T5-AFS-896

Constrained Optimization and Lagrange Multiplier Methods

This is a volume in
COMPUTER SCIENCE AND APPLIED MATHEMATICS
A Series of Monographs and Textbooks

Editor: WERNER RHEINBOLDT

A complete list of titles in this series appears at the end of this volume.

Constrained Optimization and Lagrange Multiplier Methods

DIMITRI P. BERTSEKAS
LABORATORY FOR INFORMATION
AND DECISION SYSTEMS
DEPARTMENT OF ELECTRICAL ENGINEERING
AND COMPUTER SCIENCE
MASSACHUSETTS INSTITUTE OF TECHNOLOGY
CAMBRIDGE, MASSACHUSETTS

ACADEMIC PRESS, INC.
Harcourt Brace Jovanovich, Publishers
San Diego New York Berkeley Boston
London Sydney Tokyo Toronto

ACADEMIC PRESS, INC.
1250 Sixth Avenue, San Diego, California 92101

United Kingdom Edition published by
ACADEMIC PRESS, INC. (LONDON) LTD.
24/28 Oval Road, London NW1 7DX

Library of Congress Cataloging in Publication Data

Bertsekas, Dimitri P.
Constrained optimization and Lagrange multiplier
methods.

(Computer science and applied mathematics)
Bibliography: p.
Includes index.
1. Mathematical optimization. 2. Multipliers
(Mathematical analysis) I. Title. II. Series.
QA402.5.B46 519.4 81-17612
ISBN 0-12-093480-9 AACR2

PRINTED IN THE UNITED STATES OF AMERICA

88 89 90 91 92 10 9 8 7 6 5 4 3 2

To Teli and Taki

Contents

Preface

The area of Lagrange multiplier methods for constrained minimization has undergone a radical transformation starting with the introduction of augmented Lagrangian functions and methods of multipliers in 1968 by Hestenes and Powell. The initial success of these methods in computational practice motivated further efforts aimed at understanding and improving their properties. At the same time their discovery provided impetus and a new perspective for reexamination of Lagrange multiplier methods proposed and nearly abandoned several years earlier. These efforts, aided by fresh ideas based on exact penalty functions, have resulted in a variety of interesting methods utilizing Lagrange multiplier iterations and competing with each other for solution of different classes of problems.

This monograph is the outgrowth of the author's research involvement in the area of Lagrange multiplier methods over a nine-year period beginning in early 1972. It is aimed primarily toward researchers and practitioners of mathematical programming algorithms, with a solid background in introductory linear algebra and real analysis.

Considerable emphasis is placed on the method of multipliers which, together with its many variations, may be viewed as a primary subject of the monograph. Chapters 2, 3, and 5 are devoted to this method. A large portion of Chapter 1 is devoted to unconstrained minimization algorithms on which the method relies. The developments on methods of multipliers serve as a good introduction to other Lagrange multiplier methods examined in Chapter 4.

Several results and algorithms were developed as the monograph was being written and have not as yet been published in journals. These include the algorithm for minimization subject to simple constraints (Section 1.5), the improved convergence and rate-of-convergence results of Chapter 2, the first stepsize rule of Section 2.3.1, the unification of the exact penalty methods of DiPillo and Grippo, and Fletcher, and their relationship with Newton's method (Section 4.3), the globally convergent Newton and quasi-Newton methods based on differentiable exact penalty functions (Section 4.5.2), and the methodology for solving large-scale separable integer programming problems of Section 5.6.

The line of development of the monograph is based on the author's conviction that solving practical nonlinear optimization problems efficiently (or at all) is typically a challenging undertaking and can be accomplished only through a thorough understanding of the underlying theory. This is true even if a polished packaged optimization program is used, but more so when the problem is large enough or important enough to warrant the development of a specialized algorithm. Furthermore, it is quite common in practice that methods are modified, combined, and extended in order to construct an algorithm that matches best the features of the particular problem at hand, and such modifications require a full understanding of the theoretical foundations of the method utilized. For these reasons, we place primary emphasis on the principles underlying various methods and the analysis of their convergence and rate-of-convergence properties. We also provide extensive guidance on the merits of various types of methods but, with a few exceptions, do not provide any algorithms that are specified to the last level of detail.

The monograph is based on the collective works of many researchers as well as my own. Of those people whose work had a substantial influence on my thinking and contributed in an important way to the monograph I would like to mention J. D. Buys, G. DiPillo, L. Dixon, R. Fletcher, T. Glad, L. Grippo, M. Hestenes, D. Luenberger, O. Mangasarian, D. Q. Mayne, E. Polak, B. T. Poljak, M. J. D. Powell, B. Pschenichny, R. T. Rockafellar, and R. Tapia. My research on methods of multipliers began at Stanford University. My interaction there with Daniel Gabay, Barry Kort, and David Luenberger had a lasting influence on my subsequent work on the subject. The material of Chapter 5 in particular is largely based on the results of my direct collaboration with Barry Kort. The material of Section 5.6 is based on work on electric power system scheduling at Alphatech, Inc. where I collaborated with Greg Lauer, Tom Posbergh, and Nils R. Sandell, Jr.

Finally, I wish to acknowledge gratefully the research support of the National Science Foundation, and the expert typing of Margaret Flaherty, Leni Gross, and Rosalie J. Bialy.

Chapter 1

Introduction

1.1 General Remarks

Two classical nonlinear programming problems are the equality constrained problem

$$\text{(ECP)} \qquad \text{minimize} \quad f(x) \qquad \text{subject to} \quad h(x) = 0$$

and its inequality constrained version

$$\text{(ICP)} \qquad \text{minimize} \quad f(x) \qquad \text{subject to} \quad g(x) \le 0,$$

where $f\colon R^n \to R$, $h\colon R^n \to R^m$, $g\colon R^n \to R^r$ are given functions. Computational methods for solving these problems became the subject of intensive investigation during the late fifties and early sixties. We discuss three of the approaches that were pursued.

The first approach was based on the idea of iterative descent within the confines of the constraint set. Given a feasible point x_k, a direction d_k was chosen satisfying the descent condition $\nabla f(x_k)'d_k < 0$ and the condition

$x_k + \alpha d_k$: feasible for all α positive and sufficiently small. A search along the line $\{x_k + \alpha d_k | \alpha > 0\}$ produced a new feasible point $x_{k+1} = x_k + \alpha_k d_k$ satisfying $f(x_{k+1}) < f(x_k)$. This led to various classes of *feasible direction methods* with which the names of Frank–Wolfe, Zoutendijk, Rosen, Goldstein, and Levitin–Poljak are commonly associated. These methods, together with their more sophisticated versions, enjoyed considerable success and still continue to be very popular for problems with linear constraints. On the other hand, feasible direction methods by their very nature were unable to handle problems with nonlinear equality constraints, and some of them were inapplicable or otherwise not well suited for handling nonlinear inequality constraints as well. A number of modifications were proposed for treating nonlinear equality constraints, but these involved considerable complexity and detracted substantially from the appeal of the descent idea.

A second approach was based on the possibility of solving the system of equations and (possibly) inequalities which constitute necessary conditions for optimality for the optimization problem. For (ECP), these conditions are

$$\nabla_x L(x, \lambda) = \nabla f(x) + \nabla h(x)\lambda = 0, \tag{1a}$$

$$\nabla_\lambda L(x, \lambda) = h(x) = 0, \tag{1b}$$

where L is the (ordinary) Lagrangian function

$$L(x, \lambda) = f(x) + \lambda' h(x).$$

A distinguishing feature of this approach is that the Lagrange multiplier λ is treated on an equal basis with the vector x. Iterations are carried out simultaneously on x and λ, by contrast with the descent approach where only x is iterated upon and the Lagrange multiplier plays no direct role. For this reason algorithms of this type are sometimes called *Lagrangian methods*. Several methods of this type were considered in Arrow *et al.* (1958). In addition to Newton's method for solving system (1), a gradient method was also proposed under the condition that the *local convexity assumption*

$$\nabla_{xx}^2 L(x^*, \lambda^*) > 0 \tag{2}$$

holds at a solution (x^*, λ^*). It was noted, however, by Arrow and Solow (1958) that if the local convexity assumption did not hold, then (ECP) could be replaced by the equivalent problem

$$\begin{aligned} &\text{minimize} \quad f(x) + \tfrac{1}{2}c|h(x)|^2 \\ &\text{subject to} \quad h(x) = 0, \end{aligned} \tag{3}$$

where c is a scalar and $|\cdot|$ denotes Euclidean norm. If c is taken sufficiently large, then the local convexity condition can be shown to hold for problem (3) under fairly mild conditions. The idea of focusing attention on the necessary conditions rather than the original problem also attracted considerable attention in optimal control where the necessary conditions can often be formulated as a two-point boundary value problem. However, it quickly became evident that the approach had some fundamental limitations, mainly the lack of a good mechanism to enforce convergence when far from a solution, and the difficulty of some of the methods to distinguish between local minima and local maxima.

A third approach was based on elimination of constraints through the use of *penalty functions*. For example the quadratic penalty function method (Fiacco and McCormick, 1968) for (ECP) consists of sequential unconstrained minimization of the form

$$\text{(4)} \qquad \begin{aligned} &\text{minimize} \quad f(x) + \tfrac{1}{2}c_k|h(x)|^2 \\ &\text{subject to} \quad x \in R^n, \end{aligned}$$

where $\{c_k\}$ is a positive scalar sequence with $c_k < c_{k+1}$ for all k and $c_k \to \infty$. The sequential minimization process yields

$$\text{(5)} \qquad \lim_{c_k \to \infty} \inf_{x \in R^n} \{f(x) + \tfrac{1}{2}c_k|h(x)|^2\}.$$

On the other hand, the optimal value of (ECP) can be written as

$$\text{(6)} \qquad \inf_{x \in R^n} \lim_{c_k \to \infty} \{f(x) + \tfrac{1}{2}c_k|h(x)|^2\},$$

and hence the success of the penalty method hinges on the equality of the expressions (5) and (6), i.e., the validity of interchanging "lim" and "inf." This interchange is indeed valid under mild assumptions (basically continuity of f and h—see Chapter 2). Lagrange multipliers play no direct role in this method but it can be shown under rather mild assumptions that the sequence $\{c_k h(x_k)\}$, where x_k solves problem (4), converges to a Lagrange multiplier of the problem. Despite their considerable disadvantages [mainly slow convergence and ill-conditioning when solving problem (4) for large values of c_k], penalty methods were widely accepted in practice. The reasons can be traced to the simplicity of the approach, its ability to handle nonlinear constraints, as well as the availability of very powerful unconstrained minimization methods for solving problem (4).

The main idea of the descent approach also made its appearance in a dual context whereby an ascent method is used to maximize the *dual functional* for (ECP) given by

$$d(\lambda) = \inf_x \{f(x) + \lambda' h(x)\} = \inf_x L(x, \lambda).$$

In the simplest such method one minimizes $L(\cdot, \lambda_k)$ (perhaps in a local sense) over x for a sequence of multiplier vectors $\{\lambda_k\}$. This sequence is generated by

$$\lambda_{k+1} = \lambda_k + \alpha h(x_k), \tag{7}$$

where x_k is a minimizing point of $L(\cdot, \lambda_k)$ and α is a stepsize scalar parameter. It is possible to show under the appropriate assumptions (see Section 2.6) that $h(x_k) = \nabla d(\lambda_k)$, so (7) is actually a steepest ascent iteration for maximizing the dual functional d. Such methods have been called *primal–dual methods.* Actually the dual functional and the method itself make sense only under fairly restrictive conditions including either the local convexity assumption (2) or other types of convexity conditions. The method is also often hampered by slow convergence. Furthermore in many cases it is difficult to know a priori an appropriate range for the stepsize α. For this reason primal–dual methods of the type just described initially found application only in the limited class of convex or locally convex problems where minimization of $L(\cdot, \lambda_k)$ can be carried out very efficiently due to special structure involving, for example, separable objective and constraint functions (Everett, 1963).

Starting around 1968, a number of researchers have proposed a new class of methods, called *methods of multipliers,* in which the penalty idea is merged with the primal–dual and Lagrangian philosophy. In the original method of multipliers, proposed by Hestenes (1969) and Powell (1969), the quadratic penalty term is added not to the objective function f of (ECP) but rather to the Lagrangian function $L = f + \lambda' h$ thus forming the *augmented Lagrangian* function

$$L_c(x, \lambda) = f(x) + \lambda' h(x) + \tfrac{1}{2}c|h(x)|^2. \tag{8}$$

A sequence of minimizations of the form

$$\begin{aligned} &\text{minimize} \quad L_{c_k}(x, \lambda_k) \\ &\text{subject to} \quad x \in R^n \end{aligned} \tag{9}$$

is performed where $\{c_k\}$ is a sequence of positive penalty parameters. The multiplier sequence $\{\lambda_k\}$ is generated by the iteration

$$\lambda_{k+1} = \lambda_k + c_k h(x_k), \tag{10}$$

where x_k is a solution of problem (9). The initial vector λ_0 is selected a priori, and the sequence $\{c_k\}$ may be either preselected or generated during the computation according to some scheme.

One may view the method just described within the context of penalty function methods. If $c_k \to \infty$ and the generated sequence $\{\lambda_k\}$ turns out to be bounded, then the method is guaranteed to yield in the limit the optimal value of (ECP), provided sufficient assumptions are satisfied which guarantee the validity of interchange of "lim" and "inf" in the expression

$$\lim_{c_k \to \infty} \inf_x \{f(x) + \lambda_k' h(x) + \tfrac{1}{2} c_k |h(x)|^2\},$$

similarly as for the penalty method considered earlier.

Another point of view (see Chapter 2) is based on the fact that iteration (10) is a steepest ascent iteration for maximizing the dual functional

$$d_{c_k}(\lambda) = \inf_x \{f(x) + \lambda' h(x) + \tfrac{1}{2} c_k |h(x)|^2\},$$

which corresponds to the problem

$$\begin{aligned} &\text{minimize} \quad f(x) + \tfrac{1}{2} c_k |h(x)|^2 \\ &\text{subject to} \quad h(x) = 0. \end{aligned}$$

As noted earlier, if c_k is sufficiently large, this problem has locally convex structure, so the primal–dual viewpoint is applicable.

It turns out that, by combining features of the penalty and the primal–dual approach, the method of multipliers actually moderates the disadvantages of both. As we shall see in the next chapter, convergence in the method of multipliers can usually be attained *without the need to increase* c_k *to infinity* thereby alleviating the ill-conditioning problem that plagues the penalty method. In addition *the multiplier iteration* (10) *tends to converge to a Lagrange multiplier vector much faster than iteration* (7) *of the primal–dual method, or the sequence* $\{c_k h(x_k)\}$ *in the penalty method*. Because of these attractive characteristics, the method of multipliers and its subsequently developed variations have emerged as a very important class of constrained minimization methods. A great deal of research has been directed toward their analysis and understanding. Furthermore their discovery provided impetus for reexamination of Lagrangian methods proposed and nearly abandoned many years ago. These efforts aided by fresh ideas based on penalty functions and duality have resulted in a variety of interesting methods utilizing Lagrange multiplier iterations and competing with each other for solution of different classes of problems.

The purpose of this monograph is to provide a rather thorough analysis of these Lagrange multiplier methods starting with the quadratic method of multipliers for (ECP) just described. This method is the subject of Chapter 2. In Chapter 3, the method is extended to handle problems with both equality and inequality constraints. In addition the Lagrange multiplier approach is utilized to construct algorithms for solution of nondifferentiable and minimax problems. In Chapter 4, we consider a variety of Lagrangian methods and analyze their local and global convergence properties. Finally, in Chapter 5, we explore the possibility of using a penalty function other than quadratic, and we analyze multiplier methods as applied to convex programming problems.

1.2 Notation and Mathematical Background

The purpose of this section is to provide a selective list of mathematical definitions, notations, and results that will be frequently used. For detailed expositions, the reader should consult texts on linear algebra and real analysis.

Algebraic Notions

We denote by R the real line and by R^n the space of all n-dimensional vectors. Intervals of real numbers or extended real numbers are denoted as usual by bracket–parentheses notation. For example for $a \in R$ or $a = -\infty$ and $b \in R$ or $b = +\infty$ we write $(a, b] = \{x \mid a < x \le b\}$. Given any subset $S \subset R$ which is bounded above (below), we denote by $\sup S$ ($\inf S$) the least upper bound (greatest lower bound) of S. If S is unbounded above (below) we write $\sup S = \infty$ ($\inf S = -\infty$). In our notation, *every vector is considered to be a column vector*. The transpose of an $m \times n$ matrix A is denoted A'. A vector $x \in R^n$ will be treated as an $n \times 1$ matrix, and thus x' denotes a $1 \times n$ matrix or row vector. If $x_1, \ldots, x_n$ are the coordinates of a vector $x \in R^n$, we write $x = (x_1, x_2, \ldots, x_n)$. We also write

$$x \ge 0 \quad \text{if} \quad x_i \ge 0, \quad i = 1, \ldots, n,$$

$$x \le 0 \quad \text{if} \quad x_i \le 0, \quad i = 1, \ldots, n.$$

A symmetric $n \times n$ matrix A will be said to be *positive semidefinite* if $x'Ax \ge 0$ for all $x \in R^n$. In this case we write

$$A \ge 0.$$

We say that A is *positive definite* if $x'Ax > 0$ for all $x \ne 0$, and write

$$A > 0.$$

When we say that A is positive (semi)definite we implicitly assume that it is symmetric. A symmetric $n \times n$ matrix A has n real eigenvalues $\gamma_1, \gamma_2, \ldots, \gamma_n$ and n nonzero real eigenvectors $e_1, e_2, \ldots, e_n$ which are mutually orthogonal. It can be shown that

$$\gamma x'x \le x'Ax \le \Gamma x'x \qquad \forall\, x \in R^n, \tag{1}$$

where

$$\gamma = \min\{\gamma_1, \ldots, \gamma_n\}, \qquad \Gamma = \max\{\gamma_1, \ldots, \gamma_n\}.$$

For x equal to the eigenvector corresponding to Γ (γ), the inequality on the right (left) in (1) becomes equality. It follows that $A > 0$ ($A \geq 0$), if and only if the eigenvalues of A are positive (nonnegative).

If A is positive definite, there exists a unique positive definite matrix the square of which equals A. This is the matrix that has the same eigenvectors as A and has as eigenvalues the square roots of the eigenvalues of A. We denote this matrix by $A^{1/2}$.

Let A and B be square matrices and C be a matrix of appropriate dimension. The very useful equation

$$(A + CBC')^{-1} = A^{-1} - A^{-1}C(B^{-1} + C'A^{-1}C)^{-1}C'A^{-1}$$

holds provided all the inverses appearing above exist. The equation can be verified by multiplying the right-hand side by $(A + CBC')$ and showing that the product is the identity.

Consider a partitioned square matrix M of the form

$$M = \begin{bmatrix} A & B \\ C & D \end{bmatrix}.$$

There holds

$$M^{-1} = \begin{bmatrix} Q & -QBD^{-1} \\ -D^{-1}CQ & D^{-1} + D^{-1}CQBD^{-1} \end{bmatrix},$$

where

$$Q = (A - BD^{-1}C)^{-1},$$

provided all the inverses appearing above exist. The proof is obtained by multiplying M with the expression for M^{-1} given above and verifying that the product yields the identity matrix.

Topological Notions

We shall use throughout the standard Euclidean norm in R^n denoted $|\cdot|$; i.e., for a vector $x \in R^n$, we write

$$|x| = \sqrt{x'x}.$$

The Euclidean norm of an $m \times n$ matrix A will be denoted also $|\cdot|$. It is given by

$$|A| = \max_{x \neq 0} \frac{|Ax|}{|x|} = \max_{x \neq 0} \frac{\sqrt{x'A'Ax}}{\sqrt{x'x}}.$$

In view of (1), we have

$$|A| = \sqrt{\text{max eigenvalue}(A'A)}.$$

If A is symmetric, then if $\lambda_1, \ldots, \lambda_n$ are its (real) eigenvalues, the eigenvalues of A^2 are $\lambda_1^2, \ldots, \lambda_n^2$, and we obtain

$$|A| = \max\{|\lambda_1|, \ldots, |\lambda_n|\}.$$

A sequence of vectors $x_0, x_1, \ldots, x_k, \ldots$, in R^n, denoted $\{x_k\}$, is said to converge to a limit vector x if $|x_k - x| \to 0$ as $k \to \infty$ (that is, if given $\varepsilon > 0$, there is an N such that for all $k \geq N$ we have $|x_k - x| < \varepsilon$). If $\{x_k\}$ converges to x we write $x_k \to x$ or $\lim_{k\to\infty} x_k = x$. Similarly for a sequence of $m \times n$ matrices $\{A_k\}$, we write $A_k \to A$ or $\lim_{k\to\infty} A_k = A$ if $|A_k - A| \to 0$ as $k \to \infty$. Convergence of both vector and matrix sequences is equivalent to convergence of each of the sequences of their coordinates or elements.

Given a sequence $\{x_k\}$, the subsequence $\{x_k | k \in K\}$ corresponding to an infinite index set K is denoted $\{x_k\}_K$. A vector x is said to be a *limit point* of a sequence $\{x_k\}$ if there is a subsequence $\{x_k\}_K$ which converges to x.

A sequence of real numbers $\{r_k\}$ which is monotonically nondecreasing (nonincreasing), i.e., satisfies $r_k \leq r_{k+1}$ ($r_k \geq r_{k+1}$) for all k, must either converge to a real number or be unbounded above (below) in which case we write $\lim_{k\to\infty} r_k = +\infty$ ($\lim_{k\to\infty} r_k = -\infty$). Given any bounded sequence of real numbers $\{r_k\}$, we may consider the sequence $\{s_k\}$ where $s_k = \sup\{r_i | i \geq k\}$. Since this sequence is monotonically nonincreasing and bounded, it must have a limit called the *limit superior* of $\{r_k\}$ and denoted by $\limsup_{k\to\infty} r_k$. We define similarly the *limit inferior* of $\{r_k\}$ and denote it by $\liminf_{k\to\infty} r_k$. If $\{r_k\}$ is unbounded above, we write $\limsup_{k\to\infty} r_k = +\infty$, and if it is unbounded below, we write $\liminf_{k\to\infty} r_k = -\infty$.

Open, Closed, and Compact Sets

For a vector $x \in R^n$ and a scalar $\varepsilon > 0$, we denote the open sphere centered at x with radius $\varepsilon > 0$ by $S(x; \varepsilon)$; i.e.,

$$S(x; \varepsilon) = \{z \,|\, |z - x| < \varepsilon\}. \tag{2}$$

For a subset $X \subset R^n$ and a scalar $\varepsilon > 0$, we write by extension of the preceding notation

$$S(X; \varepsilon) = \{z \,|\, |z - x| < \varepsilon \text{ for some } x \in X\}. \tag{3}$$

A subset S of R^n is said to be *open*, if for every vector $x \in S$ one can find an $\varepsilon > 0$ such that $S(x; \varepsilon) \subset S$. If S is open and $x \in S$, then S is said to be a *neighborhood* of x. The *interior* of a set $S \subset R^n$ is the set of all $x \in S$ for which there exists $\varepsilon > 0$ such that $S(x; \varepsilon) \subset S$. A set S is *closed* if and only if its complement in R^n is open. Equivalently S is closed if and only if every convergent sequence $\{x_k\}$ with elements in S converges to a point which also belongs to S. A subset S of R^n is said to be *compact* if and only if it is both closed and bounded (i.e., it is closed and for some $M > 0$ we have $|x| \leq M$ for all $x \in S$). A set S is compact if and only if every sequence $\{x_k\}$ with elements in S has at least one limit point which belongs to S. Another important fact is that if $S_0, S_1, \ldots, S_k, \ldots$ is a sequence of nonempty compact sets in R^n such that $S_k \supset S_{k+1}$ for all k then the intersection $\bigcap_{k=0}^{\infty} S_k$ is a nonempty and compact set.

Continuous Functions

A function f mapping a set $S_1 \subset R^n$ into a set $S_2 \subset R^m$ is denoted by $f: S_1 \to S_2$. The function f is said to be *continuous* at $x \in S_1$ if $f(x_k) \to f(x)$ whenever $x_k \to x$. Equivalently f is continuous at x if given $\varepsilon > 0$ there is a $\delta > 0$ such that $|y - x| < \delta$ and $y \in S_1$ implies $|f(y) - f(x)| < \varepsilon$. The function f is said to be continuous over S_1 (or simply continuous) if it is continuous at every point $x \in S_1$. If S_1, S_2, and S_3 are sets and $f_1: S_1 \to S_2$ and $f_2: S_2 \to S_3$ are functions, the function $f_2 \cdot f_1: S_1 \to S_3$ defined by $(f_2 \cdot f_1)(x) = f_2[f_1(x)]$ is called the *composition* of f_1 and f_2. If $f_1: R^n \to R^m$ and $f_2: R^m \to R^p$ are continuous, then $f_2 \cdot f_1$ is also continuous.

Differentiable Functions

A real-valued function $f: X \to R$ where $X \subset R^n$ is an open set is said to be *continuously differentiable* if the partial derivatives $\partial f(x)/\partial x_1, \ldots, \partial f(x)/\partial x_n$ exist for each $x \in X$ and are continuous functions of x over X. In this case we write $f \in C^1$ over X. More generally we write $f \in C^p$ over X for a function $f: X \to R$, where $X \subset R^n$ is an open set if all partial derivatives of order p exist and are continuous as functions of x over X. If $f \in C^p$ over R^n, we simply write $f \in C^p$. If $f \in C^1$ on X, the *gradient* of f at a point $x \in X$ is defined to be the column vector

$$\nabla f(x) = \begin{bmatrix} \dfrac{\partial f(x)}{\partial x_1} \\ \vdots \\ \dfrac{\partial f(x)}{\partial x_n} \end{bmatrix}.$$

If $f \in C^2$ over X, the *Hessian* of f at x is defined to be the symmetric $n \times n$ matrix having $\partial^2 f(x)/\partial x_i \, \partial x_j$ as the ijth element

$$\nabla^2 f(x) = \left[\frac{\partial^2 f(x)}{\partial x_i \, \partial x_j}\right].$$

If $f: X \to R^m$ where $X \subset R^n$, then f will be alternatively represented by the column vector of its component functions $f_1, f_2, \ldots, f_m$

$$f(x) = \begin{bmatrix} f_1(x) \\ \vdots \\ f_m(x) \end{bmatrix}.$$

If X is open, we write $f \in C^p$ on X if $f_1 \in C^p, f_2 \in C^p, \ldots, f_m \in C^p$ on X. We shall use the notation

$$\nabla f(x) = [\nabla f_1(x) \cdots \nabla f_m(x)].$$

Thus, the $n \times m$ matrix ∇f has as columns the gradients $\nabla f_1(x), \ldots, \nabla f_m(x)$ and is the transpose of the Jacobian matrix of the function f.

On occasion we shall need to consider gradients of functions with respect to some of the variables only. The notation will be as follows:

If $f: R^{n+r} \to R$ is a real-valued function of (x, y) where $x = (x_1, \ldots, x_n) \in R^n$, $y = (y_1, \ldots, y_r) \in R^r$, we write

$$\nabla_x f(x, y) = \begin{bmatrix} \dfrac{\partial f(x, y)}{\partial x_1} \\ \vdots \\ \dfrac{\partial f(x, y)}{\partial x_n} \end{bmatrix}, \qquad \nabla_y f(x, y) = \begin{bmatrix} \dfrac{\partial f(x, y)}{\partial y_1} \\ \vdots \\ \dfrac{\partial f(x, y)}{\partial y_r} \end{bmatrix},$$

$$\nabla_{xx} f(x, y) = \left[\frac{\partial f(x, y)}{\partial x_i\, \partial x_j}\right], \qquad \nabla_{xy} f(x, y) = \left[\frac{\partial f(x, y)}{\partial x_i\, \partial y_j}\right],$$

$$\nabla_{yy} f(x, y) = \left[\frac{\partial f(x, y)}{\partial y_i\, \partial y_j}\right].$$

If $f: R^{n+r} \to R^m$, $f = (f_1, f_2, \ldots, f_m)$, we write

$$\nabla_x f(x, y) = [\nabla_x f_1(x, y) \cdots \nabla_x f_m(x, y)],$$
$$\nabla_y f(x, y) = [\nabla_y f_1(x, y) \cdots \nabla_y f_m(x, y)].$$

For $h: R^r \to R^m$ and $g: R^n \to R^r$, consider the function $f: R^n \to R^m$ defined by

$$f(x) = h[g(x)].$$

Then if $h \in C^p$ and $g \in C^p$, we also have $f \in C^p$. The *chain rule* of differentiation is stated in terms of our notation as

$$\nabla f(x) = \nabla g(x) \nabla h[g(x)].$$

Mean Value Theorems and Taylor Series Expansions

Let $f: X \to R$, and $f \in C^1$ over the open set $X \subset R^n$. Assume that X contains the line segment connecting two points $x, y \in X$. The *mean value theorem* states that there exists a scalar α with $0 < \alpha < 1$ such that

$$f(y) = f(x) + \nabla f[x + \alpha(y - x)]'(y - x).$$

If in addition $f \in C^2$, then there exists a scalar α with $0 < \alpha < 1$ such that

$$f(y) = f(x) + \nabla f(x)'(y - x) + \tfrac{1}{2}(y - x)'\nabla^2 f[x + \alpha(y - x)](y - x).$$

Let $f: X \to R^m$ and $f \in C^1$ on the open set $X \subset R^n$. Assume that X contains the line segment connecting two points $x, y \in X$. The *first-order Taylor series expansion* of f around x is given by the equation

$$f(y) = f(x) + \int_0^1 \nabla f[x + \alpha(y - x)]'(y - x)\, d\alpha.$$

If in addition $f \in C^2$ on X, then we have the *second-order Taylor series expansion*

$$f(y) = f(x) + \nabla f(x)'(y - x) + \int_0^1 \left(\int_0^\xi (y - x)'\nabla^2 f[x + \alpha(y - x)](y - x)\, d\alpha \right) d\xi.$$

Implicit Function Theorems

Consider a system of n equations in $m + n$ variables

$$h(x, y) = 0,$$

where $h: R^{m+n} \to R^n$, $x \in R^m$, and $y \in R^n$. Implicit function theorems address the question whether one may solve the system of equations for the vector y in terms of the vector x, i.e., whether there exists a function ϕ, called the *implicit function*, such that $h[x, \phi(x)] = 0$. The following classical implicit function theorem asserts that this is possible in a local sense, i.e., in a neighborhood of a solution $(\bar{x}, \bar{y})$, provided the gradient matrix of h with respect to y is nonsingular.

Implicit Function Theorem 1: Let S be an open subset of R^{m+n}, and $h: S \to R^n$ be a function such that for some $p \geq 0$, $h \in C^p$ over S, and assume that $\nabla_y h(x, y)$ exists and is continuous on S. Let $(\bar{x}, \bar{y}) \in S$ be a vector such that $h(\bar{x}, \bar{y}) = 0$ and the matrix $\nabla_y h(\bar{x}, \bar{y})$ is nonsingular. Then there exist scalars $\varepsilon > 0$ and $\delta > 0$ and a function $\phi: S(\bar{x}; \varepsilon) \to S(\bar{y}; \delta)$ such that $\phi \in C^p$ over $S(\bar{x}; \varepsilon)$, $\bar{y} = \phi(\bar{x})$, and $h[x, \phi(x)] = 0$ for all $x \in S(\bar{x}; \varepsilon)$. The function

ϕ is unique in the sense that if $x \in S(\bar{x}; \varepsilon)$, $y \in S(\bar{y}; \delta)$, and $h(x, y) = 0$, then $y = \phi(x)$. Furthermore, if $p \geq 1$, then for all $x \in S(\bar{x}; \varepsilon)$

$$\nabla \phi(x) = -\nabla_x h[x, \phi(x)][\nabla_y h[x, \phi(x)]]^{-1}.$$

We shall also need the following implicit function theorem. It is a special case of a more general theorem found in Hestenes (1966). The notation (3) is used in the statement of the theorem.

Implicit Function Theorem 2: Let S be an open subset of R^{m+n}, $\bar{X}$ be a compact subset of R^m, and $h: S \to R^n$ be a function such that for some $p \geq 0$, $h \in C^p$ on S. Assume that $\nabla_y h(x, y)$ exists and is continuous on S. Assume that $\bar{y} \in R^n$ is a vector such that $(\bar{x}, \bar{y}) \in S$, $h(\bar{x}, \bar{y}) = 0$, and the matrix $\nabla_y h(\bar{x}, \bar{y})$ is nonsingular for all $\bar{x} \in \bar{X}$. Then there exist scalars $\varepsilon > 0$, $\delta > 0$, and a function $\phi: S(\bar{X}; \varepsilon) \to S(\bar{y}; \delta)$ such that $\phi \in C^p$ on $S(\bar{X}; \varepsilon)$, $\bar{y} = \phi(\bar{x})$ for all $\bar{x} \in \bar{X}$, and $h[x, \phi(x)] = 0$ for all $x \in S(\bar{X}; \varepsilon)$. The function ϕ is unique in the sense that if $x \in S(\bar{X}; \varepsilon)$, $y \in S(\bar{y}; \delta)$, and $h(x, y) = 0$, then $y = \phi(x)$. Furthermore, if $p \geq 1$, then for all $x \in S(\bar{X}; \varepsilon)$

$$\nabla \phi(x) = -\nabla_x h[x, \phi(x)][\nabla_y h[x, \phi(x)]]^{-1}.$$

When $\bar{X}$ consists of a single vector $\bar{x}$, the two implicit function theorems coincide.

Convexity

A set $S \subset R^n$ is said to be *convex* if for every $x, y \in S$ and $\alpha \in [0, 1]$ we have $\alpha x + (1 - \alpha) y \in S$. A function $f: S \to R$ is said to be *convex over the convex set* S if for every $x, y \in S$ and $\alpha \in [0, 1]$ we have

$$f[\alpha x + (1 - \alpha) y] \leq \alpha f(x) + (1 - \alpha) f(y).$$

If f is convex and $f \in C^1$ over an open convex set S, then

$$f(y) \geq f(x) + \nabla f(x)'(y - x) \qquad \forall\, x, y \in S. \tag{4}$$

If in addition $f \in C^2$ over S, then $\nabla^2 f(x) \geq 0$ for all $x \in S$. Conversely, if $f \in C^1$ over S and (4) holds, or if $f \in C^2$ over S and $\nabla^2 f(x) \geq 0$ for all $x \in S$, then f is convex over S.

Rate of Convergence Concepts

In minimization algorithms we are often interested in the speed with which various algorithms converge to a limit. Given a sequence $\{x_k\} \subset R^n$ with $x_k \to x^*$, the typical approach is to measure speed of convergence in terms of an *error function* $e: R^n \to R$ satisfying $e(x) \geq 0$ for all $x \in R^n$ and $e(x^*) = 0$. Typical choices are

$$e(x) = |x - x^*|, \qquad e(x) = |f(x) - f(x^*)|,$$

where f is the objective function of the problem. The sequence $\{e(x_k)\}$ is then compared with standard sequences. In our case, we compare $\{e(x_k)\}$ with geometric progressions of the form

$$r_k = q\beta^k,$$

where $q > 0$ and $\beta \in (0, 1)$ are some scalars, and with sequences of the form

$$r_k = q\beta^{p^k},$$

where $q > 0$, $\beta \in (0, 1)$, and $p > 1$ are some scalars. There is no reason for selecting these particular sequences for comparison other than the fact that they represent a sufficiently wide class which is adequate and convenient for our purposes. Our approach has much in common with that of Ortega and Rheinboldt (1970), except that we do not emphasize the distinction between Q and R linear or superlinear convergence.

Let us introduce some terminology:

Definition: Given two scalar sequences $\{e_k\}$ and $\{r_k\}$ with

$$0 \le e_k, \qquad 0 \le r_k, \qquad e_k \to 0, \qquad r_k \to 0,$$

we say that $\{e_k\}$ *converges faster than* $\{r_k\}$ if there exists an index $\bar{k} \ge 0$ such that

$$0 \le e_k \le r_k \qquad \forall\, k \ge \bar{k}.$$

We say that $\{e_k\}$ *converges slower than* $\{r_k\}$ if there exists an index $\bar{k} \ge 0$ such that

$$0 \le r_k \le e_k \qquad \forall\, k \ge \bar{k}.$$

Definition: Consider a scalar sequence $\{e_k\}$ with $e_k \ge 0$, $e_k \to 0$. The sequence $\{e_k\}$ is said to converge *at least linearly with convergence ratio* β, where $0 < \beta < 1$, if it converges faster than all geometric progressions of the form $q\bar{\beta}^k$ where $q > 0$, $\bar{\beta} \in (\beta, 1)$. It is said to converge *at most linearly with convergence ratio* β, where $0 < \beta < 1$, if it converges slower than all geometric progressions of the form $q\bar{\beta}^k$, where $q > 0$, $\bar{\beta} \in (0, \beta)$. It is said to converge *linearly with convergence ratio* β, where $0 < \beta < 1$, if it converges both at least and at most linearly with convergence ratio β. It is said to converge *superlinearly* or *sublinearly* if it converges faster or slower, respectively, then every sequence of the form $q\beta^k$, where $q > 0$, $\beta \in (0, 1)$.

Examples: (1) The following sequences all converge linearly with convergence ratio β:

$$q\beta^k, \quad q\left(\beta + \frac{1}{k}\right)^k, \quad q\left(\beta - \frac{1}{k}\right)^k, \quad q\beta^{k+(1/k)},$$

where $q > 0$ and $\beta \in (0, 1)$. This fact follows either by straightforward verification of the definition or by making use of Proposition 1.1 below.

(2) Let $0 < \beta_1 < \beta_2 < 1$, and consider the sequence $\{e_k\}$ defined by

$$e_{2k} = \beta_1^k \beta_2^k, \qquad e_{2k+1} = \beta_1^{k+1} \beta_2^k.$$

Then clearly $\{e_k\}$ converges at least linearly with convergence ratio β_2 and at most linearly with convergence ratio β_1. Actually $\{e_k\}$ can be shown to converge linearly with convergence ratio $\sqrt{\beta_1 \beta_2}$ a fact that can be proved by making use of the next proposition.

(3) The sequence $\{1/k\}$ converges sublinearly and every sequence of the form $q\beta^{p^k}$, where $q > 0$, $\beta \in (0, 1)$, $p > 1$, can be shown to converge superlinearly. Again these facts follow by making use of the proposition below.

Proposition 1.1: Let $\{e_k\}$ be a scalar sequence with $e_k \geq 0$, $e_k \to 0$. Then the following hold true:

(a) The sequence $\{e_k\}$ converges at least linearly with convergence ratio $\beta \in (0, 1)$ if and only if

$$\limsup_{k \to \infty} e_k^{1/k} \leq \beta. \tag{5}$$

It converges at most linearly with convergence ratio $\beta \in (0, 1)$ if and only if

$$\liminf_{k \to \infty} e_k^{1/k} \geq \beta. \tag{6}$$

It converges linearly with convergence ratio $\beta \in (0, 1)$ if and only if

$$\lim_{k \to \infty} e_k^{1/k} = \beta. \tag{7}$$

(b) If $\{e_k\}$ converges faster (slower) than some geometric progression of the form $q\beta^k$, $q > 0$, $\beta \in (0, 1)$, then it converges at least (at most) linearly with convergence ratio β.

(c) Assume that $e_k \neq 0$ for all k, and denote

$$\beta_1 = \liminf_{k \to \infty} \frac{e_{k+1}}{e_k}, \qquad \beta_2 = \limsup_{k \to \infty} \frac{e_{k+1}}{e_k}.$$

If $0 < \beta_1 < \beta_2 < 1$, then $\{e_k\}$ converges at least linearly with convergence ratio β_1 and at most linearly with convergence ratio β_2.

(d) Assume that $e_k \neq 0$ for all k and that

$$\lim_{k \to \infty} \frac{e_{k+1}}{e_k} = \beta.$$

If $0 < \beta < 1$, then $\{e_k\}$ converges linearly with convergence ratio β. If $\beta = 0$, then $\{e_k\}$ converges superlinearly. If $\beta = 1$, then $\{e_k\}$ converges sublinearly.

Proof: (a) If (5) holds, then for every $\bar{\beta} \in (\beta, 1)$ there exists a $\bar{k} \geq 0$ such that $e_k \leq \bar{\beta}^k$ for all $k \geq \bar{k}$. Since $\{\bar{\beta}^k\}$ converges faster than every sequence of the form $q\tilde{\beta}^k$, with $q > 0$, $\tilde{\beta} \in (\bar{\beta}, 1)$, the same is true for $\{e_k\}$. Since $\bar{\beta}$ can be taken arbitrarily close to β, it follows that $\{e_k\}$ converges at least linearly with convergence ratio β. Conversely if $\{e_k\}$ converges at least linearly with convergence ratio β, we have for every $\bar{\beta} \in (\beta, 1)$, $e_k \leq \bar{\beta}^k$ for all k sufficiently large. Hence, $\limsup_{k \to \infty} e_k^{1/k} \leq \bar{\beta}$. Since $\bar{\beta}$ can be taken arbitrarily close to β, (5) follows. An entirely similar argument proves the statement concerning (6). The statement regarding (7) is obtained by combining the two statements concerning (5) and (6).

(b) If $e_k \leq (\geq) q\beta^k$ for all k sufficiently large then $e_k^{1/k} \leq (\geq) q^{1/k}\beta$ and $\limsup_{k \to \infty} (\liminf_{k \to \infty}) e_k^{1/k} \leq (\geq) \beta$. Hence, by part (a), $\{e_k\}$ converges at least (at most) linearly with convergence ratio β.

(c) For every $\bar{\beta}_2 \in (\beta_2, 1)$, there exists $\bar{k} \geq 0$ such that

$$e_{k+1}/e_k \leq \bar{\beta}_2 \qquad \forall\, k \geq \bar{k}.$$

Hence, $e_{\bar{k}+m} \leq \bar{\beta}_2^m e_{\bar{k}}$ and $e_{\bar{k}+m}^{1/(\bar{k}+m)} \leq \bar{\beta}_2^{m/(\bar{k}+m)} e_{\bar{k}}^{1/(\bar{k}+m)}$. Taking the limit superior as $m \to \infty$, we obtain

$$\limsup_{k \to \infty} e_k^{1/k} \leq \bar{\beta}_2.$$

Since $\bar{\beta}_2$ can be taken arbitrarily close to β_2 we obtain $\limsup_{k \to \infty} e_k^{1/k} \leq \beta_2$, and the result follows by part (a). Similarly we prove the result relating to β_1.

(d) If $0 < \beta < 1$, the result follows directly from part (c). If $\beta = 0$, then for any $\bar{\beta} \in (0, 1)$ we have, for some $\bar{k} \geq 0$, $e_{k+1} \leq \bar{\beta} e_k$ for all $k \geq \bar{k}$. From this, it follows that $\{e_k\}$ converges faster than $\{\bar{\beta}^k\}$, and since $\bar{\beta}$ can be taken arbitrarily close to zero, $\{e_k\}$ converges superlinearly. Similarly we prove the result concerning sublinear convergence. Q.E.D.

When $\{e_k\}$ satisfies $\limsup_{k \to \infty} e_{k+1}/e_k = \beta < 1$ as in Proposition 1.1d, we also say that $\{e_k\}$ converges *at least quotient-linearly* (*or Q-linearly*) *with convergence ratio* β. If $\beta = 0$, then we say that $\{e_k\}$ converges *Q-superlinearly*.

Most optimization algorithms which are of interest in practice produce sequences converging either linearly or superlinearly. Linear convergence is quite satisfactory for optimization algorithms provided the convergence ratio β is not very close to unity. Algorithms which may produce sequences having sublinear convergence rates are excluded from consideration in most optimization problems as computationally inefficient. Several optimization algorithms possess superlinear convergence for particular classes of problems. For this reason, it is necessary to quantify further the notion of superlinear convergence.

Definition: Consider a scalar sequence $\{e_k\}$ with $e_k \geq 0$ converging superlinearly to zero. Then $\{e_k\}$ is said to *converge at least superlinearly with order* p, where $1 < p$, if it converges faster than all sequences of the form $q\beta^{\bar{p}^k}$, where $q > 0$, $\beta \in (0, 1)$, and $\bar{p} \in (1, p)$. It is said to converge *at most superlinearly with order* p, where $1 < p$, if it converges slower than all sequences of the form $q\beta^{\bar{p}^k}$, where $q > 0$, $\beta \in (0, 1)$, and $\bar{p} > p$. It is said to *converge superlinearly with order* p, where $p > 1$, if it converges both at least and at most superlinearly with order p.

We have the following proposition, the proof of which is similar to the one of Proposition 1.1 and is left as an exercise to the reader.

Proposition 1.2: Let $\{e_k\}$ be a scalar sequence with $e_k \geq 0$ and $e_k \to 0$. Then the following hold true:

(a) The sequence $\{e_k\}$ converges at least superlinearly with order $p > 1$ if and only if

$$\lim_{k\to\infty} e_k^{1/\bar{p}^k} = 0 \qquad \forall\, \bar{p} \in (1, p).$$

It converges at most superlinearly with order $p > 1$ if and only if

$$\lim_{k\to\infty} e_k^{1/\bar{p}^k} = 1 \qquad \forall\, \bar{p} > p.$$

(b) If $\{e_k\}$ converges faster (slower) than some sequence of the form $q\beta^{p^k}$, where $q > 0$, $\beta \in (0, 1)$, and $p > 1$, then it converges at least (at most) superlinearly with order p.

(c) Assume that $e_k \neq 0$ for all k. If for some $p > 1$, we have

$$\limsup_{k\to\infty} \frac{e_{k+1}}{e_k^p} < \infty,$$

then $\{e_k\}$ converges at least superlinearly with order p. If

$$\liminf_{k\to\infty} \frac{e_{k+1}}{e_k^p} > 0,$$

then $\{e_k\}$ converges at most superlinearly with order p.

If

$$\limsup_{k\to\infty} \frac{e_{k+1}}{e_k^p} < \infty,$$

as in Proposition 1.2c, then we say that $\{e_k\}$ converges *at least Q-superlinearly with order* p.

Cholesky Factorization

Let $A = [a_{ij}]$ be an $n \times n$ positive definite matrix and let us denote by A_i the *leading principal submatrix of A of order i*, $i = 1, \ldots, n$, where

$$A_i = \begin{bmatrix} a_{11} & a_{12} & \cdots & a_{1i} \\ a_{21} & a_{22} & \cdots & a_{2i} \\ \vdots & \vdots & & \vdots \\ a_{i1} & a_{i2} & \cdots & a_{ii} \end{bmatrix}.$$

It is easy to show that each of the submatrices A_i is a positive definite matrix. Indeed for any $y \in R^i$, $y \neq 0$, we have by positive definiteness of A

$$y'A_i y = [y' \quad 0]A\begin{bmatrix} y \\ 0 \end{bmatrix} > 0,$$

which implies that A_i is positive definite.

The matrices A_i satisfy

$$A_1 = [a_{11}],$$

$$\text{(8)} \qquad A_i = \begin{bmatrix} A_{i-1} & \alpha_i \\ \alpha_i' & a_{ii} \end{bmatrix}, \qquad i = 2, \ldots, n,$$

where α_i is the column vector in R^{i-1} given by

$$\text{(9)} \qquad \alpha_i = \begin{bmatrix} a_{1i} \\ \vdots \\ a_{i-1,i} \end{bmatrix}.$$

We now show that A can be written as

$$A = LL',$$

where L is a unique lower triangular matrix and L' is the transpose of L—an upper triangular matrix. This factorization of A is called the *Cholesky factorization.*

The Cholesky factorization may be obtained by successively factoring the principal submatrices A_i as

$$\text{(10)} \qquad A_i = L_i L_i', \qquad i = 1, 2, \ldots, n.$$

We have

$$A_1 = L_1 L_1', \qquad L_1 = [\sqrt{a_{11}}].$$

Direct calculation using (8) yields that if $A_{i-1} = L_{i-1}L'_{i-1}$, then we also have $A_i = L_i L'_i$, where

$$L_i = \begin{bmatrix} L_{i-1} & 0 \\ l'_i & \lambda_{ii} \end{bmatrix}, \tag{11}$$

$$l_i = L_{i-1}^{-1}\alpha_i, \tag{12}$$

$$\lambda_{ii} = \sqrt{a_{ii} - l'_i l_i}, \tag{13}$$

and α_i is given by (9). Thus, to show that the factorization given above is valid, it will be sufficient to show that

$$a_{ii} - l'_i l_i > 0,$$

and thus λ_{ii} is well defined as a real number from (13). Indeed define $b = A_{i-1}^{-1}\alpha_i$. Then because A_i is positive definite, we have

$$\begin{aligned} 0 < [b' \quad -1]A_i\begin{bmatrix} b \\ -1 \end{bmatrix} &= b'A_{i-1}b - 2b'\alpha_i + a_{ii} \\ &= b'\alpha_i - 2b'\alpha_i + a_{ii} = a_{ii} - b'\alpha_i \\ &= a_{ii} - \alpha'_i A_{i-1}^{-1}\alpha_i = a_{ii} - \alpha'_i(L_{i-1}L'_{i-1})^{-1}\alpha_i \\ &= a_{ii} - (L_{i-1}^{-1}\alpha_i)'(L_{i-1}^{-1}\alpha_i) = a_{ii} - l'_i l_i. \end{aligned}$$

Thus, λ_{ii} as defined by (13) is well defined as a positive real number. In order to show uniqueness of the factorization, a similar induction argument may be used. The matrix A_1 has a unique factorization, and if A_{i-1} has a unique factorization $A_{i-1} = L_{i-1}L'_{i-1}$, then L_i is uniquely determined by the requirement $A_i = L_i L'_i$ and Eqs. (8)–(13).

In practice the Cholesky factorization is computed via the algorithm (10)–(13) or some other essentially equivalent algorithm. Naturally the vectors l_i in (12) are computed by solving the triangular system

$$L_{i-1}l_i = \alpha_i$$

rather than by inverting the matrix L_{i-1}. For large n the process requires approximately $n^3/6$ multiplications.

1.3 Unconstrained Minimization

We provide an overview of analytical and computational methods for solution of the problem

$$\text{(UP)} \qquad \begin{aligned} &\text{minimize} \quad f(x) \\ &\text{subject to} \quad x \in R^n, \end{aligned}$$

where $f: R^n \to R$ is a given function. We say that a vector x^* is a *local minimum* for (UP) if there exists an $\varepsilon > 0$ such that

$$f(x^*) \le f(x) \qquad \forall\, x \in S(x^*; \varepsilon).$$

It is a *strict local minimum* if there exists an $\varepsilon > 0$ such that

$$f(x^*) < f(x) \qquad \forall\, x \in S(x^*; \varepsilon), \quad x \ne x^*.$$

We have the following well-known optimality conditions. Proofs may be found, for example, in Luenberger (1973).

Proposition 1.3: Assume that x^* is a local minimum for (UP) and, for some $\varepsilon > 0, f \in C^1$ over $S(x^*; \varepsilon)$. Then

$$\nabla f(x^*) = 0.$$

If in addition $f \in C^2$ over $S(x^*; \varepsilon)$, then

$$\nabla^2 f(x^*) \ge 0.$$

In what follows, we refer to a vector x^* satisfying $\nabla f(x^*) = 0$ as a *critical point.*

Proposition 1.4: Let x^* be such that, for some $\varepsilon > 0, f \in C^2$ over $S(x^*; \varepsilon)$ and

$$\nabla f(x^*) = 0, \qquad \nabla^2 f(x^*) > 0.$$

Then x^* is a strict local minimum for (UP). In fact, there exist scalars $\gamma > 0$ and $\delta > 0$ such that

$$f(x) \ge f(x^*) + \gamma |x - x^*|^2 \qquad \forall\, x \in S(x^*; \delta).$$

When x^* satisfies the assumptions of Proposition 1.4 we say that it is a *strong local minimum* for (UP).

We say that x^* is a *global minimum* for (UP) if

$$f(x^*) \le f(x) \qquad \forall\, x \in R^n.$$

Under convexity assumptions on f, we have the following necessary and sufficient condition:

Proposition 1.5: Assume that $f \in C^1$ and is convex over R^n. Then a vector x^* is a global minimum for (UP) if and only if

$$\nabla f(x^*) = 0.$$

Existence of global minima can be guaranteed under the assumptions of the following proposition which is a direct consequence of Weierstrass' theorem (a continuous function attains a global minimum over a compact set).

Proposition 1.6: If f is continuous over R^n and $f(x_k) \to \infty$ for every sequence $\{x_k\}$ such that $|x_k| \to \infty$, or, more generally, if the set $\{x \mid f(x) \leq \alpha\}$ is nonempty and compact for some $\alpha \in R$, then there exists a global minimum for (UP).

1.3.1 Convergence Analysis of Gradient Methods

We assume, without further mention throughout the remainder of Section 1.3, *that* $f \in C^1$ *over* R^n. The reader can easily make appropriate adjustments if $f \in C^1$ over an open subset of R^n only.

Most of the known iterative algorithms for solving (UP) take the form

$$x_{k+1} = x_k + \alpha_k d_k,$$

where if $\nabla f(x_k) \neq 0$, d_k is a *descent direction*, i.e., satisfies

$$d_k' \nabla f(x_k) < 0 \qquad \text{if} \quad \nabla f(x_k) \neq 0,$$

$$d_k = 0 \qquad \text{if} \quad \nabla f(x_k) = 0.$$

The scalar α_k is a positive stepsize parameter. We refer to such an algorithm as a *generalized gradient method* (or simply *gradient method*). Specific gradient methods that we shall consider include the method of steepest descent $[d_k = -\nabla f(x_k)]$ and scaled versions of it, Newton's method, the conjugate gradient method, quasi-Newton methods, and variations thereof. We shall examine several such methods in this section. For the time being, we focus on the convergence behavior of gradient methods. Rate of convergence issues will be addressed in the next subsection.

Stepsize Selection and Global Convergence

There are a number of rules for choosing the stepsize α_k [assuming $\nabla f(x_k) \neq 0$]. We list some that are used widely in practice:

(a) *Minimization rule:* Here α_k is chosen so that

$$f(x_k + \alpha_k d_k) = \min_{\alpha \geq 0} f(x_k + \alpha d_k).$$

(b) *Limited minimization rule:* A fixed number $s > 0$ is selected and α_k is chosen so that

$$f(x_k + \alpha_k d_k) = \min_{\alpha \in [0, s]} f(x_k + \alpha d_k).$$

(c) *Armijo rule:* Fixed scalars s, β, and σ with $s > 0$, $\beta \in (0, 1)$, and $\sigma \in (0, \frac{1}{2})$ are selected, and we set $\alpha_k = \beta^{m_k} s$, where m_k is the first nonnegative integer m for which

$$f(x_k) - f(x_k + \beta^m s d_k) \geq -\sigma \beta^m s \nabla f(x_k)' d_k,$$

i.e., $m = 0, 1, \ldots$ are tried successively until the inequality above is satisfied for $m = m_k$. (A variation of this rule is to use, instead of a fixed initial stepsize s, a sequence $\{s_k\}$ with $s_k > 0$ for all k. But this case can be reduced to the case of a fixed stepsize s by redefining the direction d_k to be $\tilde{d}_k = (s_k/s)d_k$.)

(d) *Goldstein rule:* A fixed scalar $\sigma \in (0, \frac{1}{2})$ is selected, and α_k is chosen to satisfy

$$\sigma \le \frac{f(x_k + \alpha_k d_k) - f(x_k)}{\alpha_k \nabla f(x_k)' d_k} \le 1 - \sigma.$$

It is possible to show that if f is bounded below there exists an interval of stepsizes α_k for which the relation above is satisfied, and there are fairly simple algorithms for finding such a stepsize through a finite number of arithmetic operations. However the Goldstein rule is primarily used in practice in conjunction with minimization rules in a scheme whereby an initial trial stepsize is chosen and tested to determine whether it satisfies the relation above. If it does, it is accepted. If not, a (perhaps approximate) line minimization is performed.

(e) *Constant stepsize:* Here a fixed stepsize $s > 0$ is selected and

$$\alpha_k = s \qquad \forall\, k.$$

The minimization and limited minimization rules must be implemented with the aid of one-dimensional line search algorithms (see, e.g., Luenberger, 1973; Avriel, 1976). In general, one cannot compute exactly the minimizing stepsize, and in practice, the line search is stopped once a stepsize α_k satisfying some termination criterion is obtained. An example of such a criterion is that α_k satisfies simultaneously

$$f(x_k) - f(x_k + \alpha_k d_k) \ge -\sigma \alpha_k \nabla f(x_k)' d_k \tag{1}$$

and

$$|\nabla f(x_k + \alpha_k d_k)' d_k| \le \beta |\nabla f(x_k)' d_k|, \tag{2}$$

where σ and β are some scalars with $\sigma \in (0, \frac{1}{2})$ and $\beta \in (\sigma, 1)$. If α_k is indeed a minimizing stepsize then $\nabla f(x_k + \alpha_k d_k)' d_k = \partial f(x_k + \alpha_k d_k)/\partial \alpha = 0$, so (2) is in effect a test on the accuracy of the minimization. Relation (1), in view of $\nabla f(x_k)' d_k < 0$, guarantees a function decrease. Usually σ is chosen very close to zero, for example $\sigma \in [10^{-5}, 10^{-1}]$, but trial and error must be relied upon for the choice of β. Sometimes (2) is replaced by the less stringent condition

$$\nabla f(x_k + \alpha_k d_k)' d_k \ge \beta \nabla f(x_k)' d_k. \tag{3}$$

The following lemma shows that under mild assumptions there is an interval of stepsizes α satisfying (1), (2) or (1), (3).

Lemma 1.7: Assume that there is a scalar M such that $f(x) \geq M$ for all $x \in R^n$, let $\sigma \in (0, \frac{1}{2})$ and $\beta \in (\sigma, 1)$, and assume that $\nabla f(x_k)'d_k < 0$. There exists an interval $[c_1, c_2]$ with $0 < c_1 < c_2$, such that every $\alpha \in [c_1, c_2]$ satisfies (1) and (2) [and hence also (1) and (3)].

Proof: Define $g(\alpha) = f(x_k + \alpha d_k)$. Note that $\partial g(\alpha)/\partial\alpha = \nabla f(x_k + \alpha d_k)'d_k$. Let $\hat{\beta}$ be such that $\sigma < \hat{\beta} < \beta$, and consider the set A defined by

$$A = \left\{\alpha \geq 0 \,\middle|\, \hat{\beta}\frac{\partial g(0)}{\partial\alpha} \leq \frac{\partial g(\alpha)}{\partial\alpha} \leq 0\right\}.$$

Since $g(\alpha)$ is bounded below and $\partial g(0)/\partial\alpha = \nabla f(x_k)'d_k < 0$ it is easily seen that A is nonempty. Let

$$\hat{\alpha} = \min\{\alpha \mid \alpha \in A\}.$$

Clearly $\hat{\alpha} > 0$ and it is easy to see using the fact $\hat{\beta} < \beta$ that

$$\frac{\partial g(\alpha)}{\partial\alpha} \leq \hat{\beta}\frac{\partial g(0)}{\partial\alpha} \leq 0, \qquad \forall\, \alpha \in [0, \hat{\alpha}], \tag{4}$$

and there exists a scalar $\delta_1 \in (0, \hat{\alpha})$ such that

$$\left|\frac{\partial g(\alpha)}{\partial\alpha}\right| = |\nabla f(x_k + \alpha d_k)'d_k| \leq \beta|\nabla f(x_k)'d_k|$$
$$= \beta\left|\frac{\partial g(0)}{\partial\alpha}\right|, \qquad \forall\, \alpha \in [\hat{\alpha} - \delta_1, \hat{\alpha} + \delta_1].$$

We have from (4)

$$g(\hat{\alpha}) = g(0) + \int_0^{\hat{\alpha}} \frac{\partial g(t)}{\partial\alpha}\,dt \leq g(0) + \hat{\beta}\hat{\alpha}\frac{\partial g(0)}{\partial\alpha} < g(0) + \sigma\hat{\alpha}\frac{\partial g(0)}{\partial\alpha},$$

or equivalently

$$f(x_k) - f(x_k + \hat{\alpha}d_k) > -\sigma\hat{\alpha}\nabla f(x_k)'d_k.$$

Hence there exists a scalar $\delta_2 \in (0, \hat{\alpha})$ such that

$$f(x_k) - f(x_k + \alpha d_k) \geq -\sigma\alpha\nabla f(x_k)'d_k, \qquad \forall\, \alpha \in [\hat{\alpha} - \delta_2, \hat{\alpha} + \delta_2].$$

Take $\delta = \min\{\delta_1, \delta_2\}$. Then for all α in the interval $[\hat{\alpha} - \delta, \hat{\alpha} + \delta]$ both inequalities (1) and (2) are satisfied. Q.E.D.

In practice a line search procedure may have to be equipped with various mechanisms that guarantee that a stepsize satisfying the termination criteria will indeed be obtained. We refer the reader to more specific literature for details. In all cases, it is important to have a reasonably good initial stepsize (or equivalently to scale the direction d_k in a reasonable manner). We discuss this in the next paragraph within the context of the Armijo rule.

The Armijo rule is very easy to implement and requires only one gradient evaluation per iteration. The process by which α_k is determined is shown in Fig. 1.1. We start with the trial point $(x_k + sd_k)$ and continue with $(x_k + \beta sd_k)$, $(x_k + \beta^2 sd_k), \ldots$ until the first time that $\beta^m s$ falls within the set of stepsizes α satisfying the desired inequality. While this set need not be an interval, it will always contain an interval of the form $[0, \delta]$ with $\delta > 0$, provided $\nabla f(x_k)'d_k < 0$. For this reason the stepsize α_k chosen by the Armijo rule is well defined and will be found after a finite number of trial evaluations of the value of f at the points $(x_k + sd_k)$, $(x_k + \beta sd_k), \ldots$. Usually σ is chosen close to zero, for example, $\sigma \in [10^{-5}, 10^{-1}]$. The scalar β is usually chosen from $\frac{1}{2}$ to 10^{-1} depending on the confidence we have on the quality of the initial stepsize s. Actually one can always take $s = 1$ and multiply the direction d_k by a scaling factor. Many methods incorporate automatic scaling of the direction d_k, which makes $s = 1$ a good stepsize choice (compare with Proposition 1.15 and the discussion on rate of convergence later in this section). If a suitable scaling factor for d_k is not known, one may use various ad hoc schemes to determine one. A simple possibility is to select a point $\bar{\alpha}$ on the line $\{x_k + \alpha d_k \mid \alpha > 0\}$, evaluate $f(x_k + \bar{\alpha} d_k)$, and perform a quadratic

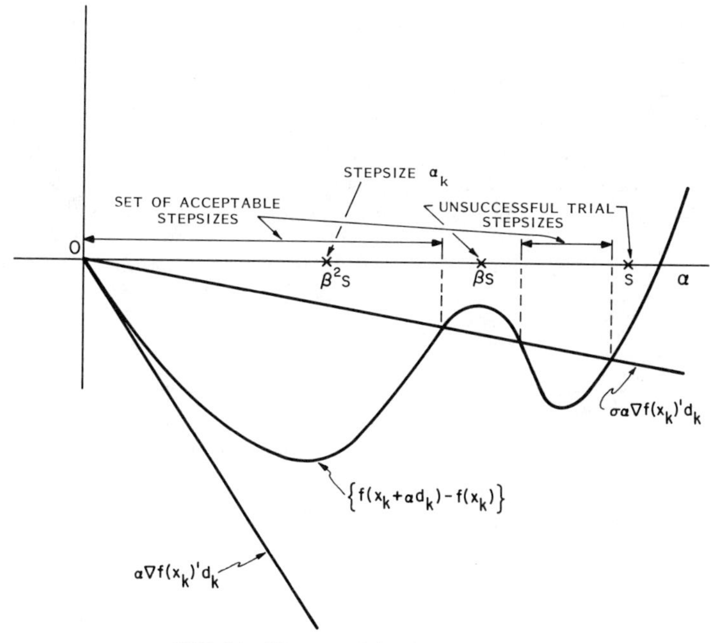

FIG. 1.1 Line search by the Armijo rule

interpolation on the basis of $f(x_k)$, $\nabla f(x_k)'d_k = \partial f(x_k + \alpha d_k)/\partial\alpha|_{\alpha=0}$, and $f(x_k + \bar{\alpha}d_k)$. If $\tilde{\alpha}$ minimizes the quadratic interpolation, d_k is replaced by $\tilde{d}_k = \tilde{\alpha}d_k$, and an initial stepsize $s = 1$ is used.

The constant stepsize rule is the simplest. It is useful in problems where evaluation of the objective function is expensive and an appropriate constant stepsize value is known or can be determined fairly easily. Interestingly enough, this is the case in the method of multipliers as we shall explain in the next chapter.

We now introduce a condition on the directions d_k of a gradient method.

Definition: Let $\{x_k\}$ be a sequence generated by a gradient method $x_{k+1} = x_k + \alpha_k d_k$. We say that the sequence $\{d_k\}$ is *uniformly gradient related* to $\{x_k\}$ if for every convergent subsequence $\{x_k\}_K$ for which

$$\lim_{\substack{k\to\infty \\ k\in K}} \nabla f(x_k) \neq 0 \tag{5}$$

there holds

$$0 < \liminf_{\substack{k\to\infty \\ k\in K}} |\nabla f(x_k)'d_k|, \qquad \limsup_{\substack{k\to\infty \\ k\in K}} |d_k| < \infty. \tag{6}$$

In words, $\{d_k\}$ is uniformly gradient related if whenever a subsequence $\{\nabla f(x_k)\}_K$ tends to a nonzero vector, the corresponding subsequence of directions d_k is bounded and does not tend to be orthogonal to $\nabla f(x_k)$. Another way of putting it is that (5) and (6) require that d_k does not become "too small" or "too large" relative to $\nabla f(x_k)$ and the angle between d_k and $\nabla f(x_k)$ does not get "too close" to $\pi/2$. Two examples of simple conditions that, if satisfied for some scalars $c_1 > 0$, $c_2 > 0$, $p_1 \geq 0$, and $p_2 \geq 0$ and all k, guarantee that $\{d_k\}$ is uniformly gradient related are

(a) $|d_k| \leq c_2 |\nabla f(x_k)|^{p_2}, \qquad c_1 |\nabla f(x_k)|^{p_1} \leq -\nabla f(x_k)'d_k;$

(b) $d_k = -D_k \nabla f(x_k)$,

with D_k a positive definite symmetric matrix satisfying

$$c_1 |\nabla f(x_k)|^{p_1} |z|^2 \leq z'D_k z \leq c_2 |\nabla f(x_k)|^{p_2} |z|^2 \qquad \forall\, z \in R^n.$$

For example, in the method of steepest descent where $D_k = I$, this condition is satisfied if we take $c_1 = c_2 = 1$, $p_1 = p_2 = 0$.

We have the following convergence result:

Proposition 1.8: Let $\{x_k\}$ be a sequence generated by a gradient method $x_{k+1} = x_k + \alpha_k d_k$ and assume that $\{d_k\}$ is uniformly gradient related and α_k is chosen by the minimization rule, or the limited minimization rule, or the Armijo rule. Then every limit point of $\{x_k\}$ is a critical point.

Proof: Consider first the Armijo rule. Assume the contrary, i.e., that $\bar{x}$ is a limit point with $\nabla f(\bar{x}) \neq 0$. Then since $\{f(x_k)\}$ is monotonically decreasing and f is continuous, it follows that $\{f(x_k)\}$ converges to $f(\bar{x})$. Hence,

$$[f(x_k) - f(x_{k+1})] \to 0.$$

By the definition of the Armijo rule, we have

$$f(x_k) - f(x_{k+1}) \geq -\sigma \alpha_k \nabla f(x_k)' d_k.$$

Hence, $\alpha_k \nabla f(x_k)' d_k \to 0$. Let $\{x_k\}_K$ be the subsequence converging to $\bar{x}$. Since $\{d_k\}$ is uniformly gradient related, we have

$$\liminf_{\substack{k \to \infty \\ k \in K}} |\nabla f(x_k)' d_k| > 0,$$

and hence,

$$\{\alpha_k\}_K \to 0.$$

Hence, by the definition of the Armijo rule, we must have for some index $\bar{k} \geq 0$

$$(7) \quad f(x_k) - f[x_k + (\alpha_k/\beta) d_k] < -\sigma(\alpha_k/\beta) \nabla f(x_k)' d_k \qquad \forall\, k \in K, \quad k \geq \bar{k};$$

i.e., the initial stepsize s will be reduced at least once for all $k \in K$, $k \geq \bar{k}$. Denote

$$p_k = d_k/|d_k|, \qquad \bar{\alpha}_k = \alpha_k |d_k|/\beta.$$

Since $\{d_k\}$ is uniformly gradient related, we have $\limsup_{k \to \infty, k \in K} |d_k| < \infty$, and it follows that

$$\{\bar{\alpha}_k\}_K \to 0.$$

Since $|p_k| = 1$ for all $k \in K$, there exists a subsequence $\{p_k\}_{\bar{K}}$ of $\{p_k\}_K$ such that $\{p_k\}_{\bar{K}} \to \bar{p}$ where $\bar{p}$ is some vector with $|\bar{p}| = 1$. From (7), we have

$$(8) \qquad \frac{f(x_k) - f(x_k + \bar{\alpha}_k p_k)}{\bar{\alpha}_k} < -\sigma \nabla f(x_k)' p_k \qquad \forall\, k \in \bar{K}, \quad k \geq \bar{k}.$$

Taking limits in (8) we obtain

$$-\nabla f(\bar{x})' \bar{p} \leq -\sigma \nabla f(\bar{x})' \bar{p} \qquad \text{or} \qquad 0 \leq (1 - \sigma) \nabla f(\bar{x})' \bar{p}.$$

Since $\sigma < 1$, we obtain

$$(9) \qquad 0 \leq \nabla f(\bar{x})' \bar{p}.$$

On the other hand, we have

$$-\nabla f(x_k)' p_k = -\nabla f(x_k)' d_k / |d_k|.$$

By taking the limit as $k \in \bar{K}$, $k \to \infty$,

$$-\nabla f(\bar{x})'\bar{p} \geq \frac{\liminf |\nabla f(x_k)'d_k|}{\limsup |d_k|} > 0,$$

which contradicts (9). This proves the result for the Armijo rule.

Consider now the minimization rule, and let $\{x_k\}_K$ converge to $\bar{x}$ with $\nabla f(\bar{x}) \neq 0$. Again we have that $\{f(x_k)\}$ decreases monotonically to $f(\bar{x})$. Let $\tilde{x}_{k+1}$ be the point generated from x_k via the Armijo rule, and let $\tilde{\alpha}_k$ be the corresponding stepsize. We have

$$f(x_k) - f(x_{k+1}) \geq f(x_k) - f(\tilde{x}_{k+1}) \geq -\sigma\tilde{\alpha}_k \nabla f(x_k)'d_k.$$

By simply replacing α_k by $\tilde{\alpha}_k$ and repeating the arguments of the earlier proof, we obtain a contradiction. In fact the line of argument just used establishes that *any stepsize rule that gives a larger reduction in objective function value at each step than the Armijo rule inherits its convergence properties.* This proves also the proposition for the limited minimization rule. Q.E.D.

Similarly the following proposition can be shown to be true. Its proof is left to the reader.

Proposition 1.9: The conclusions of Proposition 1.8 hold if $\{d_k\}$ is uniformly gradient related and α_k is chosen by the Goldstein rule or satisfies (1) and (2) for all k.

The next proposition establishes, among other things, convergence for the case of a constant stepsize.

Proposition 1.10: Let $\{x_k\}$ be a sequence generated by a gradient method $x_{k+1} = x_k + \alpha_k d_k$, where $\{d_k\}$ is uniformly gradient related. Assume that for some constant $L > 0$, we have

(10) $$|\nabla f(x) - \nabla f(y)| \leq L|x - y| \qquad \forall\, x, y \in R^n,$$

and that there exists a scalar ε such that for all k we have $d_k \neq 0$ and

(11) $$0 < \varepsilon \leq \alpha_k \leq \frac{2 - \varepsilon}{L} \frac{|\nabla f(x_k)'d_k|}{|d_k|^2}.$$

Then every limit point of $\{x_k\}$ is a critical point of f.

NOTE: If $\{d_k\}$ is such that there exist $c_1, c_2 > 0$ such that for all k we have

(12) $$-\nabla f(x_k)'d_k \geq c_1|\nabla f(x_k)|^2, \qquad c_2|\nabla f(x_k)|^2 \geq |d_k|^2,$$

then (11) is satisfied if for all k we have

$$0 < \varepsilon \le \alpha_k \le (2 - \varepsilon)c_1/Lc_2. \tag{13}$$

For steepest descent $[d_k = -\nabla f(x_k)]$ in particular, we can take $c_1 = c_2 = 1$, and the condition on the stepsize becomes

$$0 < \varepsilon \le \alpha_k \le (2 - \varepsilon)/L.$$

Proof: We have the following equality for $\alpha \ge 0$,

$$f(x_k + \alpha d_k) = f(x_k) + \alpha \nabla f(x_k)' d_k + \int_0^\alpha [\nabla f(x_k + t d_k) - \nabla f(x_k)]' d_k \, dt.$$

By using (10), we obtain

$$\begin{aligned} f(x_k + \alpha d_k) - f(x_k) &\le \alpha \nabla f(x_k)' d_k + \int_0^\alpha |\nabla f(x_k + t d_k) - \nabla f(x_k)| \, |d_k| \, dt \\ &\le \alpha \nabla f(x_k)' d_k + \int_0^\alpha t L |d_k|^2 \, dt \\ &= \alpha[-|\nabla f(x_k)' d_k| + \tfrac{1}{2}\alpha L |d_k|^2]. \end{aligned}$$

From (11), we have $\alpha_k \ge \varepsilon$ and $\frac{1}{2}\alpha_k L |d_k|^2 - |\nabla f(x_k)' d_k| \le -\frac{1}{2}\varepsilon |\nabla f(x_k)' d_k|$. Using these relations in the inequality above, we obtain

$$f(x_k) - f(x_k + \alpha_k d_k) \ge \tfrac{1}{2}\varepsilon^2 |\nabla f(x_k)' d_k|.$$

Now if a subsequence $\{x_k\}_K$ converges to a noncritical point $\bar{x}$, the above relation implies that $|\nabla f(x_k)' d_k| \to 0$. But this contradicts the fact that $\{d_k\}$ is uniformly gradient related. Hence, every limit point of $\{x_k\}$ is critical. Q.E.D.

Note that when $d_k = -D_k \nabla f(x_k)$ with D_k positive definite symmetric, relation (12) holds with

$$c_1 = \bar{\gamma}, \qquad c_2 = \bar{\Gamma}^2,$$

if the eigenvalues of D_k lie in the interval $[\bar{\gamma}, \bar{\Gamma}]$ for all k. It is also possible to show that (10) is satisfied for some $L > 0$, if $f \in C^2$ and the Hessian $\nabla^2 f$ is bounded over R^n. Unfortunately, however, it is difficult in general to obtain an estimate of L and thus in most cases the interval of stepsizes in (11) or (13) which guarantees convergence is not known a priori. Thus, experimentation with the problem at hand is necessary in order to obtain a range of stepsize values which lead to convergence. We note, however, that in the method of multipliers, it is possible to obtain a satisfactory estimate of L as will be explained in Chapter 2.

Gradient Convergence

The convergence results given so far are concerned with limit points of the sequence $\{x_k\}$. It can also be easily seen that the corresponding sequence $\{f(x_k)\}$ will converge to some value whenever $\{x_k\}$ has at least one limit point and there holds $f(x_{k+1}) \leq f(x_k)$ for all k. Concerning the sequence $\{\nabla f(x_k)\}$, we have by continuity of ∇f that if a subsequence $\{x_k\}_K$ converges to some point $\bar{x}$ then $\{\nabla f(x_k)\}_K \to \nabla f(\bar{x})$. If $\bar{x}$ is critical, then $\{\nabla f(x_k)\}_K \to 0$. More generally, we have the following result:

Proposition 1.11: Let $\{x_k\}$ be a sequence generated by a gradient method $x_{k+1} = x_k + \alpha_k d_k$, which is convergent in the sense that every limit point of sequences that it generates is a critical point of f. Then if $\{x_k\}$ is a bounded sequence, we have $\nabla f(x_k) \to 0$.

Proof: Assume the contrary, i.e., that there exists a subsequence $\{x_k\}_K$ and an $\varepsilon > 0$ such that $|\nabla f(x_k)| \geq \varepsilon$ for all $k \in K$. Since $\{x_k\}_K$ is bounded, it has at least one limit point $\bar{x}$ and we must have $|\nabla f(\bar{x})| \geq \varepsilon$. But this contradicts our hypothesis which implies that $\bar{x}$ must be critical. Q.E.D.

The proposition above forms the basis for terminating the iterations of gradient methods. Thus, computation is stopped when a point $x_{\bar{k}}$ is obtained with

$$|\nabla f(x_{\bar{k}})| \leq \varepsilon, \tag{14}$$

where ε is a small positive scalar. The point $x_{\bar{k}}$ is considered for practical purposes to be a critical point. Sometimes one terminates computation when the norm of the direction d_k becomes too small; i.e.,

$$|d_{\bar{k}}| \leq \varepsilon. \tag{15}$$

If d_k satisfies

$$c_1 |\nabla f(x_k)|^{p_1} \leq |d_k| \leq c_2 |\nabla f(x_k)|^{p_2}$$

for some positive scalars c_1, c_2, p_1, p_2, and all k, then the termination criterion (15) is of the same nature as (14). Unfortunately, it is not known a priori how small one should take ε in order to guarantee that the final point $x_{\bar{k}}$ is a "good" approximation to a stationary point. For this reason it is necessary to conduct some experimentation prior to settling on a reasonable termination criterion for a given problem, unless bounds are known (or can be estimated) for the Hessian matrix of f (see the following exercise).

Exercise: Let x^* be a local minimum of f and assume that for all x in a sphere $S(x^*; \delta)$ we have, for some $m > 0$ and $M > 0$,

$$m|z|^2 \leq z'\nabla^2 f(x)z \leq M|z|^2 \qquad \forall\, z \in R^n.$$

Then every $x \in S(x^*; \delta)$ satisfying $|\nabla f(x)| \le \varepsilon$ also satisfies

$$|x - x^*| \le \varepsilon/m, \qquad f(x) - f(x^*) \le M\varepsilon^2/2m^2.$$

Local Convergence

A weakness of the convergence results of the preceding subsection is that they do not guarantee that convergence (to a single point) of the generated sequence $\{x_k\}$ will occur. Thus, the sequence $\{x_k\}$ may have one, more than one, or no limit points at all. It is not infrequent for a gradient method to generate an unbounded sequence $\{x_k\}$. This will typically occur if the function f has no critical point or if f decreases monotonically as $|x| \to \infty$ along some directions. However $\{x_k\}$ will have at least one limit point if the set $\{x \mid f(x) \le f(x_0)\}$ is bounded or more generally if $\{x_k\}$ is a bounded sequence.

On the other hand, practical experience suggests that a sequence generated by a gradient method will rarely have more than one critical limit point. This is not very surprising since the generated sequence of function values $\{f(x_k)\}$ is monotonically nonincreasing and will always converge to a finite value whenever $\{x_k\}$ has at least one limit point. Hence, any two critical limit points, say $\bar{x}$ and $\tilde{x}$, of the sequence $\{x_k\}$ must simultaneously satisfy $\nabla f(\bar{x}) = \nabla f(\tilde{x}) = 0$ and $f(\bar{x}) = f(\tilde{x}) = \lim_{k\to\infty} f(x_k)$. These relations are unlikely to hold if the critical points of f are "isolated" points. One may also prove that if f has a *finite* number of critical points and the Armijo rule or the limited minimization rule is used in connection with a gradient method with uniformly gradient-related direction sequence $\{d_k\}$, then the generated sequence $\{x_k\}$ will converge to a unique critical point provided that $\{x_k\}$ is a bounded sequence. We leave this as an exercise for the reader.

The following proposition may also help to explain to some extent why sequences generated by gradient methods tend to have unique limit points. It states that strong local minima tend to attract gradient methods.

Proposition 1.12: Let $f \in C^2$ and $\{x_k\}$ be a sequence satisfying $f(x_{k+1}) \le f(x_k)$ for all k and generated by a gradient method $x_{k+1} = x_k + \alpha_k d_k$ which is convergent in the sense that every limit point of sequences that it generates is a critical point of f. Assume that there exist scalars $s > 0$ and $c > 0$ such that for all k there holds $\alpha_k \le s$ and $|d_k| \le c|\nabla f(x_k)|$. Then for every local minimum x^* of f with $\nabla^2 f(x^*) > 0$, there exists an open set L containing x^* such that if $x_{\bar{k}} \in L$ for some $\bar{k} \ge 0$ then $x_k \in L$ for all $k \ge \bar{k}$ and $\{x_k\} \to x^*$. Furthermore, given any scalar $\varepsilon > 0$, the set L can be chosen so that $L \subset S(x^*; \varepsilon)$.

NOTE: The condition $\alpha_k \le s$ is satisfied for the Armijo rule and the limited minimization rule. The condition $|d_k| \le c|\nabla f(x_k)|$ is satisfied if $d_k = -D_k \nabla f(x_k)$ with the eigenvalues of D_k uniformly bounded from above.

Proof: Let x^* be a local minimum with $\nabla^2 f(x^*)$ positive definite. Then there exists $\bar{\varepsilon} > 0$ such that for all x with $|x - x^*| \leq \bar{\varepsilon}$, the matrix $\nabla^2 f(x)$ is also positive definite. Denote

$$\gamma = \min_{\substack{|x - x^*| \leq \bar{\varepsilon} \\ |z| = 1}} z'\nabla^2 f(x)z, \qquad \Gamma = \max_{\substack{|x - x^*| \leq \bar{\varepsilon} \\ |z| = 1}} z'\nabla^2 f(x)z.$$

We have $\gamma > 0$ and $\Gamma > 0$. Consider the open set

$$L = \{x \,|\, |x - x^*| < \bar{\varepsilon},\ f(x) < f(x^*) + \tfrac{1}{2}\gamma[\bar{\varepsilon}/(1 + sc\Gamma)]^2\}.$$

We claim that if $x_{\bar{k}} \in L$ for some $\bar{k} \geq 0$ then $x_k \in L$ for all $k \geq \bar{k}$ and furthermore $x_k \to x^*$.

Indeed if $x_{\bar{k}} \in L$ then by using Taylor's theorem, we have

$$\tfrac{1}{2}\gamma|x_{\bar{k}} - x^*|^2 \leq f(x_{\bar{k}}) - f(x^*) < \tfrac{1}{2}\gamma[\bar{\varepsilon}/(1 + sc\Gamma)]^2$$

from which we obtain

$$|x_{\bar{k}} - x^*| < \bar{\varepsilon}/(1 + cs\Gamma). \tag{16}$$

On the other hand, we have

$$\begin{aligned} |x_{\bar{k}+1} - x^*| &= |x_{\bar{k}} - x^* + \alpha_{\bar{k}} d_{\bar{k}}| \leq |x_{\bar{k}} - x^*| + \alpha_{\bar{k}}|d_{\bar{k}}| \\ &\leq |x_{\bar{k}} - x^*| + sc|\nabla f(x_{\bar{k}})|. \end{aligned}$$

By using Taylor's theorem, we have $|\nabla f(x_{\bar{k}})| \leq \Gamma|x_{\bar{k}} - x^*|$ and substituting in the inequality above, we obtain

$$|x_{\bar{k}+1} - x^*| \leq (1 + sc\Gamma)|x_{\bar{k}} - x^*|.$$

By combining this relation with (16), we obtain

$$|x_{\bar{k}+1} - x^*| < \bar{\varepsilon}.$$

Furthermore, using the hypothesis $f(x_{k+1}) \leq f(x_k)$ for all k, we have

$$f(x_{\bar{k}+1}) \leq f(x_{\bar{k}}) < f(x^*) + \tfrac{1}{2}\gamma[\bar{\varepsilon}/(1 + sc\Gamma)]^2.$$

It follows from the above two inequalities that $x_{\bar{k}+1} \in L$ and similarly $x_k \in L$ for all $k \geq \bar{k}$. Let $\bar{L}$ be the closure of L. Since $\bar{L}$ is a compact set, the sequence $\{x_k\}$ will have at least one limit point which by assumption must be a critical point of f. Now the only critical point of f within $\bar{L}$ is the point x^* (since f is strictly convex within $\bar{L}$). Hence $x_k \to x^*$. Finally given any $\varepsilon > 0$, we can choose $\bar{\varepsilon} \leq \varepsilon$ in which case $L \subset S(x^*; \varepsilon)$. Q.E.D.

Rate of Convergence—Quadratic Objective Function

The second major question relating to the behavior of a gradient method concerns the speed (or rate) of convergence of generated sequences $\{x_k\}$. The mere fact that x_k converges to a critical point x^* will be of little value in practice unless the points x_k are reasonably close to x^* after relatively few iterations. Thus, the study of the rate of convergence of an algorithm or a class of algorithms not only provides useful information regarding computational efficiency, but also delineates what in most cases are the dominant criteria for selecting one algorithm in favor of others for solving a particular problem.

Most of the important characteristics of gradient methods are revealed by investigation of the case where the objective function is quadratic. Indeed, assume that a gradient method is applied to minimization of a function $f: R^n \to R$, $f \in C^2$, and it generates a sequence $\{x_k\}$ converging to a strong local minimum x^* where

$$\nabla f(x^*) = 0, \qquad \nabla^2 f(x^*) > 0.$$

Then we have, by Taylor's Theorem,

$$f(x) = f(x^*) + \tfrac{1}{2}(x - x^*)'\nabla^2 f(x^*)(x - x^*) + o(|x - x^*|^2),$$

where $o(|x - x^*|^2)/|x - x^*|^2 \to 0$ as $x \to x^*$. This implies that f can be accurately approximated near x^* by the quadratic function

$$f(x^*) + \tfrac{1}{2}(x - x^*)'\nabla^2 f(x^*)(x - x^*).$$

We thus expect that rate-of-convergence results obtained through analysis of the case where the objective function is the quadratic function above have direct analogs to the general case. The validity of this conjecture can indeed be established by rigorous analysis and has been substantiated by extensive numerical experimentation.

Consider the quadratic function

$$f(x) = \tfrac{1}{2}(x - x^*)'Q(x - x^*)$$

and the gradient method

$$x_{k+1} = x_k - \alpha_k D_k g_k, \tag{17}$$

where

$$g_k = \nabla f(x_k) = Q(x_k - x^*). \tag{18}$$

We assume that Q and D_k are positive definite and symmetric. Let

$$M_k = \text{max eigenvalue of } (D_k^{1/2} Q D_k^{1/2}),$$
$$m_k = \text{min eigenvalue of } (D_k^{1/2} Q D_k^{1/2}).$$

We have the following proposition:

Proposition 1.13: Consider iteration (17), and assume that α_k is chosen according to the minimization rule

$$f(x_k - \alpha_k D_k g_k) = \min_{\alpha \geq 0} f(x_k - \alpha D_k g_k).$$

Then

$$f(x_{k+1}) \le \left(\frac{M_k - m_k}{M_k + m_k}\right)^2 f(x_k). \tag{19}$$

Proof: The result clearly holds if $g_k = 0$, so we assume $g_k \neq 0$. We first compute the minimizing stepsize α_k. We have

$$\begin{aligned}(d/d\alpha) f(x_k - \alpha D_k g_k) &= -g_k' D_k Q(x_k - \alpha D_k g_k - x^*) \\ &= -g_k' D_k g_k + \alpha g_k' D_k Q D_k g_k.\end{aligned}$$

Hence, by setting this derivative equal to zero, we obtain

$$\alpha_k = g_k' D_k g_k / g_k' D_k Q D_k g_k. \tag{20}$$

We have, using (17) and (20),

$$\begin{aligned}f(x_{k+1}) &= f(x_k - \alpha_k D_k g_k) = \tfrac{1}{2}(x_k - x^* - \alpha_k D_k g_k)' Q(x_k - x^* - \alpha_k D_k g_k) \\ &= \tfrac{1}{2}(x_k - x^*)' Q(x_k - x^*) + \tfrac{1}{2}\alpha_k^2 g_k' D_k Q D_k g_k - \alpha_k g_k' D_k Q(x_k - x^*) \\ &= f(x_k) + \tfrac{1}{2}\alpha_k^2 g_k' D_k Q D_k g_k - \alpha_k g_k' D_k g_k,\end{aligned}$$

and finally

$$f(x_{k+1}) = f(x_k) - \frac{1}{2}\frac{(g_k' D_k g_k)^2}{g_k' D_k Q D_k g_k}. \tag{21}$$

Also we have

$$\begin{aligned}f(x_k) &= \tfrac{1}{2}(x_k - x^*)' Q(x_k - x^*) \\ &= \tfrac{1}{2}(x_k - x^*)' Q D_k^{1/2}(D_k^{1/2} Q D_k^{1/2})^{-1} D_k^{1/2} Q(x_k - x^*) \\ &= \tfrac{1}{2} g_k' D_k^{1/2}(D_k^{1/2} Q D_k^{1/2})^{-1} D_k^{1/2} g_k.\end{aligned} \tag{22}$$

Setting $y_k = D_k^{1/2} g_k$, $L_k = D_k^{1/2} Q D_k^{1/2}$, and using (21) and (22), we obtain

$$\begin{aligned}f(x_{k+1}) &= f(x_k) - \frac{(y_k' y_k)^2}{(y_k' L_k y_k)(y_k' L_k^{-1} y_k)} f(x_k) \\ &= \left[1 - \frac{(y_k' y_k)^2}{(y_k' L_k y_k)(y_k' L_k^{-1} y_k)}\right] f(x_k).\end{aligned} \tag{23}$$

We shall now need the following lemma, a proof of which can be found in Luenberger (1973, p. 151).

Lemma (Kantorovich Inequality): Let L be a positive definite symmetric $n \times n$ matrix. Then for any vector $y \in R^n$, $y \neq 0$, there holds

$$\frac{(y'y)^2}{(y'Ly)(y'L^{-1}y)} \ge \frac{4Mm}{(M + m)^2},$$

where M and m are the largest and smallest eigenvalues of L.

Returning to the proof of the proposition, we have by using Kantorovich's inequality in (23)

$$f(x_{k+1}) \le \left[1 - \frac{4M_k m_k}{(M_k + m_k)^2}\right] f(x_k) = \left(\frac{M_k - m_k}{M_k + m_k}\right)^2 f(x_k). \qquad \text{Q.E.D.}$$

From (19), we obtain, assuming $g_k \neq 0$ for all k,

$$\limsup_{k\to\infty} \frac{f(x_{k+1})}{f(x_k)} \le \limsup_{k\to\infty} \left(\frac{M_k - m_k}{M_k + m_k}\right)^2 \triangleq \beta.$$

If $\beta < 1$ (as will be the case if $\{m_k/M_k\}$ is bounded away from zero), it follows that $\{f(x_k)\}$ converges at least Q-linearly with convergence ratio β (see Section 1.2). If $\beta = 0$, then the convergence rate is *superlinear*. If $\beta < 1$, then the sequence $\{f(x_{k+1})\}$ is majorized for all k sufficiently large by any geometric progression of the form $q\bar{\beta}^k$, where $q > 0$, $\bar{\beta} > \beta$ (see Section 1.2). If γ is the minimum eigenvalue of Q, we have

$$\tfrac{1}{2}\gamma |x_k - x^*|^2 \le f(x_k)$$

so the same conclusion can be drawn for the sequence $\{|x_k - x^*|^2\}$. Relation (19) also indicates that the iteration $x_{k+1} = x_k - \alpha_k D_k g_k$ yields a large relative reduction in objective function value if $M_k/m_k \sim 1$. This shows that in order to achieve fast convergence, one should select D_k so that the eigenvalues of $D_k^{1/2} Q D_k^{1/2}$ are close together, such as when $D_k \sim Q^{-1}$, and this is the main motivation for introducing the matrix D_k instead of taking $D_k \equiv I$. If in particular $D_k = Q^{-1}$, then we obtain $M_k = m_k = 1$ and, from (19), $f(x_{k+1}) = 0$ which implies $x_{k+1} = x^*$; i.e., convergence to the minimum is attained in a single iteration.

When the ratio M_k/m_k is much larger than unity, then (19) indicates that convergence can be very slow. Actually, the speed of convergence of $\{x_k\}$ depends strongly on the starting point x_0. However, if D_k is constant, it is possible to show that there always exist "worst" starting points for which (19) is satisfied with equality for all k. [The reader may wish to verify this by considering the case $D_k \equiv I, f(x) = \frac{1}{2}\sum_{i=1}^n \gamma_i x_i^2$, where $0 < \gamma_1 \le \gamma_2 \le \cdots \le \gamma_n$, and the starting point $x_0 = (\gamma_1^{-1}, 0, \ldots, 0, \gamma_n^{-1})$.]

Similar convergence rate results can be obtained for the case of the limited minimization rule. For example, notice that from (20), we obtain

$$\alpha_k = y_k' y_k / y' D_k^{1/2} Q D_k^{1/2} y_k,$$

where $y_k = D_k^{1/2} g_k$. Hence, we have $\alpha_k \le 1/m_k$, and (19) also holds when α_k is chosen by the limited minimization rule

$$f(x_k - \alpha_k D_k g_k) = \min_{0 \le \alpha \le s} f(x_k - \alpha D_k g_k)$$

provided that

$$s \geq 1/m_k, \qquad k = 0, 1, \ldots .$$

Qualitatively, similar results are also obtained when other stepsize rules are used, such as a constant stepsize. We have the following proposition:

Proposition 1.14: Consider the iteration $x_{k+1} = x_k - \alpha_k D_k g_k$. For all $\alpha_k \geq 0$ and k, we have

$$(24) \qquad (x_{k+1} - x^*)' D_k^{-1}(x_{k+1} - x^*) \\ \leq \max\{|1 - \alpha_k m_k|^2, |1 - \alpha_k M_k|^2\}(x_k - x^*)' D_k^{-1}(x_k - x^*).$$

Furthermore, the right-hand side of (24) is minimized when

$$(25) \qquad \alpha_k = 2/(m_k + M_k),$$

and with this choice of α_k, we obtain

$$(26) \quad (x_{k+1} - x^*)' D_k^{-1}(x_{k+1} - x^*) \leq \left(\frac{M_k - m_k}{M_k + m_k}\right)^2 (x_k - x^*)' D_k^{-1}(x_k - x^*).$$

Proof: We have

$$x_{k+1} - x^* = x_k - x^* - \alpha_k D_k g_k = x_k - x^* - \alpha_k D_k Q(x_k - x^*).$$

A straightforward calculation yields

$$(x_{k+1} - x^*)' D_k^{-1}(x_{k+1} - x^*) \\ = (x_k - x^*)' D_k^{-1/2}(I - \alpha_k D_k^{1/2} Q D_k^{1/2})^2 D_k^{-1/2}(x_k - x^*).$$

Hence,

$$(x_{k+1} - x^*)' D_k^{-1}(x_{k+1} - x^*) \leq A_k^2 (x_k - x^*)' D_k^{-1}(x_k - x^*),$$

where A_k is the maximum eigenvalue of $G_k = (I - \alpha_k D_k^{1/2} Q D_k^{1/2})$. The eigenvalues of G_k are $1 - \alpha_k e_i(D_k^{1/2} Q D_k^{1/2})$, $i = 1, \ldots, n$, where $e_i(D_k^{1/2} Q D_k^{1/2})$ is the ith eigenvalue of $D_k^{1/2} Q D_k^{1/2}$. From this we obtain by an elementary calculation

$$|A_k| = \max\{|1 - \alpha_k m_k|, |1 - \alpha_k M_k|\},$$

and (24) follows. The verification of the fact that α_k as given by (25) minimizes the right-hand side of (24) is elementary and is left to the reader. Q.E.D.

The result shows that if $D_k = D$ for all k where D is positive definite and

$$\limsup_{k \to \infty} \max\{|1 - \alpha_k m|^2, |1 - \alpha_k M|^2\} = \beta,$$

where m, M are the smallest and largest eigenvalues of $(D^{1/2} Q D^{1/2})$, then $\{(x_k - x^*)' D^{-1}(x_k - x^*)\}$ converges at least linearly with convergence ratio β provided $0 < \beta < 1$. If $c > 0$ is the smallest eigenvalue of D^{-1} and Γ is the largest eigenvalue of Q, we have

$$(c/\Gamma) f(x_k) \le \tfrac{1}{2} c |x_k - x^*|^2 \le \tfrac{1}{2}(x_k - x^*)' D^{-1} (x_k - x^*).$$

Hence, if $0 < \beta < 1$, we have that $\{f(x_k)\}$ and $\{|x_k - x^*|^2\}$ will also converge faster than linearly with convergence ratio β. The important point is that [compare with (26)]

$$\left(\frac{M - m}{M + m}\right)^2 \le \beta,$$

and hence if M/m is much larger than unity, again the convergence rate can be very slow even if the optimal stepsize $\alpha_k = 2/(m_k + M_k)$ (which is generally unknown) were to be utilized. From this, it follows again that D_k should be chosen as close as possible to Q^{-1} so that $M_k \sim m_k \sim 1$. Notice that if D_k has indeed been so chosen, then (25) shows that the stepsize $\alpha_k = 1$ is a good choice. This fact also follows from (20), which shows that when $D_k \sim Q^{-1}$ then the minimizing stepsize is near unity.

Rate of Convergence—Nonquadratic Objective Function

One can show that our main conclusions on rate of convergence carry over to the nonquadratic case for sequences converging to strong local minima.

Let $f \in C^2$ and consider the gradient method

$$x_{k+1} = x_k - \alpha_k D_k \nabla f(x_k), \tag{27}$$

where D_k is positive definite symmetric. Consider a generated sequence $\{x_k\}$ and assume that

$$x_k \to x^*, \qquad \nabla f(x^*) = 0, \qquad \nabla^2 f(x^*) > 0, \tag{28}$$

and that $x_k \neq x^*$ for all k. Then it is possible to show the following:

(a) If α_k is chosen by the line minimization rule there holds

$$\limsup_{k\to\infty} \frac{f(x_{k+1}) - f(x^*)}{f(x_k) - f(x^*)} \le \limsup_{k\to\infty} \left(\frac{M_k - m_k}{M_k + m_k}\right)^2, \tag{29}$$

where M_k and m_k are the largest and smallest eigenvalues of $D_k^{1/2} \nabla^2 f(x^*) D_k^{1/2}$.

(b) There holds

$$\limsup_{k\to\infty} \frac{(x_{k+1} - x^*)' D_k^{-1} (x_{k+1} - x^*)}{(x_k - x^*)' D_k^{-1} (x_k - x^*)} \le \limsup_{k\to\infty} \max\{|1 - \alpha_k m_k|^2, |1 - \alpha_k M_k|^2\}.$$

The proof of these facts involves essentially a repetition of the proofs of Propositions 1.13 and 1.14. However, the details are somewhat more technical and will not be given.

When $D_k \to \nabla^2 f(x^*)^{-1}$, then (29) shows that the convergence rate of $\{f(x_k) - f(x^*)\}$ is superlinear. A somewhat more general version of this result for the case of the Armijo rule is given by the following proposition:

Proposition 1.15: Consider a sequence $\{x_k\}$ generated by (27) and satisfying (28). Assume further that $\nabla f(x_k) \neq 0$ for all k and

$$\lim_{k\to\infty} \frac{|[D_k - \nabla^2 f(x^*)^{-1}]\nabla f(x_k)|}{|\nabla f(x_k)|} = 0. \tag{30}$$

Then if α_k is chosen by means of the Armijo rule with initial stepsize $s = 1$, we have

$$\lim_{k\to\infty} \frac{|x_{k+1} - x^*|}{|x_k - x^*|} = 0,$$

and hence $\{|x_k - x^*|\}$ converges superlinearly. Furthermore, there exists an integer $\bar{k} \geq 0$ such that we have $\alpha_k = 1$ for all $k \geq \bar{k}$ (i.e., eventually no reduction of the initial stepsize will be taking place).

Proof: We first prove that there exists a $\bar{k} \geq 0$ such that for all $k \geq \bar{k}$ we have $\alpha_k = 1$. By the mean value theorem we have

$$f(x_k) - f[x_k - D_k\nabla f(x_k)] = \nabla f(x_k)'D_k\nabla f(x_k) - \tfrac{1}{2}\nabla f(x_k)'D_k\nabla^2 f(\bar{x}_k)D_k\nabla f(x_k),$$

where $\bar{x}_k$ is a point on the line segment joining x_k and $x_k - D_k\nabla f(x_k)$. It will be sufficient to show that for k sufficiently large we have

$$\nabla f(x_k)'D_k\nabla f(x_k) - \tfrac{1}{2}\nabla f(x_k)'D_k\nabla^2 f(\bar{x}_k)D_k\nabla f(x_k) \geq \sigma\nabla f(x_k)'D_k\nabla f(x_k)$$

or equivalently, by defining $p_k = \nabla f(x_k)/|\nabla f(x_k)|$,

$$(1 - \sigma)p_k'D_k p_k \geq \tfrac{1}{2}p_k'D_k\nabla^2 f(\bar{x}_k)D_k p_k. \tag{31}$$

From (28), (30), we obtain $D_k\nabla f(x_k) \to 0$. Hence, $x_k - D_k\nabla f(x_k) \to x^*$, and it follows that $\bar{x}_k \to x^*$ and $\nabla^2 f(\bar{x}_k) \to \nabla^2 f(x^*)$. Now (30) is written as

$$D_k p_k = [\nabla^2 f(x^*)]^{-1}p_k + \beta_k,$$

where $\{\beta_k\}$ denotes a vector sequence with $\beta_k \to 0$. By using the above relation and the fact that $\nabla^2 f(\bar{x}_k) \to \nabla^2 f(x^*)$, we may write (31) as

$$(1 - \sigma)p_k'[\nabla^2 f(x^*)]^{-1}p_k \geq \tfrac{1}{2}p_k'[\nabla^2 f(x^*)]^{-1}p_k + \gamma_k,$$

where $\{\gamma_k\}$ is some scalar sequence with $\gamma_k \to 0$. Thus (31) is equivalent to

$$(\tfrac{1}{2} - \sigma)p_k'[\nabla^2 f(x^*)]^{-1}p_k \geq \gamma_k.$$

Since $\frac{1}{2} - \sigma > 0$, $|p_k| = 1$, and $\nabla^2 f(x^*) > 0$, the above relation holds for k sufficiently large, and we have $\alpha_k = 1$ for $k \geq \bar{k}$ where $\bar{k}$ is some index.

To show superlinear convergence we write, for $k \geq \bar{k}$,

$$x_{k+1} - x^* = x_k - x^* - D_k \nabla f(x_k). \tag{32}$$

We have, from (30) and for some sequence $\{\delta_k\}$ with $\delta_k \to 0$,

$$D_k \nabla f(x_k) = \nabla^2 f(x^*)^{-1} \nabla f(x_k) + |\nabla f(x_k)| \delta_k. \tag{33}$$

From Taylor's theorem we obtain

$$\nabla f(x_k) = \nabla^2 f(x^*)(x_k - x^*) + o(|x_k - x^*|)$$

from which

$$[\nabla^2 f(x^*)]^{-1} \nabla f(x_k) = x_k - x^* + o(|x_k - x^*|),$$
$$|\nabla f(x_k)| = O(|x_k - x^*|).$$

Using the above two relations in (33), we obtain

$$D_k \nabla f(x_k) = x_k - x^* + o(|x_k - x^*|)$$

and (32) becomes

$$x_{k+1} - x^* = o(|x_k - x^*|),$$

from which

$$\lim_{k \to \infty} \frac{|x_{k+1} - x^*|}{|x_k - x^*|} = \lim_{k \to \infty} \frac{o(|x_k - x^*|)}{|x_k - x^*|} = 0. \qquad \text{Q.E.D.}$$

We note that one can prove that Eq. (30) is equivalent to

$$\lim_{k \to \infty} \frac{|[D_k^{-1} - \nabla^2 f(x^*)] D_k \nabla f(x_k)|}{|D_k \nabla f(x_k)|} = 0 \tag{34}$$

assuming (28) holds. Equation (34) has been used by Dennis and Moré (1974) in the analysis of quasi-Newton methods and is sometimes called the *Dennis–Moré condition* (see also McCormick and Ritter, 1972).

A slight modification of the proof of Proposition 1.15 shows also that its conclusion holds if α_k is chosen by means of the Goldstein rule with initial trial stepsize equal to unity. Furthermore for all k sufficiently large, we shall have $\alpha_k = 1$ (i.e., the initial stepsize will be acceptable after a certain index).

Several additional results relating to the convergence rate of gradient methods are possible. The main guideline which consistently emerges from this analysis (and which has been supported by extensive numerical experience) is that *in order to achieve fast convergence of the iteration*

$$x_{k+1} = x_k - \alpha_k D_k \nabla f(x_k),$$

one should try to choose the matrices D_k as close as possible to $[\nabla^2 f(x^*)]^{-1}$ so that the corresponding maximum and minimum eigenvalues of $D_k^{1/2}\nabla^2 f(x^*)D_k^{1/2}$ satisfy $M_k \sim 1$ and $m_k \sim 1$. This fact holds true for all stepsize rules that we have examined. Furthermore, when $M_k \sim 1$ and $m_k \sim 1$, the initial stepsize $s = 1$ is a good choice for the Armijo rule and other related rules or as a starting point for one-dimensional minimization procedures in minimization stepsize rules.

Spacer Steps in Descent Algorithms

Often in optimization problems, we utilize complex descent algorithms in which the rule used to determine the next point may depend on several previous points or on the iteration index k. Some of the conjugate direction algorithms to be examined in the next chapter are of this type. Other algorithms may represent a combination of different methods and switch from one method to the other in a manner which may either be prespecified or may depend on the progress of the algorithm. Such combinations are usually introduced in order to improve speed of convergence or reliability. However, their convergence analysis can become extremely complicated. It is thus often of value to know that if in such algorithms one inserts, perhaps irregularly but infinitely often, an iteration of a convergent algorithm such as steepest descent, then the theoretical convergence properties of the overall algorithm are quite satisfactory. Such an iteration will be referred to as a *spacer step*. The related convergence result is given in the following proposition. The only requirement imposed on the iterations of the algorithm other than the spacer steps is that they do not increase the value of the objective function.

Proposition 1.16: Consider a sequence $\{x_k\}$ such that

$$f(x_{k+1}) \le f(x_k) \qquad \forall\, k = 0, 1, \ldots.$$

Assume that there exists an infinite set K of nonnegative integers for which we have

$$x_{k+1} = x_k + \alpha_k d_k \qquad \forall\, k \in K,$$

where $\{d_k\}_K$ is uniformly gradient related and α_k is chosen by the minimization rule, or the limited minimization rule, or the Armijo rule. Then every limit point of the subsequence $\{x_k\}_K$ is a critical point.

The proof requires a simple modification of the proof of Proposition 1.8 and is left to the reader. Notice that if f is a convex function, it is possible to strengthen the conclusion of the proposition and assert that every limit point of the whole sequence $\{x_k\}$ is a global minimum of f.

1.3.2 Steepest Descent and Scaling

Consider the steepest descent method

$$x_{k+1} = x_k - \alpha_k \nabla f(x_k),$$

and assume that $f \in C^2$. We saw in the previous section that the convergence rate depends on the eigenvalue structure of the Hessian matrix $\nabla^2 f$. This structure in turn depends strongly on the particular choice of variables x used to define the problem. A different choice may change substantially the convergence rate.

Let T be an invertible $n \times n$ matrix. We can then represent points in R^n either by the vector x which enters in the objective function $f(x)$, or by the vector y, where

$$Ty = x. \tag{35}$$

Then the problem of minimizing f is equivalent to the problem

$$\begin{aligned} &\text{minimize} \quad h(y) \triangleq f(Ty) \\ &\text{subject to} \quad y \in R^n. \end{aligned} \tag{36}$$

If y^* is a local minimum of h, the vector $x^* = Ty^*$ is a local minimum of f.

Now steepest descent for problem (36) takes the form

$$y_{k+1} = y_k - \alpha_k \nabla h(y_k) = y_k - \alpha_k T' \nabla f(Ty_k). \tag{37}$$

Multiplying both sides by T and using (35) we obtain the iteration in terms of the x variables

$$x_{k+1} = x_k - \alpha_k TT' \nabla f(x_k).$$

Setting $D = TT'$, we obtain the following scaled version of steepest descent

$$x_{k+1} = x_k - \alpha_k D \nabla f(x_k) \tag{38}$$

with D being a positive definite symmetric matrix. The convergence rate of (37) or equivalently (38), however, is governed by the eigenvalue structure of $\nabla^2 h$ rather than of $\nabla^2 f$. We have $\nabla^2 h(y) = T' \nabla^2 f(Ty) T$, and if T is symmetric and positive definite, then $T = D^{1/2}$ and

$$\nabla^2 h(y) = D^{1/2} \nabla^2 f(x) D^{1/2}.$$

When $D \sim [\nabla^2 f(x)]^{-1}$, we obtain $\nabla^2 h(y) \sim I$, and the problem of minimizing h becomes well scaled and can be solved efficiently by steepest descent. This is consistent with the rate of convergence results of the previous section.

The more general iteration

$$x_{k+1} = x_k - \alpha_k D_k \nabla f(x_k)$$

with D_k positive definite may be viewed as a scaled version of the steepest descent method where at each iteration we use different scaling for the variables. Good scaling is obtained when $D_k \sim [\nabla^2 f(x^*)]^{-1}$, where x^* is a local minimum to which the method is assumed to converge ultimately. Since $\nabla^2 f(x^*)$ is unavailable, often we use $D_k = [\nabla^2 f(x_k)]^{-1}$ or $D = [\nabla^2 f(x_0)]^{-1}$, where these matrices are positive definite. This type of scaling results in modified forms of Newton's method. A less complicated form of scaling is obtained when D is chosen to be diagonal of the form

$$D = \begin{bmatrix} d^1 & & & 0 \\ & d^2 & & \\ & & \ddots & \\ 0 & & & d^n \end{bmatrix}$$

with

$$d^i \sim [\partial^2 f(x_0)/(\partial x^i)^2]^{-1}, \qquad i = 1, \ldots, n;$$

i.e., the Hessian matrix is approximated by a diagonal matrix. The approximate inverse second derivatives d^i are obtained either analytically or by finite differences of first derivatives at the starting point x_0. It is also possible to update the scaling factors d^i periodically. The scaled version of steepest descent takes the form

$$x_{k+1}^i = x_k^i - \alpha_k d^i \, \partial f(x_k)/\partial x^i, \qquad i = 1, \ldots, n.$$

While such simple scaling schemes are not guaranteed to improve the convergence rate of steepest descent, in many cases they can result in spectacular improvements. An additional advantage when using the simple diagonal scaling device described above is that usually the initial stepsize $s = 1$ will work well for the Armijo rule, thus eliminating the need for determining a range of good initial stepsize choices by experimentation.

1.3.3 Newton's Method and Its Modifications

Newton's method consists of the iteration

$$x_{k+1} = x_k - \alpha_k [\nabla^2 f(x_k)]^{-1} \nabla f(x_k), \tag{39}$$

assuming that $[\nabla^2 f(x_k)]^{-1}$ exists and that the Newton direction

$$d_k = -[\nabla^2 f(x_k)]^{-1} \nabla f(x_k)$$

is a direction of descent (i.e., $d_k' \nabla f(x_k) < 0$). This direction is obtained as the solution of the linear system of equations

$$\nabla^2 f(x_k) d_k = -\nabla f(x_k).$$

As explained in the section on scaling, one may view this iteration as a scaled version of steepest descent where the "optimal" scaling matrix $D_k = [\nabla^2 f(x_k)]^{-1}$ is utilized. It is also worth mentioning that *Newton's method is "scale-free"* in the sense that the method cannot be affected by a change in coordinate system as is the case with steepest descent (Section 1.3.2). Indeed if we consider a linear invertible transformation of variables $x = Ty$, then Newton's method in the space of the variables y is written as

$$y_{k+1} = y_k - \alpha_k[\nabla_{yy}^2 f(Ty_k)]^{-1}\nabla_y f(Ty_k) = y_k - \alpha_k T^{-1}\nabla^2 f(Ty_k)^{-1}\nabla f(Ty_k),$$

and by applying T to both sides of this equation we recover (39).

When the Armijo rule is utilized with initial stepsize $s = 1$, then no reduction of the stepsize will be necessary near convergence to a strong minimum, as shown in Proposition 1.15. Thus, near convergence the method takes the form

$$x_{k+1} = x_k - [\nabla^2 f(x_k)]^{-1}\nabla f(x_k), \tag{40}$$

which will be referred to as the *pure form of Newton's method.* A valuable interpretation of this iteration is obtained by observing that x_{k+1} as given above minimizes the second-order Taylor's series expansion of f around x_k given by

$$\tilde{f}_k(x) = f(x_k) + \nabla f(x_k)'(x - x_k) + \tfrac{1}{2}(x - x_k)'\nabla^2 f(x_k)(x - x_k).$$

Indeed by setting the derivative of $\tilde{f}_k$ equal to zero, we obtain

$$\nabla\tilde{f}_k(x) = \nabla f(x_k) + \nabla^2 f(x_k)(x - x_k) = 0.$$

The solution of this equation is x_{k+1} as given by Eq. (40). It follows that when f is positive definite quadratic the pure form of Newton's method yields the unique minimum of f in a single iteration. Thus, one expects that iteration (40) will have a fast rate of convergence. This is substantiated by the following result which applies to Newton's method for solving systems of equations:

Proposition 1.17: Consider a mapping g: $R^n \to R^n$, and let $\varepsilon > 0$ and x^* be such that $g \in C^1$ on $S(x^*; \varepsilon)$, $g(x^*) = 0$, and $\nabla g(x^*)$ is invertible. Then there exists a $\delta > 0$ such that if $x_0 \in S(x^*; \delta)$, the sequence $\{x_k\}$ generated by the iteration

$$x_{k+1} = x_k - [\nabla g(x_k)']^{-1} g(x_k)$$

is well defined, converges to x^*, and satisfies $x_k \in S(x^*; \delta)$ for all k. Furthermore, if $x_k \neq x^*$ for all k, then

$$\lim_{k\to\infty} \frac{|x_{k+1} - x^*|}{|x_k - x^*|} = 0; \tag{41}$$

i.e., $\{|x_k - x^*|\}$ converges Q-superlinearly. In addition given any $r > 0$, there exists a $\delta_r > 0$ such that if $x_k \in S(x^*; \delta_r)$, then

$$|x_{k+1} - x^*| \le r|x_k - x^*|, \tag{42}$$

$$|g(x_{k+1})| \le r|g(x_k)|. \tag{43}$$

If we assume further that for some $L > 0$ and $M > 0$, we have

$$|\nabla g(x)' - \nabla g(y)'| \le L|x - y| \qquad \forall\, x, y \in S(x^*; \varepsilon), \tag{44a}$$

$$|[\nabla g(x)']^{-1}| \le M, \qquad \forall\, x \in S(x^*; \varepsilon), \tag{44b}$$

then

$$|x_{k+1} - x^*| \le \tfrac{1}{2}LM|x_k - x^*|^2 \qquad \forall\, k = 0, 1, \ldots,$$

and $\{|x_k - x^*|\}$ converges Q-superlinearly with order at least two.

Proof: Let $\delta \in (0, \varepsilon)$ and $M > 0$ be such that $[\nabla g(x)]^{-1}$ exists for all $x \in S(x^*; \delta)$ and

$$|[\nabla g(x)']^{-1}| \le M \qquad \forall\, x \in S(x^*; \delta). \tag{45}$$

If $x_k \in S(x^*; \delta)$, we have

$$g(x_k) = \int_0^1 \nabla g[x^* + t(x_k - x^*)]'\, dt(x_k - x^*)$$

from which

(46)

$$\begin{aligned}
x_{k+1} - x^* &= x_k - x^* - [\nabla g(x_k)']^{-1} g(x_k) \\
&= [\nabla g(x_k)']^{-1}[\nabla g(x_k)'(x_k - x^*) - g(x_k)] \\
&= [\nabla g(x_k)']^{-1}\left[\nabla g(x_k)' - \int_0^1 \nabla g[x^* + t(x_k - x^*)]'\, dt\right](x_k - x^*) \\
&= [\nabla g(x_k)']^{-1} \int_0^1 \{\nabla g(x_k)' - \nabla g[x^* + t(x_k - x^*)]'\}\, dt(x_k - x^*).
\end{aligned}$$

By continuity of ∇g, we can take δ sufficiently small to ensure that

$$|\nabla g(x)' - \nabla g(y)'| < \tfrac{1}{2}M^{-1} \qquad \forall\, x, y \in S(x^*; \delta). \tag{47}$$

Then from (45), (46), and (47), we obtain

(48)

$$|x_{k+1} - x^*| \le |[\nabla g(x_k)']^{-1}| \int_0^1 |\nabla g(x_k)' - \nabla g[x^* + t(x_k - x^*)]'|\, dt\, |x_k - x^*|$$

and

$$|x_{k+1} - x^*| < \tfrac{1}{2}|x_k - x^*|.$$

It follows that if $x_0 \in S(x^*;\delta)$ then $x_k \in S(x^*;\delta)$ for all k and $x_k \to x^*$. Equation (41) then follows from (48).

We have

$$g_i(x) = \nabla g_i(\tilde{x}_i)'(x - x^*) \qquad \forall\, i = 1, \ldots, n,$$

where $\tilde{x}_i$ is a vector lying in the line segment connecting x and x^*. Therefore by denoting $\nabla g(\tilde{x})$ the matrix with columns $\nabla g_i(\tilde{x}_i)$, we have

$$|g(x)|^2 = (x - x^*)'\nabla g(\tilde{x})\nabla g(\tilde{x})'(x - x^*).$$

Choose $\delta_1 > 0$ sufficiently small so that $\nabla g(\tilde{x})\nabla g(\tilde{x})'$ is positive definite for all x with $|x - x^*| \leq \delta_1$, and let $\Lambda > 0$ and $\lambda > 0$ be upper and lower bounds to the eigenvalues of $[\nabla g(\tilde{x})\nabla g(\tilde{x})']^{1/2}$ for $x \in S(x^*;\delta_1)$. Then

$$\begin{aligned}\lambda^2|x - x^*|^2 &\leq (x - x^*)'\nabla g(\tilde{x})\nabla g(\tilde{x})'(x - x^*)\\ &\leq \Lambda^2|x - x^*|^2 \qquad \forall\, x \in S(x^*;\delta_1).\end{aligned}$$

Hence, we have

$$\lambda|x - x^*| \leq |g(x)| \leq \Lambda|x - x^*| \qquad \forall\, x \in S(x^*;\delta_1).$$

Now from (48), it follows easily that given any $r > 0$, we can find a $\delta_r \in (0, \delta_1]$ such that if $x_k \in S(x^*;\delta_r)$, then

$$|x_{k+1} - x^*| \leq (\lambda r/\Lambda)|x_k - x^*| \leq r|x_k - x^*|,$$

thereby showing (42). Combining the last two inequalities we also obtain

$$|g(x_{k+1})| \leq r|g(x_k)| \qquad \forall\, x \in S(x^*;\delta_r),$$

and (43) is proved.

If (44a) and (44b) hold, then from (48) we have

$$|x_{k+1} - x^*| \leq M\int_0^1 Lt|x_k - x^*|\,dt\,|x_k - x^*| = \frac{ML}{2}|x_k - x^*|^2.$$

Q.E.D.

For $g(x) = \nabla f(x)$, the result of the proposition applies to the pure form of Newton's method (40). Extensive computational experience suggests that the fast convergence rate indicated in the proposition is indeed realized in a practical setting. On the other hand, Newton's method in its pure form has several serious drawbacks. First, the inverse $[\nabla^2 f(x_k)]^{-1}$ may fail to exist, in which case the method breaks down. This may happen, for example, if f is linear within some region in which case $\nabla^2 f = 0$. Second, iteration (40) is not a descent method in the sense that it may easily happen that $f(x_{k+1}) > f(x_k)$. Third, the method tends to be attracted by local maxima just as much as it is attracted by local minima. This is evident from Proposition 1.17 where it is assumed that $\nabla g(x^*)$ is invertible but not necessarily positive definite.

For these reasons, it is necessary to modify the pure form of Newton's method (40) in order to convert it to a reliable minimization algorithm. There are several schemes by means of which this can be accomplished. All these schemes convert iteration (40) into a gradient method with a uniformly gradient-related direction sequence, while guaranteeing that whenever the algorithm gets sufficiently close to a point x^* satisfying the second-order sufficiency conditions, then the algorithm assumes the pure form (40) and achieves the attendant fast convergence rate.

First Modification Scheme: This method consists of the iteration

$$x_{k+1} = x_k + \alpha_k d_k,$$

where α_k is chosen by the Armijo rule with initial stepsize unity ($s = 1$), and d_k is chosen by

$$d_k = -[\nabla^2 f(x_k)]^{-1}\nabla f(x_k), \tag{49}$$

if $[\nabla^2 f(x_k)]^{-1}$ exists and

$$\nabla f(x_k)'[\nabla^2 f(x_k)]^{-1}\nabla f(x_k) \geq c_1 |\nabla f(x_k)|^{p_1}, \tag{50}$$

$$c_2 |\nabla f(x_k)| \geq |[\nabla^2 f(x_k)]^{-1}\nabla f(x_k)|^{p_2}, \tag{51}$$

while otherwise

$$d_k = -D\nabla f(x_k).$$

The matrix D is some positive definite symmetric scaling matrix. The scalars c_1, c_2, p_1, and p_2 satisfy

$$c_1 > 0, \qquad c_2 > 0, \qquad p_1 > 2, \qquad \text{and} \qquad p_2 > 1.$$

In practice c_1 should be very small, say 10^{-5}, c_2 should be very large, say 10^5, and p_1 and p_2 can be chosen equal to three and two, respectively.

It is clear, from Proposition 1.8, that a sequence $\{d_k\}$ generated by the scheme above is uniformly gradient related and hence the resulting algorithm is convergent in the sense that every limit point of a sequence that it generates is a critical point of f. Now consider the algorithm near a local minimum x^* satisfying

$$\nabla f(x^*) = 0, \qquad \nabla^2 f(x^*) > 0.$$

Then it is easy to see that for x_k close enough to x^*, the Hessian $\nabla^2 f(x_k)$ will be invertible and the tests (50) and (51) will be passed. Thus, d_k will be the Newton direction (49) for all x_k sufficiently close to x. Furthermore, from Propositions 1.12 and 1.15, we shall have $x_k \to x^*$, and the stepsize

α_k will equal unity. Hence, if x_k is sufficiently close to x^*, then $x_k \to x^*$, and the pure form of Newton's method will be employed after some index, thus achieving the fast convergence rate indicated in Proposition 1.17.

A variation of this modification scheme is given by the iteration

$$x_{k+1} = x_k + \alpha_k[\alpha_k d_k^N - (1 - \alpha_k)D\nabla f(x_k)],$$

where D is a positive definite matrix and d_k^N is the Newton direction

$$d_k^N = -[\nabla^2 f(x_k)]^{-1}\nabla f(x_k),$$

if $[\nabla^2 f(x_k)]^{-1}$ exists. Otherwise $d_k^N = -D\nabla f(x_k)$. The stepsize α_k is chosen by an Armijo-type rule with initial stepsize unity whereby $\alpha_k = \beta^{m_k}$ and m_k is the first nonnegative integer m for which

$$f(x_k) - f[x_k + \beta^m d_k(\beta^m)] \geq -\sigma\beta^m |\nabla f(x_k)|^2,$$

where $\sigma \in (0, \frac{1}{2})$, $\beta \in (0, 1)$, and

$$d_k(\beta^m) = \beta^m d_k^N - (1 - \beta^m)D\nabla f(x_k).$$

This is a line search along the curve of points of the form

$$z_\alpha = \alpha[\alpha d_k^N - (1 - \alpha)D\nabla f(x_k)]$$

with $\alpha \in [0, 1]$. For $\alpha = 1$ we obtain the Newton direction, while as $\alpha \to 0$ the vector z_α/α tends to the (scaled) steepest descent direction $-D\nabla f(x_k)$. Assuming σ is chosen sufficiently small, one can prove similar convergence and rate of convergence results as the ones stated earlier for this modified version of Newton's method.

Second Modification Scheme: Since calculation of the Newton direction d_k involves solution of the system of linear equations

$$\nabla^2 f(x_k)d_k = -\nabla f(x_k),$$

it is natural to compute d_k by attempting to form the Cholesky factorization of $\nabla^2 f(x_k)$ (see the preceding section). During the factorization process, one can detect whether $\nabla^2 f(x_k)$ is either nonpositive definite or nearly singular, in which case $\nabla^2 f(x_k)$ is replaced by a positive definite matrix of the form $F_k = \nabla^2 f(x_k) + E_k$, where E_k is a diagonal matrix. The elements of E_k are introduced sequentially during the factorization process, so that at the end we obtain F_k in the form $F_k = L_k L_k'$, where L_k is lower triangular. Subsequently d_k is obtained as the solution of the system of equations $L_k L_k' d_k = -\nabla f(x_k)$, and the next point x_{k+1} is determined from $x_{k+1} = x_k + \alpha_k d_k$, where α_k is chosen according to the Armijo rule. The matrix E_k is such that the sequence $\{d_k\}$ is uniformly gradient related. Furthermore, $E_k = 0$ when x_k is close enough to a point x^* satisfying the second-order sufficiency conditions for optimality. Thus, near such a point, the method is again identical to the pure form of Newton's method and achieves the corresponding superlinear convergence rate. The precise mechanization of the scheme is as follows.

Let $c > 0$, $\mu > 0$, and $p > 0$ denote fixed scalars and let a_{ij}^k denote the elements of $\nabla^2 f(x_k)$. Consider the $i \times i$ lower triangular matrices L_k^i, $i = 1, \ldots, n$, defined recursively by the following modified Cholesky factorization process (compare with Section 1.2):

$$L_k^1 = \begin{cases} \sqrt{a_{11}^k} & \text{if } a_{11}^k > 0 \quad \text{and} \quad \sqrt{a_{11}^k} \geq c|\nabla f(x_k)|^p, \\ \mu & \text{otherwise,} \end{cases}$$

$$L_k^i = \begin{bmatrix} L_k^{i-1} & 0 \\ l_i^{k'} & \lambda_{ii}^k \end{bmatrix}, \qquad i = 2, \ldots, n,$$

where

$$l_i^k = (L_k^{i-1})^{-1} a_i^k, \qquad a_i^k = \begin{bmatrix} a_{1i}^k \\ \vdots \\ a_{i-1,i}^k \end{bmatrix},$$

$$\lambda_{ii}^k = \begin{cases} \sqrt{a_{ii}^k - l_i^{k'} l_i^k} & \text{if } \ a_{ii}^k > l_i^{k'} l_i^k \quad \text{and} \quad \sqrt{a_{ii}^k - l_i^{k'} l_i^k} \geq c|\nabla f(x_k)|^p, \\ \mu & \text{otherwise.} \end{cases}$$

Then the direction d_k is determined from

$$L_k L_k' d_k = -\nabla f(x_k),$$

where $L_k = L_k^n$. The next point x_{k+1} is determined from

$$x_{k+1} = x_k + \alpha_k d_k,$$

where α_k is chosen by the Armijo rule with initial stepsize $s = 1$ whenever $\nabla^2 f(x_k) = L_k L_k'$.

Some trial and error may be necessary in order to determine appropriate values for c, μ, and p. Usually, one takes c very small so that the Newton direction will be modified as infrequently as possible. The value of μ should be considerably larger than that of c in order that the matrix $L_k L_k'$ is not nearly singular. A choice $0 < p \leq 1$ is usually satisfactory. Sometimes one takes $p = 0$, although in this case the theoretical convergence rate properties of the algorithm depend on the value of c.

The following facts may be verified for the algorithm described above:

(a) The direction sequence $\{d_k\}$ is uniformly gradient related, and hence the resulting algorithm is convergent in the sense that every limit point of $\{x_k\}$ is a critical point of f.

(b) For each point x^* satisfying $\nabla f(x^*) = 0$ and $\nabla^2 f(x^*) > 0$, there exists a scalar $\varepsilon > 0$ such that if $|x_k - x^*| < \varepsilon$ then $L_k L_k' = \nabla^2 f(x_k)$; i.e., the Newton direction will not be modified, and furthermore the stepsize α_k will equal unity. Thus, when sufficiently close to such a point x^*, the algorithm assumes the pure form of Newton's method and converges to x^* with superlinear convergence rate.

There is another interesting modification scheme that can be used when $\nabla^2 f(x_k)$ is indefinite. In this case one can use, instead of the direction $[\nabla^2 f(x_k)]^{-1}\nabla f(x_k)$, a descent direction which is also *a direction of negative curvature*, i.e., a d_k such that $\nabla f(x_k)' d_k < 0$ and $d_k' \nabla^2 f(x_k) d_k < 0$. This can be done in a numerically stable and efficient manner via a form of triangular factorization of $\nabla^2 f(x_k)$. For a detailed presentation we refer to Fletcher and Freeman (1977), Moré and Sorensen (1979), and Goldfarb (1980).

Periodic Reevaluation of the Hessian

Finally, we mention that a Newton-type method, which in many cases is considerably more efficient computationally than those described above, is obtained if the Hessian matrix $\nabla^2 f$ is recomputed every p iterations ($p \geq 2$) rather than at every iteration. This method in unmodified form is given by

$$x_{k+1} = x_k - \alpha_k D_k \nabla f(x_k),$$

where

$$D_{ip+j} = [\nabla^2 f(x_{ip})]^{-1}, \qquad j = 0, 1, \ldots, p-1, \quad i = 0, 1, \ldots.$$

A significant advantage of this method when coupled with the second modification scheme described above is that the Cholesky factorization of $\nabla^2 f(x_{ip})$ is obtained at the ipth iteration and is subsequently used for a total of p iterations in the computation of the direction of search. This reduction in computational burden per iteration is achieved at the expense of what is usually a small or imperceptible degradation in speed of convergence.

Approximate Newton Methods

One of the main drawbacks of Newton's method in its pure or modified forms is the need to solve a system of linear equations in order to obtain the descent direction at each iteration. We have so far implicitly assumed that this system will be solved by some version of the Gaussian elimination method which requires a finite number of arithmetic operations $[O(n^3)]$. On the other hand, if the dimension n is large, the amount of calculation required for exact solution of the Newton system can be prohibitive and one may have to be satisfied with only an approximate solution of this system. This approach is often used in fact for solving large linear systems of equations where in some cases an adequate approximation to the solution can be obtained by iterative methods such as successive overrelaxation (SOR) much faster than the exact solution can be obtained by Gaussian elimination. The fact that Gaussian elimination can solve the system in a finite number of arithmetic operations while this is not guaranteed by SOR methods can be quite irrelevant, since the computational cost of finding the exact solution can be entirely prohibitive.

Another possibility is to solve the Newton system approximately by using the conjugate gradient method to be presented in the next section. More generally any system of the form $Hd = -g$, where H is a positive definite symmetric $n \times n$ matrix and $g \in R^n$ can be solved by the conjugate gradient method by converting it to the quadratic optimization problem

$$\begin{aligned} &\text{minimize} \quad \tfrac{1}{2}d'Hd + g'd \\ &\text{subject to} \quad d \in R^n. \end{aligned}$$

It will be seen in the next section that actually the conjugate gradient method solves this problem exactly in at most n iterations. However this fact is not particularly relevant since for the type of problems where the use of the conjugate gradient method makes sense, the dimension n is very large and the main hope is that only a few conjugate gradient steps will be necessary in order to obtain a good approximation to the solution.

For the purposes of unconstrained optimization, an important property of any approximate method of solving a system of the form $H_k d = -\nabla f(x_k)$, where H_k is positive definite, is that the approximate direction $\bar{d}$ obtained is a descent direction, i.e., it satisfies $\nabla f(x_k)'\bar{d} < 0$. This will be automatically satisfied if the approximate method used is a descent method for solving the quadratic optimization problem

$$\begin{aligned} &\text{minimize} \quad \tfrac{1}{2}d'H_k d + \nabla f(x_k)'d \\ &\text{subject to} \quad d \in R^n, \end{aligned}$$

and the starting point $d_0 = 0$ is used, for the descent property implies

$$\tfrac{1}{2}\bar{d}'H_k\bar{d} + \nabla f(x_k)'\bar{d} < \tfrac{1}{2}d_0'H_k d_0 + \nabla f(x_k)'d_0 = 0,$$

or $\nabla f(x_k)'\bar{d} < -\tfrac{1}{2}\bar{d}'H_k\bar{d} < 0$. As will be seen in the next section, the conjugate gradient method has this property.

Conditions on the accuracy of the approximate solution $\bar{d}$ that ensure linear or superlinear rate of convergence in connection with approximate methods are given in Dembo *et al.* (1980). Generally speaking if $H_k \to \nabla^2 f(x_k)$ and the approximate Newton directions d_k satisfy

$$\lim_{k \to \infty} \frac{|H_k d_k + \nabla f(x_k)|}{|\nabla f(x_k)|} \to 0,$$

the superlinear convergence rate property of the method to a strong local minimum is maintained (compare with Proposition 1.15). Approximate Newton methods based on the conjugate gradient method are applied to large scale nonlinear multicommodity flow problems in Bertsekas and Gafni (1981).

1.3.4 Conjugate Direction and Conjugate Gradient Methods

Conjugate direction methods are motivated by a desire to accelerate the convergence rate of steepest descent while avoiding the overhead and evaluation of second derivatives associated with Newton's method. Conjugate direction methods are typically analyzed for the purely quadratic problem

$$\text{(52)} \qquad \begin{aligned} &\text{minimize} \quad f(x) = \tfrac{1}{2}x'Qx \\ &\text{subject to} \quad x \in R^n, \end{aligned}$$

where $Q > 0$, which they can solve in at most n iterations (see Proposition 1.18 that follows). It is then argued that the general problem can be approximated near a strong local minimum by a quadratic problem. One therefore expects that conjugate direction methods, suitably modified, should work well for the general problem—a conjecture that has been substantiated by analysis as well as practical experience.

Definition: Given a positive definite $n \times n$ matrix Q, we say that a collection of nonzero vectors $d_1, \ldots, d_k \in R^n$ is mutually *Q-conjugate* if for all i and j with $i \neq j$ we have $d_i'Qd_j = 0$.

It is clear that *if $d_1, \ldots, d_k$ are mutually Q-conjugate then they are linearly independent*, since if, for example, we had for scalars $\alpha_1, \ldots, \alpha_{k-1}$

$$d_k = \alpha_1 d_1 + \cdots + \alpha_{k-1} d_{k-1},$$

then

$$d_k'Qd_k = \alpha_1 d_k'Qd_1 + \cdots + \alpha_{k-1} d_k'Qd_{k-1} = 0,$$

which is impossible since $d_k \neq 0$, and Q is positive definite.

Given a collection of mutually Q-conjugate directions $d_0, \ldots, d_{n-1}$, we define the corresponding *conjugate direction method* for solving problem (52) by

$$\text{(53)} \qquad x_{k+1} = x_k + \alpha_k d_k, \qquad k = 0, 1, \ldots, n-1,$$

where x_0 is a given vector in R^n and α_k is defined by the line minimization rule

$$\text{(54)} \qquad f(x_k + \alpha_k d_k) = \min_{\alpha} f(x_k + \alpha d_k).$$

We shall employ in what follows in this and the next section the notation

$$g_k = \nabla f(x_k) = Qx_k.$$

We have the following result:

Proposition 1.18: If $x_1, x_2, \ldots, x_n$ are the vectors generated by the conjugate direction method (53), we have

$$g'_{k+1} d_i = 0 \qquad \forall\, i = 0, \ldots, k. \tag{55}$$

Furthermore, for $k = 0, 1, \ldots, n-1$, x_{k+1} minimizes f over the linear manifold

$$M_k = \{z \,|\, z = x_0 + \gamma_0 d_0 + \cdots + \gamma_k d_k,\ \gamma_0, \ldots, \gamma_k \in R\},$$

and hence x_n minimizes f over R^n.

Proof: By (54), we have

$$\partial f(x_i + \alpha_i d_i)/\partial \alpha = g'_{i+1} d_i = 0, \qquad i = 0, \ldots, n-1,$$

so we need only verify (55) for $i = 0, 1, \ldots, k-1$. We have, for $i = 0, 1, \ldots, k-1$,

$$g'_{k+1} d_i = x'_{k+1} Q d_i = \left(x_{i+1} + \sum_{j=i+1}^{k} \alpha_j d_j \right)' Q d_i = x'_{i+1} Q d_i = g'_{i+1} d_i = 0.$$

To show the last part of the proposition, we must show that

$$\partial f(x_0 + \gamma_0 d_0 + \cdots + \gamma_k d_k)/\partial \gamma_i \big|_{\substack{\gamma_0 = \alpha_0 \\ \vdots \\ \gamma_k = \alpha_k}} = 0 \qquad \forall\, i = 0, \ldots, k$$

or

$$g'_{k+1} d_i = 0 \qquad \forall\, i = 0, \ldots, k,$$

which is (55). Q.E.D.

It is easy to visualize the result of Proposition 1.18 for the case where $Q = I$, for in this case, the surfaces of equal cost of f are concentric spheres, and the notion of Q-conjugacy reduces to usual orthogonality. By elementary geometry or a simple algebraic argument, we have that minimization along n orthogonal directions leads to the global minimum of f, i.e., the center of the spheres. The case of a general positive definite Q can actually be reduced to the case where $Q = I$ by means of a scaling transformation. By setting $y = Q^{1/2}x$, the problem becomes min $\{\frac{1}{2}|y|^2 \,|\, y \in R^n\}$. If $w_0, \ldots, w_{n-1}$ are any set of orthogonal nonzero vectors in R^n, the algorithm

$$y_{k+1} = y_k + \alpha_k w_k, \qquad k = 0, 1, \ldots, n-1,$$

where α_k minimizes $\frac{1}{2}|y_k + \alpha w_k|^2$ over α, terminates in at most n steps at $y_n = 0$. To pass back to the x-coordinate system, we multiply this equation by $Q^{-1/2}$ and obtain

$$x_{k+1} = x_k + \alpha_k d_k, \qquad k = 0, 1, \ldots, n-1,$$

where $d_k = Q^{-1/2} w_k$. Since $w_i' w_j = 0$ for $i \neq j$, we obtain $d_i' Q d_j = 0$ for $i \neq j$; i.e., the directions $d_0, \ldots, d_{n-1}$ are Q-conjugate. This argument can be reversed and shows that the collection of conjugate direction methods for the problem $\min\{\frac{1}{2}x'Qx \,|\, x \in R^n\}$ is in one-to-one correspondence with the set of methods for solving the problem $\min\{\frac{1}{2}|y|^2 \,|\, y \in R^n\}$, which consist of successive minimization along n orthogonal directions.

Given any set of linearly independent vectors $\xi_0, \ldots, \xi_{n-1}$, we can construct a set of mutually Q-conjugate directions $d_0, \ldots, d_{n-1}$ as follows. Set

$$d_0 = \xi_0, \tag{56}$$

and for $i = 1, 2, \ldots, n-1$, define successively

$$d_i = \xi_i + \sum_{j=0}^{i-1} c_{ij} d_j, \tag{57}$$

where the coefficients c_{ij} are chosen so that d_i is Q-conjugate to the previous directions $d_{i-1}, \ldots, d_0$. This will be so if, for $k = 0, \ldots, i-1$,

$$d_i' Q d_k = \xi_i' Q d_k + \sum_{j=0}^{i-1} c_{ij} d_j' Q d_k = 0. \tag{58}$$

If previous coefficients were chosen so that $d_0, \ldots, d_{i-1}$ are Q-conjugate, then we have $d_j' Q d_k = 0$ if $j \neq k$, and (58) yields

$$c_{ij} = -\xi_i' Q d_j / d_j' Q d_j \qquad \forall\, i = 1, 2, \ldots, n-1, \quad j = 0, 1, \ldots, i-1. \tag{59}$$

Thus the set of directions $d_0, \ldots, d_{n-1}$ defined by (56), (57) and (59) is Q-conjugate, and (56) and (57) show also that, for $i = 0, \ldots, n-1$, we have

$$(\text{subspace spanned by } d_0, \ldots, d_i) = (\text{subspace spanned by } \xi_0, \ldots, \xi_i). \tag{60}$$

We now define the most important conjugate direction method.

The Conjugate Gradient Method

The conjugate gradient method is obtained by the procedure described above by taking $\xi_0 = -g_0, \ldots, \xi_{n-1} = -g_{n-1}$. More specifically, starting at x_0 with $g_0 \neq 0$, we use g_0 as our first conjugate direction, i.e., $d_0 = -g_0$. We find $x_1 = x_0 + \alpha_0 d_0$ by line search and obtain our second direction

d_1 using the procedure defined by (56), (57) and (59) with $\xi_0 = -g_0$ and $\xi_1 = -g_1$. This yields, from (57) and (59),

$$d_1 = -g_1 + \frac{g_1' Q d_0}{d_0' Q d_0} d_0. \tag{61}$$

By using the equation

$$g_1 - g_0 = Q(x_1 - x_0) = \alpha_0 Q d_0,$$

we can write (61) as

$$d_1 = -g_1 + \frac{g_1'(g_1 - g_0)}{d_0'(g_1 - g_0)} d_0.$$

By repeating the process with $\xi_0 = -g_0$, $\xi_1 = -g_1, \ldots,$ and $\xi_k = -g_k$, we obtain at the $(k + 1)$st step

$$d_k = -g_k + \sum_{j=0}^{k-1} \frac{g_k' Q d_j}{d_j' Q d_j} d_j$$

from which

$$d_k = -g_k + \sum_{j=0}^{k-1} \frac{g_k'(g_{j+1} - g_j)}{d_j'(g_{j+1} - g_j)} d_j. \tag{62}$$

By using the fact that the subspace spanned by $g_0, \ldots, g_{k-1}$ is also the subspace spanned by $d_0, \ldots, d_{k-1}$ [compare with (60)] and the relation $g_k' d_j = 0$ for $j = 0, \ldots, k - 1$ (Proposition 1.18), we obtain

$$g_k' g_j = 0, \qquad j = 0, \ldots, k - 1,$$

so (62) reduces to the simple formula

$$d_k = -g_k + \beta_k d_{k-1}, \tag{63}$$

with

$$\beta_k = \frac{g_k'(g_k - g_{k-1})}{d_{k-1}'(g_k - g_{k-1})}. \tag{64}$$

Note that by using the facts $g_k' g_j = g_k' d_j = 0, j = 0, \ldots, k - 1$, and $d_{k-1} = -g_{k-1} + \beta_{k-1} d_{k-2}$, we see that the coefficient β_k of (64) can also be written as

$$\beta_k = \frac{g_k'(g_k - g_{k-1})}{g_{k-1}' g_{k-1}} = \frac{g_k' g_k}{g_{k-1}' g_{k-1}}. \tag{65}$$

An important observation from (63) and (64) is that *in order to generate the direction d_k one need only know the current and previous gradients g_k and*

g_{k-1} and the previous direction d_{k-1}. This fact is particularly significant when the method is extended to nonquadratic problems.

Scaled Conjugate Gradient Method

This method, also referred to as the *preconditioned conjugate gradient method*, is really the conjugate gradient method implemented in a new coordinate system. Suppose we make a change of variables, as in Section 1.3.2, $x = Ty$, where T is a symmetric invertible $n \times n$ matrix, and apply the conjugate gradient method to the equivalent problem

$$\begin{aligned} &\text{minimize} \quad h(y) = f(Ty) = \tfrac{1}{2}y'TQTy \\ &\text{subject to} \quad y \in R^n. \end{aligned}$$

The method is described by [compare with (63) and (65)]

$$y_{k+1} = y_k + \alpha_k \tilde{d}_k, \tag{66}$$

where α_k is obtained by line minimization and $\tilde{d}_k$ is generated by

$$\tilde{d}_0 = -\nabla h(y_0), \qquad \tilde{d}_k = -\nabla h(y_k) + \beta_k \tilde{d}_{k-1}, \qquad k = 1, 2, \ldots, n, \tag{67}$$

where

$$\beta_k = \frac{\nabla h(y_k)'\nabla h(y_k)}{\nabla h(y_{k-1})'\nabla h(y_{k-1})}. \tag{68}$$

Setting $x_k = Ty_k$, $\nabla h(y_k) = Tg_k$, $d_k = T\tilde{d}_k$, and $H = T^2$, we obtain from (66)–(68) the equivalent method

$$x_{k+1} = x_k + \alpha_k d_k, \tag{69}$$

$$d_0 = -Hg_0, \qquad d_k = -Hg_k + \beta_k d_{k-1}, \qquad k = 1, \ldots, n, \tag{70}$$

where

$$\beta_k = g_k' H g_k / g_{k-1}' H g_{k-1}. \tag{71}$$

Since $\nabla^2 h(y) = TQT$, we have that $\tilde{d}_0, \ldots, \tilde{d}_{n-1}$ are (TQT)-conjugate, and in view of $d_k = T\tilde{d}_k$, we have that $d_0, \ldots, d_{n-1}$ are Q-conjugate. By carrying further this line of argument we see that

$$g_k' H g_j = g_k' d_j = 0 \qquad \forall\, j = 0, \ldots, k-1,$$

and x_k minimizes f over the linear manifold

$$\begin{aligned} M_k &= \{z \mid z = x_0 + \gamma_0 d_0 + \cdots + \gamma_{k-1} d_{k-1}, \gamma_0, \ldots, \gamma_{k-1} \in R\} \\ &= \{z \mid z = x_0 + \gamma_0 H g_0 + \cdots + \gamma_{k-1} H g_{k-1}, \gamma_0, \ldots, \gamma_{k-1} \in R\}. \end{aligned}$$

The motivation for employing scaling typically stems from a desire to improve the speed of convergence of the method within an n-iteration cycle (see the following analysis). This in turn may be important even for a quadratic problem if n is large.

Rate of Convergence of the Conjugate Gradient Method

There are a number of results relating to the convergence rate of the conjugate gradient method applied to quadratic problems. We describe a particular result due to Luenberger (1973).

Consider an algorithm of the form

$$\begin{aligned} x_1 &= x_0 + \gamma_{00} g_0, \\ x_2 &= x_0 + \gamma_{10} g_0 + \gamma_{11} g_1, \\ &\vdots \\ x_{k+1} &= x_0 + \gamma_{k0} g_0 + \cdots + \gamma_{kk} g_k, \end{aligned} \tag{72}$$

where γ_{ij} are arbitrary scalars. Since $g_i = Qx_i$, we have that for suitable scalars ζ_{ki} the algorithm above can be written for all k

$$\begin{aligned} x_{k+1} &= x_0 + \zeta_{k0} Q x_0 + \zeta_{k1} Q^2 x_0 + \cdots + \zeta_{kk} Q^{k+1} x_0 \\ &= [I + QP_k(Q)]x_0, \end{aligned}$$

where P_k is a polynomial of degree k. Among all algorithms of the form (72), the conjugate gradient method is optimal in the sense that for every k, it minimizes $f(x_{k+1})$ over all sets of coefficients $\gamma_{k0}, \ldots, \gamma_{kk}$. It follows from the equation above that in the conjugate gradient method we have, for every k,

$$f(x_{k+1}) = \min_{P_k} \tfrac{1}{2} x_0' Q[I + QP_k(Q)]^2 x_0. \tag{73}$$

Let $\lambda_1, \ldots, \lambda_n$ be the eigenvalues of Q, and let $e_1, \ldots, e_n$ be corresponding orthogonal eigenvectors normalized so that $|e_i| = 1$. Since $e_1, \ldots, e_n$ form a basis, any vector $x_0 \in R^n$ can be written as

$$x_0 = \sum_{i=1}^{n} \zeta_i e_i$$

for some scalars ζ_i. Since

$$Qx_0 = \sum_{i=1}^{n} \zeta_i Q e_i = \sum_{i=1}^{n} \zeta_i \lambda_i e_i,$$

we have, using the orthogonality of $e_1, \ldots, e_n$ and the fact that $|e_i| = 1$,

$$f(x_0) = \frac{1}{2} x_0' Q x_0 = \frac{1}{2} \left(\sum_{i=1}^{n} \zeta_i e_i \right)' \left(\sum_{i=1}^{n} \zeta_i \lambda_i e_i \right) = \frac{1}{2} \sum_{i=1}^{n} \lambda_i \zeta_i^2.$$

Applying the same process to (73), we obtain for *any* polynomial P_k of degree k

$$f(x_{k+1}) \leq \frac{1}{2} \sum_{i=1}^{n} [1 + \lambda_i P_k(\lambda_i)]^2 \lambda_i \zeta_i^2,$$

and it follows that

$$f(x_{k+1}) \leq \max_i \, [1 + \lambda_i P_k(\lambda_i)]^2 f(x_0) \qquad \forall\, P_k, k. \tag{74}$$

One can use this relationship for different choices of polynomials P_k to obtain a number of convergence rate results. We provide one such result.

Proposition 1.19: Assume that Q has $n - k$ eigenvalues in an interval $[a, b]$ with $a > 0$, and the remaining k eigenvalues are greater than b. Then for every x_0, the vector x_{k+1} generated after $(k + 1)$ steps of the conjugate gradient method satisfies

$$f(x_{k+1}) \leq \left(\frac{b - a}{b + a}\right)^2 f(x_0). \tag{75}$$

This relation also holds for the scaled conjugate gradient method (69)–(71) if the eigenvalues of Q are replaced by those of $H^{1/2}QH^{1/2}$.

Proof: Let $\lambda_1, \lambda_2, \ldots, \lambda_k$ be the eigenvalues of Q that are greater than b and consider the polynomial P_k defined by

$$1 + \lambda P_k(\lambda) = \frac{2}{(a + b)\lambda_1 \cdots \lambda_k}\left(\frac{a + b}{2} - \lambda\right)(\lambda_1 - \lambda) \cdots (\lambda_k - \lambda). \tag{76}$$

Since $1 + \lambda_i P_k(\lambda_i) = 0$ we have, using (74), (76), and a simple calculation,

$$\begin{aligned} f(x_{k+1}) &\leq \max_{a \leq \lambda \leq b} [1 + \lambda P_k(\lambda)]^2 f(x_0) \\ &\leq \max_{a \leq \lambda \leq b} \frac{[\lambda - \frac{1}{2}(a + b)]^2}{[\frac{1}{2}(a + b)]^2} f(x_0) = \left(\frac{b - a}{b + a}\right)^2 f(x_0). \end{aligned} \qquad \text{Q.E.D.}$$

An immediate consequence of the proposition is that if the eigenvalues of Q take only k distinct values then the conjugate gradient method will find the minimum of the quadratic function f in at most k iterations. (Simply take $a = b$ in the proposition.) Another interesting possibility, arising for example in some optimal control problems, is when Q has the form

$$Q = M + \sum_{i=1}^{k} v_i v_i', \tag{77}$$

where M is positive definite symmetric, and v_i are some vectors in R^n. We have the following result, the proof of which we leave as an exercise for the reader.

Exercise: Show that if Q is of the form (77), then the vector x_{k+1} generated after $(k+1)$ steps of the conjugate gradient method satisfies

$$f(x_{k+1}) \le \left(\frac{b-a}{b+a}\right)^2 f(x_0),$$

where a and b are the smallest and largest eigenvalues of M. Show also that the vector x_{k+1} generated by the scaled conjugate gradient method with $H = M^{-1}$ minimizes f. [*Hint:* Use the interlocking eigenvalues lemma of Luenberger (1973, p. 202).]

The $(k+1)$-step scaled conjugate gradient method is particularly interesting when Q is of the form (77), k is small relative to n, and systems of equations involving M can be solved easily (see Bertsekas, 1974a).

We also leave the following strengthened version of Proposition 1.19 as an exercise to the reader.

Exercise (Hessian with Clustered Eigenvalues): Assume that Q has all its eigenvalues concentrated at k intervals of the form

$$[z_i - \delta_i, z_i + \delta_i], \qquad i = 1, \ldots, k,$$

where we assume that $\delta_i \ge 0$, $i = 1, \ldots, k$, $0 < z_1 - \delta_1$, and

$$0 < z_1 < z_2 < \cdots < z_k, \qquad z_i + \delta_i \le z_{i+1} - \delta_{i+1}, \qquad i = 1, \ldots, k-1.$$

Show that the vector x_{k+1} generated after $(k+1)$ steps of the conjugate gradient method satisfies

$$f(x_{k+1}) \le R f(x_0),$$

where

$$R = \max\left\{\frac{\delta_1^2}{z_1^2}, \frac{\delta_2^2(z_2+\delta_2-z_1)^2}{z_1^2 z_2^2}, \frac{\delta_3^2(z_3+\delta_3-z_1)^2(z_3+\delta_3-z_2)^2}{z_1^2 z_2^2 z_3^2}, \ldots, \frac{\delta_k^2(z_k+\delta_k-z_1)^2\cdots(z_k+\delta_k-z_{k-1})^2}{z_1^2 z_2^2 \cdots z_k^2}\right\}.$$

The Conjugate Gradient Method Applied to Nonquadratic Problems

The conjugate gradient method can be applied to the not necessarily quadratic problem

$$\begin{aligned} &\text{minimize} \quad f(x) \\ &\text{subject to} \quad x \in R^n. \end{aligned}$$

It takes the form

$$x_{k+1} = x_k + \alpha_k d_k, \tag{78}$$

where α_k is obtained by line search

$$f(x_k + \alpha_k d_k) = \min_{\alpha} f(x_k + \alpha d_k), \tag{79}$$

and d_k is generated by

$$d_k = -\nabla f(x_k) + \beta_k d_{k-1}. \tag{80}$$

The two most common ways to compute β_k are

$$\beta_k = \frac{\nabla f(x_k)'\nabla f(x_k)}{\nabla f(x_{k-1})'\nabla f(x_{k-1})} \tag{81}$$

and

$$\beta_k = \frac{\nabla f(x_k)'[\nabla f(x_k) - \nabla f(x_{k-1})]}{\nabla f(x_{k-1})'\nabla f(x_{k-1})}. \tag{82}$$

The use of (81) has been suggested by Fletcher and Reeves (1964) while the use of (82) was proposed by Polak and Ribiere (1969), Poljak (1969a), and Sorenson (1969). The direction d_k generated by (80) will be a direction of descent in either case. To see this, note that if $\nabla f(x_k) \neq 0$, then

$$\nabla f(x_k)'d_k = -|\nabla f(x_k)|^2 + \beta_k \nabla f(x_k)'d_{k-1} = -|\nabla f(x_k)|^2 < 0,$$

since $\nabla f(x_k)'d_{k-1} = 0$ in view of (79). However, while these two formulas, along with several others, are equivalent when the method is applied to a quadratic problem, this is no more true in the general case. Extensive computational experience has established that the use of (82) results in much more efficient computation than the use of (81). A heuristic reason that can be given is that due to nonquadratic terms in the objective function and possibly inaccurate line searches, conjugacy of the generated directions is progressively lost and a situation may be created where the method temporarily "jams" in the sense that the generated direction d_k is nearly orthogonal to the gradient $\nabla f(x_k)$. When this occurs, then $\nabla f(x_{k+1}) \simeq \nabla f(x_k)$. In that case β_{k+1}, generated by (82), will be nearly zero and the next direction d_{k+1}, generated by (80), will be close to $-\nabla f(x_{k+1})$ thereby breaking the jam. This is not the case when (81) is used. A more detailed explanation of this phenomenon is given by Powell (1977).

Regardless of the formula for computing the scalar β_k, one must deal with the loss of conjugacy that results from nonquadratic terms in the objective function. The conjugate gradient method is often employed in problems where the number of variables n is large, and it is not unusual for the method to start generating nonsensical and inefficient directions of search after a few iterations. For this reason it is important to operate the method in cycles of conjugate direction steps given by (80), with the first step in the cycle being a steepest descent step. Some possible restarting policies are:

(a) Restart with a steepest descent step n iterations after the preceding restart.

(b) Restart with a steepest descent step k iterations after the preceding restart with $k < n$. This is recommended when the problem has special structure so that the resulting method has good convergence rate (compare with Proposition 1.19 and the following discussion).

(c) Restart with a steepest descent step n iterations after the preceding restart or if

$$|\nabla f(x_k)'\nabla f(x_{k-1})| > \gamma |\nabla f(x_{k-1})|^2, \tag{83}$$

where γ is a scalar with $0 < \gamma < 1$, whichever comes first. Relation (83) is a test on loss of conjugacy, for if the generated directions were indeed conjugate then we would have $\nabla f(x_k)'\nabla f(x_{k-1}) = 0$. This procedure was suggested by Powell (1977) who recommended the choice of $\gamma = 0.2$.

Note that in all these restart procedures the steepest descent iteration serves as a spacer step and guarantees global convergence (Proposition 1.16). If the scaled version of the conjugate gradient method is used, then a scaled steepest descent iteration is used to restart a cycle. The scaling matrix may change at the beginning of a cycle but should remain unchanged during the cycle. Another possibility, stemming from a suggestion of Beale (1972), is to use the last direction generated in a cycle as the first direction in the new conjugate direction cycle instead of using steepest descent. We refer to papers by Powell (1977) and Shanno (1978a,b) for a discussion of this possibility.

An important practical issue relates to the line search accuracy that is necessary for efficient computation. An elementary calculation shows that if line search is carried out to the extent that

$$\nabla f(x_k)'d_{k-1} < |\nabla f(x_{k-1})|^2,$$

then d_k, generated by (80) and (81), satisfies $\nabla f(x_k)'d_k < 0$ and is a direction of descent. On the other hand, a much more accurate line search may be necessary in order to keep loss of direction conjugacy and deterioration of rate of convergence within a reasonable level. At the same time, insisting on a very accurate line search can be computationally expensive. Considerable research has been directed towards clarifying these questions, and several implementations of the conjugate gradient method with inexact line search have been proposed by Klessig and Polak (1972), Lenard (1973,

1976), and Powell (1977). Among recent works, Shanno (1978a,b) suggests a rather imprecise line search coupled with a method for computing conjugate gradient directions which views each iteration as a memoryless quasi-Newton step. This method appears relatively insensitive to line search errors and yields descent directions under essentially no restriction on line search accuracy.

1.3.5 Quasi-Newton Methods

Quasi-Newton methods are descent methods of the form

$$x_{k+1} = x_k + \alpha_k d_k, \tag{84}$$

$$d_k = -D_k \nabla f(x_k), \tag{85}$$

where D_k is a positive definite matrix adjusted during the course of the computation in a way that (84) tends to approximate Newton's method. The stepsize α_k is determined by one of the stepsize rules of Section 1.3.1. The popularity of the most successful of these methods stems from the fact that they tend to exhibit a fast rate of convergence while avoiding the second derivative calculations associated with Newton's method.

There is a large variety of quasi-Newton methods, but we shall restrict ourselves to the so-called *Broyden class of quasi-Newton algorithms* where D_{k+1} is obtained from D_k and the vectors

$$p_k = x_{k+1} - x_k \tag{86}$$

$$q_k = \nabla f(x_{k+1}) - \nabla f(x_k), \tag{87}$$

by means of the equation

$$D_{k+1} = D_k + \frac{p_k p_k'}{p_k' q_k} - \frac{D_k q_k q_k' D_k}{q_k' D_k q_k} + \zeta_k \tau_k v_k v_k'; \tag{88}$$

where

$$v_k = \frac{p_k}{p_k' q_k} - \frac{D_k q_k}{\tau_k} \tag{89}$$

$$\tau_k = q_k' D_k q_k \tag{90}$$

the scalars ζ_k satisfy, for all k,

$$0 \le \zeta_k \le 1, \tag{91}$$

and D_0 is an arbitrary positive definite matrix. If $\zeta_k \equiv 0$, one obtains the *Davidon–Fletcher–Powell (DFP) method* (Davidon, 1959; Fletcher and Powell, 1963), which is historically the first quasi-Newton method. If $\zeta_k \equiv 1$, one obtains the *Broyden–Fletcher–Goldfarb–Shanno (BFGS) method*

(Broyden, 1970; Fletcher, 1970; Goldfarb, 1970; Shanno, 1970) for which there is growing evidence that it is the best general purpose quasi-Newton method currently available.

We first show that under a mild assumption the matrices D_k generated by (88) are positive definite. This is a most important property, since it guarantees that the search direction d_k is a direction of descent.

Proposition 1.20: If D_k is positive definite, $\nabla f(x_{k+1}) \neq 0$, and the stepsize α_k is chosen so that x_{k+1} satisfies

$$\nabla f(x_k)'d_k < \nabla f(x_{k+1})'d_k \tag{92}$$

(or equivalently $p_k' q_k > 0$), then D_{k+1} given by (88) is well defined and is positive definite.

Proof: First note that (92) implies that $q_k \neq 0$ and

$$p_k' q_k = \alpha_k d_k' [\nabla f(x_{k+1}) - \nabla f(x_k)] > 0. \tag{93}$$

Thus all denominator terms in (88), (89), and (90) are nonzero, and D_{k+1} is well defined.

Now for any $z \neq 0$, we have

$$z' D_{k+1} z = z' D_k z + \frac{(z' p_k)^2}{p_k' q_k} - \frac{(q_k' D_k z)^2}{q_k' D_k q_k} + \zeta_k \tau_k (v_k' z)^2. \tag{94}$$

Define $a = D_k^{1/2} z$, $b = D_k^{1/2} q_k$, and write (94) as

$$z' D_{k+1} z = \frac{|a|^2 |b|^2 - (a'b)^2}{|b|^2} + \frac{(z' p_k)^2}{p_k' q_k} + \zeta_k \tau_k (v_k' z)^2. \tag{95}$$

From (90), (91), (93), and the Cauchy–Schwarz inequality we have that all the terms on the right-hand side of (95) are nonnegative. In order that $z' D_{k+1} z > 0$, it will suffice to show that we cannot have simultaneously

$$|a|^2 |b|^2 = (a'b)^2 \qquad \text{and} \qquad z' p_k = 0.$$

Indeed if $|a|^2 |b|^2 = (a'b)^2$, we must have $a = \lambda b$ for some $\lambda \neq 0$ or $z = \lambda q_k$, so if $z' p_k = 0$, we must have $q_k' p_k = 0$, which is impossible by (93). Q.E.D.

Note that if D_k is positive definite, we have $\nabla f(x_k)' d_k < 0$, so in order to satisfy condition (92), it is sufficient to carry out the line search to a point where

$$|\nabla f(x_{k+1})' d_k| < |\nabla f(x_k)' d_k|.$$

If α_k is determined by the line minimization rule, then $\nabla f(x_{k+1})' d_k = 0$ and (92) is certainly satisfied.

A most interesting property of the Broyden class of algorithms is that when applied to the positive definite quadratic function

$$f(x) = \tfrac{1}{2}x'Qx,$$

with the stepsize α_k determined by line minimization, they generate a Q-conjugate direction sequence, while simultaneously constructing the inverse Hessian Q^{-1} after n iterations. This is the subject of the next proposition.

Proposition 1.21: Let $\{x_k\}$ and $\{d_k\}$ be sequences generated by the algorithm (84)–(90) applied to minimization of the positive definite quadratic function $f(x) = \frac{1}{2}x'Qx$ with α_k chosen by

$$f(x_k + \alpha_k d_k) = \min_{\alpha} f(x_k + \alpha d_k). \tag{96}$$

Assume none of the vectors $x_0, \ldots, x_{n-1}$ is optimal. Then

(a) The vectors $d_0, \ldots, d_{n-1}$ are mutually Q-conjugate.
(b) There holds

$$D_n = Q^{-1}.$$

Proof: It will be sufficient to show that for all k

$$d_i'Qd_j = 0, \qquad 0 \le i < j \le k, \tag{97}$$

$$D_{k+1}q_i = D_{k+1}Qp_i = p_i, \qquad 0 \le i \le k. \tag{98}$$

Equation (97) proves (a). Equation (98) proves (b), since for $k = n - 1$ it shows that $p_0, \ldots, p_{n-1}$ are eigenvectors of D_nQ corresponding to unity eigenvalue. Since $p_i = \alpha_i d_i$ and $d_0, \ldots, d_{n-1}$ are Q-conjugate, it follows that the eigenvectors $p_0, \ldots, p_{n-1}$ are linearly independent and therefore D_nQ equals the identity.

We first verify that for all k

$$D_{k+1}q_k = D_{k+1}Qp_k = p_k. \tag{99}$$

From (88), we have

$$D_{k+1}q_k = D_kq_k + \frac{p_kp_k'q_k}{p_k'q_k} - \frac{D_kq_kq_k'D_kq_k}{q_k'D_kq_k} + \zeta_k\tau_kv_kv_k'q_k = p_k + \zeta_k\tau_kv_kv_k'q_k.$$

An elementary calculation shows that $v_k'q_k = 0$, and (99) follows.

We now show (97) and (98) simultaneously by induction. For $k = 0$ there is nothing to show for (97), while (98) holds in view of (99). Assuming that (97) and (98) hold for k, we prove them for $k + 1$. We have, for $i < k$,

$$\nabla f(x_{k+1}) = \nabla f(x_{i+1}) + Q(p_{i+1} + \cdots + p_k). \tag{100}$$

Using (96), (97), (100), the fact $p_i = \alpha_i d_i$, and the fact $p_k' \nabla f(x_{k+1}) = 0$, we obtain

$$p_i' \nabla f(x_{k+1}) = p_i' \nabla f(x_{i+1}) = 0, \qquad 0 \le i < k + 1.$$

Hence from (98),

$$p_i' Q D_{k+1} \nabla f(x_{k+1}) = 0, \qquad 0 \le i < k + 1,$$

and since $p_i = \alpha_i d_i$, $d_{k+1} = -D_{k+1} \nabla f(x_{k+1})$, we obtain

$$d_i' Q d_{k+1} = 0, \qquad 0 \le i < k + 1.$$

This proves (97) for $k + 1$.

From the induction hypothesis (98) and (97), we have

$$(101) \qquad q_{k+1}' D_{k+1} q_i = q_{k+1}' D_{k+1} Q p_i = q_{k+1}' p_i = p_{k+1}' Q p_i = 0, \qquad 0 \le i \le k.$$

Using (88), (89), (97), (101), and a straightforward calculation, we have, for $0 \le i \le k$,

$$\begin{aligned} D_{k+2} q_i &= D_{k+1} q_i + \frac{p_{k+1} p_{k+1}' q_i}{p_{k+1}' q_{k+1}} - \frac{D_{k+1} q_{k+1} q_{k+1}' D_{k+1} q_i}{q_{k+1}' D_{k+1} q_{k+1}} \\ &\quad + \zeta_{k+1} \tau_{k+1} v_{k+1} v_{k+1}' q_i \\ &= D_{k+1} q_i = p_i. \end{aligned}$$

Taking into account (99), we have a proof of (98) for $k + 1$. Q.E.D.

It is also interesting to note that *the sequence* $\{x_k\}$ *in Proposition* 1.21 *is identical to the one that would be generated by the scaled conjugate gradient method with scaling matrix* $H = D_0$; i.e., for $k = 0, 1, \ldots, n - 1$, the vector x_{k+1} minimizes f over the linear manifold

$$M_k = \{z \mid z = x_0 + \gamma_0 D_0 \nabla f(x_0) + \cdots + \gamma_k D_0 \nabla f(x_k), \gamma_0, \ldots, \gamma_k \in R\}.$$

This can be proved for the case where $D_0 = I$ by verifying by induction that for all k there exist scalars β_{ij}^k such that

$$D_k = I + \sum_{i=0}^{k} \sum_{j=0}^{k} \beta_{ij}^k \nabla f(x_i) \nabla f(x_j)'.$$

Therefore, for some scalars b_i^k and all k, we have

$$d_k = -D_k \nabla f(x_k) = \sum_{i=0}^{k} b_i^k \nabla f(x_i).$$

Hence, for all i, x_{i+1} lies on the manifold

$$M_i = \{z \mid z = x_0 + \gamma_0 \nabla f(x_0) + \cdots + \gamma_i \nabla f(x_i), \gamma_0, \ldots, \gamma_i \in R\},$$

and since the algorithm is a conjugate direction method the result follows using Proposition 1.18. The proof for the case where $D_0 \neq I$ follows by making a transformation of variables so that in the transformed space the initial matrix is the identity. A consequence of this result is that *any algorithm in Broyden's class employing line minimization generates identical sequences of points for the case of a quadratic objective function. This is also true even for a nonquadratic objective function* (Dixon, 1972a,b) which is a rather surprising result. Thus the choice of the scalar ζ_k makes a difference only if the line minimization is inaccurate.

Computational Aspects of Quasi-Newton Methods

Consider now the case of a nonquadratic problem. Even though the quasi-Newton method (84)–(90) is equivalent to the conjugate gradient method for quadratic problems, it has certain advantages which manifest themselves in the presence of inaccurate line search and nonquadratic terms in the objective function. The first advantage is that when line search is accurate the algorithm (84)–(90) not only tends to generate conjugate directions but also constructs an approximation to the inverse Hessian matrix which tends to be more accurate as the algorithm progresses. As a result, near convergence to a strong local minimum, it tends to approximate Newton's method thereby attaining a fast convergence rate. This fact is suggested by Proposition 1.21 and has also been established analytically by Powell (1971) [for a proof, see also Polak (1971)]. It is significant that this property does not depend on the starting matrix D_0, and as a result it is not usually necessary to periodically restart the method with a steepest descent-type step—something that is essential for the conjugate gradient method. A second advantage over the conjugate gradient method is that quasi-Newton methods are not as sensitive to accuracy in the line search. This has been verified by extensive computational experience and can be substantiated to some extent by analysis (see Broyden *et al.*, 1973). One reason that can be given is that, under essentially no restriction on the line search accuracy, the quasi-Newton method (84)–(90) generates positive definite matrices D_k and hence directions of descent (Proposition 1.20).

In an effort to compare further the conjugate gradient method and quasi-Newton methods, we consider their computational requirements per iteration. The kth iteration of the conjugate gradient method requires computation of the objective function and its gradient (perhaps several times in view of the employment of line search) together with $O(n)$† multiplications to compute the conjugate direction d_k and next point x_{k+1}. A

† In this context $O(n)$ multiplications means that there is an integer M such that the number of multiplications per iteration is bounded by Mn, where n is the dimension of the problem.

quasi-Newton method requires roughly the same amount of computation for function and gradient evaluations together with $O(n^2)$ multiplications to compute the matrix D_k and next point x_{k+1}. If the computation time necessary for a function and gradient evaluation is larger or comparable to $O(n^2)$ multiplications, the quasi-Newton method requires only slightly more computation per iteration than the conjugate gradient method and holds the edge in view of its other advantages mentioned earlier. In problems where a function and gradient evaluation requires computation time much less than $O(n^2)$ multiplications, the conjugate gradient method is preferable. For example in optimal control problems where typically n is very large (over 100 and often over 1000) and a function and gradient evaluation typically requires $O(n)$ multiplications, the conjugate gradient method is preferred. In general, both methods require less computation per iteration than Newton's method which requires a function, gradient, and Hessian evaluation, as well as $O(n^3)$ multiplications at each step. This is counterbalanced by the faster speed of convergence of Newton's method. The case for Newton's method is strengthened if periodic reevaluation of the Hessian is employed since each step that utilizes a previously evaluated (and factored) Hessian requires only $O(n^2)$ multiplications. The same is true if the problem has special structure that can be exploited to compute the Newton direction efficiently. For example in optimal control problems, Newton's method typically requires $O(n)$ multiplications per iteration versus $O(n^2)$ multiplications for quasi-Newton methods.

Finally, we note that multiplying the initial matrix D_0 by a positive scaling factor can have a significant beneficial effect on the behavior of the algorithm. A popular choice is to compute

$$\tilde{D}_0 = (p_0' q_0 / q_0' D_0 q_0) D_0 \tag{102}$$

once the vector x_1 (and hence also p_0 and q_0) has been obtained, and use $\tilde{D}_0$ in place of D_0 in computing D_1. The rationale for this is explained in Luenberger (1973). Among other things it can be shown that if the initial scaling (102) is used, then the condition number M_k/m_k, where

$$M_k = \text{max eigenvalue of } (D_k^{1/2} Q D_k^{1/2}),$$

$$m_k = \text{min eigenvalue of } (D_k^{1/2} Q D_k^{1/2}),$$

is not increased (and is usually decreased) at each iteration (compare with the discussion on rate of convergence in Section 1.3.1). Sometimes it is beneficial to scale D_k even after the first iteration by the factor $p_k' q_k / q_k' D_k q_k$ and this has given rise to the class of self-scaling quasi-Newton algorithms due to Oren and Luenberger [see Oren and Luenberger (1974), Oren (1973, 1974), Oren and Spedicato (1976)].

1.3.6 Methods Not Requiring Evaluation of Derivatives

All the gradient methods examined so far in Section 1.3 require calculation of at least the gradient $\nabla f(x_k)$ and possibly the Hessian matrix $\nabla^2 f(x_k)$ at each generated point x_k. In many problems, these derivatives are either not available in explicit form or else are given by very complicated expressions and hence their evaluation requires excessive computation time. In such cases, it is possible to use the same algorithms as earlier with all unavailable derivatives approximated by finite differences. Thus, second derivatives may be approximated by the *forward difference formula*

$$\frac{\partial^2 f(x_k)}{\partial x^i \partial x^j} \sim \frac{1}{h}\left[\frac{\partial f(x_k + he_j)}{\partial x^i} - \frac{\partial f(x_k)}{\partial x^i}\right] \tag{103}$$

or the *central difference formula*

$$\frac{\partial^2 f(x_k)}{\partial x^i \partial x^j} \sim \frac{1}{2h}\left[\frac{\partial f(x_k + he_j)}{\partial x^i} - \frac{\partial f(x_k - he_j)}{\partial x^i}\right]. \tag{104}$$

In these relations, h is a small positive scalar and e_j is the jth unit vector (jth column of the identity matrix). Similarly first derivatives may be approximated by

$$\partial f(x_k)/\partial x^i \sim (1/h)[f(x_k + he_i) - f(x_k)] \tag{105}$$

or by

$$\partial f(x_k)/\partial x^i \sim (1/2h)[f(x_k + he_i) - f(x_k - he_i)]. \tag{106}$$

The central difference formula has the disadvantage that it requires twice as much computation as the forward difference formula. However, it is much more accurate. By forming the corresponding Taylor series expansions, it may be seen that the absolute value of the error between the approximation and the actual derivatives is $O(h)$ for the forward difference formula while it is $O(h^2)$ for the central difference formula. In some cases the same value of h can be used for all partial derivatives, but in other cases, particularly when the problem is poorly scaled, it is essential to use a different value of h for each partial derivative.

From the point of view of reducing the approximation error (or truncation error), it is advantageous to choose the finite difference interval h as small as possible. Unfortunately there is a limit to the amount that h can be reduced due to the significant cancellation error, which occurs when quantities of similar magnitude are subtracted by the computer. Cancellation error is particularly evident in the approximate formulas (105) and (106) near a critical point where ∇f is nearly zero.

Practical experience suggests that a good policy is to keep the scalar h for each derivative at a *fixed* value which balances the truncation error against the cancellation error. When second derivatives are approximated by finite differences of first derivatives in discretized versions of Newton's method, practical experience suggests that extreme accuracy is not very important in terms of speed of convergence. For this reason, exclusive use of the forward difference formula (103) is advisable in most cases. By contrast, when first derivatives are approximated by finite differences of function values, the approximation can become poor near a critical point and can vitally affect the convergence characteristics of the algorithm if the forward difference formula (105) is used exclusively. A good practical rule is to use the forward difference formula (105) until the absolute value of the corresponding approximate derivative becomes less than a certain tolerance; i.e.,

$$|(1/h)[f(x_k + he_i) - f(x_k)]| \leq \varepsilon,$$

where $\varepsilon > 0$ is some small prespecified scalar. At that point a switch to the central difference formula is made; i.e., the formula (106) is used whenever the inequality above is satisfied. This has been suggested by Gill and Murray (1972). An extensive discussion of implementation of gradient methods based on finite difference approximations can be found in Gill *et al.* (1981).

There are several other algorithms for minimizing differentiable functions without the explicit use of derivatives, the most interesting of which, at least from the theoretical point of view, are coordinate descent methods. For a discussion of these and other nonderivative methods we refer the reader to Avriel (1976), Brent (1972), Luenberger (1973), Polak (1971), Powell (1964, 1973), Sargent and Sebastian (1973), and Zangwill (1967a, 1969).

1.4 Constrained Minimization

We consider the problem

$$\text{(CP)} \qquad \begin{array}{ll} \text{minimize} & f(x) \\ \text{subject to} & x \in X, \end{array}$$

where $f: R^n \to R$ is a given function and X is a given subset of R^n. We say that a vector $x^* \in X$ is a *local minimum* for (CP) if there exists an $\varepsilon > 0$ such that

$$f(x^*) \leq f(x) \qquad \forall\, x \in S(x^*; \varepsilon), \quad x \in X.$$

It is a *strict local minimum* if there exists an $\varepsilon > 0$ such that

$$f(x^*) < f(x) \qquad \forall\, x \in S(x^*; \varepsilon), \quad x \in X, \quad x \neq x^*.$$

It is a *global minimum* if

$$f(x^*) \le f(x) \qquad \forall\, x \in X.$$

We have the following optimality conditions for the case where X is a convex set. Proofs may be found in the sources given at the end of the chapter.

Proposition 1.22: Assume that X is a convex set and for some $\varepsilon > 0$ and $x^* \in X$, $f \in C^1$ over $S(x^*; \varepsilon)$.

(a) If x^* is a local minimum for (CP), then

$$\nabla f(x^*)'(x - x^*) \ge 0 \qquad \forall\, x \in X. \tag{1}$$

(b) If f is in addition convex over X and (1) holds, then x^* is a global minimum for (CP).

We shall be mostly interested in optimality conditions for problems where the constraint set X is described by equality and inequality constraints.

Equality Constrained Problems

We consider first the following equality constrained problem

(ECP)
$$\text{minimize} \quad f(x)$$
$$\text{subject to} \quad h(x) = 0,$$

where $f\colon R^n \to R$ and $h\colon R^n \to R^m$ are given functions and $m \le n$. The components of h are denoted $h_1, \ldots, h_m$.

Definition: Let x^* be a vector such that $h(x^*) = 0$ and, for some $\varepsilon > 0$, $h \in C^1$ on $S(x^*; \varepsilon)$. We say that x^* is a *regular point* if the gradients $\nabla h_1(x^*), \ldots, \nabla h_m(x^*)$ are linearly independent.

Consider the *Lagrangian* function $L\colon R^{n+m} \to R$ defined by

$$L(x, \lambda) = f(x) + \lambda' h(x).$$

We have the following classical results (see, e.g., Luenberger, 1973).

Proposition 1.23: Let x^* be a local minimum for (ECP), and assume that, for some $\varepsilon > 0$, $f \in C^1$, $h \in C^1$ on $S(x^*; \varepsilon)$, and x^* is a regular point. Then there exists a unique vector $\lambda^* \in R^m$ such that

$$\nabla_x L(x^*, \lambda^*) = 0. \tag{2}$$

If in addition $f \in C^2$ and $h \in C^2$ on $S(x^*; \varepsilon)$ then

$$z' \nabla_{xx}^2 L(x^*, \lambda^*) z \ge 0 \qquad \forall\, z \in R^n \quad \text{with} \quad \nabla h(x^*)' z = 0. \tag{3}$$

Proposition 1.24: Let x^* be such that $h(x^*) = 0$ and, for some $\varepsilon > 0$, $f \in C^2$ and $h \in C^2$ on $S(x^*; \varepsilon)$. Assume that there exists a vector $\lambda^* \in R^m$ such that

$$\nabla_x L(x^*, \lambda^*) = 0 \tag{4}$$

and

$$z' \nabla_{xx}^2 L(x^*, \lambda^*) z > 0 \qquad \forall\, z \neq 0 \quad \text{with} \quad \nabla h(x^*)' z = 0. \tag{5}$$

Then x^* is a strict local minimum for (ECP).

It is instructive to provide a proof of Proposition 1.24 that utilizes concepts that will be of interest later in the analysis of multiplier methods. We have the following lemma:

Lemma 1.25: Let P be a symmetric $n \times n$ matrix and Q a positive semidefinite symmetric $n \times n$ matrix. Assume that $x'Px > 0$ for all $x \neq 0$ satisfying $x'Qx = 0$. Then there exists a scalar c such that

$$P + cQ > 0.$$

Proof: Assume the contrary. Then for every integer k, there exists a vector x_k with $|x_k| = 1$ such that

$$x_k' P x_k + k x_k' Q x_k \leq 0. \tag{6}$$

The sequence $\{x_k\}$ has a subsequence $\{x_k\}_K$ converging to a vector $\bar{x}$ with $|\bar{x}| = 1$. Taking the limit superior in (6), we obtain

$$\bar{x}' P \bar{x} + \limsup_{\substack{k \to \infty \\ k \in K}} (k x_k' Q x_k) \leq 0. \tag{7}$$

Since $x_k' Q x_k \geq 0$, (7) implies that $\{x_k' Q x_k\}_K$ converges to zero and hence $\bar{x}' Q \bar{x} = 0$. From the hypothesis it then follows that $\bar{x}' P \bar{x} > 0$ and this contradicts (7). Q.E.D.

Consider now a vector x^* satisfying the sufficiency assumptions of Proposition 1.24. By Lemma 1.25 it follows that there exists a scalar $\bar{c}$ such that

$$\nabla_{xx}^2 L(x^*, \lambda^*) + \bar{c} \nabla h(x^*) \nabla h(x^*)' > 0. \tag{8}$$

Let us introduce the so-called, *augmented Langrangian function*, L_c: $R^{n+m+1} \to R$ defined by

$$L_c(x, \lambda) = f(x) + \lambda' h(x) + \tfrac{1}{2} c |h(x)|^2. \tag{9}$$

We have, by a straightforward calculation,

$$(10)\quad \nabla_x L_c(x, \lambda) = \nabla f(x) + \nabla h(x)[\lambda + ch(x)],$$

$$(11)\quad \nabla^2_{xx} L_c(x, \lambda) = \nabla^2 f(x) + \sum_{i=1}^{m} [\lambda_i + ch_i(x)]\nabla^2 h_i(x) + c\nabla h(x)\nabla h(x)'.$$

Therefore, using also (8), we have, for all $c \geq \bar{c}$,

$$(12)\quad \nabla_x L_c(x^*, \lambda^*) = \nabla_x L(x^*, \lambda^*) = 0,$$

$$(13)\quad \nabla^2_{xx} L_c(x^*, \lambda^*) = \nabla^2_{xx} L(x^*, \lambda^*) + c\nabla h(x^*)\nabla h(x^*)' > 0.$$

Now by using Proposition 1.4 and the preceding discussion, we obtain the following result:

Proposition 1.26: Under the sufficiency assumptions of Proposition 1.24, there exist scalars $\bar{c}, \gamma > 0$, and $\delta > 0$ such that

$$(14)\quad L_c(x, \lambda^*) \geq L_c(x^*, \lambda^*) + \gamma|x - x^*|^2 \qquad \forall\, x \in S(x^*; \delta), \quad c \geq \bar{c}.$$

Notice that from (9) and (14), we obtain

$$f(x) \geq f(x^*) + \gamma|x - x^*|^2 \qquad \forall\, x \in S(x^*; \varepsilon), \quad h(x) = 0,$$

which implies that x^* is a strict local minimum for (ECP). Thus a proof of Proposition 1.24 has been obtained.

The next proposition yields a valuable sensitivity interpretation of Lagrange multipliers. We shall need the following lemma:

Lemma 1.27: Let x^* be a local minimum for (ECP) which is a regular point and together with its associated Lagrange multiplier vector λ^* satisfies the sufficiency assumptions of Proposition 1.24. Then the $(n + m) \times (n + m)$ matrix

$$(15)\quad J = \begin{bmatrix} \nabla^2_{xx} L(x^*, \lambda^*) & \nabla h(x^*) \\ \nabla h(x^*)' & 0 \end{bmatrix}$$

is nonsingular.

Proof: If J were singular, there would exist $y \in R^n$ and $z \in R^m$ not both zero such that (y, z) is in the nullspace of J or equivalently

$$(16)\quad \nabla^2_{xx} L(x^*, \lambda^*)y + \nabla h(x^*)z = 0,$$

$$(17)\quad \nabla h(x^*)'y = 0.$$

Premultiplying (16) by y' and using (17), we obtain

$$y'\nabla^2_{xx} L(x^*, \lambda^*)y = 0.$$

Hence $y = 0$, for otherwise the sufficiency assumption is violated. It follows that $\nabla h(x^*)z = 0$, which in view of the fact that $\nabla h(x^*)$ has rank m implies $z = 0$. This contradicts the fact that y and z cannot be both zero. Q.E.D.

Proposition 1.28: Let the assumptions of Lemma 1.27 hold. Then there exists a scalar $\delta > 0$ and continuously differentiable functions $x(\cdot): S(0;\delta) \to R^n$, $\lambda(\cdot): S(0;\delta) \to R^m$ such that $x(0) = x^*$, $\lambda(0) = \lambda^*$, and for all $u \in S(0;\delta)$, $\{x(u), \lambda(u)\}$ are a local minimum–Lagrange multiplier pair for the problem

$$\text{(18)} \qquad \text{minimize} \quad f(x)$$

$$\text{subject to} \quad h(x) = u.$$

Furthermore,

$$\nabla_u f[x(u)] = -\lambda(u) \qquad \forall\, u \in S(0;\delta).$$

Proof: Consider the system of equations in (x, λ, u):

$$\nabla f(x) + \nabla h(x)\lambda = 0, \qquad h(x) - u = 0.$$

It has the solution $(x^*, \lambda^*, 0)$. Furthermore the Jacobian of the system with respect to (x, λ) at this solution is the invertible matrix J of (15). Hence by the implicit function theorem (Section 1.2), there exists a $\delta > 0$ and functions $x(\cdot) \in C^1$, $\lambda(\cdot) \in C^1$ on $S(0;\delta)$ such that

$$\text{(19)} \quad \nabla f[x(u)] + \nabla h[x(u)]\lambda(u) = 0, \qquad h[x(u)] = u \qquad \forall\, u \in S(0;\delta).$$

For u sufficiently close to $u = 0$, the vectors $x(u)$, $\lambda(u)$ satisfy the sufficiency conditions for problem (18) in view of the fact that they satisfy them by assumption for $u = 0$. Hence δ can be chosen so that $\{x(u), \lambda(u)\}$ are a local minimum–Lagrange multiplier pair for problem (18).

Now from (19), we have

$$\nabla_u x(u)\nabla f[x(u)] + \nabla_u x(u)\nabla h[x(u)]\lambda(u) = 0$$

or

$$\text{(20)} \qquad \nabla_u f[x(u)] = -\nabla_u x(u)\nabla h[x(u)]\lambda(u).$$

By differentiating the relation $h[x(u)] = u$, we obtain

$$\text{(21)} \qquad I = \nabla_u h[x(u)] = \nabla_u x(u)\nabla h[x(u)].$$

Combining (20) and (21), we have

$$\nabla_u f[x(u)] = -\lambda(u),$$

which was to be proved. Q.E.D.

Inequality Constraints

Consider now the case of a problem involving both equality and inequality constraints

$$\text{(NLP)} \qquad \text{minimize} \quad f(x)$$
$$\text{subject to} \quad h(x) = 0, \qquad g(x) \le 0,$$

where $f: R^n \to R$, $h: R^n \to R^m$, $g: R^n \to R^r$ are given functions and $m \le n$. The components of g are denoted by $g_1, \ldots, g_r$. We first generalize the definition of a regular point. For any vector x satisfying $g(x) \le 0$, we denote

$$A(x) = \{j \mid g_j(x) = 0, j = 1, \ldots, r\}. \tag{22}$$

Definition: Let x^* be a vector such that $h(x^*) = 0$, $g(x^*) \le 0$ and, for some $\varepsilon > 0$, $h \in C^1$ and $g \in C^1$ on $S(x^*; \varepsilon)$. We say that x^* is a *regular point* if the gradients $\nabla h_1(x^*), \ldots, \nabla h_m(x^*)$ and $\nabla g_j(x^*)$, $j \in A(x^*)$, are linearly independent.

Define the Lagrangian function $L: R^{n+m+r} \to R$ for (NLP) by

$$L(x, \lambda, \mu) = f(x) + \lambda' h(x) + \mu' g(x).$$

We have the following optimality conditions paralleling those for equality constrained problems (see, e.g., Luenberger, 1973).

Proposition 1.29: Let x^* be a local minimum for (NLP) and assume that, for some $\varepsilon > 0$, $f \in C^1$, $h \in C^1$, $g \in C^1$ on $S(x^*; \varepsilon)$, and x^* is a regular point. Then there exist unique vectors $\lambda^* \in R^m$, $\mu^* \in R^r$ such that

$$\nabla_x L(x^*, \lambda^*, \mu^*) = 0, \tag{23}$$

$$\mu_j^* \ge 0, \qquad \mu_j^* g_j(x^*) = 0 \qquad \forall j = 1, \ldots, r. \tag{24}$$

If in addition $f \in C^2$, $h \in C^2$, and $g \in C^2$ on $S(x^*; \varepsilon)$, then for all $z \in R^n$ satisfying $\nabla h(x^*)'z = 0$ and $\nabla g_j(x^*)'z = 0$, $j \in A(x^*)$, we have

$$z' \nabla_{xx}^2 L(x^*, \lambda^*, \mu^*) z \ge 0. \tag{25}$$

Proposition 1.30: Let x^* be such that $h(x^*) = 0$, $g(x^*) \le 0$, and, for some $\varepsilon > 0$, $f \in C^2$, $h \in C^2$, and $g \in C^2$ on $S(x^*; \varepsilon)$. Assume that there exist vectors $\lambda^* \in R^m$, $\mu^* \in R^r$ such that

$$\nabla_x L(x^*, \lambda^*, \mu^*) = 0, \tag{26}$$

$$\mu_j^* \ge 0, \qquad \mu_j^* g_j(x^*) = 0 \qquad \forall j = 1, \ldots, r, \tag{27}$$

and for every $z \neq 0$ satisfying $\nabla h(x^*)'z = 0$, $\nabla g_j(x^*)'z \leq 0$, for all $j \in A(x^*)$, and $\nabla g_j(x^*)'z = 0$, for all $j \in A(x^*)$ with $\mu_j^* > 0$, we have

$$z'\nabla_{xx}^2 L(x^*, \lambda^*, \mu^*)z > 0. \tag{28}$$

Then x^* is a strict local minimum for (NLP).

Optimality Conditions via Conversion to the Equality Constrained Case

Some of the results for inequality constraints may also be proved by using the results for equality constraints *provided we assume that* $f, h_i, g_j \in C^2$. In this approach, we convert the inequality constrained problem (NLP) into a problem which involves exclusively equality constraints and then use the results for (ECP) to obtain necessary conditions, sufficiency conditions, and a sensitivity result for (NLP).

Consider the equality constrained problem

$$\begin{aligned} &\text{minimize} \quad f(x) \\ &\text{subject to} \quad h_1(x) = 0, \dots, h_m(x) = 0, \\ &\qquad g_1(x) + z_1^2 = 0, \dots, g_r(x) + z_r^2 = 0, \end{aligned} \tag{29}$$

where we have introduced additional variables $z_1, \dots, z_r$. It is clear that (NLP) and problem (29) are equivalent in the sense that x^* is a local minimum for problem (NLP) if and only if $(x^*, [-g_1(x^*)]^{1/2}, \dots, [-g_r(x^*)]^{1/2})$ is a local minimum for (29). By introducing the vector $z = (z_1, \dots, z_r)$ and the functions

$$\begin{aligned} \bar{f}(x, z) &= f(x), \\ \bar{h}_i(x, z) &= h_i(x), \qquad i = 1, \dots, m, \\ \bar{g}_j(x, z) &= g_j(x) + z_j^2, \qquad j = 1, \dots, r, \end{aligned}$$

problem (29) may be written as

$$\begin{aligned} &\text{minimize} \quad \bar{f}(x, z) \\ &\text{subject to} \quad \bar{h}_i(x, z) = 0, \quad \bar{g}_j(x, z) = 0, \quad i = 1, \dots, m, \quad j = 1, \dots, r. \end{aligned} \tag{30}$$

Let x^* be a local minimum for our original problem (NLP) as well as a regular point. Then (x^*, z^*), where $z^* = (z_1^*, \dots, z_r^*)$, $z_j^* = [-g_j(x^*)]^{1/2}$, is a local minimum for problem (30). In addition (x^*, z^*) is a regular point since the gradients

$$\nabla \bar{h}_i(x^*, z^*) = \begin{bmatrix} \nabla h_i(x^*) \\ 0 \end{bmatrix}, \qquad i = 1, \ldots, m,$$

$$\nabla \bar{g}_j(x^*, z^*) = \begin{bmatrix} \nabla g_j(x^*) \\ 0 \\ \vdots \\ 0 \\ 2z_j^* \\ 0 \\ \vdots \\ 0 \end{bmatrix}, \qquad j = 1, \ldots, r,$$

can be easily verified to be linearly independent when x^* is a regular point. By the necessary conditions for equality constraints (Proposition 1.23), there exist Lagrange multipliers $\lambda_1^*, \ldots, \lambda_m^*, \mu_1^*, \ldots, \mu_r^*$ such that

$$\nabla \bar{f}(x^*, z^*) + \sum_{i=1}^{m} \lambda_i^* \nabla \bar{h}_i(x^*, z^*) + \sum_{j=1}^{r} \mu_j^* \nabla \bar{g}_j(x^*, z^*) = 0.$$

In view of the form of the gradients of $\bar{f}$, $\bar{h}_i$, and $\bar{g}_j$, the condition above is equivalent to

$$\text{(31a)} \qquad \nabla f(x^*) + \sum_{i=1}^{m} \lambda_i^* \nabla h_i(x^*) + \sum_{j=1}^{r} \mu_j^* \nabla g_j(x^*) = 0,$$

$$\text{(31b)} \qquad 2\mu_j^*[-g_j(x^*)]^{1/2} = 0, \qquad j = 1, \ldots, r.$$

The last equation implies $\mu_j^* = 0$ for all $j \notin A(x^*)$ and may also be written as

$$\text{(32)} \qquad \mu_j^* g_j(x^*) = 0, \qquad j = 1, \ldots, r.$$

The second-order necessary condition for problem (30) is applicable, in view of our assumption $f, h_i, g_j \in C^2$ which in turn implies $\bar{f}, \bar{h}_i, \bar{g}_j \in C^2$. It yields

$$\text{(33)} \qquad [y', v'] \left[\begin{array}{c|ccc} \nabla_{xx}^2 L(x^*, \lambda^*, \mu^*) & & 0 & \\ \hline & 2\mu_1^* & & 0 \\ 0 & & \ddots & \\ & 0 & & 2\mu_r^* \end{array}\right] \begin{bmatrix} y \\ v \end{bmatrix} \geq 0$$

for all $y \in R^n$, $v = (v_1, \ldots, v_r) \in R^r$ satisfying

$$(34) \qquad \nabla h(x^*)'y = 0, \qquad \nabla g_j(x^*)'y + 2z_j^* v_j = 0, \qquad j = 1, \ldots, r.$$

By setting $v_j = 0$ for $j \in A(x^*)$ and taking into account the fact $\mu_j^* = 0$ for $j \notin A(x^*)$ [compare with (32)] we obtain, from (33) and (34),

$$(35) \qquad y'\nabla_{xx}^2 L(x^*, \lambda^*, \mu^*)y \geq 0,$$

$$\forall\ y, \text{ with } \nabla h(x^*)'y = 0, \quad \nabla g_j(x^*)'y = 0, \quad j \in A(x^*).$$

For every j with $z_j^* = 0$, we may choose $y = 0$, $v_j \neq 0$, and $v_k = 0$, for $k \neq j$, in (33) to obtain

$$(36) \qquad \mu_j^* \geq 0.$$

Relations (31), (32), (35), and (36) represent all the necessary conditions of Proposition 1.29. Thus we have obtained a proof of Proposition 1.29 (under the assumption f, h_i, $g_j \in C^2$) based on the transformation of the inequality constrained problem (NLP) to the equality constrained problem (29).

The transformation described above may also be used to derive a set of sufficiency conditions for (NLP) which are somewhat weaker than those of Proposition 1.30.

Proposition 1.31: Let x^* be such that $h(x^*) = 0$, $g(x^*) \leq 0$, and, for some $\varepsilon > 0$, $f \in C^2$, $h \in C^2$, and $g \in C^2$ on $S(x^*; \varepsilon)$. Assume that there exist vectors $\lambda^* \in R^m$, $\mu^* \in R^r$ satisfying

$$\nabla_x L(x^*, \lambda^*, \mu^*) = 0,$$

$$\mu_j^* \geq 0, \qquad \mu_j^* g_j(x^*) = 0, \qquad j = 1, \ldots, r,$$

as well as the *strict complementarity condition*

$$\mu_j^* > 0 \qquad \text{if} \quad j \in A(x^*).$$

Assume further that for all $y \neq 0$ satisfying $\nabla h(x^*)'y = 0$ and $\nabla g_j(x^*)'y = 0$, for all $j \in A(x^*)$, we have

$$y'\nabla_{xx}^2 L(x^*, \lambda^*, \mu^*)y > 0.$$

Then x^* is a strict local minimum for (NLP).

Proof: From (31), (33), and (34), we see that our assumptions imply that the sufficiency conditions of Proposition 1.24 are satisfied for (x^*, z^*) and λ^*, μ^*, where $z^* = ([-g_1(x^*)]^{1/2}, \ldots, [-g_r(x^*)]^{1/2})$ for problem (29). Hence (x^*, z^*) is a strict local minimum for problem (29) and it follows that x^* is a strict local minimum of f subject to $h(x) = 0$, and $g(x) \leq 0$. Q.E.D.

We formalize some of the arguments in the preceding discussion in the following proposition.

Proposition 1.32: If the sufficiency conditions for (NLP) of Proposition 1.31 hold, then the sufficiency conditions of Proposition 1.24 are satisfied for problem (29). If in addition x^* is a regular point for (NLP), then (x^*, z^*), where $z^* = ([-g_1(x^*)]^{1/2}, \ldots, [-g_r(x^*)]^{1/2})$, is a regular point for problem (29).

Linear Constraints

The preceding necessary conditions rely on a regularity assumption on the local minimum x^* to assert the existence of a unique Lagrange multiplier vector. When x^* is not regular, there are two possibilities. Either there does not exist a Lagrange multiplier vector or there exists an infinity of such vectors. There are a number of assumptions other than regularity that guarantee the existence of a Lagrange multiplier vector. A very useful one is linearity of the constraint functions as in the following proposition.

Proposition 1.33: Let x^* be a local minimum for the problem

$$\text{minimize} \quad f(x)$$
$$\text{subject to} \quad a_j'x - b_j \le 0, \qquad j = 1, \ldots, r,$$

where $f: R^n \to R$, $b \in R^r$, and $a_j \in R^n$, $j = 1, \ldots, r$. Assume that, for some $\varepsilon > 0$, $f \in C^1$ on $S(x^*; \varepsilon)$. Then there exists a vector $\mu^* = (\mu_1^*, \ldots, \mu_r^*)$ such that

$$\nabla f(x^*) + \sum_{j=1}^{r} \mu_j^* a_j = 0,$$
$$\mu_j^* \ge 0, \qquad \mu_j^*(a_j'x^* - b_j) = 0, \qquad j = 1, \ldots, r.$$

Sufficiency Conditions under Convexity Assumptions

Consider the convex programming problem

$$\text{(37)} \qquad \text{minimize} \quad f(x)$$
$$\text{subject to} \quad g(x) \le 0,$$

where we assume that the functions f and $g_1, \ldots, g_r$ are convex and differentiable over R^n. Then every local minimum is global, and the necessary optimality conditions of Proposition 1.29 are also sufficient as stated in the following proposition.

Proposition 1.34: Assume that f and $g_1, \ldots, g_r$ are convex and continuously differentiable functions on R^n. Let $x^* \in R^n$ and $\mu^* \in R^r$ satisfy

$$\nabla f(x^*) + \nabla g(x^*)\mu^* = 0,$$

$$g(x^*) \le 0, \qquad \mu_j^* \ge 0, \qquad \mu_j^* g_j(x^*) = 0, \qquad j = 1, \ldots, r.$$

Then x^* is a global minimum of problem (37).

1.5 Algorithms for Minimization Subject to Simple Constraints

There is a large number of algorithms of the feasible direction type for minimization of differentiable functions subject to linear constraints. A survey of some of the most popular ones may be found in the volume edited by Gill and Murray (1974), and computational results may be found in the paper by Lenard (1979). In this section, we shall focus on a new class of methods that is well suited for problems with simple inequality constraints such as those that might arise in methods of multipliers and differentiable exact penalty methods, where the simple constraints are not eliminated by means of a penalty but rather are treated directly (cf. Sections 2.4 and 4.3). We shall restrict ourselves exclusively to problems involving lower and/or upper bounds on the variables, but there are extensions of the class of algorithms presented that handle problems with general linear constraints (see Bertsekas, 1980c).

Consider the problem

$$\text{(SCP)} \qquad \text{minimize } f(x) \quad \text{subject to } x \ge 0,$$

where $f\colon R^n \to R$ is a continuously differentiable function. By applying Proposition 1.22, we obtain the following necessary conditions for optimality of a vector $x^* \ge 0$.

$$\text{(1a)} \qquad \partial f(x^*)/\partial x^i = 0 \quad \text{if} \quad x^{*i} > 0, \quad i = 1, \ldots, n,$$

$$\text{(1b)} \qquad \partial f(x^*)/\partial x^i \ge 0 \quad \text{if} \quad x^{*i} = 0, \quad i = 1, \ldots, n.$$

An equivalent way of writing these conditions is

$$\text{(2)} \qquad x^* = [x^* - \alpha \nabla f(x^*)]^+,$$

where α is any positive scalar and $[\cdot]^+$ denotes projection on the positive orthant; i.e., for every $z = (z^1, \ldots, z^n)$,

$$\text{(3)} \qquad [z]^+ = \begin{bmatrix} \max\{0, z^1\} \\ \vdots \\ \max\{0, z^n\} \end{bmatrix}.$$

If a vector $x^* \geq 0$ satisfies (1), we say that it is a *critical point* with respect to (SCP).

Equation (2) motivates the following extension of the steepest descent method

(4) $$x_{k+1} = [x_k - \alpha_k \nabla f(x_k)]^+, \qquad k = 0, 1, \ldots,$$

where α_k is a positive scalar stepsize. There are a number of rules for choosing α_k that guarantee that limit points of sequences generated by iteration (4) satisfy the necessary condition (1) (Goldstein, 1964, 1974; Levitin and Poljak, 1965; McCormick, 1969; Bertsekas, 1974c). The rate of convergence of iteration (4) is however at best linear for general problems. We shall provide Newton-like generalizations of iteration (4) which preserve its basic simplicity while being capable of superlinear convergence.

Consider an iteration of the form

(5) $$x_{k+1} = [x_k - \alpha_k D_k \nabla f(x_k)]^+, \qquad k = 0, 1, \ldots,$$

where D_k is a positive definite symmetric matrix and α_k is chosen by search along the arc of points

(6) $$x_k(\alpha) = [x_k - \alpha D_k \nabla f(x_k)]^+, \qquad \alpha \geq 0.$$

It is easy to construct examples (see Fig. 1.2) where an arbitrary choice of the matrix D_k leads to situations where it is impossible to reduce the value

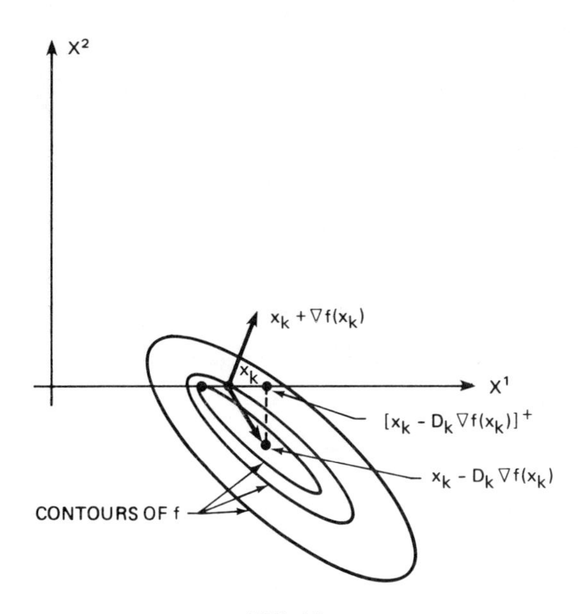

FIG. 1.2

of the objective by suitable choice of the stepsize α (i.e., $f[x_k(\alpha)] \geq f(x_k)$ $\forall\, \alpha \geq 0$). The following proposition identifies a class of matrices D_k for which an objective reduction is possible. Define, for all $x \geq 0$,

$$I^+(x) = \{i \mid x^i = 0, \partial f(x)/\partial x^i > 0\}. \tag{7}$$

We say that a symmetric matrix D with elements d^{ij} is *diagonal with respect to a subset of indices* $I \subset \{1, 2, \ldots, n\}$, if

$$d^{ij} = 0 \qquad \forall\, i \in I, \quad j = 1, 2, \ldots, n, \quad j \neq i. \tag{8}$$

Proposition 1.35: Let $x \geq 0$ and D be a positive definite symmetric matrix which is diagonal with respect to $I^+(x)$, and denote

$$x(\alpha) = [x - \alpha D \nabla f(x)]^+ \qquad \forall\, \alpha \geq 0. \tag{9}$$

(a) The vector x is a critical point with respect to (SCP), if and only if

$$x = x(\alpha) \qquad \forall\, \alpha \geq 0.$$

(b) If x is not a critical point with respect to (SCP), there exists a scalar $\bar{\alpha} > 0$ such that

$$f[x(\alpha)] < f(x) \qquad \forall\, \alpha \in (0, \bar{\alpha}]. \tag{10}$$

Proof: Assume without loss of generality that for some integer r, we have

$$I^+(x) = \{r + 1, \ldots, n\}.$$

Then D has the form

$$D = \left[\begin{array}{c|ccc} \bar{D} & & 0 & \\ \hline & d^{r+1} & & 0 \\ 0 & & \ddots & \\ & 0 & & d^n \end{array}\right], \tag{11}$$

where $\bar{D}$ is positive definite and $d^i > 0$, $i = r + 1, \ldots, n$.

Denote

$$p = D \nabla f(x). \tag{12}$$

(a) Assume x is a critical point. Then, using (1), (7),

$$\partial f(x)/\partial x^i = 0 \qquad \forall\, i = 1, \ldots, r$$
$$\partial f(x)/\partial x^i > 0, \qquad x^i = 0 \qquad \forall\, i = r + 1, \ldots, n.$$

These relations and the positivity of d^i, $i = r + 1, \ldots, n$, imply that

$$p^i = 0 \qquad \forall\, i = 1, \ldots, r,$$
$$p^i > 0 \qquad \forall\, i = r + 1, \ldots, n.$$

Since $x^i(\alpha) = [x^i - \alpha p^i]^+$ and $x^i = 0$ for $i = r + 1, \ldots, n$, it follows that $x^i(\alpha) = x^i$, for all i, and $\alpha \geq 0$.

Conversely assume that $x = x(\alpha)$ for all $\alpha \geq 0$. Then we must have

$$p^i = 0 \qquad \forall\, i = 1, \ldots, n \quad \text{with} \quad x^i > 0,$$

$$p^i \geq 0 \qquad \forall\, i = 1, \ldots, n \quad \text{with} \quad x^i = 0.$$

Now by definition of $I^+(x)$, we have that if $x^i = 0$ and $i \notin I^+(x)$, then $\partial f(x)/\partial x^i \leq 0$. This together with the relations above imply

$$\sum_{i=1}^{r} p^i \frac{\partial f(x)}{\partial x^i} \leq 0.$$

Since, by (11) and (12),

$$\begin{bmatrix} p_1 \\ \vdots \\ p_r \end{bmatrix} = \bar{D} \begin{bmatrix} \partial f(x)/\partial x^1 \\ \vdots \\ \partial f(x)/\partial x^r \end{bmatrix}$$

and $\bar{D}$ is positive definite, it follows that

$$p^i = \partial f(x)/\partial x^i = 0 \qquad \forall\, i = 1, \ldots, r.$$

Since, for $i = r + 1, \ldots, n$, $\partial f(x)/\partial x^i > 0$, and $x^i = 0$, we obtain that x is a critical point.

(b) For $i = r + 1, \ldots, n$, we have $\partial f(x)/\partial x^i > 0$, $x^i = 0$, and, from (11) and (12), $p^i > 0$. Since $x^i(\alpha) = [x^i - \alpha p^i]^+$, we obtain

$$x^i = x^i(\alpha) = 0 \qquad \forall\, \alpha \geq 0, \quad i = r + 1, \ldots, n. \tag{13}$$

Consider the sets of indices

$$I_1 = \{i \mid x^i > 0 \quad \text{or} \quad x^i = 0 \quad \text{and} \quad p^i < 0, \quad i = 1, \ldots, r\}, \tag{14}$$

$$I_2 = \{i \mid x^i = 0 \quad \text{and} \quad p^i \geq 0, \quad i = 1, \ldots, r\}. \tag{15}$$

Let

$$\alpha_1 = \sup\{\alpha \geq 0 \mid x^i - \alpha p^i \geq 0, i \in I_1\}. \tag{16}$$

Note that, in view of the definition of I_1, α_1 is either positive or $+\infty$. Define the vector $\bar{p}$ with coordinates

$$\bar{p}^i = \begin{cases} p^i & \text{if} \quad i \in I_1, \\ 0 & \text{if} \quad i \in I_2 \quad \text{or} \quad i = r + 1, \ldots, n. \end{cases} \tag{17}$$

In view of (13)–(16), we have

$$x(\alpha) = x - \alpha \bar{p} \qquad \forall\, \alpha \in (0, \alpha_1). \tag{18}$$

In view of (15) and the definition of $I^+(x)$, we have

(19) $$\partial f(x)/\partial x^i \leq 0 \qquad \forall\, i \in I_2,$$

and hence

(20) $$\sum_{i \in I_2} \frac{\partial f(x)}{\partial x^i} p^i \leq 0.$$

Now using (17) and (20), we have

(21) $$\nabla f(x)'\bar{p} = \sum_{i \in I_1} \frac{\partial f(x)}{\partial x^i} p^i \geq \sum_{i=1}^{r} \frac{\partial f(x)}{\partial x^i} p^i.$$

Since x is not a critical point, by part (a) and (18), we must have $x \neq x(\alpha)$ for some $\alpha > 0$, and hence also in view of (13), $p^i \neq 0$ for some $i \in \{1, \ldots, r\}$. In view of the positive definiteness of $\bar{D}$ and (11) and (12), it follows that

$$\sum_{i=1}^{r} \frac{\partial f(x)}{\partial x^i} p^i > 0.$$

It follows, from (21), that

$$\nabla f(x)'\bar{p} > 0.$$

Combining this relation with (18) and the fact that $\alpha_1 > 0$, it follows that $\bar{p}$ is a feasible descent direction at x and there exists a scalar $\bar{\alpha} > 0$ for which the desired relation (10) is satisfied. Q.E.D.

Based on Proposition 1.35, we are led to the conclusion that the matrix D_k in the iteration

$$x_{k+1} = [x_k - \alpha_k D_k \nabla f(x_k)]^+$$

should be chosen diagonal with respect to a subset of indices that contains

$$I^+(x_k) = \{i \,|\, x_k^i = 0, \partial f(x_k)/\partial x^i > 0\}.$$

Unfortunately, the set $I^+(x_k)$ exhibits an undesirable discontinuity at the boundary of the constraint set whereby given a sequence $\{x_k\}$ of interior points that converges to a boundary point $\bar{x}$, all the sets $I^+(x_k)$ may be strictly smaller than the set $I^+(\bar{x})$. This causes difficulties in proving convergence of the algorithm and may have an adverse effect on its rate of convergence. (This phenomenon is quite common in feasible direction algorithms and is referred to as zigzagging or jamming.) For this reason, we shall employ certain enlargements of the sets $I^+(x_k)$ with the aim of bypassing these difficulties.

The algorithm that we describe utilizes a scalar $\varepsilon > 0$ (typically small), a fixed† *diagonal* positive definite matrix M (for example, the identity), and two parameters $\beta \in (0, 1)$ and $\sigma \in (0, \frac{1}{2})$ that will be used in connection with an Armijo-like stepsize rule. An initial vector $x_0 \geq 0$ is chosen and at the kth iteration of the algorithm, we have a vector $x_k \geq 0$. Denote

$$w_k = |x_k - [x_k - M\nabla f(x_k)]^+|, \qquad \varepsilon_k = \min\{\varepsilon, w_k\}.$$

(Actually there are several other possibilities for defining the scalar ε_k as can be seen by examination of the proof of the subsequent proposition. It is also possible to use a separate scalar ε_k^i for each coordinate.)

$(k + 1)$st Iteration of the Algorithm

We select a positive definite symmetric matrix D_k which is diagonal with respect to the set I_k^+ given by

$$I_k^+ = \{i \mid 0 \leq x_k^i \leq \varepsilon_k, \partial f(x_k)/\partial x^i > 0\}. \tag{22}$$

Denote

$$p_k = D_k \nabla f(x_k), \tag{23}$$

$$x_k(\alpha) = [x_k - \alpha p_k]^+ \qquad \forall\, \alpha \geq 0. \tag{24}$$

Then x_{k+1} is given by

$$x_{k+1} = x_k(\alpha_k), \tag{25}$$

where

$$\alpha_k = \beta^{m_k}, \tag{26}$$

and m_k is the first nonnegative integer m such that

$$f(x_k) - f[x_k(\beta^m)] \geq \sigma\left\{\beta^m \sum_{i \notin I_k^+} \frac{\partial f(x_k)}{\partial x^i} p_k^i + \sum_{i \in I_k^+} \frac{\partial f(x_k)}{\partial x^i} [x_k^i - x_k^i(\beta^m)]\right\}. \tag{27}$$

The stepsize rule (26) and (27) is quite similar to the Armijo rule of Section 1.3. We have chosen a unity initial stepsize, but any other positive initial stepsize can be incorporated in the matrix D_k, so this choice involves no loss of generality. The results that follow can also be proved if

$$\sum_{i \notin I_k^+} \frac{\partial f(x_k)}{\partial x^i} p_k^i$$

† Actually the results that follow can also be proved if the fixed matrix M is replaced by a sequence of diagonal positive definite matrices $\{M_k\}$ with diagonal elements that are bounded above and away from zero.

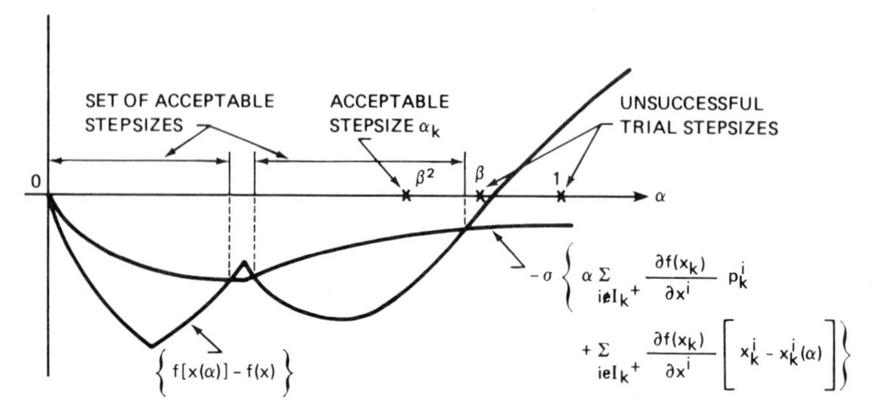

FIG. 1.3 Line search by the Armijo-like rule (26), (27).

in (27) is replaced by $\gamma_k \sum_{i \notin I_k^+} [\partial f(x_k)/\partial x^i] p_k^i$, where $\gamma_k = \min\{1, \bar{\alpha}_k\}$ and $\bar{\alpha}_k = \sup\{\alpha \,|\, x_k^i - \alpha p_k^i \geq 0 \; \forall\, i \notin I_k^+\}$. Other variations of the stepsize rule are also possible. The process of determining the stepsize α_k is illustrated in Fig. 1.3. When I_k^+ is empty, the right-hand side of (27) becomes $\sigma \beta^m \nabla f(x_k)' p_k$ and is identical to the corresponding expression of the Armijo rule for unconstrained minimization. Note that, for all k, $I_k^+ \supset I^+(x_k)$ so D_k is diagonal with respect to $I^+(x_k)$. It is possible to show that for all $m \geq 0$, the right-hand side of (27) is nonnegative and is positive if and only if x_k is not a critical point. Indeed since D_k is positive definite and diagonal with respect to I_k^+, we have

$$\sum_{i \notin I_k^+} \frac{\partial f(x_k)}{\partial x^i} p_k^i \geq 0 \qquad \forall\, k = 0, 1, \ldots, \tag{28}$$

while for all $i \in I_k^+$, in view of the fact $\partial f(x^k)/\partial x^i > 0$, we have $p_k^i > 0$, and hence

$$x_k^i - x_k^i(\alpha) \geq 0 \qquad \forall\, \alpha \geq 0, \quad i \in I_k^+, \quad k = 0, 1, \ldots,$$

$$\frac{\partial f(x_k)}{\partial x^i} [x_k^i - x_k^i(\alpha)] \geq 0 \qquad \forall\, \alpha \geq 0, \quad i \in I_k^+, \quad k = 0, 1, \ldots. \tag{29}$$

This shows that the right-hand side of (27) is nonnegative. If x_k is not critical, then it is easily seen [compare also with the proof of Proposition 1.35(b)] that one of the inequalities (28) or (29) is strict for $\alpha > 0$ so the right-hand side of (27) is positive for all $m \geq 0$. A slight modification of the proof of Proposition 1.35(b) also shows that if x_k is not a critical point, then (27) will be satisfied for all m sufficiently large so the stepsize α_k is well defined and can be determined via a finite number of arithmetic operations. If x_k is a critical point then, by Proposition 1.35(a), we have $x_k = x_k(\alpha)$ for all $\alpha \geq 0$.

Furthermore the argument given in the proof of Proposition 1.35(a) shows that

$$\sum_{i \notin I_k^+} \frac{\partial f(x_k)}{\partial x^i} p_k^i = 0,$$

so both terms in the right-hand side of (27) are zero. Since also $x_k = x_k(\alpha)$ for all $\alpha \geq 0$, it follows that (27) is satisfied for $m = 0$ thereby implying that

$$x_{k+1} = x_k(1) = x_k \quad \text{if} \quad x_k \text{ is critical.}$$

In conclusion the algorithm is well defined, decreases the value of the objective function at each iteration k for which x_k is not a critical point, and essentially terminates if x_k is critical. We proceed to analyze its convergence and rate of convergence properties. To this end, we shall make use of the following two assumptions:

Assumption (A): *The gradient ∇f is Lipschitz continuous on each bounded set of R^n; i.e., given any bounded set $S \subset R^n$ there exists a scalar L (depending on S) such that*

$$|\nabla f(x) - \nabla f(y)| \leq L|x - y| \qquad \forall\, x, y \in S. \tag{30}$$

Assumption (B): *There exist positive scalars λ_1 and λ_2 and nonnegative integers q_1 and q_2, such that*

$$\lambda_1 w_k^{q_1} |z|^2 \leq z' D_k z \leq \lambda_2 w_k^{q_2} |z|^2 \qquad \forall\, z \in R^n, \quad k = 0, 1, \ldots, \tag{31}$$

where

$$w_k = |x_k - [x_k - M\nabla f(x_k)]^+|.$$

Assumption (A) is not essential for the result of Proposition 1.36 that follows but simplifies its proof. It is satisfied for just about every problem likely to appear in practice. For example, it is satisfied when f is twice differentiable, as well as when f is an augmented Lagrangian of the type considered in Chapter 3 for problems involving twice differentiable functions. Assumption (B) is a condition of the type utilized in connection with unconstrained minimization algorithms (compare with the discussion preceding Proposition 1.8). When $q_1 = q_2 = 0$, relation (31) takes the form

$$\lambda_1 |z|^2 \leq z' D_k z \leq \lambda_2 |z|^2 \qquad \forall\, z \in R^n, \quad k = 0, 1, \ldots, \tag{32}$$

and simply says that the eigenvalues of D_k are uniformly bounded above and away from zero.

Proposition 1.36: Under Assumptions (A) and (B) above, every limit point of a sequence $\{x_k\}$ generated by iteration (25) is a critical point with respect to (SCP).

Proof: Assume the contrary; i.e., there exists a subsequence $\{x_k\}_K$ converging to a vector $\bar{x}$ which is not critical. Since $\{f(x_k)\}$ is decreasing and f is continuous, it follows that $\{f(x_k)\}$ converges to $f(\bar{x})$ and therefore

$$[f(x_k) - f(x_{k+1})] \to 0.$$

Since each of the sums in the right-hand side of (27) is nonnegative [compare with (28) and (29)], we must have

$$\alpha_k \sum_{i \notin I_k^+} \frac{\partial f(x_k)}{\partial x^i} p_k^i \to 0, \tag{33}$$

$$\sum_{i \in I_k^+} \frac{\partial f(x_k)}{\partial x^i} [x_k^i - x_k^i(\alpha_k)] \to 0. \tag{34}$$

Also since $\bar{x}$ is not critical and M is positive definite and diagonal, we have clearly $|\bar{x} - [\bar{x} - M\nabla f(\bar{x})]^+| \neq 0$, so (31) implies that the eigenvalues of $\{D_k\}_K$ are uniformly bounded above and away from zero. In view of the fact that D_k is diagonal with respect to I_k^+, it follows that there exist positive scalars $\bar{\lambda}_1$ and $\bar{\lambda}_2$ such that, for all $k \in K$ that are sufficiently large,

$$0 < \bar{\lambda}_1 \, \partial f(x_k)/\partial x^i \leq p_k^i \leq \bar{\lambda}_2 \, \partial f(x_k)/\partial x^i \qquad \forall \, i \in I_k^+, \tag{35}$$

$$\bar{\lambda}_1 \sum_{i \notin I_k^+} \left| \frac{\partial f(x_k)}{\partial x^i} \right|^2 \leq \sum_{i \notin I_k^+} p_k^i \frac{\partial f(x_k)}{\partial x^i} \leq \bar{\lambda}_2 \sum_{i \notin I_k^+} \left| \frac{\partial f(x_k)}{\partial x^i} \right|^2. \tag{36}$$

We shall show that our hypotheses so far lead to the conclusion that

$$\liminf_{\substack{k \to \infty \\ k \in K}} \alpha_k = 0. \tag{37}$$

Indeed since $\bar{x}$ is not a critical point, there must exist an index i such that either

$$\bar{x}^i > 0 \qquad \text{and} \qquad \partial f(\bar{x})/\partial x^i \neq 0 \tag{38}$$

or

$$\bar{x}^i = 0 \qquad \text{and} \qquad \partial f(\bar{x})/\partial x^i < 0. \tag{39}$$

If $i \notin I_k^+$ for an infinite number of indices $k \in K$, then (37) follows from (33), (36), (38), and (39). If $i \in I_k^+$ for an infinite number of indices $k \in K$, then for all those indices we must have $\partial f(x_k)/\partial x^i > 0$, so (39) cannot hold. Therefore, from (38),

$$\bar{x}^i > 0 \qquad \text{and} \qquad \partial f(\bar{x})/\partial x^i > 0. \tag{40}$$

Since, for all $k \in K$ for which $i \in I_k^+$ [compare with (29)], we have

$$\sum_{j \in I_k^+} \frac{\partial f(x_k)}{\partial x^j}[x_k^j - x_k^j(\alpha_k)] \geq \frac{\partial f(x_k)}{\partial x^i}[x_k^i - x_k^i(\alpha_k)] \geq 0,$$

it follows from (34) and (40) that

$$\lim_{\substack{k \to \infty \\ k \in K}} [x_k^i - x_k^i(\alpha_k)] = 0.$$

Using the above relation, (35), and (40), we obtain (37).

We shall complete the proof by showing that $\{\alpha_k\}_K$ is bounded away from zero thereby contradicting (37). Indeed in view of (31), the subsequences $\{x_k\}_K$, $\{p_k\}_K$, and $\{x_k(\alpha)\}_K$, $\alpha \in [0, 1]$, are uniformly bounded, so by Assumption (A) there exists a scalar $L > 0$ such that, for all $t \in [0, 1]$, $\alpha \in [0, 1]$, and $k \in K$, we have

$$|\nabla f(x_k) - \nabla f[x_k - t[x_k - x_k(\alpha)]]| \leq tL|x_k - x_k(\alpha)|. \tag{41}$$

For all $k \in K$ and $\alpha \in [0, 1]$, we have

$$f[x_k(\alpha)] = f(x_k) + \nabla f(x_k)'[x_k(\alpha) - x_k]$$
$$+ \int_0^1 \{\nabla f(x_k) - \nabla f[x_k - t[x_k - x_k(\alpha)]]\}' \, dt[x_k - x_k(\alpha)],$$

so

$$\begin{aligned} f(x_k) - f[x_k(\alpha)] &= \nabla f(x_k)'[x_k - x_k(\alpha)] \\ &\quad + \int_0^1 \{\nabla f[x_k - t[x_k - x_k(\alpha)]] - \nabla f(x_k)\}' \, dt[x_k - x_k(\alpha)] \\ &\geq \nabla f(x_k)'[x_k - x_k(\alpha)] \\ &\quad - \int_0^1 |\nabla f[x_k - t[x_k - x_k(\alpha)]] - \nabla f(x_k)| \, dt|x_k - x_k(\alpha)|, \end{aligned}$$

and finally, by using (41),

$$f(x_k) - f[x_k(\alpha)] \geq \nabla f(x_k)'[x_k - x_k(\alpha)] - \tfrac{1}{2}L|x_k - x_k(\alpha)|^2. \tag{42}$$

For $i \in I_k^+$, we have $x_k^i(\alpha) = [x_k^i - \alpha p_k^i]^+ \geq x_k^i - \alpha p_k^i$ and $p_k^i > 0$, so $0 \leq x_k^i - x_k^i(\alpha) \leq \alpha p_k^i$. It follows, using (35), that

$$\sum_{i \in I_k^+} |x_k^i - x_k^i(\alpha)|^2 \leq \alpha \sum_{i \in I_k^+} p_k^i[x_k^i - x_k^i(\alpha)] \leq \bar{\lambda}_2 \alpha \sum_{i \in I_k^+} \frac{\partial f(x_k)}{\partial x^i}[x_k^i - x_k^i(\alpha)]. \tag{43}$$

Consider the sets

$$I_{1,k} = \{i \mid \partial f(x_k)/\partial x^i > 0, i \notin I_k^+\}, \qquad I_{2,k} = \{i \mid \partial f(x_k)/\partial x^i \leq 0, i \notin I_k^+\}.$$

For all $i \in I_{1,k}$ we must have $x_k^i > \varepsilon_k$ for otherwise we would have $i \in I_k^+$. Since $|\bar{x} - [\bar{x} - M\nabla f(\bar{x})]^+| \neq 0$, we must have $\liminf_{k\to\infty, k\in K} \varepsilon_k > 0$ and $\varepsilon_k > 0$ for all k. Let $\bar{\varepsilon} > 0$ be such that $\bar{\varepsilon} \leq \varepsilon_k$ for all $k \in K$, and let B be such that $|p_k^i| \leq B$ for all i and $k \in K$. Then, for all $\alpha \in [0, \bar{\varepsilon}/B]$, we have $x_k^i(\alpha) = x_k^i - \alpha p_k^i$, so it follows that

$$(44) \qquad \sum_{i\in I_{1,k}} \frac{\partial f(x_k)}{\partial x^i}[x_k^i - x_k^i(\alpha)] = \alpha \sum_{i\in I_{1,k}} \frac{\partial f(x_k)}{\partial x^i} p_k^i \qquad \forall\, \alpha \in \left[0, \frac{\bar{\varepsilon}}{B}\right].$$

Also, for all $\alpha \geq 0$, we have $x_k^i - x_k^i(\alpha) \leq \alpha p_k^i$, and since $\partial f(x_k)/\partial x^i \leq 0$, for all $i \in I_{2,k}$, we obtain

$$(45) \qquad \sum_{i\in I_{2,k}} \frac{\partial f(x_k)}{\partial x^i}[x_k^i - x_k^i(\alpha)] \geq \alpha \sum_{i\in I_{2,k}} \frac{\partial f(x_k)}{\partial x^i} p_k^i.$$

Combining (44) and (45), we obtain

$$(46) \qquad \sum_{i\notin I_k^+} \frac{\partial f(x_k)}{\partial x^i}[x_k^i - x_k^i(\alpha)] \geq \alpha \sum_{i\notin I_k^+} \frac{\partial f(x_k)}{\partial x^i} p_k^i \qquad \forall\, \alpha \in \left[0, \frac{\bar{\varepsilon}}{B}\right].$$

For all $\alpha \geq 0$, we also have

$$|x_k^i - x_k^i(\alpha)| \leq \alpha |p_k^i| \qquad \forall\, i = 1, \ldots, n.$$

Furthermore, it is easily seen, using Assumption (B), that there exists $\lambda > 0$ such that

$$\sum_{i\notin I_k^+} (p_k^i)^2 \leq \lambda \sum_{i\notin I_k^+} \frac{\partial f(x_k)}{\partial x^i} p_k^i \qquad \forall\, k \in K.$$

Using the last two relations, we obtain, for all $\alpha \geq 0$,

$$(47) \qquad \sum_{i\notin I_k^+} |x_k^i - x_k^i(\alpha)|^2 \leq \alpha^2 \lambda \sum_{i\notin I_k^+} \frac{\partial f(x_k)}{\partial x^i} p_k^i \qquad \forall\, k \in K.$$

We now combine (42), (43), (46), and (47) to obtain, for all $\alpha \in [0, \bar{\varepsilon}/B]$ with $\alpha \leq 1$ and $k \in K$,

$$(48) \qquad f(x_k) - f[x_k(\alpha)] \geq \left(\alpha - \frac{\alpha^2 \lambda L}{2}\right) \sum_{i\notin I_k^+} \frac{\partial f(x_k)}{\partial x^i} p_k^i$$
$$+ (1 - \tfrac{1}{2}\alpha\bar{\lambda}_2 L) \sum_{i\in I_k^+} \frac{\partial f(x_k)}{\partial x^i}[x_k^i - x_k^i(\alpha)].$$

Suppose α is chosen so that

$$(49) \quad 0 \leq \alpha \leq \bar{\varepsilon}/B, \qquad 1 - \tfrac{1}{2}\alpha\lambda L \geq \sigma, \qquad 1 - \tfrac{1}{2}\alpha\bar{\lambda}_2 L \geq \sigma, \qquad \alpha \leq 1,$$

or equivalently

$$0 \le \alpha \le \min\left\{\frac{\bar{\varepsilon}}{B}, \frac{2(1-\sigma)}{\lambda L}, \frac{2(1-\sigma)}{\bar{\lambda}_2 L}, 1\right\}. \tag{50}$$

Then we have from (48) and (49), for all $k \in K$,

$$f(x_k) - f[x_k(\alpha)] \ge \sigma\left\{\alpha \sum_{i \notin I_k^+} \frac{\partial f(x_k)}{\partial x^i} p_k^i + \sum_{i \in I_k^+} \frac{\partial f(x_k)}{\partial x^i}[x_k^i - x_k^i(\alpha)]\right\}.$$

This means that if (50) is satisfied with $\beta^m = \alpha$, then the inequality (27) of the Armijo-like rule will be satisfied. It follows from the way the stepsize is reduced that α_k satisfies

$$\alpha_k \ge \beta \min\left\{\frac{\bar{\varepsilon}}{B}, \frac{2(1-\sigma)}{\lambda L}, \frac{2(1-\sigma)}{\bar{\lambda}_2 L}, 1\right\} \qquad \forall\, k \in K. \tag{51}$$

This contradicts (37) and proves the proposition. Q.E.D.

We now focus attention at a local minimum x^* satisfying the following second-order sufficiency conditions which are in fact the ones of Proposition 1.31 applied to (SCP), as the reader can easily verify. For all $x \ge 0$, we denote by $A(x)$ the set of indices of active constraints at x; i.e.,

$$A(x) = \{i \,|\, x^i = 0\} \qquad \forall\, x \ge 0. \tag{52}$$

Assumption (C): *The local minimum x^* of* (SCP) *is such that, for some $\delta > 0$, f is twice continuously differentiable in the open sphere $S(x^*;\ \delta)$ and there exist positive scalars m_1 and m_2 such that*

$$m_1|z|^2 \le z'\nabla^2 f(x)z \le m_2|z|^2 \qquad \forall\, x \in S(x^*;\delta) \quad \text{and} \quad z \ne 0, \quad \text{such that} \quad z^i = 0 \quad \forall\, i \in A(x^*). \tag{53}$$

Furthermore,

$$\partial f(x^*)/\partial x^i > 0 \qquad \forall\, i \in A(x^*). \tag{54}$$

The following proposition demonstrates an important property of the algorithm, namely, that under mild conditions *it is attracted by a local minimum x^* satisfying Assumption* (C) *and identifies the set of active constraints at x^* in a finite number of iterations. Thus, if the algorithm converges to x^*, then after a finite number of iterations it is equivalent to an unconstrained optimization method restricted on the subspace of active constraints at x^*.* This property is instrumental in proving superlinear convergence of the algorithm when the portion of D_k, corresponding to the indices $i \notin I_k^+$, is chosen in a way that approximates the inverse of the portion of the Hessian of f corresponding to these same indices.

Proposition 1.37: Let x^* be a local minimum of (SCP) satisfying Assumption (C), and let Assumption (B) hold in the stronger form whereby, in addition to (31), it is assumed that there exists a scalar $\bar{\lambda}_1 > 0$ such that the diagonal elements d_k^{ii} of the matrices D_k satisfy

$$\bar{\lambda}_1 \leq d_k^{ii} \qquad \forall\, k = 0, 1, \ldots, i \in I_k^+. \tag{55}$$

There exists a scalar $\bar{\delta} > 0$ such that if $\{x_k\}$ is a sequence generated by iteration (25) and for some index $\bar{k}$, we have

$$|x_{\bar{k}} - x^*| \leq \bar{\delta},$$

then $\{x_k\}$ converges to x^*, and we have

$$I_k^+ = A(x_k) = A(x^*) \qquad \forall\, k \geq \bar{k} + 1.$$

Proof: Since f is twice differentiable on $S(x^*; \delta)$, it follows that there exist scalars $L > 0$ and $\delta_1 \in (0, \delta]$ such that for all x and $\bar{x}$ with $|x - x^*| \leq \delta_1$ and $|\bar{x} - x^*| \leq \delta_1$, we have

$$|\nabla f(x) - \nabla f(\bar{x})| \leq L|x - \bar{x}|.$$

Also for x_k sufficiently close to x^*, the scalar

$$w_k = |x_k - [x_k - Mf(x_k)]^+|$$

is arbitrarily close to zero while, in view of (54), we have

$$\left[x_k^i - m^i \frac{\partial f(x_k)}{\partial x^i}\right]^+ = 0 \qquad \forall\, i \in A(x^*),$$

where m^i is the ith diagonal element of M. It follows that, for x_k sufficiently close to x^*, we have

$$x_k^i \leq w_k = \varepsilon_k < \varepsilon \qquad \forall\, i \in A(x^*), \tag{56}$$

while

$$x_k^i > \varepsilon_k \qquad \forall\, i \notin A(x^*). \tag{57}$$

Since, by Assumption (C), $\partial f(x_k)/\partial x^i > 0$ for all $i \in A(x^*)$ and x_k sufficiently close to x^*, (56) and (57) imply that there exists $\delta_2 \in (0, \delta_1]$ such that

$$A(x^*) = I_k^+ \qquad \forall\, k \quad \text{such that} \quad |x_k - x^*| \leq \delta_2. \tag{58}$$

Also there exist scalars $\bar{\varepsilon} > 0$ and $\delta_3 \in (0, \delta_2]$ such that

$$x_k^i > \bar{\varepsilon} \qquad \forall\, i \notin A(x^*) \quad \text{and} \quad k \text{ such that } |x_k - x^*| \leq \delta_3. \tag{59}$$

By repeating the argument in the proof of Proposition 1.36 that led to (51), we find that there exists a scalar $\bar{\alpha} > 0$ such that

(60) $$\alpha_k \geq \bar{\alpha} \qquad \forall\, k \quad \text{such that} \quad |x_k - x^*| \leq \delta_3.$$

By using (55) and (58), it follows that

(61) $$0 < \bar{\lambda}_1\, \partial f(x_k)/\partial x^i \leq p_k^i \qquad \forall\, i \in A(x^*) \quad \text{and} \quad k \quad \text{such that} \quad |x_k - x^*| \leq \delta_3,$$

while, by Assumption (B), there exists a scalar $\lambda > 0$ such that

(62) $$\sum_{i \notin A(x^*)} |p_k^i|^2 \leq \lambda \sum_{i \notin A(x^*)} \left| \frac{\partial f(x_k)}{\partial x^i} \right|^2 \qquad \forall\, k \quad \text{such that} \quad |x_k - x^*| \leq \delta_3.$$

Since $\partial f(x^*)/\partial x^i > 0$ for all $i \in A(x^*)$ and $\partial f(x^*)/\partial x^i = 0$ for all $i \notin A(x^*)$, it follows from (58)–(62) that there exists a scalar $\delta_4 \in (0, \delta_3]$ such that

(63) $$A(x^*) = A(x_{k+1}) \qquad \forall\, k \quad \text{such that} \quad |x_k - x^*| \leq \delta_4$$

and

(64) $$|x_{k+1} - x^*| \leq \delta_3 \qquad \forall\, k \quad \text{such that} \quad |x_k - x^*| \leq \delta_4.$$

In view of (58), we obtain, from (63) and (64),

(65) $$A(x^*) = A(x_{k+1}) = I_{k+1}^+ \qquad \forall\, k \quad \text{such that} \quad |x_k - x^*| \leq \delta_4.$$

Thus when $|x_k - x^*| \leq \delta_4$, we have $|x_{k+1} - x^*| \leq \delta_3$, $A(x^*) = A(x_{k+1})$, and the $(k + 1)$th iteration of the algorithm reduces to an iteration of an unconstrained minimization algorithm on the subspace of active constraints at x^* to which Proposition 1.12 applies. From this proposition, it follows that there exists an open set $N(x^*)$ containing x^* such that $N(x^*) \subset S(x^*; \delta_4)$ and with the property that if $x_{k+1} \in N(x^*)$ and $A(x_{k+1}) = A(x^*)$, then $x_{k+2} \in N(x^*)$ and, by (63), $A(x_{k+2}) = A(x^*)$. This argument can be repeated and shows that if for some $\bar{k} \geq 0$ we have

$$x_{\bar{k}} \in N(x^*), \qquad A(x_{\bar{k}}) = A(x^*),$$

then $\{x_k\} \to x^*$ and

$$x_k \in N(x^*), \qquad A(x_k) = A(x^*) \qquad \forall\, k \geq \bar{k}.$$

To complete the proof, it is sufficient to show that there exists $\bar{\delta} > 0$ such that if $|x_k - x^*| \leq \bar{\delta}$ then $x_{k+1} \in N(x^*)$ and $A(x_{k+1}) = A(x^*)$. Indeed by repeating the argument that led to (63) and (64), we find that given any $\tilde{\delta} > 0$ there exists a $\bar{\delta} > 0$ such that if $|x_k - x^*| \leq \bar{\delta}$, then

$$|x_{k+1} - x^*| \leq \tilde{\delta}, \qquad A(x_{k+1}) = A(x^*).$$

By taking $\tilde{\delta}$ sufficiently small so that $S(x^*; \tilde{\delta}) \subset N(x^*)$ the proof is completed. Q.E.D.

Under the assumptions of Proposition 1.37, we see that if the algorithm converges to a local minimum x^* satisfying Assumption (C) then it reduces eventually to an unconstrained minimization method restricted to the subspace

$$S^* = \{x \mid x^i = 0, \ \forall\, i \in A(x^*)\}.$$

Furthermore, as shown in the proof of Proposition 1.37 [compare with (58)], for some index $\bar{k}$, we shall have

$$I_k^+ = A(x^*) \qquad \forall\, k \geq \bar{k}. \tag{66}$$

This shows that if the portion of the matrix D_k corresponding to the indices $i \notin I_k^+$ is chosen to be the inverse of the Hessian of f with respect to the indices $i \notin I_k^+$, then the algorithm eventually reduces to Newton's method restricted to the subspace S^*.

More specifically, by rearranging indices if necessary, assume without loss of generality that

$$I_k^+ = \{r_k + 1, \ldots, n\}, \tag{67}$$

where r_k is some integer. Then D_k has the form

$$D_k = \left[\begin{array}{c|ccc} \bar{D}_k & & 0 & \\ \hline & d^{r_k+1} & & 0 \\ & & \ddots & \\ & 0 & & d^n \end{array}\right], \tag{68}$$

where $d_k^i > 0$, $i = r_k + 1, \ldots, n$, and $\bar{D}_k$ can be an *arbitrary* positive definite matrix. Suppose we choose $\bar{D}_k$ to be the inverse of the Hessian of f with respect to the indices $i = 1, \ldots, r_k$; i.e., the elements $[\bar{D}_k^{-1}]_{ij}$ of $\bar{D}_k^{-1}$ are

$$[\bar{D}_k^{-1}]_{ij} = \partial^2 f(x_k)/\partial x^i\, \partial x^j \qquad \forall\, i, j \notin I_k^+. \tag{69}$$

By Assumption (C), $\nabla^2 f(x^*)$ is positive definite on S^*, so it follows from (66) that this choice is well defined and satisfies the assumption of Proposition 1.37 for k sufficiently large. Since the conclusion of this proposition asserts that the method eventually reduces to Newton's method restricted to the subspace S^*, a superlinear convergence rate result follows. This type of argument can be used to construct a number of Newton-like and quasi-Newton methods and prove corresponding convergence and rate of convergence results. We state one of the simplest such results regarding a Newton-like algorithm which is well suited for problems where f is strictly convex and twice differentiable. Its proof follows simply from the preceding discussion and Propositions 1.15 and 1.17 and is left to the reader.

Proposition 1.38: Let f be convex and twice continuously differentiable. Assume that (SCP) has a unique optimal solution x^* satisfying Assumption (C), and there exist positive scalars m_1 and m_2, such that

$$m_1|z|^2 \le z'\nabla^2 f(x)z \le m_2|z|^2 \qquad \forall\, z \in \{x \mid f(x) \le f(x_0)\}.$$

Assume also that in the algorithm (22)–(27), the matrix D_k is given by $D_k = H_k^{-1}$, where H_k is the matrix with elements H_k^{ij} given by

$$H_k^{ij} = \begin{cases} 0 & \text{if } \; i \ne j \; \text{ and either } \; i \in I_k^+ \; \text{ or } \; j \in I_k^+, \\ \partial^2 f(x_k)/\partial x^i\, \partial x^j & \text{otherwise.} \end{cases}$$

Then the sequence $\{x_k\}$ generated by iteration (25) converges to x^*, and the rate of convergence of $\{|x_k - x^*|\}$ is superlinear (of order at least two if $\nabla^2 f$ is Lipschitz continuous in a neighborhood of x^*).

It is worth noting that when $f(x)$ is a positive definite quadratic function, the algorithm of Proposition 1.38 finds the unique solution x^* in a finite number of iterations, assuming x^* satisfies Assumption (C).

An additional property of the algorithm of Proposition 1.38 is that after a finite number of iterations and once the set of binding constraints is identified, the initial unity stepsize is accepted by the Armijo rule. Computational experience with the algorithm suggests that this is also true for most iterations even before the set of binding constraints is identified. In some cases, however, it may be necessary to reduce the initial unity stepsize several times before a sufficient reduction in objective function value is effected. A typical situation where this may occur is when the scalar $\tilde{\gamma}_k$ defined by

$$\tilde{\gamma}_k = \min\{1, \tilde{\alpha}_k\}, \qquad \tilde{\alpha}_k = \sup\{\alpha \mid x_k^i - \alpha p_k^i \ge 0,\, x_k^i > 0,\, i \notin I_k^+\}$$

is much smaller than unity. Under these circumstances a nonbinding constraint that was not included in the set I_k^+ becomes binding after a small movement along the arc $\{x_k(\alpha) \mid \alpha \ge 0\}$ and it may happen that the objective function value increases as α becomes larger than $\tilde{\gamma}_k$. To correct such a situation, it may be useful to modify the Armijo rule so that if after a fixed number r of trial stepsizes $1, \beta, \ldots, \beta^{r-1}$ have failed to pass the Armijo rule test, then $\tilde{\gamma}_k$ is computed and, if it is smaller than β^{r-1}, it is used as the next trial stepsize.

Another (infrequent) situation, where the algorithm of Proposition 1.38 can exhibit a large number of stepsize reductions and slow convergence when far from the optimum, arises sometimes if the set of indices

$$\tilde{I}_k^+ = \{i \mid 0 \le x_k^i \le \varepsilon_k,\, p_k^i > 0\}, \tag{70}$$

where $p_k = D_k \nabla f(x_k)$, is strictly larger than the set I_k^+ of (22). (Note that, under the assumptions of Proposition 1.38, we always have $I_k^+ \subset \tilde{I}_k^+$ with equality holding in a neighborhood of the optimal solution x^*.) Under these circumstances, the initial motion along the arc $\{x(\alpha) \mid \alpha \geq 0\}$ may be along a search direction that is not a Newton direction on any subspace. A possible remedy for this difficulty is to combine the Armijo rule with some form of line minimization rule.

Extension to Upper and Lower Bounds

The algorithm (22)–(27) described so far in this section can be easily extended to handle problems of the form

$$\begin{aligned} &\text{minimize} \quad f(x) \\ &\text{subject to} \quad b_1 \leq x \leq b_2, \end{aligned}$$

where b_1 and b_2 are given vectors of lower and upper bounds. The set I_k^+ is replaced by

$$\begin{aligned} I_k^\# = \{i \mid\ & b_1^i \leq x_k^i \leq b_1^i + \varepsilon_k \text{ and } \partial f(x_k)/\partial x^i > 0 \\ &\text{or } b_2^i - \varepsilon_k \leq x_k^i \leq b_2^i \text{ and } \partial f(x_k)/\partial x^i < 0\}, \end{aligned}$$

and the definition of $x_k(\alpha)$ is changed to

$$x_k(\alpha) = [x_k - \alpha D_k \nabla f(x_k)]^\#,$$

where for all $z \in R^n$ we denote by $[z]^\#$ the vector with coordinates

$$[z]^{\# i} = \begin{cases} b_2^i & \text{if } b_2^i \leq z^i, \\ z^i & \text{if } b_1^i < z^i < b_2^i, \\ b_1^i & \text{if } z^i \leq b_1^i. \end{cases}$$

The scalar ε_k is given by

$$\varepsilon_k = \min\{\varepsilon, |x_k - [x_k - M\nabla f(x_k)]^\# |\}.$$

The matrix D_k is positive definite and diagonal with respect to $I_k^\#$, and M is a fixed diagonal positive definite matrix. The iteration is given by

$$x_{k+1} = x_k(\alpha_k),$$

where α_k is chosen by the Armijo rule (26), (27) with $[x_k^i - x_k^i(\beta^m)]^+$ replaced by $[x_k^i - x_k^i(\beta^m)]^\#$.

Similar extensions of the basic algorithm can be provided for problems where only some of the variables x^i are simply constrained by upper and/or lower bounds.

1.6 Notes and Sources

Notes on Section 1.2: The proof of the second implicit function theorem may be found in Hestenes (1966, p. 23). The theorem itself is apparently due to Bliss (Hestenes, personal communication).

Notes on Section 1.3: The convergence analysis of gradient methods given here stems from the papers of Goldstein (1962, 1966), and bears similarity with the corresponding analysis in Ortega and Rheinboldt (1970). Some other influential works in this area are Armijo (1966), Wolfe (1969), and Daniel (1971). Zangwill (1969) and Polak (1971) have proposed general convergence theories for optimization algorithms. The gradient method with constant stepsize was first analyzed by Poljak (1963). Proposition 1.12 is thought to be new. The linear convergence rate results stem from Kantorovich (1945) and Poljak (1963), while the superlinear rate results stem from Goldstein and Price (1967). For convergence rate analysis of the steepest descent method near local minima with singular Hessian, see Dunn (1981b). The spacer step theorem (Proposition 1.16) is due to Zangwill (1969). For an extensive analysis and references on Newton-like methods, see Ortega and Rheinboldt (1970). The modification scheme for Newton's method based on the Cholesky factorization is related to one due to Murray (1972).

Conjugate direction methods were originally developed in Hestenes and Stiefel (1952). Extensive presentations may be found in Faddeev and Faddeeva (1963), Luenberger (1973), and Hestenes (1980). Scaled $(k + 1)$-step conjugate gradient methods for problems with Hessian matrix of the form

$$Q = M + \sum_{i=l}^{k} v_i v_i'$$

were first proposed in Bertsekas (1974a). For further work on this subject, see Oren (1978).

Extensive surveys of quasi-Newton methods can be found in Avriel (1976), Broyden (1972), and Dennis and Moré (1977).

Notes on Section 1.4: Presentations of optimality conditions for constrained optimization can be found in many sources including Fiacco and McCormick (1968), Mangasarian (1969), Cannon *et al.* (1970), Luenberger (1973), and Avriel (1976). For a development of optimality conditions based on the notion of augmentability, which is intimately related to methods of multipliers, see Hestenes (1975).

Notes on Section 1.5: The methods in this section are new and were developed while the monograph was being written. Extensions to general linear constraints may be found in Bertsekas (1980c). The methods are particularly well suited for large scale problems with many simple constraints. An example is nonlinear multicommodity flow problems arising in communication and transportation networks (see Bertsekas and Gafni, 1981). The constrained version of the Armijo rule (26), (27) is based on a similar rule first proposed in Bertsekas (1974c). The main advantage that the methods of this section offer over methods based on active set strategies [compare with Gill and Murray (1974) and Ritter (1973)] is that there is no limit to the number of constraints that can be added or dropped from the active set in a single iteration, and this is significant for problems of large dimension. At the same time, there is no need to solve a quadratic programming problem at each iteration as in the Newton and quasi-Newton methods of Levitin and Poljak (1965), Garcia-Palomares (1975), and Brayton and Cullum (1979).

Chapter 2

The Method of Multipliers for Equality Constrained Problems

The main idea in the methods to be examined in this chapter is to approximate a constrained minimization problem by a problem which is considerably easier to solve. Naturally by solving an approximate problem, we can only expect to obtain an approximate solution of the original problem. However, if we can construct a sequence of approximate problems which "converges" in a well-defined sense to the original problem, then hopefully the corresponding sequence of approximate solutions will yield in the limit a solution of the original problem.

It may appear odd at first sight that we would prefer solving a sequence of minimization problems rather than a single problem. However, in practice only a finite number of approximate problems need to be solved in order to obtain what would be an acceptable approximate solution of the original problem. Furthermore, usually each approximate problem need not be solved itself exactly but rather only approximately. In addition, one may utilize efficiently information obtained from each approximate problem in the solution of the next approximate problem.

2.1 The Quadratic Penalty Function Method

The basic idea in penalty methods is to eliminate some or all of the constraints and add to the objective function a penalty term which prescribes a high cost to infeasible points. Associated with these methods is a parameter c, which determines the severity of the penalty and as a consequence the extent to which the resulting unconstrained problem approximates the original constrained problem. As c takes higher values, the approximation becomes increasingly accurate. In this chapter, we restrict attention to the popular quadratic penalty function. Other penalty functions will be considered in Chapter 5.

Throughout this section we consider the problem

$$\text{(1)} \qquad \begin{aligned} &\text{minimize} \quad f(x) \\ &\text{subject to} \quad x \in X,\ h(x) = 0, \end{aligned}$$

where $f\colon R^n \to R$, $h\colon R^n \to R^m$ are given functions and X is a given subset of R^n. We assume throughout that *problem* (1) *has at least one feasible solution.*

For any scalar c, let us define the *augmented Lagrangian* function $L_c\colon R^n \times R^m \to R$ by

$$\text{(2)} \qquad L_c(x, \lambda) = f(x) + \lambda' h(x) + \tfrac{1}{2}c|h(x)|^2.$$

We refer to c as the *penalty parameter* and to λ as the *multiplier vector* (or simply multiplier).

The quadratic penalty method consists of solving a sequence of problems of the form

$$\text{(3)} \qquad \begin{aligned} &\text{minimize} \quad L_{c_k}(x, \lambda_k) \\ &\text{subject to} \quad x \in X, \end{aligned}$$

where $\{\lambda_k\}$ is a *bounded* sequence in R^m and $\{c_k\}$ is a penalty parameter sequence satisfying

$$0 < c_k < c_{k+1} \qquad \forall\, k, \quad c_k \to \infty.$$

In the original version of the penalty method the multipliers λ_k are taken to be equal to zero,

$$\lambda_k = 0 \qquad \forall\, k = 0, 1, \ldots,$$

and the method depends for its success on sequentially increasing the penalty parameter to infinity. We shall see later in this chapter that it is possible to improve considerably the performance of the method (under certain assumptions) by employing nonzero multipliers λ_k and by updating them in an intelligent manner after each minimization of the form (3).

In this section, however, we concentrate on the effect of the penalty parameter, and we make no assumption on $\{\lambda_k\}$ other than boundedness.

The rationale for the penalty method is based on the fact that when $\{\lambda_k\}$ is bounded and $c_k \to \infty$, then the term

$$\lambda_k' h(x) + \tfrac{1}{2}c_k |h(x)|^2,$$

which is added to the objective function, tends to infinity if $h(x) \neq 0$ and equals zero if $h(x) = 0$. Thus, if we define the function $\tilde{f}: R^n \to (-\infty, +\infty]$ by

$$\tilde{f}(x) = \begin{cases} f(x) & \text{if } \ h(x) = 0, \\ \infty & \text{if } \ h(x) \neq 0, \end{cases}$$

the optimal value of the original problem can be written as

$$f^* = \inf_{\substack{h(x)=0 \\ x \in X}} f(x) = \inf_{x \in X} \tilde{f}(x) = \inf_{x \in X} \lim_{k \to \infty} L_{c_k}(x, \lambda_k). \tag{4}$$

On the other hand, the penalty method determines, via the sequence of minimizations (3),

$$\bar{f} = \lim_{k \to \infty} \inf_{x \in X} L_{c_k}(x, \lambda_k). \tag{5}$$

Thus, in order for the penalty method to be successful, the original problem should be such that the interchange of "lim" and "inf" in (4) and (5) is valid. The following proposition guarantees the validity of the interchange, under mild assumptions, and constitutes the basic convergence result for the penalty method.

Proposition 2.1: Assume that f and h are continuous functions and X is a closed set. For $k = 0, 1, \ldots$, let x_k be a global minimum of the problem

$$\begin{aligned} &\text{minimize} \quad L_{c_k}(x, \lambda_k) \\ &\text{subject to} \quad x \in X, \end{aligned} \tag{6}$$

where $\{\lambda_k\}$ is bounded and $0 < c_k < c_{k+1}$ for all k, $c_k \to \infty$. Then every limit point of the sequence $\{x_k\}$ is a global minimum of f subject to $x \in X$, $h(x) = 0$.

Proof: Let $\bar{x}$ be a limit point of $\{x_k\}$. We have by definition of x_k

$$L_{c_k}(x_k, \lambda_k) \leq L_{c_k}(x, \lambda_k) \qquad \forall\, x \in X. \tag{7}$$

Let f^* denote the optimal value of the original problem. We have

$$f^* = \inf_{\substack{h(x)=0 \\ x \in X}} f(x) = \inf_{\substack{h(x)=0 \\ x \in X}} L_{c_k}(x, \lambda_k).$$

Hence, by taking the infimum of the right-hand side of (7) over $x \in X$, $h(x) = 0$, we obtain

$$L_{c_k}(x_k, \lambda_k) = f(x_k) + \lambda_k' h(x_k) + \tfrac{1}{2}c_k |h(x_k)|^2 \leq f^*.$$

The sequence $\{\lambda_k\}$ is bounded and hence it has a limit point $\bar{\lambda}$. Without loss of generality, we may assume $\lambda_k \to \bar{\lambda}$. By taking the limit superior in the relation above and by using the continuity of f and h, we obtain

$$f(\bar{x}) + \bar{\lambda}' h(\bar{x}) + \limsup_{k \to \infty} \tfrac{1}{2}c_k |h(x_k)|^2 \leq f^*. \tag{8}$$

Since $|h(x_k)|^2 \geq 0$, $c_k \to \infty$, it follows that we must have $h(x_k) \to 0$ and

$$h(\bar{x}) = 0, \tag{9}$$

for otherwise the limit superior in the left-hand side of (8) will equal $+\infty$. Since X is a closed set we also obtain that $\bar{x} \in X$. Hence $\bar{x}$ is feasible, and

$$f^* \leq f(\bar{x}). \tag{10}$$

Using (8), (9), and (10), we obtain

$$f^* + \limsup_{k \to \infty} \tfrac{1}{2}c_k |h(x_k)|^2 \leq f(\bar{x}) + \limsup_{k \to \infty} \tfrac{1}{2}c_k |h(x_k)|^2 \leq f^*.$$

Hence,

$$\lim_{k \to \infty} \tfrac{1}{2}c_k |h(x_k)|^2 = 0$$

and

$$f(\bar{x}) = f^*,$$

which proves that $\bar{x}$ is a global minimum for problem (1). Q.E.D.

The proposition shown above has several weaknesses. First, it assumes that the problem

$$\begin{aligned} &\text{minimize} \quad L_{c_k}(x, \lambda_k) \\ &\text{subject to} \quad x \in X \end{aligned}$$

has a global minimum. This may not be true, even if the original problem (1) has a global minimum. As an example, consider the scalar problem

$$\begin{aligned} &\text{minimize} \quad -x^4 \\ &\text{subject to} \quad x = 0. \end{aligned}$$

This problem has, of course, a unique global minimum—the point $x^* = 0$. We have

$$L_{c_k}(x, \lambda_k) = -x^4 + \lambda_k x + \tfrac{1}{2}c_k x^2.$$

Clearly, $\inf_x L_{c_k}(x, \lambda_k) = -\infty$, and $L_{c_k}(x, \lambda_k)$ has no global minimum for every c_k and λ_k. This example shows a weakness of the penalty method and focuses attention at a situation where some care should be exercised. One should choose the order of growth of the penalty function in such a way that the augmented Lagrangian has a minimum. For instance, in the example above if we use a penalty function of the form

$$\tfrac{1}{2}c\,|h(x)|^2 + |h(x)|^\rho,$$

where $\rho > 4$, then $L_{c_k}(x, \lambda_k)$ has a global minimum for every λ_k and $c_k > 0$. We shall consider such penalty functions in Chapter 5. If one cannot find a suitable penalty function, an alternative is to impose additional artificial constraints on the problem so that the constraint set X is compact. Then $L_{c_k}(x, \lambda_k)$ will attain a global minimum over X by Weierstrass' theorem. Another possibility is to replace the objective function f by an equivalent objective which is bounded below, such as $e^{f(x)}$, although this is seldom recommended as it tends to introduce numerical difficulties.

A second weakness of Proposition 2.1 is that it relates exclusively to global (as opposed to local) minima of both the original problem and the augmented Lagrangian. The following proposition remedies the situation somewhat. We first introduce a definition:

Definition: A nonempty set $X^* \subset R^n$ is said to be *an isolated set of local minima* of problem (1) if each point in X^* is a local minimum of problem (1) and, for some $\varepsilon > 0$, the set

$$X_\varepsilon^* = \{x \mid |x - x^*| \le \varepsilon \text{ for some } x^* \in X^*\} \tag{11}$$

contains no local minima of problem (1) other than the points of X^*.

Note that a strict local minimum may be viewed as an isolated set of local minima consisting of a single point.

Proposition 2.2: Let f and h be continuous functions, X be a closed set, $\{\lambda_k\}$ be bounded, and $0 < c_k < c_{k+1}$ for all k, $c_k \to \infty$. Assume that X^* is an isolated set of local minima of problem (1) which is compact. Then there exists a subsequence $\{x_k\}_K$ converging to a point $x^* \in X^*$ such that x_k is a local minimum for the problem

$$\begin{aligned} &\text{minimize} \quad L_{c_k}(x, \lambda_k) \\ &\text{subject to} \quad x \in X \end{aligned} \tag{12}$$

for each $k \in K$. Furthermore, if X^* consists of a single point x^*, there exists a sequence $\{x_k\}$ and an integer $\bar{k} \ge 0$ such that $x_k \to x^*$ and x_k is a local minimum of problem (12) for all $k \ge \bar{k}$.

Proof: Consider the set

$$X_{\varepsilon'}^* = \{x \,|\, |x - x^*| \le \varepsilon' \text{ for some } x^* \in X^*\},$$

where $0 < \varepsilon' < \varepsilon$ and ε is as in (11). The compactness of X^* implies that $X_{\varepsilon'}^*$ is also compact, and hence the problem

$$\begin{aligned} &\text{minimize} \quad L_{c_k}(x, \lambda_k) \\ &\text{subject to} \quad x \in X_{\varepsilon'}^* \cap X \end{aligned}$$

has a global minimum x_k by Weierstrass' theorem. By Proposition 1.1, every limit point of $\{x_k\}$ is a global minimum of the problem

$$\begin{aligned} &\text{minimize} \quad f(x) \\ &\text{subject to} \quad x \in X_{\varepsilon'}^* \cap X, \qquad h(x) = 0. \end{aligned}$$

Furthermore, each global minimum of the problem above must belong to X^* by the definition of $X_{\varepsilon'}^*$. It follows that there is a subsequence $\{x_k\}_{K'}$ converging to a point $x^* \in X^*$. Let $K = \{k \in K' \,|\, |x_k - x^*| < \varepsilon'\}$. Then K is an infinite subset of the integers, and x_k is a local minimum of problem (12) for each $k \in K$. The argument given above proves also the last part of the proposition. Q.E.D.

Both Propositions 2.1 and 2.2 assume implicitly that a method is available that can find a local or global minimum of the augmented Lagrangian. On the other hand, unconstrained minimization methods are usually terminated when the gradient of the objective function is sufficiently small, but not necessarily zero. In particular, when $X = R^n$ and $f, h \in C^1$ the algorithm for solving the unconstrained problem

$$\begin{aligned} &\text{minimize} \quad L_{c_k}(x, \lambda_k) \\ &\text{subject to} \quad x \in R^n \end{aligned}$$

will typically be terminated at a point x_k satisfying

$$|\nabla_x L_{c_k}(x_k, \lambda_k)| \le \varepsilon_k,$$

where ε_k is some small scalar. We address this situation in the next proposition, where it is shown in addition that one can usually obtain as a by-product of the computation a Lagrange multiplier vector.

Proposition 2.3: Assume that $X = R^n$ and $f, h \in C^1$. For $k = 0, 1, \ldots,$ let x_k satisfy

$$|\nabla_x L_{c_k}(x_k, \lambda_k)| \le \varepsilon_k,$$

where $\{\lambda_k\}$ is bounded, $0 < c_k < c_{k+1}$ for all k, $c_k \to \infty$, and $0 \le \varepsilon_k$ for all k, $\varepsilon_k \to 0$. Assume that a subsequence $\{x_k\}_K$ converges to a vector x^* such that $\nabla h(x^*)$ has rank m. Then for some vector λ^*, we have

$$\{\lambda_k + c_k h(x_k)\}_K \to \lambda^*,$$

$$\nabla f(x^*) + \nabla h(x^*)\lambda^* = 0, \qquad h(x^*) = 0.$$

Proof: Define for all k

$$\tilde{\lambda}_k = \lambda_k + c_k h(x_k).$$

We have

$$\begin{aligned}\nabla_x L_{c_k}(x_k, \lambda_k) &= \nabla f(x_k) + \nabla h(x_k)[\lambda_k + c_k h(x_k)] \\ &= \nabla f(x_k) + \nabla h(x_k)\tilde{\lambda}_k = \nabla_x L_0(x_k, \tilde{\lambda}_k),\end{aligned}$$

and, for all k such that $\nabla h(x_k)$ has rank m, it follows that

$$\tilde{\lambda}_k = [\nabla h(x_k)'\nabla h(x_k)]^{-1}\nabla h(x_k)'[\nabla_x L_{c_k}(x_k, \lambda_k) - \nabla f(x_k)].$$

Since $\nabla_x L_{c_k}(x_k, \lambda_k) \to 0$, it follows that

$$\{\tilde{\lambda}_k\}_K \to \lambda^* \triangleq -[\nabla h(x^*)'\nabla h(x^*)]^{-1}\nabla h(x^*)'\nabla f(x^*),$$

and

$$\nabla_x L_0(x^*, \lambda^*) = 0.$$

Since $\{\lambda_k\}$ is bounded and $\{\lambda_k + c_k h(x_k)\}_K \to \lambda^*$, it follows that $\{c_k h(x_k)\}_K$ is bounded. Since $c_k \to \infty$, we must have $h(x^*) = 0$. Q.E.D.

Proposition 2.3 relates to the case where we utilize a method for unconstrained minimization which aims at finding for each k a critical point of the augmented Lagrangian. Assuming that the kth unconstrained minimization is terminated when $|\nabla_x L_{c_k}(x_k, \lambda_k)| \le \varepsilon_k$ where $\varepsilon_k \to 0$, there are three possibilities:

(a) The method breaks down, because for some k a vector x_k satisfying $|\nabla_x L_{c_k}(x_k, \lambda_k)| \le \varepsilon_k$ cannot be found.

(b) A sequence $\{x_k\}$ with $|\nabla_x L_{c_k}(x_k, \lambda_k)| \le \varepsilon_k$ for all k is found, but it either has no limit points, or for each of its limit points x^* the matrix $\nabla h(x^*)$ has linearly dependent columns.

(c) A sequence $\{x_k\}$ with $|\nabla_x L_{c_k}(x_k, \lambda_k)| \le \varepsilon_k$ for all k is found and it has a limit point x^* with $\nabla h(x^*)$ having rank m. This point x^* together with λ^*—the corresponding limit point of $\{\lambda_k + c_k h(x_k)\}$—satisfies the first-order conditions for optimality.

Possibility (a) will usually occur if $L_{c_k}(\cdot, \lambda_k)$ is unbounded below as discussed following Proposition 2.1.

Possibility (b) usually occurs when $L_{c_k}(\cdot, \lambda_k)$ is bounded below, but the original problem has no feasible solution. Typically then the penalty term dominates as $k \to \infty$, and the method usually converges to an infeasible vector x^* which is a critical point of the function $|h(x)|^2$. This means that $\nabla h(x^*)h(x^*) = 0$ implying that $\nabla h(x^*)$ does not have rank m. However possibility (b) may also occur even if the original problem has a feasible solution. A typical example is when $f(x) \triangleq 0$ and $\lambda_k \equiv 0$. Then for all $c_k > 0$, a vector x^* is a critical point of $L_{c_k}(\cdot, 0)$ if and only if it is a critical point of the function $|h(x)|^2$, or equivalently if and only if $\nabla h(x^*)h(x^*) = 0$. If $\nabla h(x^*)$ does not have rank m, it is possible that $\nabla h(x^*)h(x^*) = 0$ while $h(x^*) \neq 0$. This can occur regardless of whether there is a feasible solution. We provide an example of such a situation.

Example: Let $n = 2$, $m = 2$, $f(x) \triangleq 0$,

$$h_1(x_1, x_2) = x_1^2 - 3, \qquad h_2(x_1, x_2) = -2x_1 + x_2^2.$$

The vectors $(\sqrt{3}, \pm\sqrt{2\sqrt{3}})$ are the only feasible solutions. On the other hand for the infeasible vector $x^* = (-1, 0)$ we have, for all $c > 0$,

$$\nabla_x L_c(x^*, 0) = c\nabla h(x^*)h(x^*) = c\begin{bmatrix} -2 & -2 \\ 0 & 0 \end{bmatrix}\begin{bmatrix} -2 \\ 2 \end{bmatrix} = \begin{bmatrix} 0 \\ 0 \end{bmatrix}.$$

Possibility (c) is the normal case where the unconstrained minimization algorithm terminates successfully for each k and $\{x_k\}$ converges to a feasible vector which is also a regular point. It is possible of course that $\{x_k\}$ converges to a local minimum x^* which is not a regular point as shown by Proposition 1.2. In this case, if there is no Lagrange multiplier vector corresponding to x^*, the sequence $\{\lambda_k + c_k h(x_k)\}$ diverges and has no limit point.

Extensive practical experience has shown that the penalty function method is on the whole quite reliable and usually converges to at least a local minimum of the original problem. Whenever it fails, this is usually due to the fact that unconstrained minimization of $L_{c_k}(x, \lambda_k)$ becomes increasingly ill-conditioned as $c_k \to \infty$. We proceed to discuss this in what follows in this section. In Section 2.2, we shall show how, by introducing suitable updating formulas for the multipliers λ_k, the difficulties due to ill-conditioning can be significantly alleviated, and in fact it might not even be necessary to have $c_k \to \infty$ in order to induce convergence.

The Problem of Ill-Conditioning

Since the penalty method is based on the solution of problems of the form

$$\text{(13)} \qquad \begin{array}{ll} \text{minimize} & L_{c_k}(x, \lambda_k) \\ \text{subject to} & x \in X, \end{array}$$

it is natural to inquire about the degree of difficulty in solving such problems. When $X = R^n$ and $f, h \in C^2$, the degree of difficulty for solving problem (13) depends on the eigenvalue structure of the Hessian matrix $\nabla_{xx}^2 L_{c_k}(x_k, \lambda_k)$. We have

$$\nabla_{xx}^2 L_{c_k}(x_k, \lambda_k) = \nabla^2 f(x_k) + \sum_{i=1}^{m} [\lambda_k + c_k h(x_k)]_i \nabla^2 h_i(x_k) + c_k \nabla h(x_k) \nabla h(x_k)'.$$

By using the notation $\tilde{\lambda}_k = \lambda_k + c_k h(x_k)$, we can write

$$\nabla_{xx}^2 L_{c_k}(x_k, \lambda_k) = \nabla_{xx}^2 L_0(x_k, \tilde{\lambda}_k) + c_k \nabla h(x_k) \nabla h(x_k)'. \tag{14}$$

The minimum eigenvalue $\gamma(x_k, \lambda_k, c_k)$ of $\nabla_{xx}^2 L_{c_k}(x_k, \lambda_k)$ satisfies

$$\gamma(x_k, \lambda_k, c_k) = \min_{z \neq 0} \frac{z' \nabla_{xx}^2 L_{c_k}(x_k, \lambda_k) z}{z'z} \leq \min_{\substack{z \neq 0 \\ \nabla h(x_k)'z = 0}} \frac{z' \nabla_{xx}^2 L_0(x, \tilde{\lambda}_k) z}{z'z}, \tag{15}$$

where we assume that $m < n$, and hence there exists a vector $z \neq 0$ with $\nabla h(x_k)'z = 0$. The maximum eigenvalue $\Gamma(x_k, \lambda_k, c_k)$ of $\nabla_{xx}^2 L_{c_k}(x_k, \lambda_k)$ satisfies

$$\begin{aligned} \Gamma(x_k, \lambda_k, c_k) &= \max_{z \neq 0} \frac{z' \nabla_{xx}^2 L_{c_k}(x_k, \lambda_k) z}{z'z} \\ &\geq \min_{z \neq 0} \frac{z' \nabla_{xx}^2 L_0(x_k, \tilde{\lambda}_k) z}{z'z} + c_k \max_{z \neq 0} \frac{z' \nabla h(x_k) \nabla h(x_k)' z}{z'z}. \end{aligned} \tag{16}$$

If $\{x_k\}$ converges to a local minimum x^* which is a regular point with associated Lagrange multiplier vector λ^*, then, by Proposition 2.3, we have $\tilde{\lambda}_k \to \lambda^*$. Since $\nabla h(x^*) \neq 0$, it follows, from (15), (16), that

$$\lim_{k \to \infty} \frac{\Gamma(x_k, \lambda_k, c_k)}{\gamma(x_k, \lambda_k, c_k)} = \infty.$$

In other words, the condition number of problem (13) becomes progressively worse and tends to infinity as $k \to \infty$.

A conclusion that can be drawn from the above analysis is that *for high values of the penalty parameter c_k the corresponding unconstrained optimization problem becomes ill-conditioned* and hence difficult to solve. For example, steepest descent is out of the question as a possible solution method. Even Newton's method can encounter significant difficulties if c_k is very high, and the starting point for minimizing $L_{c_k}(\cdot, \lambda_k)$ is not near a solution.

The ill-conditioning associated with the unconstrained minimization problems (13) is a basic characteristic feature of penalty methods and represents the overriding factor in determining the manner in which these methods are operated. Ill-conditioning can be overcome only by using for each k a starting point for the unconstrained minimization routine which is close to a minimizing point of $L_{c_k}(\cdot, \lambda_k)$. Usually, one adopts as a starting point the last point x_{k-1} of the previous minimization. In order for x_{k-1} to be near a minimizing point of $L_{c_k}(\cdot, \lambda_k)$, it is necessary that c_k is close to c_{k-1}. This in turn implies that the rate of increase of the penalty parameter c_k should be relatively small. If c_k is increased at a fast rate, then convergence of the method (i.e., of the sequence $\{x_k\}$) is faster, albeit at the expense of ill-conditioning. In practice, one must operate the method in a way which balances the benefit of fast convergence with the evil of ill-conditioning. Usually, a sequence $\{c_k\}$ satisfying $c_{k+1} = \beta c_k$ with $\beta \in [4, 10]$ works well. There is no safe guideline as to what is a suitable value for c_0, so one may have to resort to trial and error in order to determine the value of this parameter.

2.2 The Original Method of Multipliers

Consider the equality constrained problem

(ECP)
$$\text{minimize} \quad f(x)$$
$$\text{subject to} \quad h(x) = 0,$$

where f: $R^n \to R$, h: $R^n \to R^m$ are given functions. The components of h are denoted by $h_1, \ldots, h_m$. For any scalar c, consider also the augmented Lagrangian function

$$L_c(x, \lambda) = f(x) + \lambda' h(x) + \tfrac{1}{2}c|h(x)|^2.$$

Throughout this section, we shall assume that x^* is a local minimum satisfying the following second-order sufficiency condition (compare with Proposition 1.24).

Assumption (S): The vector x^* is a strict local minimum and a regular point of (ECP), and $f, h \in C^2$ on some open sphere centered at x^*. Furthermore x^* together with its associated Lagrange multiplier vector λ^* satisfies

$$z'\nabla_{xx}^2 L_0(x^*, \lambda^*)z > 0,$$

for all $z \neq 0$ with $\nabla h(x^*)'z = 0$.

A formal description of the typical step of the original version of the method of multipliers (Hestenes, 1969; Powell, 1969) is as follows:

Given a multiplier vector λ_k and a penalty parameter c_k, we minimize $L_{c_k}(\cdot, \lambda_k)$ over R^n thereby obtaining a vector x_k. We then set

$$\lambda_{k+1} = \lambda_k + c_k h(x_k), \tag{1}$$

we choose a penalty parameter $c_{k+1} \geq c_k$ and repeat the process.

The initial vector λ_0 is chosen arbitrarily, and the sequence $\{c_k\}$ may be either preselected or determined on the basis of results obtained during the algorithmic process.

The description given above is not meant to be precise, but is rather aimed at providing a starting point for the analysis that follows. The reader can view, for the time being, the method of multipliers simply as the penalty function method where the multipliers λ_k are determined by using the updating formula (1).

2.2.1 *Geometric Interpretation*

We provide a geometric interpretation of the method of multipliers which motivates the subsequent convergence analysis. Consider the *primal functional* p of (ECP) defined by

$$p(u) = \min_{h(x)=u} f(x),$$

where the minimization is understood to be local in an open sphere within which x^* is the unique local minimum of (ECP). We specify p more precisely later, but for the moment we shall use this informal definition. Clearly $p(0) = f(x^*)$, and, from Proposition 1.28, we have $\nabla p(0) = -\lambda^*$. We can break down the minimization of $L_c(\cdot, \lambda)$ into two stages, first minimizing over all x such that $h(x) = u$ with u fixed, and then minimizing over all u so that

$$\begin{aligned}\min_x L_c(x, \lambda) &= \min_u \min_{h(x)=u} \{f(x) + \lambda' h(x) + \tfrac{1}{2}c|h(x)|^2\} \\ &= \min_u \{p(u) + \lambda' u + \tfrac{1}{2}c|u|^2\},\end{aligned}$$

where the minimization above is understood to be local in a neighborhood of $u = 0$. This minimization can be interpreted as shown in Fig. 2.1. The minimum is attained at the point $u(\lambda, c)$ for which the gradient of $p(u) + \lambda' u + \frac{1}{2}c|u|^2$ is zero, or equivalently

$$\nabla\{p(u) + \tfrac{1}{2}c|u|^2\}|_{u=u(\lambda,c)} = -\lambda.$$

Thus the minimizing point $u(\lambda, c)$ is obtained as shown in Fig. 2.1. We have also

$$\min_x L_c(x, \lambda) - \lambda' u(\lambda, c) = p[u(\lambda, c)] + \tfrac{1}{2}c|u(\lambda, c)|^2,$$

so the tangent hyperplane to the graph of $p(u) + \frac{1}{2}c|u|^2$ at $u(\lambda, c)$ (which has "slope" $-\lambda$) intersects the vertical axis at the value $\min_x L_c(x, \lambda)$ as shown in Fig. 2.1. It can be seen that if c is sufficiently large then $p(u) + \lambda' u + \frac{1}{2}c|u|^2$ is convex in a neighborhood of the origin. Furthermore, the

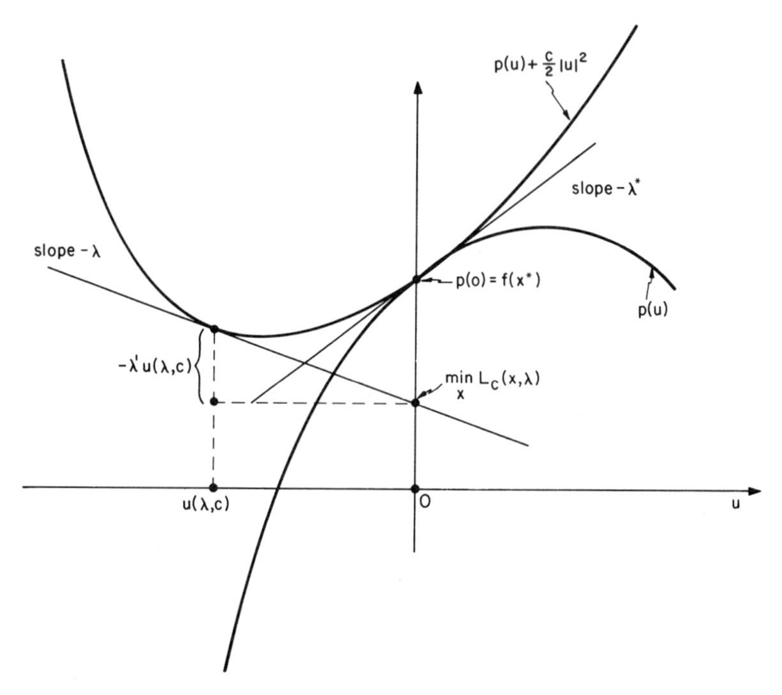

FIG. 2.1 Geometric interpretation of minimization of the Augmented Lagrangian

value $\min_x L_c(x, \lambda)$ is close to $p(0) = f(x^*)$ for values of λ close to λ^* and large values of c.

Figure 2.2 provides a geometric interpretation of the multiplier iteration (1). To understand this figure, note that if x_k minimizes $L_{c_k}(\cdot, \lambda_k)$, then by the analysis above the vector u_k given by $u_k = h(x_k)$ minimizes $p(u) + \lambda_k' u + \frac{1}{2} c_k |u|^2$. Hence,

$$\nabla\{p(u) + \tfrac{1}{2} c_k |u|^2\}|_{u=u_k} = -\lambda_k,$$

and

$$\nabla p(u_k) = -(\lambda_k + c_k u_k) = -[\lambda_k + c_k h(x_k)].$$

It follows that, for the next multiplier λ_{k+1}, we have

$$\lambda_{k+1} = \lambda_k + c_k h(x_k) = -\nabla p(u_k),$$

as shown in Fig. 2.2. The figure shows that if λ_k is sufficiently close to λ^* and/or c_k is sufficiently large, the next multiplier λ_{k+1} will be closer to λ^* than λ_k is. In fact if $p(u)$ is linear, convergence to λ^* will be achieved in one iteration. If $\nabla^2 p(0) = 0$, the convergence is very fast. Furthermore it is not

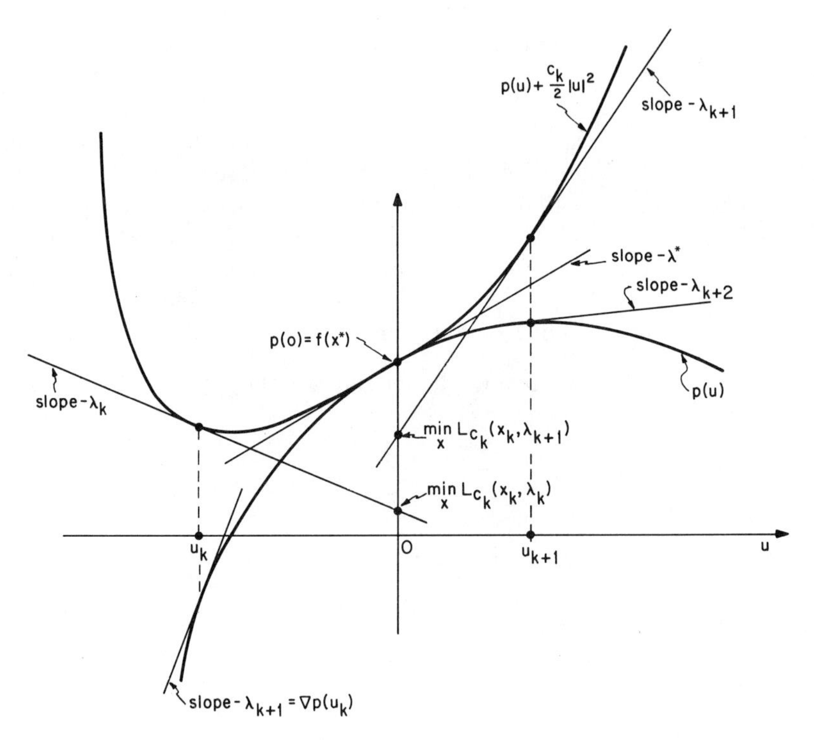

FIG. 2.2 Geometric interpretation of the first-order multiplier iteration

necessary to have $c_k \to \infty$ in order to obtain convergence but merely to have that c_k exceeds some threshold level after some index. We proceed to make these observations precise.

2.2.2 *Existence of Local Minima of the Augmented Lagrangian*

As in the case of the penalty method, it is natural to inquire whether local minima of the augmented Lagrangian exist, and if so how their distance from local minima of the original problem is affected by the values of the multiplier λ and the penalty parameter c. To this end, we focus on the local minimum x^* satisfying Assumption (S) together with its corresponding Lagrange multiplier λ^*. We have, for any scalar c,

(2) $\nabla_x L_c(x^*, \lambda^*) = \nabla f(x^*) + \nabla h(x^*)[\lambda^* + ch(x^*)] = \nabla_x L_0(x^*, \lambda^*) = 0$

and [compare with (14) in the previous section]

$$\nabla_{xx}^2 L_c(x^*, \lambda^*) = \nabla_{xx}^2 L_0(x^*, \lambda^*) + c\nabla h(x^*)\nabla h(x^*)'.$$

Lemma 1.25 and Assumption (S) imply that there exists a scalar $\bar{c}$ such that

$$\nabla^2_{xx} L_c(x^*, \lambda^*) > 0 \qquad \forall\, c \geq \bar{c}. \tag{3}$$

From (2) and (3), we have that x^* is a strict local minimum of $L_c(\cdot, \lambda^*)$ for all $c \geq \bar{c}$. It is thus reasonable to infer that if λ is close enough to λ^*, there should exist a local minimum of $L_c(\cdot, \lambda)$ close to x^* for every $c \geq \bar{c}$. Proposition 2.2 suggests that this will also be true even if λ is far from λ^* provided c is sufficiently large. The following proposition makes this idea precise. It also provides estimates on the proximity of the local minimum of $L_c(\cdot, \lambda)$ to x^* and the corresponding Lagrange multiplier estimate to λ^*.

Proposition 2.4: Assume (S) holds and let $\bar{c}$ be a positive scalar such that

$$\nabla^2_{xx} L_{\bar{c}}(x^*, \lambda^*) > 0. \tag{4}$$

There exist positive scalars δ, ε, and M such that:

(a) For all (λ, c) in the set $D \subset R^{m+1}$ defined by

$$D = \{(\lambda, c) \,|\, |\lambda - \lambda^*| < \delta c, \bar{c} \leq c\}, \tag{5}$$

the problem

$$\begin{aligned} &\text{minimize} \quad L_c(x, \lambda) \\ &\text{subject to} \quad x \in S(x^*; \varepsilon) \end{aligned} \tag{6}$$

has a unique solution denoted $x(\lambda, c)$. The function $x(\cdot, \cdot)$ is continuously differentiable in the interior of D, and, for all $(\lambda, c) \in D$, we have

$$|x(\lambda, c) - x^*| \leq M |\lambda - \lambda^*| / c. \tag{7}$$

(b) For all $(\lambda, c) \in D$, we have

$$|\tilde{\lambda}(\lambda, c) - \lambda^*| \leq M |\lambda - \lambda^*| / c, \tag{8}$$

where

$$\tilde{\lambda}(\lambda, c) = \lambda + c h[x(\lambda, c)]. \tag{9}$$

(c) For all $(\lambda, c) \in D$, the matrix $\nabla^2_{xx} L_c[x(\lambda, c), \lambda]$ is positive definite and the matrix $\nabla h[x(\lambda, c)]$ has rank m.

Proof: For $c > 0$, consider the system of equations in $(x, \tilde{\lambda}, \lambda, c)$

$$\nabla f(x) + \nabla h(x)\tilde{\lambda} = 0, \qquad h(x) + (\lambda - \tilde{\lambda})/c = 0. \tag{10}$$

By introducing the variables $t \in R^m$, $\gamma \in R$ defined by

$$t = (\lambda - \lambda^*)/c, \qquad \gamma = 1/c, \tag{11}$$

we can write system (10) as

$$\nabla f(x) + \nabla h(x)\tilde{\lambda} = 0, \qquad h(x) + t + \gamma\lambda^* - \gamma\tilde{\lambda} = 0. \tag{12}$$

For $t = 0$ and $\gamma \in [0, 1/\bar{c}]$, system (12) has the solution $x = x^*$ and $\tilde{\lambda} = \lambda^*$. The Jacobian with respect to $(x, \tilde{\lambda})$ at such a solution is

$$\begin{bmatrix} \nabla_{xx}^2 L_0(x^*, \lambda^*) & \nabla h(x^*) \\ \nabla h(x^*)' & -\gamma I \end{bmatrix}, \tag{13}$$

where I is the identity matrix. We show that matrix (13) is invertible for all $\gamma \in [0, 1/\bar{c}]$. By Lemma 1.27, this is true for $\gamma = 0$. To show this fact for $\gamma \in (0, 1/\bar{c}]$, suppose that for some $z \in R^n$ and $w \in R^m$, we have

$$\begin{bmatrix} \nabla_{xx}^2 L_0(x^*, \lambda^*) & \nabla h(x^*) \\ \nabla h(x^*)' & -\gamma I \end{bmatrix} \begin{bmatrix} z \\ w \end{bmatrix} = 0 \tag{14}$$

or equivalently

$$\nabla_{xx}^2 L_0(x^*, \lambda^*)z + \nabla h(x^*)w = 0, \tag{15}$$

$$\nabla h(x^*)'z - \gamma w = 0. \tag{16}$$

Substituting the value of w from (16) into (15), we obtain

$$[\nabla_{xx}^2 L_0(x^*, \lambda^*) + (1/\gamma)\nabla h(x^*)\nabla h(x^*)']z = 0.$$

For $\gamma = 1/c$ with $c \geq \bar{c}$, this yields $\nabla_{xx}^2 L_c(x^*, \lambda^*)z = 0$, and since $\nabla_{xx}^2 L_c(x^*, \lambda^*) > 0$ for $c \geq \bar{c}$, we obtain $z = 0$. From (16), we also obtain $w = 0$. Thus if (14) holds, we must have $z = 0$ and $w = 0$, and it follows that matrix (13) is invertible for all $\gamma \in [0, 1/\bar{c}]$.

We now apply the second implicit function theorem of Section 1.2, to the system (12), where we identify the compact set $K = \{(0, \gamma) \mid \gamma \in [0, 1/\bar{c}]\}$ with the set $\bar{X}$ of that theorem. It follows that there exist $\varepsilon > 0$ and $\delta > 0$ and unique continuously differentiable functions $\hat{x}(t, \gamma)$ and $\hat{\lambda}(t, \gamma)$ defined on $S(K; \delta)$ such that $(|\hat{x}(t, \gamma) - x^*|^2 + |\hat{\lambda}(t, \gamma) - \lambda^*|^2)^{1/2} < \varepsilon$ for all $(t, \gamma) \in S(K; \delta)$ and satisfying

$$\nabla f[\hat{x}(t, \gamma)] + \nabla h[\hat{x}(t, \gamma)]\hat{\lambda}(t, \gamma) = 0, \tag{17}$$

$$h[\hat{x}(t, \gamma)] + t + \gamma\lambda^* - \gamma\hat{\lambda}(t, \gamma) = 0. \tag{18}$$

Clearly δ and ε can be chosen so that in addition $\nabla h[\hat{x}(t, \gamma)]$ has rank m and

$$\nabla_{xx}^2 L_0[\hat{x}(t, \gamma), \hat{\lambda}(t, \gamma)] + c\nabla h[\hat{x}(t, \gamma)]\nabla h[\hat{x}(t, \gamma)]' > 0$$

for $(t, \gamma) \in S(K; \delta), c \geq \bar{c}$. For $c \geq \bar{c}$ and $|\lambda - \lambda^*| < \delta c$, define

$$x(\lambda, c) = \hat{x}\left(\frac{\lambda - \lambda^*}{c}, \frac{1}{c}\right), \qquad \tilde{\lambda}(\lambda, c) = \hat{\lambda}\left(\frac{\lambda - \lambda^*}{c}, \frac{1}{c}\right).$$

Then in view of (11), (17), and (18), we obtain, for $(\lambda, c) \in D$,

$$\nabla f[x(\lambda, c)] + \nabla h[x(\lambda, c)]\tilde{\lambda}(\lambda, c) = 0,$$

$$\tilde{\lambda}(\lambda, c) = \lambda + ch[x(\lambda, c)],$$

$$\nabla_{xx}^2 L_0[x(\lambda, c), \tilde{\lambda}(\lambda, c)] + c\nabla h[x(\lambda, c)]\nabla h[x(\lambda, c)]' = \nabla_{xx}^2 L_c[x(\lambda, c), \lambda] > 0.$$

Thus, the proposition is proved except for (7) and (8).

In order to show (7) and (8), we differentiate (17) and (18) with respect to t and γ. We obtain, via a straightforward calculation,

$$\begin{bmatrix} \nabla_t \hat{x}(t, \gamma)' & \nabla_\gamma \hat{x}(t, \gamma)' \\ \nabla_t \hat{\lambda}(t, \gamma)' & \nabla_\gamma \hat{\lambda}(t, \gamma)' \end{bmatrix} = A(t, \gamma) \begin{bmatrix} 0 & 0 \\ -I & \hat{\lambda}(t, \gamma) - \lambda^* \end{bmatrix}, \tag{19}$$

where

$$A(t, \gamma) = \begin{bmatrix} \nabla_{xx}^2 L_0[\hat{x}(t, \gamma), \hat{\lambda}(t, \gamma)] & \nabla h[\hat{x}(t, \gamma)] \\ \nabla h[\hat{x}(t, \gamma)]' & -\gamma I \end{bmatrix}^{-1}. \tag{20}$$

We have for all (t, γ), such that $|t| < \delta$ and $\gamma \in [0, 1/\bar{c}]$,

$$\begin{bmatrix} \hat{x}(t, \gamma) - x^* \\ \hat{\lambda}(t, \gamma) - \lambda^* \end{bmatrix} = \begin{bmatrix} \hat{x}(t, \gamma) - \hat{x}(0, 0) \\ \hat{\lambda}(t, \gamma) - \hat{\lambda}(0, 0) \end{bmatrix} = \int_0^1 A(\zeta t, \zeta\gamma) \begin{bmatrix} 0 & 0 \\ -I & \hat{\lambda}(\zeta t, \zeta\gamma) - \lambda^* \end{bmatrix} \begin{bmatrix} t \\ \gamma \end{bmatrix} d\zeta. \tag{21}$$

Since matrix (13) is invertible for all $\gamma \in [0, 1/\bar{c}]$, it follows that, for δ sufficiently small, $A(t, \gamma)$ is uniformly bounded on $\{(t, \gamma) \mid |t| < \delta, \gamma \in [0, 1/\bar{c}]\}$. Let μ be such that $|A(t, \gamma)| \le \mu$ for all $|t| < \delta$, $\gamma \in [0, 1/\bar{c}]$, and if necessary take δ sufficiently small to ensure that $\mu\delta < 1$. Then, from (21), we obtain

$$(|\hat{x}(t, \gamma) - x^*|^2 + |\hat{\lambda}(t, \gamma) - \lambda^*|^2)^{1/2} \le \mu\left(|t| + \max_{0 \le \zeta \le 1} |\hat{\lambda}(\zeta t, \zeta\gamma) - \lambda^*|\gamma\right). \tag{22}$$

From this, it follows that, for all (t, γ) with $|t| < \delta$, $\gamma \in [0, 1/\bar{c}]$, and $\gamma < \delta$,

$$|\hat{\lambda}(t, \gamma) - \lambda^*| \le \mu|t| + \mu\gamma \max_{0 \le \zeta \le 1} |\hat{\lambda}(\zeta t, \zeta\gamma) - \lambda^*|.$$

Using the inequality above with $\zeta t, \zeta\gamma, \zeta \in [0, 1]$ in place of t, γ, we obtain

$$\max_{0 \le \zeta \le 1} |\hat{\lambda}(\zeta t, \zeta\gamma) - \lambda^*| \le \frac{\mu}{1 - \mu\gamma}|t|. \tag{23}$$

Combining (22) and (23), we obtain, for all (t, γ) with $|t| < \delta$, $\gamma \in [0, 1/\bar{c}]$, and $\gamma < \delta$,

$$(|\hat{x}(t, \gamma) - x^*|^2 + |\hat{\lambda}(t, \gamma) - \lambda^*|^2)^{1/2} \le \left(\mu + \frac{\mu^2\gamma}{1 - \mu\gamma}\right)|t| \le \frac{\mu}{1 - \mu\delta}|t|.$$

By taking δ sufficiently small if necessary, we obtain

$$(|\hat{x}(t, \gamma) - x^*|^2 + |\hat{\lambda}(t, \gamma) - \lambda^*|^2)^{1/2} \leq 2\mu|t|.$$

By using (11) and writing $x(\lambda, c) = \hat{x}(t, \gamma)$ and $\tilde{\lambda}(\lambda, c) = \hat{\lambda}(t, \gamma)$, we have that, for all (λ, c) with $|\lambda - \lambda^*| < \delta c$ and $c > \max\{\bar{c}, 1/\delta\}$, there holds

$$|x(\lambda, c) - x^*| \leq 2\mu|\lambda - \lambda^*|/c, \qquad |\tilde{\lambda}(\lambda, c) - \lambda^*| \leq 2\mu|\lambda - \lambda^*|/c.$$

Thus (7) and (8) hold with $M = 2\mu$ for all (λ, c) with $|\lambda - \lambda^*| < \delta c$ and $c > \max\{\bar{c}, 1/\delta\}$. Because $x(\cdot, \cdot)$ is continuously differentiable, we can also find an M so that (7) and (8) hold for all (λ, c) with $|\lambda - \lambda^*| < \delta c$ and $\bar{c} \leq c \leq \max\{\bar{c}, 1/\delta\}$. This completes the proof. Q.E.D.

Figure 2.3 shows the set D of (5) within which the conclusions of Proposition 2.4 are valid. It can be seen that, for any λ, there exists a c_λ such that (λ, c) belongs to D for every $c \geq c_\lambda$. The estimate δc on the allowable distance of λ from λ^* [compare with (5)] grows linearly with c. In particular problems, the actual allowable distance may grow at a higher than linear rate, and in fact it is possible that for every λ and $c > 0$ there exists a unique global minimum of $L_c(\cdot, \lambda)$. (Take for instance the scalar problem $\min\{\frac{1}{2}x^2 \,|\, x = 0\}$.) The following example shows however that the estimate of a linear order of growth cannot be improved.

Example: Let $n = m = 1$, and consider the problem

$$\text{minimize} \quad -x^p$$
$$\text{subject to} \quad x = 0,$$

where p is an even integer with $p > 2$. We have $x^* = \lambda^* = 0$ and Assumption (S) is satisfied. We have

$$L_c(x, \lambda) = -x^p + \lambda x + \tfrac{1}{2}c|x|^2,$$
$$\nabla_x L_c(x, \lambda) = -px^{p-1} + \lambda + cx,$$
$$\nabla_{xx}^2 L_c(x, \lambda) = -p(p-1)x^{p-2} + c.$$

A straightforward calculation shows that

$$\nabla_{xx}^2 L_c(x, \lambda) > 0 \Leftrightarrow |x| < [c/p(p-1)]^{1/(p-2)},$$
$$\nabla_x L_c(x, \lambda) = 0 \Leftrightarrow \lambda = x(px^{p-2} - c).$$

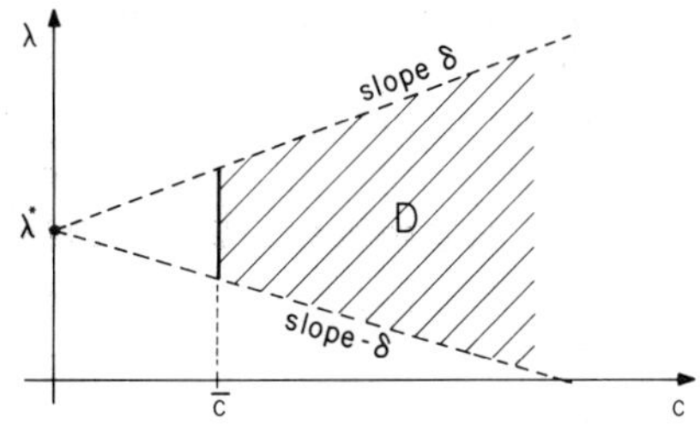

FIG. 2.3 Region D of pairs (λ, c) for which the method of multipliers is defined

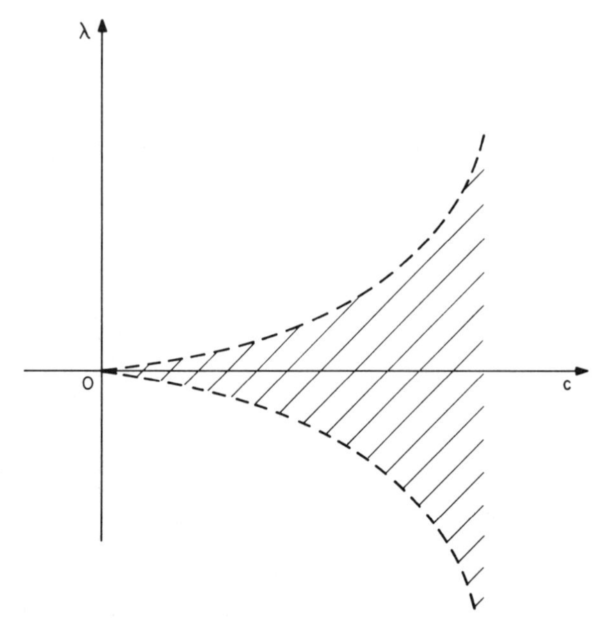

FIG. 2.4 Form of the set (24)

Using these relations it can be verified that $L_c(\cdot, \lambda)$ has a unique local minimum $x(\lambda, c)$ with $\nabla_{xx}^2 L_c[x(\lambda, c), \lambda] > 0$ for all (λ, c) in the set

$$\left\{(\lambda, c) \,\middle|\, |\lambda| < \frac{p-2}{p-1}\left[\frac{1}{p(p-1)}\right]^{1/(p-2)} c^{(p-1)/(p-2)}, c > 0\right\}, \tag{24}$$

shown in Fig. 2.4. The order of growth on the allowable distance of λ from λ^* is $(p-1)/(p-2)$ and tends to unity as p increases. Thus, we cannot demonstrate, in general, a better than linear order of growth of the allowable distance of λ from λ^* as c increases.

Proposition 2.4 can yield both a convergence and a rate-of-convergence result for the multiplier iteration

$$\lambda_{k+1} = \lambda_k + c_k h(x_k).$$

It shows that if the generated sequence $\{\lambda_k\}$ is bounded [this can be enforced if necessary by leaving λ_k unchanged if $\lambda_k + c_k h(x_k)$ does not belong to a prespecified bounded open set known to contain λ^*], the penalty parameter c_k is sufficiently large after a certain index, and after that index minimization of $L_{c_k}(\cdot, \lambda_k)$ yields the local minimum $x_k = x(\lambda_k, c_k)$ closest to x^*, then we obtain $x_k \to x^*$, $\lambda_k \to \lambda^*$. However the threshold level for the penalty parameter is unknown thus far. We try to characterize this level and obtain a sharper convergence and rate of convergence result. The primal functional plays a key role in this respect, so we first analyze its properties.

2.2.3 The Primal Functional

Consider the system of equations in (x, λ, u)

$$\nabla f(x) + \nabla h(x)\lambda = 0, \qquad h(x) - u = 0.$$

It has the solution $(x^*, \lambda^*, 0)$. By the standard implicit function theorem, there exists a $\delta > 0$ and functions $x(\cdot) \in C^1$ and $\lambda(\cdot) \in C^1$, such that $x(0) = x^*$ and $\lambda(0) = \lambda^*$, and, for $|u| < \delta$,

$$\nabla f[x(u)] + \nabla h[x(u)]\lambda(u) = 0, \qquad h[x(u)] - u = 0. \tag{25}$$

Furthermore, for some $\varepsilon > 0$, we have $|x(u) - x^*| < \varepsilon$ and $|\lambda(u) - \lambda^*| < \varepsilon$ for $|u| < \delta$. The function $p: S(0; \delta) \to R$ given by

$$p(u) = f[x(u)] \qquad \forall\, u \in S(0; \delta)$$

is referred to as the *primal functional* corresponding to x^*. In view of (S), we can take δ and ε sufficiently small so that $x(u)$ is actually a local minimum of the problem of minimizing $f(x)$ subject to $h(x) = u$. Thus, an equivalent definition of p is given by

$$p(u) = f[x(u)] = \min\{f(x) \mid h(x) = u, x \in S(x^*; \varepsilon)\}. \tag{26}$$

From Proposition 1.28, we have

$$\nabla p(u) = -\lambda(u) \qquad \forall\, u \in S(0; \delta). \tag{27}$$

Differentiating (25), we obtain

$$\nabla_u x(u) \nabla_{xx}^2 L_0[x(u), \lambda(u)] + \nabla_u \lambda(u) \nabla h[x(u)]' = 0, \tag{28}$$

$$\nabla_u x(u) \nabla h[x(u)] = I. \tag{29}$$

From (29), we have, for any $c \in R$,

$$c\nabla_u x(u) \nabla h[x(u)] \nabla h[x(u)]' = c\nabla h[x(u)]', \tag{30}$$

and by adding (28) and (30), we obtain

$$\begin{aligned} \nabla_u x(u)\{\nabla_{xx}^2 L_0[x(u), \lambda(u)] &+ c\nabla h[x(u)]\nabla h[x(u)]'\} \\ &+ [\nabla_u \lambda(u) - cI]\nabla h[x(u)]' = 0. \end{aligned}$$

From this, we obtain, for every c for which the inverse below exists,

$$\begin{aligned} \nabla_u x(u) &+ [\nabla_u \lambda(u) - cI]\nabla h[x(u)]' \\ &\times \{\nabla_{xx}^2 L_0[x(u), \lambda(u)] + c\nabla h[x(u)]\nabla h[x(u)]'\}^{-1} = 0. \end{aligned}$$

Multiplying both sides with $\nabla h[x(u)]$ and using (27) and (29), we obtain

$$cI + \nabla^2 p(u) = \{\nabla h[x(u)]'\{\nabla_{xx}^2 L_0[x(u), \lambda(u)] + c\nabla h[x(u)]\nabla h[x(u)]'\}^{-1}\nabla h[x(u)]\}^{-1}. \tag{31}$$

Equation (31) holds for all u with $|u| < \delta$ and for all c for which the inverse above exists. For $u = 0$, we obtain

$$\nabla^2 p(0) = \{\nabla h(x^*)'[\nabla_{xx}^2 L_c(x^*, \lambda^*)]^{-1}\nabla h(x^*)\}^{-1} - cI, \tag{32}$$

for any c for which $\nabla_{xx}^2 L_c(x^*, \lambda^*)$ is invertible. If $[\nabla_{xx}^2 L_0(x^*, \lambda^*)]^{-1}$ exists, then

$$\nabla^2 p(0) = \{\nabla h(x^*)'[\nabla_{xx}^2 L_0(x^*, \lambda^*)]^{-1}\nabla h(x^*)\}^{-1}. \tag{33}$$

The following proposition shows that the threshold level for the penalty parameter in Proposition 2.4 can be characterized in terms of the eigenvalues of the matrix $\nabla^2 p(0)$. In the next section, we shall see that the rate of convergence of the method of multipliers can also be characterized in terms of these eigenvalues.

Proposition 2.5: Let (S) hold. For any scalar c, we have

$$\nabla_{xx}^2 L_c(x^*, \lambda^*) > 0 \Leftrightarrow c > \max\{-e_1, \ldots, -e_m\} \Leftrightarrow \nabla^2 p(0) + cI > 0, \tag{34}$$

where $e_1, \ldots, e_m$ are the eigenvalues of $\nabla^2 p(0)$.

Proof: Since the eigenvalues of $\nabla^2 p(0) + cI$ are $e_i + c$, $i = 1, \ldots, m$, the condition $c > \max\{-e_1, \ldots, -e_m\}$ is equivalent to

$$\nabla^2 p(0) + cI > 0. \tag{35}$$

If $\nabla_{xx}^2 L_c(x^*, \lambda^*) > 0$, then from (32) it follows that (35) also holds. Conversely if (35) holds, then, using (27), we have that $u = 0$ is a strong local minimum of $p(u) + \lambda^{*\prime}u + \frac{1}{2}c|u|^2$. It follows that, for some $\delta_1 > 0$, $\gamma > 0$, and all u with $|u| < \delta_1$,

$$p(u) + \lambda^{*\prime}u + \tfrac{1}{2}c|u|^2 \geq p(0) + \tfrac{1}{2}\gamma|u|^2.$$

Hence using (26), we have that there is an $\varepsilon > 0$ such that

$$\min_{x \in S(x^*; \varepsilon)} \left\{ f(x) + \lambda^{*\prime}h(x) + \frac{c - \gamma}{2}|h(x)|^2 \right\} = f(x^*).$$

As a result $\nabla_{xx}^2 L_0(x^*, \lambda^*) + (c - \gamma)\nabla h(x^*)\nabla h(x^*)' \geq 0$ or

$$\nabla_{xx}^2 L_c(x^*, \lambda^*) \geq \gamma\nabla h(x^*)\nabla h(x^*)'. \tag{36}$$

It follows that $\nabla^2_{xx} L_c(x^*, \lambda^*) \geq 0$. If there exists a $z \neq 0$ such that

$$z'\nabla^2_{xx} L_c(x^*, \lambda^*)z = 0,$$

then, from (36), $\nabla h(x^*)'z = 0$ and using (S) we have $z'\nabla^2_{xx} L_0(x^*,\lambda^*)z > 0$ or equivalently $z'\nabla^2_{xx} L_c(x^*, \lambda^*)z > 0$. Thus, we must have $z'\nabla^2_{xx} L_c(x^*, \lambda^*)z > 0$ for all $z \neq 0$, and $\nabla^2_{xx} L_c(x^*, \lambda^*)$ must be positive definite. Q.E.D.

An alternative proof of Proposition 2.5 can be given by making use of (32) and the result of the following exercise.

Exercise: Let Q be a symmetric $n \times n$ matrix and L be a subspace of R^n. Assume that

$$z'Qz > 0 \qquad \forall\, z \in L, \quad z \neq 0.$$

Then

$$Q > 0 \Leftrightarrow Q^{-1} \text{ exists and } w'Q^{-1}w > 0 \qquad \forall\, w \in L^\perp, \quad w \neq 0,$$

where $L^\perp$ is the orthogonal complement of L. *Hint*: Consider a set of basis vectors for $L^\perp$ and let B be the matrix having as columns these vectors. Show that $\min\{\frac{1}{2}x'Qx \,|\, B'x = u\} = \frac{1}{2}u'(B'Q^{-1}B)^{-1}u$.

It is interesting to interpret the condition (35) in terms of the "penalized" primal functional $p_c(u) = p(u) + \frac{1}{2}c|u|^2$ of Figs. 2.1 and 2.2. Relation (34) can be written as

$$\nabla^2_{xx} L_c(x^*, \lambda^*) > 0 \Leftrightarrow \nabla^2 p_c(0) > 0, \tag{37}$$

so *we have* $\nabla^2_{xx} L_c(x^*, \lambda^*) > 0$ *if and only if* p_c *is convex and has positive definite Hessian in a neighborhood of* $u = 0$.

2.2.4 *Convergence Analysis*

We shall obtain a convergence and rate-of-convergence result for the method of multipliers which is sharper than the one implied by Proposition 2.4. To this end, we need the following intermediate result.

Proposition 2.6: Assume (S) holds and let $\bar{c}$ and δ be as in Proposition 2.4. For all (λ, c) in the set D defined by (5), there holds

$$\tilde{\lambda}(\lambda, c) - \lambda^* = \int_0^1 N_c[\lambda^* + \zeta(\lambda - \lambda^*)](\lambda - \lambda^*)\, d\zeta, \tag{38}$$

where for all $(\lambda, c) \in D$, the $m \times m$ matrix N_c is given by

$$N_c(\lambda) = I - c\nabla h[x(\lambda, c)]'\{\nabla^2_{xx} L_c[x(\lambda, c), \lambda]\}^{-1}\nabla h[x(\lambda, c)] \tag{39}$$

and I is the identity matrix.

Proof: Going back to the proof of Proposition 2.4 we have, by using the matrix inversion formula of Section 1.2, that for $\gamma > 0$ the matrix $A(t, \gamma)$, defined by (20), is given by

$$A(t,\gamma) = \begin{bmatrix} Q(t,\gamma) & \gamma^{-1}Q(t,\gamma)\nabla h[\hat{x}(t,\gamma)] \\ \gamma^{-1}\nabla h[\hat{x}(t,\gamma)]'Q(t,\gamma) & -\gamma^{-1}I + \gamma^{-2}\nabla h[\hat{x}(t,\gamma)]'Q(t,\gamma)\nabla h[\hat{x}(t,\gamma)] \end{bmatrix},$$

where

$$Q(t,\gamma) = \{\nabla^2_{xx}L_0[\hat{x}(t,\gamma), \hat{\lambda}(t,\gamma)] + \gamma^{-1}\nabla h[\hat{x}(t,\gamma)]\nabla h[\hat{x}(t,\gamma)]'\}^{-1}.$$

Using (19) we have, for any $\gamma \in (0, 1/\bar{c}]$,

$$\begin{aligned}\hat{\lambda}(t,\gamma) - \lambda^* &= \hat{\lambda}(t,\gamma) - \hat{\lambda}(0,\gamma) = \int_0^1 \nabla_t \hat{\lambda}(\zeta t, \gamma)' t \, d\zeta \\ &= \int_0^1 \{\gamma^{-1}I - \gamma^{-2}\nabla h[\hat{x}(\zeta t,\gamma)]'Q(\zeta t,\gamma)\nabla h[\hat{x}(\zeta t,\gamma)]\} t \, d\zeta.\end{aligned}$$

Substituting $\gamma = 1/c$, $t = (\lambda - \lambda^*)/c$, $x(\lambda, c) = \hat{x}(t,\gamma)$, and $\tilde{\lambda}(\lambda, c) = \hat{\lambda}(t,\gamma) = \lambda + ch[x(\lambda, c)]$, we obtain the result. Q.E.D.

We can now show our main convergence and rate-of-convergence result.

Proposition 2.7: Assume (S) holds, and let $\bar{c}$ and δ be as in Proposition 2.4. Denote by $e_1, \ldots, e_m$ the eigenvalues of the matrix $\nabla^2 p(0)$ given by (32) or (33). Assume also that

$$\bar{c} > \max\{-2e_1, \ldots, -2e_m\}, \tag{40}$$

(or equivalently that $\nabla^2 p(0) + \frac{1}{2}\bar{c}I > 0$). Then there exists a scalar δ_1 with $0 < \delta_1 \le \delta$ such that if $\{c_k\}$ and λ_0 satisfy

$$|\lambda_0 - \lambda^*|/c_0 < \delta_1, \qquad \bar{c} \le c_k \le c_{k+1} \qquad \forall\, k = 0, 1, \ldots, \tag{41}$$

then the sequence $\{\lambda_k\}$ generated by

$$\lambda_{k+1} = \lambda_k + c_k h[x(\lambda_k, c_k)] \tag{42}$$

is well defined† and we have $\lambda_k \to \lambda^*$ and $x(\lambda_k, c_k) \to x^*$. Furthermore if $\limsup_{k\to\infty} c_k = c^* < \infty$ and $\lambda_k \ne \lambda^*$ for all k, there holds

$$\limsup_{k\to\infty} \frac{|\lambda_{k+1} - \lambda^*|}{|\lambda_k - \lambda^*|} \le \max_{i=1,\ldots,m} \left| \frac{e_i}{e_i + c^*} \right|, \tag{43}$$

while if $c_k \to \infty$ and $\lambda_k \ne \lambda^*$ for all k there holds

$$\lim_{k\to\infty} \frac{|\lambda_{k+1} - \lambda^*|}{|\lambda_k - \lambda^*|} = 0. \tag{44}$$

† By this we mean that, for all k, (λ_k, c_k) belongs to the set D of (5), and hence $x(\lambda_k, c_k)$ is well defined.

Proof: Consider the matrix N_c of (39). We have

$$N_c(\lambda^*) = I - c\nabla h(x^*)'[\nabla^2_{xx} L_c(x^*, \lambda^*)]^{-1}\nabla h(x^*).$$

Using (32), we obtain

$$N_c(\lambda^*) = I - c[\nabla^2 p(0) + cI]^{-1}.$$

If $\mu_1(c), \ldots, \mu_m(c)$ are the eigenvalues of $N_c(\lambda^*)$, we have

$$\mu_i(c) = 1 - \frac{c}{e_i + c} = \frac{e_i}{e_i + c}, \qquad i = 1, \ldots, m.$$

For all (λ, c) in the set D of (5), we can rewrite (39) as

$$N_c(\lambda) = I - \nabla h[x(\lambda, c)]'\{c^{-1}\nabla^2_{xx} L_0[x(\lambda, c), \tilde{\lambda}(\lambda, c)] + \nabla h[x(\lambda, c)]\nabla h[x(\lambda, c)]'\}^{-1}\nabla h[x(\lambda, c)].$$

By using the result of Proposition 2.4 and the above expression, it is easy to see that given any $\varepsilon_1 > 0$ there exists a $\delta_1 \in (0, \delta]$ such that, for all (λ, c) with $|\lambda - \lambda^*|/c < \delta_1$, $\bar{c} \le c$, we have

$$|N_c(\lambda)| \le |N_c(\lambda^*)| + \varepsilon_1 = \max_{i=1,\ldots,m} |\mu_i(c)| + \varepsilon_1 = \max_{i=1,\ldots,m} \left|\frac{e_i}{e_i + c}\right| + \varepsilon_1.$$

Using (38), we obtain for all these pairs (λ, c)

$$|\tilde{\lambda}(\lambda, c) - \lambda^*| \le \left(\max_{i=1,\ldots,m} \left|\frac{e_i}{e_i + c}\right| + \varepsilon_1\right)|\lambda - \lambda^*|. \tag{45}$$

From (40) and (41), we have $\max_{i=1,\ldots,m} |e_i/(e_i + c)| < 1$, so by choosing ε_1 sufficiently small we have for some $\rho \in (0, 1)$ and all (λ, c) with $|\lambda - \lambda^*|/c < \delta_1$, $\bar{c} \le c$,

$$|\tilde{\lambda}(\lambda, c) - \lambda^*| \le \rho|\lambda - \lambda^*|.$$

This combined with (7) and (41) shows that $\lambda_k \to \lambda^*$ and $x(\lambda_k, c_k) \to x^*$. The rate-of-convergence estimates (43) and (44) follow from (45) and the preceding argument. Q.E.D.

The region D_1 of initial multiplier-penalty parameter pairs (λ_0, c_0) for which convergence is attained according to Proposition 2.7 is shown schematically in Fig. 2.5. It can be seen that a poor choice of λ_0 can be compensated by a choice of sufficiently high c_0. Furthermore, if for some k the algorithm generates a pair (λ_k, c_k) that lies in the shaded region of Fig. 2.5, convergence of λ_k to λ^* is guaranteed.

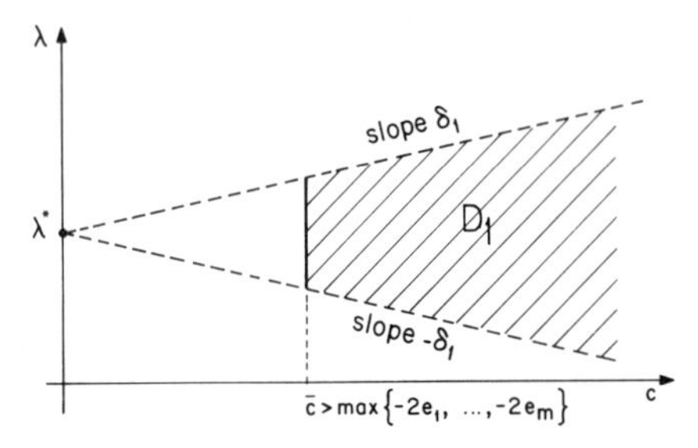

FIG. 2.5 Region of convergence of the first-order multiplier iteration

Regarding the threshold value $\bar{c}$, note that:

(a) If $\nabla^2 p(0) > 0$, which by Proposition 2.5 is equivalent to the local convexity condition

$$\nabla_{xx}^2 L_0(x^*, \lambda^*) > 0,$$

then *any positive $\bar{c}$ can serve as a threshold level.* The same is true even if $\nabla^2 p(0) \geq 0$. We shall reencounter this result in the context of convex programming problems in Chapter 5.

(b) If $\nabla^2 p(0)$ has a negative eigenvalue, then any $\bar{c}$ satisfying $\bar{c} > \max\{-e_1, \ldots, -e_m\}$ is sufficient for $\nabla_{xx}^2 L_{\bar{c}}(x^*, \lambda^*) > 0$ to hold (Proposition 2.5). However, it is necessary to take $\bar{c} > 2\max\{-e_1, \ldots, -e_m\}$ in order to induce convergence [compare with (40)]. The reason for this can be understood by examination of Fig. 2.2, where it can be seen that to achieve convergence the "penalized" primal functional p_c must have at least as much "positive curvature" as the "negative curvature" of p.

Regarding rate of convergence, we see from (43) and (44) that we have at least *Q-linear convergence if $\{c_k\}$ is bounded* and *superlinear convergence if either $\{c_k\}$ is unbounded or* $\nabla^2 p(0) = 0$. These rate-of-convergence results cannot be improved, since for any dimensions n and m, it is possible to construct a problem with a quadratic objective function and linear equality constraints and a starting point λ_0 for which, if $c_k = c^*$ for all k, relation (43) holds as an equality. The reader can verify this by first considering the scalar problems $\min\{x^2 \,|\, x = 0\}$ and $\min\{-x^2 \,|\, x = 0\}$ and then by constructing a related example for the general case. Note that *the rate of convergence improves as c_k increases.* The case where $\nabla^2 p(0) = 0$ is not as uninteresting as might appear at first sight. From (27) and (28), we have

$$\nabla_{xx}^2 L_0(x^*, \lambda^*)\nabla h(x^*) = 0 \Rightarrow \nabla^2 p(0) = 0.$$

Thus if the objective and all the constraint functions have zero curvature on the subspace

$$\{z \,|\, z = \nabla h(x^*)w, w \in R^m\},$$

which is orthogonal to the manifold tangent to the constraints at x^*, then we have $\nabla^2 p(0) = 0$ and a superlinear rate of convergence. A related situation occurs in linear programming problems, for which in fact we shall show in Chapter 5 that the corresponding method of multipliers converges in a finite number of iterations.

It should be noted that our convergence analysis is contingent upon the generation of the points $x(\lambda_k, c_k)$ by the unconstrained minimization method employed, at least for all k after a certain index. These points are, by Proposition 2.4, well defined as local minima of $L_{c_k}(\cdot, \lambda_k)$ closest to x^*. Naturally $L_{c_k}(\cdot, \lambda_k)$ may have other local minima to which the unconstrained minimization method may be attracted. Thus, unless the unconstrained minimization method stays after some index in the neighborhood of the same local minimum x^* of (ECP), our convergence analysis is not applicable. On the other hand, as extensive computational experience has shown, the usual practice of using the point x_k obtained from the kth minimization as a starting point for the $(k + 1)$th minimization tends to produce sequences $\{x_k\}$ that are close to one and the same local minimum x^* of (ECP). As a result, in the great majority of practical cases, our analysis applies and provides an accurate measure of the convergence behavior of the method of multipliers.

Dependence of Convergence Results on the Assumptions

A careful examination of the convergence results obtained reveals that they depend to a large extent on the fact that the primal functional p is twice continuously differentiable in a neighborhood of $u = 0$. This in turn is guaranteed under the sufficiency assumption (S). When this assumption is relaxed the primal functional need not be twice differentiable, and this can have a substantial effect on the convergence and rate-of-convergence properties of the method of multipliers. Some simple examples illustrate these points.

Example 1: Consider the problem $\min\{-|x|^\rho \,|\, x = 0\}$, where $1 < \rho < 2$. Then for any $c > 0$, one can find a neighborhood of $x^* = 0$ within which $L_c(\cdot, \lambda)$ does not have a local minimum for every value of λ. Here Assumption (S) is violated. The situation can be corrected by using a nonquadratic penalty function of the form $\phi(t) = |t|^{\rho'}$ or $\phi(t) = |t|^{\rho'} + \frac{1}{2}t^2$, where ρ' satisfies $1 < \rho' < \rho$ (see Chapter 5).

Example 2: Consider the problem $\min\{|x|^\rho \,|\, x = 0\}$, where $1 < \rho < 2$. Here again Assumption (S) is violated, but it will be shown in Chapter 5 that the method of multipliers converges to $\lambda^* = 0$ for any starting λ_0 and any nondecreasing penalty parameter sequence $\{c_k\}$. When $\{c_k\}$ is bounded and $\lambda_0 \neq \lambda^*$, the sequence $\{|\lambda_k - \lambda^*|\}$ converges *sublinearly*. This can be verified directly by the reader (see also the analysis of Chapter 5).

The choice of a quadratic penalty function also has a substantial effect on the convergence rate. If a different penalty function is chosen, then the convergence rate can become sublinear or superlinear. We provide some examples below. A general convergence rate theory for nonquadratic penalty functions will be provided in Chapter 5 in the context of a convex programming problem.

Example 3: Consider the scalar problem $\min\{\frac{1}{2}x^2 \,|\, x = 0\}$, with $x^* = 0$ and $\lambda^* = 0$, and the generalized method of multipliers (see Chapter 5) consisting of sequential unconstrained minimization of

$$L_{c_k}(x, \lambda_k) = \tfrac{1}{2}x^2 + \lambda_k x + (1/c_k)\phi(c_k x),$$

followed by multiplier iterations of the form

$$\lambda_{k+1} = \lambda_k + \nabla\phi(c_k x_k).$$

Here $\phi(t)$ is the penalty function, and $\nabla\phi$ is its first derivative. If $\phi(t) = \frac{1}{3}|t|^3$, $c = 1$, and $\lambda \leq 0$, the minimizing point of $L_1(x, \lambda)$ is

$$x(\lambda, 1) = \tfrac{1}{2}(-1 + \sqrt{1 - 4\lambda}).$$

For a starting point $\lambda_0 < 0$ and $c_k \equiv 1$, the multiplier iteration takes the form

$$\lambda_{k+1} = \lambda_k + [\tfrac{1}{2}(-1 + \sqrt{1 - 4\lambda_k})]^2 = \tfrac{1}{2}(1 - \sqrt{1 - 4\lambda_k}).$$

It can be verified that $\lambda_k \to \lambda^* = 0$ and $\lim_{k\to\infty} |\lambda_{k+1} - \lambda^*|/|\lambda_k - \lambda^*| = 1$. Thus, *sublinear* convergence occurs.

If instead we use the penalty function $\phi(t) = \frac{2}{3}|t|^{3/2}$, then, for $\lambda \leq 0$,

$$x(\lambda, 1) = \tfrac{1}{4}[-1 + \sqrt{1 - 4\lambda}]^2.$$

If $\lambda_0 < 0$ and $c_k \equiv 1$, the multiplier iteration takes the form

$$\lambda_{k+1} = \lambda_k + \tfrac{1}{2}(-1 + \sqrt{1 - 4\lambda_k}).$$

It can be verified that $\lambda_k \to \lambda^* = 0$ and $\lim_{k\to\infty} |\lambda_{k+1} - \lambda^*|/|\lambda_k - \lambda^*|^2 = 1$. Thus the convergence is *superlinear* of order two.

We finally point out that if we use the (nondifferentiable) penalty function $\phi(t) = |t|$, then the augmented Lagrangian $L_c(x, \lambda)$ is minimized at $x^* = 0$ for all λ and c such that $|\lambda| < c$. Thus the method is *exact* in the sense that the optimal solution x^* can be obtained by a single minimization of $L_c(x, \lambda)$. This type of method will be studied in Sections 4.1 and 5.5.

2.2.5 *Comparison with the Penalty Method—Computational Aspects*

Since Proposition 2.4 applies to both the method of multipliers and the penalty method where the iteration

$$\lambda_{k+1} = \lambda_k + c_k h(x_k) \tag{46}$$

is not employed, it provides a natural vehicle for comparison of these methods. From (7), it follows that, in the penalty method where $\lambda_k \equiv$ const, it is ordinarily necessary to increase c_k to infinity. It follows, from Proposition 2.6, that *it is not necessary to increase c_k to infinity in order to induce convergence in the method of multipliers.* This is an important advantage, since it results in elimination or at least moderation of the ill-conditioning problem. A second important advantage of the method of multipliers is that *its convergence rate is considerably better than that of the penalty method.* This can be seen by comparing the convergence rate of the two methods as given by the estimates (7), (8) and (43), (44). While in the method of multipliers, the rate of convergence is linear or superlinear, in the penalty method, the rate of convergence is much worse and essentially depends on the rate at which the penalty parameter is increased. This advantage in speed of convergence has been verified in many computational studies, where a consistent reduction in computation time ranging from 80 to 30% has been reported when the multipliers were updated via (46) over the case where λ_k was kept constant. For illustration purposes, we provide the following example, which is trivial in terms of computational complexity but nonetheless is representative of the computational savings resulting from employment of the multiplier iteration (46).

Example: Consider the two-dimensional problem

$$\text{minimize} \quad \tfrac{1}{2}[(x^1)^2 + \tfrac{1}{3}(x^2)^2]$$

$$\text{subject to} \quad x^1 + x^2 = 1.$$

The augmented Lagrangian is given by

$$L_{c_k}(x, \lambda_k) = \tfrac{1}{2}[(x^1)^2 + \tfrac{1}{3}(x^2)^2] + \lambda_k(x^1 + x^2 - 1) + \tfrac{1}{2}c_k(x^1 + x^2 - 1)^2.$$

Minimization of $L_{c_k}(\cdot, \lambda_k)$ yields

$$x_k^1 = \frac{c_k - \lambda_k}{1 + 4c_k}, \qquad x_k^2 = \frac{3(c_k - \lambda_k)}{1 + 4c_k}.$$

The optimal solution is $x^* = (0.25, 0.75)$, and the corresponding Lagrange multiplier is $\lambda^* = -0.25$. In Table 2.1, we show the results of the computation for the penalty method where $\lambda_k = 0$ for all k and for the method of multipliers, where

$$\lambda_{k+1} = \lambda_k + c_k(x_k^1 + x_k^2 - 1), \qquad \lambda_0 = 0.$$

TABLE 2.1

Computational Results with the Penalty Method and the First-Order Method of Multipliers.

	$c_k = 0.1 \times 2^k$				$c_k = 0.1 \times 4^k$				$c_k = 0.1 \times 8^k$			
	Penalty		Multiplier		Penalty		Multiplier		Penalty		Multiplier	
k	x_k^1	x_k^2	x_k^1	x_k^2	x_k^1	x_k^2	x_k^1	x_k^2	x_k^1	x_k^2	x_k^1	x_k^2
0	0.0714	0.2142	0.0714	0.2142	0.0714	0.2142	0.0714	0.2142	0.0714	0.2142	0.0714	0.2142
1	0.1111	0.3333	0.1507	0.4523	0.1538	0.4615	0.1813	0.5439	0.1904	0.5714	0.2074	0.6224
2	0.1538	0.4615	0.2118	0.6355	0.2162	0.6486	0.2407	0.7221	0.2406	0.7218	0.2824	0.7452
3	0.1904	0.5714	0.2409	0.7227	0.2406	0.7218	0.2496	0.7489	0.2487	0.7463	0.2499	0.7499
4	0.2162	0.6486	0.2487	0.7463	0.2475	0.7427	0.2499	0.7499	0.2498	0.7495		
5	0.2318	0.6956	0.2499	0.7497	0.2493	0.7481			0.2499	0.7499		
6	0.2406	0.7218	0.2499	0.7499	0.2498	0.7495						
7	0.2452	0.7356			0.2499	0.7498						
8	0.2475	0.7427			0.2499	0.7499						
9	0.2487	0.7463										
10	0.2493	0.7481										
11	0.2496	0.7490										
12	0.2498	0.7495										
13	0.2499	0.7497										
14	0.2499	0.7498										
15	0.2499	0.7499										

Notice that the method of multipliers requires a smaller number of minimizations to obtain the solution. The number of minimizations required for both methods decreases when the penalty parameter is increased at a faster rate. However, the effects of ill-conditioning are felt more under these circumstances when the unconstrained minimization is carried out numerically.

An important practical question is how one should select the initial multiplier λ_0 and the penalty parameter sequence. Clearly, in view of (8), any prior knowledge should be utilized to select λ_0 as close as possible to λ^*. The main considerations to be kept in mind for selecting the penalty parameter sequence are as follows:

(a) The parameter c_k should eventually become larger than the threshold level necessary to bring to bear the positive features of the multiplier iteration.

(b) The initial parameter c_0 should not be too large to the point that ill-conditioning results in the first unconstrained minimization.

(c) The parameter c_k is not increased too fast to the point that too much ill-conditioning is forced upon the unconstrained minimization routine too early.

(d) The parameter c_k is not increased too slowly, at least in the early minimizations, to the extent that the multiplier iteration has poor convergence rate.

These considerations are to some extent contradictory, and, in addition for nonconvex problems, it is difficult to know a priori the corresponding threshold level for the penalty parameter. A scheme that usually works well in practice is one whereby a moderate value of c_0 is chosen (if necessary by some preliminary experimentation), and subsequent values of c_k are monotonically increased via the equation $c_{k+1} = \beta c_k$, where β is a scalar with $\beta > 1$. Typical choices are $\beta \in [4, 10]$. In this way, the threshold level for multiplier convergence will eventually be exceeded.

Another reasonable parameter adjustment scheme is to increase c_k by multiplication with a factor $\beta > 1$ only if the constraint violation as measured by $|h[x(\lambda_k, c_k)]|$ is not decreased by a factor $\gamma < 1$ over the previous minimization; i.e.,

$$c_{k+1} = \begin{cases} \beta c_k & \text{if} \quad |h[x(\lambda_k, c_k)]| > \gamma |h[x(\lambda_{k-1}, c_{k-1})]|, \\ c_k & \text{if} \quad |h[x(\lambda_k, c_k)]| \leq \gamma |h[x(\lambda_{k-1}, c_{k-1})]|. \end{cases} \tag{47}$$

Choices such as $\beta = 10$ and $\gamma = 0.25$ are typically recommended. Assuming that $\{\lambda_k\}$ remains bounded, one can prove, for this scheme, that *the penalty parameter sequence* $\{c_k\}$ *will remain bounded*. To see this, suppose that $\{\lambda_k\}$ remains bounded and c_k becomes unbounded. Then after some $\bar{k}$, the parameter c_k will be sufficiently high for the estimates (7) and (8) of Proposition

2.4 to become effective and in addition $c_{k-1} > M$. Let L be a Lipschitz constant for h. We have, from (7) and (8) for all $k \geq \bar{k}$,

$$|h[x(\lambda_k, c_k)]| \leq L|x(\lambda_k, c_k) - x^*| \leq LM|\lambda_k - \lambda^*|/c_k, \tag{48}$$

$$|h[x(\lambda_{k-1}, c_{k-1})]| = \frac{|\lambda_k - \lambda_{k-1}|}{c_{k-1}} \geq \frac{|\lambda_{k-1} - \lambda^*|}{c_{k-1}} - \frac{|\lambda_k - \lambda^*|}{c_{k-1}} \tag{49}$$

$$\geq (M^{-1} - c_{k-1}^{-1})|\lambda_k - \lambda^*|.$$

Combining (48) and (49), we obtain

$$|h[x(\lambda_k, c_k)]| \leq \frac{LM}{c_k(M^{-1} - c_{k-1}^{-1})}|h[x(\lambda_{k-1}, c_{k-1})]|. \tag{50}$$

From (47) and (50), it follows that the assumption that $\{c_k\}$ is unbounded implies $c_{k+1} = c_k$, for all k sufficiently large, which is a contradiction. Hence, $\{c_k\}$ will remain bounded if the penalty parameter adjustment scheme (47) is adopted, while convergence will be achieved by virtue of enforcement of asymptotic feasibility of the constraints; i.e.,

$$\lim_{k\to\infty} |h[x(\lambda_k, c_k)]| = 0.$$

Another possibility along the same lines is to use a different penalty parameter for each constraint $h_i(x) = 0$, and to increase by a certain factor only the penalty parameters which correspond to those constraint equations for which the constraint violation as measured by $|h_i[x(\lambda_k, c_k)]|$ is not decreased by a certain factor over the previous minimization. It is to be noted that the convergence analysis given earlier can be easily modified to handle the case where a separate penalty parameter is used for each constraint.

As an example of a situation where using a different penalty parameter for each constraint can be beneficial, consider a problem with "poorly scaled" constraints such as

$$\text{minimize} \quad \tfrac{1}{2}[(x^1)^2 + (x^2)^2 + (x^3)^2]$$
$$\text{subject to} \quad x^2 = 0, \qquad 10^5 x^3 = 0.$$

We have

$$L_c(x, \lambda) = \tfrac{1}{2}[(x^1)^2 + (x^2)^2 + (x^3)^2] + \lambda^1 x^2$$
$$+ 10^5 \lambda^2 x^3 + \tfrac{1}{2}c(x^2)^2 + \tfrac{1}{2}10^{10}c(x^3)^2.$$

Clearly, minimization of $L_c(\cdot, \lambda)$ is an ill-conditioned problem. A scheme that allows a different penalty parameter for each constraint and adjusts these parameters depending on the progress made towards satisfying these constraints can partially compensate for poor scaling. As an additional measure, one can multiply initially the constraints with scaling factors that make the norms of their gradients at the starting point equal to unity. This is generally recommended as a good heuristic (but not fail-safe) technique.

We finally note that the overall efficiency of the method strongly depends on the initial choice of the multiplier λ_0. As suggested by the convergence analysis given thus far, a choice of λ_0 close to λ^* can reduce dramatically the computational requirements of the method.

2.3 Duality Framework for the Method of Multipliers

Let $\bar{c}$, δ, and ε be as in Proposition 2.4, and define for (λ, c) in the set

$$D = \{(\lambda, c) \mid |\lambda - \lambda^*| < \delta c, \bar{c} \le c\} \tag{1}$$

the *dual functional* d_c given by

$$d_c(\lambda) = \min_{x \in S(x^*; \varepsilon)} L_c(x, \lambda) = f[x(\lambda, c)] + \lambda' h[x(\lambda, c)] + \tfrac{1}{2}c|h[x(\lambda, c)]|^2. \tag{2}$$

Since $x(\cdot, c)$ is continuously differentiable (Proposition 2.4), the same is true for d_c. We compute the gradient of d_c with respect to λ. We have

$$\begin{aligned}\nabla d_c(\lambda) &= \nabla_\lambda x(\lambda, c)\{\nabla f[x(\lambda, c)] + \nabla h[x(\lambda, c)]\lambda + c\nabla h[x(\lambda, c)]h[x(\lambda, c)]\} \\ &\quad + h[x(\lambda, c)] \\ &= \nabla_\lambda x(\lambda, c)\nabla_x L_c[x(\lambda, c), \lambda] + h[x(\lambda, c)].\end{aligned}$$

Since $\nabla_x L_c[x(\lambda, c), \lambda] = 0$, we obtain

$$\nabla d_c(\lambda) = h[x(\lambda, c)]. \tag{3}$$

Since $x(\cdot, c)$ is continuously differentiable, the same is true for ∇d_c. Differentiating with respect to λ, we obtain

$$\nabla^2 d_c(\lambda) = \nabla_\lambda x(\lambda, c)\nabla h[x(\lambda, c)]. \tag{4}$$

We also have, for all (λ, c) in the set D,

$$\nabla_x L_c[x(\lambda, c), \lambda] = 0.$$

Differentiating with respect to λ, we obtain

$$\nabla_\lambda x(\lambda, c)\nabla^2_{xx} L_c[x(\lambda, c), \lambda] + \nabla^2_{\lambda x} L_c[x(\lambda, c), \lambda] = 0,$$

and since

$$\nabla^2_{\lambda x} L_c[x(\lambda, c), \lambda] = \nabla h[x(\lambda, c)]',$$

we obtain

$$\nabla_\lambda x(\lambda, c) = -\nabla h[x(\lambda, c)]'\{\nabla_{xx}^2 L_c[x(\lambda, c), \lambda]\}^{-1}.$$

Substitution in (4) yields the formula

$$\text{(5)} \qquad \nabla^2 d_c(\lambda) = -\nabla h[x(\lambda, c)]'\{\nabla_{xx}^2 L_c[x(\lambda, c), \lambda]\}^{-1}\nabla h[x(\lambda, c)].$$

Since $\nabla_{xx}^2 L_c[x(\lambda, c), \lambda] > 0$ and $\nabla h[x(\lambda, c)]$ has rank m for $(\lambda, c) \in D$ (Proposition 2.4), it follows from (5) that $\nabla^2 d_c(\lambda) < 0$ for all $(\lambda, c) \in D$. Furthermore using (3), we have, for all $c \geq \bar{c}$,

$$\nabla d_c(\lambda^*) = h[x(\lambda^*, c)] = h(x^*) = 0.$$

Thus, *for every* $c \geq \bar{c}$, λ^* *maximizes* $d_c(\lambda)$ *over* $\{\lambda \mid |\lambda - \lambda^*| < \delta c\}$. Also the multiplier iteration in view of (3) can be written as

$$\text{(6)} \qquad \lambda_{k+1} = \lambda_k + c_k \nabla d_{c_k}(\lambda_k)$$

and *represents a steepest ascent iteration for maximizing* d_{c_k}. When $c_k = c$ for all k, then (6) is the constant stepsize steepest ascent method

$$\text{(7)} \qquad \lambda_{k+1} = \lambda_k + c\nabla d_c(\lambda_k)$$

for maximizing d_c and is of the type discussed in Section 1.3.1.

2.3.1 Stepsize Analysis for the Method of Multipliers

As discussed following Proposition 1.10, the choice of stepsize in the steepest ascent method is crucial both in terms of convergence and rate of convergence. It is a rather remarkable fact that the particular stepsize c used in iteration (7) works so well. Nonetheless, it is of interest to try to compare the stepsize c with other possible stepsizes and investigate whether there exists an optimal stepsize. In order to simplify the analysis, we restrict ourselves to the case where f is a (not necessarily positive definite) quadratic function and h is a linear function, i.e., the problem

$$\begin{aligned} &\text{minimize} \quad \tfrac{1}{2}x'Qx \\ &\text{subject to} \quad Ax = b, \end{aligned}$$

where Q is a symmetric $n \times n$ matrix, A is an $m \times n$ matrix of rank m, and $b \in R^m$ is a given vector. It is assumed that this problem has a unique minimum x^*, which together with a Lagrange multiplier λ^* satisfies Assumption (S). It is a routine matter to extend the analysis to the general case under Assumption (S). If f^* is the optimal value of the problem, the function d_c is quadratic of the form [compare with (5)]

$$\text{(8)} \qquad d_c(\lambda) = -\tfrac{1}{2}(\lambda - \lambda^*)'A(Q + cA'A)^{-1}A'(\lambda - \lambda^*) + f^*$$

for all c for which $Q + cA'A > 0$. Consider, for $\lambda_k \neq \lambda^*$ and $\alpha > 0$, the iteration

$$\lambda_{k+1} = \lambda_k + \alpha \nabla d_c(\lambda_k). \tag{9}$$

By Proposition 1.14, we have

$$|\lambda_{k+1} - \lambda^*|/|\lambda_k - \lambda^*| \leq r(\alpha), \tag{10}$$

where

$$r(\alpha) = \max\{|1 - \alpha E_c|, |1 - \alpha e_c|\}, \tag{11}$$

and E_c and e_c are the maximum and minimum eigenvalues of

$$A(Q + c\, A'A)^{-1}A'.$$

Convergence occurs for

$$0 < \alpha < 2/E_c. \tag{12}$$

The optimal convergence ratio is attained for the stepsize α^*, minimizing $r(\alpha)$ over α,

$$\alpha^* = 2/(E_c + e_c), \tag{13}$$

and is given by

$$r(\alpha^*) = (E_c - e_c)/(E_c + e_c).$$

Let us assume that Q is invertible. Then, by using the matrix identity given in Section 1.2, we have

$$(I + cAQ^{-1}A')^{-1} = I - cA(Q + cA'A)^{-1}A'.$$

Thus the eigenvalues of $(AQ^{-1}A')^{-1}$ and $A(Q + cA'A)^{-1}A'$ are related by

$$(1 + c/e_i[(AQ^{-1}A')^{-1}])^{-1} = 1 - ce_i[A(Q + cA'A)^{-1}A'].$$

Let γ and Γ denote the eigenvalues of $(A'Q^{-1}A)^{-1}$ corresponding to E_c and e_c via the relation above. We have, via a straightforward calculation,

$$E_c = \frac{1}{\gamma + c}, \qquad e_c = \frac{1}{\Gamma + c}. \tag{14}$$

Note, from Eq. (33) of Section 2.2.2, that the Hessian of the primal functional is given by

$$\nabla^2 p(0) = (AQ^{-1}A')^{-1},$$

and *γ and Γ are the smallest and largest eigenvalues of* $\nabla^2 p(0)$, *respectively*.

In view of (10), (11), (12), and (14), we have that convergence occurs for

$$0 < \alpha < 2(\gamma + c), \tag{15}$$

and the convergence ratio is

$$r(\alpha) = \max\left\{\left|1 - \frac{\alpha}{\gamma + c}\right|, \left|1 - \frac{\alpha}{\Gamma + c}\right|\right\}. \tag{16}$$

In particular for $\alpha = c$, we obtain that convergence occurs if $-2\gamma < c$ and

$$r(c) = \max\left\{\left|\frac{\gamma}{\gamma + c}\right|, \left|\frac{\Gamma}{\Gamma + c}\right|\right\}, \tag{17}$$

as already derived in Proposition 2.6. Using (13), (14), and (16) we obtain, via a straightforward calculation, the expressions for the optimal stepsize α^* and the corresponding convergence ratio

$$\alpha^* = \frac{2(\Gamma + c)(\gamma + c)}{\Gamma + \gamma + 2c}, \tag{18}$$

$$r(\alpha^*) = \frac{\Gamma - \gamma}{\Gamma + \gamma + 2c}. \tag{19}$$

We note that, from (18), one can verify that we have

$$\lim_{c \to \infty} \frac{|\alpha^* - c|}{c} = 0$$

thereby implying that as c increases the ratio α^*/c tends to unity. We now distinguish two cases of interest.

CASE (*a*) ($\gamma < 0 < \Gamma$): Here we assume that $\nabla^2 p(0)$ is neither positive semidefinite nor negative semidefinite. It can be seen that we must have $-\gamma < c$ in order to guarantee $\nabla_{xx}^2 L_c(x^*, \lambda^*) > 0$ (Proposition 2.5), in which case, from (15), we see that there exist some stepsizes α which achieve convergence. However, the particular stepsize $\alpha = c$ guarantees convergence only if $-2\gamma < c$ (compare with Proposition 2.6). For values of c close to -2γ, Eq. (17) shows that the convergence ratio $r(c)$ is poor (close to one). However, as c increases, not only does the convergence ratio $r(c)$ improve but also the ratio $r(c)/r(\alpha^*)$ decreases, and in fact, from (17) and (19), we have, via an easy calculation,

$$\lim_{c \to \infty} \frac{r(c)}{r(\alpha^*)} = \max\left\{\frac{2|\gamma|}{\Gamma - \gamma}, \frac{2|\Gamma|}{\Gamma - \gamma}\right\} < 2.$$

Thus, in the case $\gamma < 0 < \Gamma$ for large values of c, we have that $r(c)$ is close to being optimal and can be improved only by a factor of at most 2 by optimal stepsize choice.

CASE (*b*) ($\gamma \le \Gamma \le 0$ *or* $0 \le \gamma \le \Gamma$): Here $\nabla^2 p(0)$ is either positive semidefinite or negative semidefinite. The case where $0 < \gamma \le \Gamma$ can often be easily recognized in practice since it corresponds to a convex programming problem. We have, from (17) and (19), that if $\Gamma \neq \gamma$

$$\lim_{c \to \infty} \frac{r(c)}{r(\alpha^*)} = \max\left\{\frac{2|\gamma|}{\Gamma - \gamma}, \frac{2|\Gamma|}{\Gamma - \gamma}\right\} \geq 2,$$

while if $\Gamma = \gamma$ we have $r(\alpha^*) = 0$. Thus, in this case, there is considerable room for improvement by alternative stepsize choice. The potential benefits become greater as the ratio γ/Γ approaches unity.

It is, of course, possible to compute at least approximately the eigenvalues of $\nabla^2 p(0)$, thereby obtaining an approximate value of the optimal stepsize. This can be done by computing the eigenvalues of

$$\nabla h[x(\lambda_k, c_k)]'\{\nabla_{xx}^2 L_{c_k}[x(\lambda_k, c_k), \lambda_k]\}^{-1} \nabla h[x(\lambda_k, c_k)]$$

and by using Eq. (31) of Section 2.2. Such an approach probably would not be worthwhile for most problems, but, in some problems which are solved repetitively with slightly varying data, it may be profitable to compute approximately these eigenvalues at least once and obtain approximate values of γ and Γ for use in the optimal stepsize formula (18).

Another possibility is based on the fact that, from (18), we obtain

$$\alpha^* = \frac{2(\Gamma + c)(\gamma + c)}{\Gamma + \gamma + 2c} = 2c\left[1 - \frac{c}{\Gamma + \gamma + 2c} + \frac{\Gamma\gamma}{c(\Gamma + \gamma + 2c)}\right],$$

and an underestimate of α^* is given by

$$\alpha^* \geq 2c[1 - c/(\Gamma + \gamma + 2c)].$$

This underestimate is quite accurate for large c. For problems that are solved repetitively with slightly varying data one can use the stepsize formula

$$\alpha_k = 2c_k[1 - c_k/(\mu + 2c_k)], \tag{20}$$

where μ is a parameter approximating $(\Gamma + \gamma)$ and determined by experimentation with a few trial runs. When $\mu = 0$, (20) yields $\alpha_k = c_k$. For c_k very large, we have $\alpha_k \simeq c_k$. For problems with convex structure where $\nabla^2 p(0) \geq 0$, we shall have $\mu \geq 0$, so the stepsize α_k of (20) satisfies

$$c_k \leq \alpha_k < 2c_k.$$

Such a stepsize lies within the interval of convergence [compare with (15)], so the corresponding method

$$\lambda_{k+1} = \lambda_k + \alpha_k h(x_k),$$

with α_k given by (20), is guaranteed to converge in the quadratic case. In fact, one may show that this is also true in the general case under Assumption

(S). We leave the proof of this fact as an exercise for the reader (see also Proposition 5.13). For problems with $\nabla^2 p(0) \le 0$, we shall have $\mu \le 0$. In order to avoid a negative α_k from (20), it is advisable to select μ so that α_k as given by (20) satisfies

$$\rho c_k \le \alpha_k \le c_k \tag{21}$$

where ρ is some small scalar with $0 < \rho < 1$. For c_k greater than the threshold level -2γ, we see that α_k lies within the interval of convergence [compare with (15)], and the resulting method is guaranteed to converge. Unfortunately it is not ordinarily easy to determine whether the condition $\nabla^2 p(0) \le 0$ holds in a given problem.

Example: Consider the three-dimensional problem

$$\text{minimize} \quad \tfrac{1}{2}\{(x^2 + x^3)^2 + (x^1 + x^3)^2 + (x^1 + x^2)^2\}$$

$$\text{subject to} \quad x^1 + x^2 + 2x^3 = 2, \qquad x^1 - x^2 = 0.$$

The optimal solution is $x^* = (0, 0, 1)$ and the corresponding Lagrange multiplier is $\lambda^* = (-1, 0)$. The optimal value is equal to unity. We show, in Table 2.2, the sequences $\{d_{c_k}(\lambda_k)\}$ generated via the iteration

$$\lambda_{k+1} = \lambda_k + \alpha_k h(x_k),$$

where α_k is given by the stepsize rule (20) for various values of μ. The starting point in all runs was $\lambda_0 = (10, -5)$. It can be seen that a value of μ between 1 and 5 improves considerably the rate of convergence over the standard stepsize ($\mu = 0$ and $\alpha_k = c_k$).

There is another way to improve the convergence rate of the method of multipliers by alternative stepsize choice when either $\nabla^2 p(0) > 0$ or $\nabla^2 p(0) < 0$. We proceed to describe it briefly for the case of the general problem (ECP) under Assumption (S). Consider the system of equations in (x, λ)

$$\nabla f(x) + \nabla h(x)\lambda = 0, \qquad h(x) = 0.$$

Using the implicit function theorem, it follows that there exists a continuously differentiable function $x(\lambda)$ defined in a neighborhood $N(\lambda^*)$ of λ^* such that $x(\lambda^*) = x^*$ and

$$\nabla f[x(\lambda)] + \nabla h[x(\lambda)]\lambda = 0, \qquad h[x(\lambda)] = 0 \qquad \forall\, \lambda \in N(\lambda^*).$$

Define

$$d(\lambda) = f[x(\lambda)] + \lambda' h[x(\lambda)] \qquad \forall\, \lambda \in N(\lambda^*).$$

TABLE 2.2

Performance of Stepsize Rule (20) for Various Values of μ

	$c_k = 0.1 \times 2^k$					$c_k = 0.1 \times 4^k$				
k	$\mu = 0$	$\mu = 1$	$\mu = 2.5$	$\mu = 5$	$\mu = 25$	$\mu = 0$	$\mu = 1$	$\mu = 2.5$	$\mu = 5$	$\mu = 25$
0	−120.6	−120.6	−120.6	−120.6	−120.6	−120.6	−120.6	−120.6	−120.6	−120.6
1	−71.42	−49.29	−47.08	−46.24	−45.53	−55.32	−38.11	−36.39	−35.74	−35.19
2	−27.73	−9.182	−7.190	−6.432	−5.787	−6.450	−0.5969	−0.2323	0.5318	0.7580
3	−5.140	0.3284	0.7284	0.8470	0.9266	0.8714	0.9984	0.9978	0.9926	0.9859
4	0.4376	0.9909	0.9999	0.9993	0.9982	0.9998	0.9999	0.9999	0.9999	0.9989
5	0.9819	0.9999		0.9999	0.9998	0.9999				0.9999
6	0.9998				0.9999					
7	0.9999									

	$c_k = 1 \times 2^k$					$c_k = 1 \times 4^k$				
k	$\mu = 0$	$\mu = 1$	$\mu = 2.5$	$\mu = 5$	$\mu = 25$	$\mu = 0$	$\mu = 1$	$\mu = 2.5$	$\mu = 5$	$\mu = 25$
0	−47.66	−47.66	−47.66	−47.66	−47.66	−47.66	−47.66	−47.66	−47.66	−47.66
1	−2.244	0.6395	0.9599	0.4040	−1.354	−0.8024	0.7997	0.9777	0.6689	−0.3079
2	0.9279	0.9996	0.9997	0.9802	0.6863	0.9939	0.9999	0.9999	0.9951	0.8872
3	0.9995	0.9999	0.9999	0.9994	0.9475	0.9999			0.9999	0.9954
4	0.9999			0.9999	0.9928					0.9999
5					0.9994					

A calculation analogous to the one given in this section following (2) shows that

$$\nabla d(\lambda) = h[x(\lambda)],$$
$$\nabla^2 d(\lambda) = -\nabla h[x(\lambda)]'\{\nabla_{xx}^2 L_0[x(\lambda), \lambda]\}^{-1}\nabla h[x(\lambda)],$$

where we assume that $\nabla_{xx}^2 L_0(x^*, \lambda^*)$ is invertible and $N(\lambda^*)$ is chosen so that $\{\nabla_{xx}^2 L_0[x(\lambda), \lambda]\}^{-1}$ exists for all $\lambda \in N(\lambda^*)$. We have

$$\nabla d(\lambda^*) = h(x^*) = 0.$$

From (33) of Section 2.2, we have that if $\nabla^2 p(0) > 0$ then $\nabla^2 d(\lambda^*) < 0$ while if $\nabla^2 p(0) < 0$ then $\nabla^2 d(\lambda^*) > 0$. Thus, if $\nabla^2 p(0) > 0$, we have that λ^* maximizes d, while if $\nabla^2 p(0) < 0$, we have that λ^* minimizes d. We mention that in the case where $\nabla^2 p(0) > 0$, the function d is the ordinary dual functional in the local duality framework of Luenberger (1973). Now when operating the method of multipliers, each time we obtain a vector $x(\lambda_k, c_k)$, we have

$$\nabla f[x(\lambda_k, c_k)] + \nabla h[x(\lambda_k, c_k)]\tilde{\lambda}_k = 0,$$

where

$$\tilde{\lambda}_k = \lambda_k + c_k h[x(\lambda_k, c_k)].$$

This means that if $\tilde{\lambda}_k$ is sufficiently close to λ^*, we have

$$\nabla d(\tilde{\lambda}_k) = h[x(\lambda_k, c_k)],$$
$$d(\tilde{\lambda}_k) = f[x(\lambda_k, c_k)] + \tilde{\lambda}_k' h[x(\lambda_k, c_k)].$$

This information on gradients and values of d can be utilized to determine a stepsize for the multiplier iteration by interpolation aimed at maximizing or minimizing d depending on whether $\nabla^2 p(0) > 0$ or $\nabla^2 p(0) < 0$, respectively. It is necessary to carry out the interpolation every second iteration so as to collect sufficient data in the intermediate iteration. We describe the typical step of this procedure.

Given λ_{2k} and c_{2k}, $k = 0, 1, \ldots$, we obtain x_{2k} and $h(x_{2k})$ by unconstrained minimization of the augmented Lagrangian, and we set

$$\lambda_{2k+1} = \lambda_{2k} + c_{2k} h(x_{2k}).$$

Similarly, we obtain x_{2k+1} and $h(x_{2k+1})$ by means of unconstrained minimization of the augmented Lagrangian. However, we now set

$$\lambda_{2k+2} = \lambda_{2k+1} + \alpha_{2k+1} h(x_{2k+1}),$$

where

$$\alpha_{2k+1} = c_{2k+1} \frac{h(x_{2k+1})'h(x_{2k})}{h(x_{2k+1})'h(x_{2k}) - |h(x_{2k+1})|^2} \tag{22}$$

This choice of stepsize α_{2k+1} is based on quadratic interpolation of $d[\lambda_{2k+1} + \alpha h(x_{2k+1})]$ based on

$$\nabla d(\lambda_{2k+1}) = h(x_{2k}), \qquad \nabla d[\lambda_{2k+1} + c_{2k+1}h(x_{2k+1})] = h(x_{2k+1}).$$

The stepsize α_{2k+1}, given by (22), is the one for which the derivative of the interpolating polynomial is zero. In the case where $\nabla^2 p(0) > 0$ $[\nabla^2 p(0) < 0]$, it may be a good idea to restrict α_{2k+1} to be less or equal to $2c_{2k+1}$ $[c_{2k+1}]$. This will guarantee convergence [compare with (15)].

It is possible to show that this stepsize procedure improves the convergence rate of the method of multipliers (see Bertsekas, 1975c), although in most cases the improvement is not spectacular. Some related computational results can be found in Bertsekas (1975a,c). On the other hand, the added computational overhead of the procedure is negligible, and it may be worth trying as a means of accelerating the rate of convergence of the basic method when Newton-type iterations to be examined in the next section are inappropriate.

2.3.2 *The Second-Order Multiplier Iteration*

In view of the interpretation of the multiplier iteration as a steepest ascent method, it is natural to consider Newton's method for maximizing the dual functional d_c which is given by

$$\lambda_{k+1} = \lambda_k - [\nabla^2 d_{c_k}(\lambda_k)]^{-1} \nabla d_{c_k}(\lambda_k). \tag{23}$$

In view of (3) and (5), this iteration can be written as

$$\lambda_{k+1} = \lambda_k + B_k^{-1} h[x(\lambda_k, c_k)], \tag{24}$$

where

$$B_k = \nabla h[x(\lambda_k, c_k)]' \{\nabla_{xx}^2 L_{c_k}[x(\lambda_k, c_k), \lambda_k]\}^{-1} \nabla h[x(\lambda_k, c_k)]. \tag{25}$$

We shall provide a convergence and rate of convergence result for iteration (24), (25). To this end, we consider Newton's method for solving, for $c \in R$, the system of necessary conditions

$$\nabla_x L_c(x, \lambda) = \nabla f(x) + \nabla h(x)[\lambda + ch(x)] = 0, \qquad h(x) = 0. \tag{26}$$

In this method, given the current iterate, say (x, λ), one obtains the next iterate $(\hat{x}, \hat{\lambda})$ as the solution of the linear system of equations (compare with Proposition 1.17)

$$\begin{bmatrix} \nabla_{xx}^2 L_c(x, \lambda) & \nabla h(x) \\ \nabla h(x)' & 0 \end{bmatrix} \begin{bmatrix} \hat{x} - x \\ \hat{\lambda} - \lambda \end{bmatrix} = - \begin{bmatrix} \nabla_x L_c(x, \lambda) \\ h(x) \end{bmatrix}. \tag{27}$$

If $\nabla^2_{xx} L_c(x, \lambda)$ is invertible and $\nabla h(x)$ has rank m, we can solve system (27) explicitly. We first write (27) as

$$\nabla^2_{xx} L_c(x, \lambda)(\hat{x} - x) + \nabla h(x)(\hat{\lambda} - \lambda) = -\nabla_x L_c(x, \lambda), \tag{28}$$

$$\nabla h(x)'(\hat{x} - x) = -h(x). \tag{29}$$

Premultiplying (28) with $\nabla h(x)'[\nabla^2_{xx} L_c(x, \lambda)]^{-1}$ and using (29), we obtain

$$\begin{aligned} -h(x) + \nabla h(x)'[\nabla^2_{xx} L_c(x, \lambda)]^{-1}\nabla h(x)(\hat{\lambda} - \lambda) \\ = -\nabla h(x)'[\nabla^2_{xx} L_c(x, \lambda)]^{-1}\nabla_x L_c(x, \lambda) \end{aligned}$$

from which

$$\begin{aligned} \hat{\lambda} = \lambda + \{\nabla h(x)'[\nabla^2_{xx} L_c(x, \lambda)]^{-1}\nabla h(x)\}^{-1} \\ \times [h(x) - \nabla h(x)'[\nabla^2_{xx} L_c(x, \lambda)]^{-1}\nabla_x L_c(x, \lambda)]. \end{aligned} \tag{30}$$

Substitution in (28) yields

$$\hat{x} = x - [\nabla^2_{xx} L_c(x, \lambda)]^{-1}\nabla_x L_c(x, \hat{\lambda}). \tag{31}$$

Returning to (24), (25) and using the fact that $\nabla_x L_c[x(\lambda, c), \lambda] = 0$, we see that *iteration* (24), (25) *is of the form* (30).

For a triple (x, λ, c) for which the matrix on the left-hand side of (27) is invertible, we denote by $\hat{x}(x, \lambda, c)$, $\hat{\lambda}(x, \lambda, c)$ the unique solution of (27) and say that $\hat{x}(x, \lambda, c)$, $\hat{\lambda}(x, \lambda, c)$ are *well defined*. Thus (24), (25) is written

$$\lambda_{k+1} = \hat{\lambda}[x(\lambda_k, c_k), \lambda_k, c_k]. \tag{32}$$

Proposition 2.8: Let c be a scalar. For every triple (x, λ, c), the vectors $\hat{x}(x, \lambda, c)$, $\hat{\lambda}(x, \lambda, c)$ are well defined if and only if the vectors $\hat{x}[x, \lambda + ch(x), 0]$, $\hat{\lambda}[x, \lambda + ch(x), 0]$ are well defined. Furthermore

$$\hat{x}(x, \lambda, c) = \hat{x}[x, \lambda + ch(x), 0], \tag{33}$$

$$\hat{\lambda}(x, \lambda, c) = \hat{\lambda}[x, \lambda + ch(x), 0]. \tag{34}$$

Proof: We have

$$\nabla^2_{xx} L_c(x, \lambda) = \nabla^2_{xx} L_0[x, \lambda + ch(x)] + c\nabla h(x)\nabla h(x)',$$

$$\nabla_x L_c(x, \lambda) = \nabla_x L_0[x, \lambda + ch(x)].$$

As a result, the system (27) can be written as

$$\begin{bmatrix} \nabla^2_{xx} L_0[x, \lambda + ch(x)] + c\nabla h(x)\nabla h(x)' & \nabla h(x) \\ \nabla h(x)' & 0 \end{bmatrix} \begin{bmatrix} \hat{x} - x \\ \hat{\lambda} - \lambda \end{bmatrix} = -\begin{bmatrix} \nabla_x L_0[x, \lambda + ch(x)] \\ h(x) \end{bmatrix}. \tag{35}$$

The second equation yields $\nabla h(x)'(\hat{x} - x) = -h(x)$, which, when substituted in the first equation, yields

$$\nabla_{xx}^2 L_0[x, \lambda + ch(x)](\hat{x} - x) - c\nabla h(x)h(x) + \nabla h(x)(\hat{\lambda} - \lambda) = -\nabla_x L_0[x, \lambda + ch(x)].$$

Thus, system (35) is equivalent to

(36)
$$\begin{bmatrix} \nabla_{xx}^2 L_0[x, \lambda + ch(x)] & \nabla h(x) \\ \nabla h(x)' & 0 \end{bmatrix} \begin{bmatrix} \hat{x} - x \\ \hat{\lambda} - \lambda - ch(x) \end{bmatrix} = -\begin{bmatrix} \nabla_x L_0[x, \lambda + ch(x)] \\ h(x) \end{bmatrix}.$$

This shows (33) and (34). Q.E.D.

In view of (34), we can write (32) [or equivalently (24), (25)] as

$$\lambda_{k+1} = \hat{\lambda}[x(\lambda_k, c_k), \tilde{\lambda}(\lambda_k, c_k), 0], \tag{37}$$

where

$$\tilde{\lambda}(\lambda_k, c_k) = \lambda_k + c_k h[x(\lambda_k, c_k)]. \tag{38}$$

This means that one can carry out the second-order multiplier iteration (24), (25) in two stages. *First execute the first-order iteration* (38) *and then the second-order iteration* (37), *which is part of Newton's iteration at* $[x(\lambda_k, c_k), \tilde{\lambda}(\lambda_k, c_k)]$ *for solving the system of necessary conditions* $\nabla f(x) + \nabla h(x)\lambda = 0$, $h(x) = 0$. Now, we know that $[x(\lambda_k, c_k), \tilde{\lambda}(\lambda_k, c_k)]$ is close to (x^*, λ^*) for (λ_k, c_k) in an appropriate region of R^{m+1} (Proposition 2.4). Therefore, using known results for Newton's method, we expect that (37) will yield a vector λ_{k+1} which is closer to λ^* than λ_k. This argument is the basis for the proof of the following two propositions.

Proposition 2.9: Assume (S) holds, and let $\bar{c}$ and δ be as in Proposition 2.4. Then, given any scalar $\gamma > 0$, there exists a scalar δ_2 with $0 < \delta_2 \leq \delta$ such that for all (λ, c) in the set D_2 defined by

$$D_2 = \{(\lambda, c) \,|\, |\lambda - \lambda^*| < \delta_2 c, \bar{c} \leq c\}, \tag{39}$$

there holds

$$|\hat{\lambda}(\lambda, c) - \lambda^*| \leq \gamma |\lambda - \lambda^*|/c, \tag{40}$$

where

$$\hat{\lambda}(\lambda, c) = \lambda + B_c(\lambda)^{-1} h[x(\lambda, c)], \tag{41}$$

$$B_c(\lambda) = \nabla h[x(\lambda, c)]'\{\nabla_{xx}^2 L_c[x(\lambda, c), \lambda]\}^{-1} \nabla h[x(\lambda, c)]. \tag{42}$$

If, in addition, $\nabla^2 f$ and $\nabla^2 h_i$, $i = 1, \ldots, m$, are Lipschitz continuous in a neighborhood of x^*, there exists a scalar M_2 such that, for all $(\lambda, c) \in D_2$, there holds

$$|\hat{\lambda}(\lambda, c) - \lambda^*| \leq M_2 |\lambda - \lambda^*|^2/c^2. \tag{43}$$

Proof: Let M be as in Proposition 2.4. Given any $\gamma > 0$, there exists an $\varepsilon > 0$ such that if $x(\lambda, c) \in S(x^*; \varepsilon)$ and $\tilde{\lambda}(\lambda, c) \in S(\lambda^*; \varepsilon)$, there holds

$$\begin{aligned} &|(\hat{x}[x(\lambda, c), \tilde{\lambda}(\lambda, c), 0], \hat{\lambda}[x(\lambda, c), \tilde{\lambda}(\lambda, c), 0]) - (x^*, \lambda^*)| \\ &\quad \leq (\gamma/\sqrt{2}M)|[x(\lambda, c), \tilde{\lambda}(\lambda, c)] - (x^*, \lambda^*)| \end{aligned} \tag{44}$$

(compare with Proposition 1.17). Take δ_2 sufficiently small so that, for $(x, \lambda) \in D_2$, we have $x(\lambda, c) \in S(x^*; \varepsilon)$ and $\tilde{\lambda}(\lambda, c) \in S(\lambda^*; \varepsilon)$ (compare with Proposition 2.4). Using (7) and (8) in Section 2.2, (44), and the fact

$$\hat{\lambda}(\lambda, c) = \hat{\lambda}[x(\lambda, c), \tilde{\lambda}(\lambda, c), 0],$$

we obtain

$$|\hat{\lambda}(\lambda, c) - \lambda^*| \leq \frac{\gamma}{\sqrt{2}M} \left(\frac{M^2 |\lambda - \lambda^*|^2}{c^2} + \frac{M^2 |\lambda - \lambda^*|^2}{c^2} \right)^{1/2} = \frac{\gamma |\lambda - \lambda^*|}{c},$$

and (40) is proved.

If $\nabla^2 f$ and $\nabla^2 h_i$ are Lipschitz continuous, then there exists an M_2 such that for $x(\lambda, c) \in S(x^*; \varepsilon)$ and $\tilde{\lambda}(\lambda, c) \in S(x^*; \varepsilon)$, we have

$$\begin{aligned} &|(\hat{x}[x(\lambda, c), \tilde{\lambda}(\lambda, c), 0], \hat{\lambda}[x(\lambda, c), \tilde{\lambda}(\lambda, c), 0]) - (x^*, \lambda^*)| \\ &\quad \leq (M_2/2M^2)|[x(\lambda, c), \tilde{\lambda}(\lambda, c)] - (x^*, \lambda^*)|^2. \end{aligned} \tag{45}$$

Using again (7) and (8) in Section 2.2, and (45), we obtain

$$|\hat{\lambda}(\lambda, c) - \lambda^*| \leq \frac{M_2}{2M^2} \left(\frac{M^2 |\lambda - \lambda^*|^2}{c^2} + \frac{M^2 |\lambda - \lambda^*|^2}{c^2} \right) = \frac{M_2 |\lambda - \lambda^*|^2}{c^2},$$

and (43) is proved. Q.E.D.

An almost immediate consequence of Proposition 2.9 is the following convergence result:

Proposition 2.10: Assume (S) holds, and let $\bar{c}$ and δ be as in Proposition 2.4. Then there exists a scalar δ_2 with $0 < \delta_2 \leq \delta$ such that if $\{c_k\}$ and λ_0 satisfy

$$|\lambda_0 - \lambda^*|/c_0 < \delta_2, \qquad \bar{c} \leq c_k \leq c_{k+1} \qquad \forall\, k = 0, 1, \ldots, \tag{46}$$

then the sequence $\{\lambda_k\}$ generated by [compare with (41), (42)]

$$\lambda_{k+1} = \hat{\lambda}(\lambda_k, c_k) \tag{47}$$

is well defined, and we have $\lambda_k \to \lambda^*$ and $x(\lambda_k, c_k) \to x^*$. Furthermore $\{|\lambda_k - \lambda^*|\}$ and $\{|x(\lambda_k, c_k) - x^*|\}$ converge superlinearly. If in addition $\nabla^2 f$ and $\nabla^2 h_i$, $i = 1, \ldots, m$, are Lipschitz continuous in a neighborhood of x^*, then $\{|\lambda_k - \lambda^*|\}$ and $\{|x(\lambda_k, c_k) - x^*|\}$ converge superlinearly with order at least two.

Proof: Take $\gamma \in (0, \bar{c})$ and let δ_2 be as in Proposition 2.9. Then $\lambda_k \to \lambda^*$ by (40) and $x(\lambda_k, c_k) \to x^*$ by Proposition 2.4. The convergence rate assertions follow from (40) and (43). Q.E.D.

It is interesting to compare the result of the proposition above with the corresponding result for the first-order iteration

$$\lambda_{k+1} = \lambda_k + c_k h[x(\lambda_k, c_k)] \tag{48}$$

(Proposition 2.7). The region of convergence of both iterations is of a similar type but *the threshold level for the penalty parameter in the first order iteration* (48) *is higher than that for the second-order iteration* (47) *if* $\nabla^2 p(0)$ *has a negative eigenvalue.* Indeed if γ is the smallest eigenvalue of $\nabla^2 p(0)$ and $\gamma < 0$, we must have $\bar{c} > 2|\gamma|$ in the first-order method and $\bar{c} > |\gamma|$ in the second-order method in order to assert convergence for λ_0 sufficiently close to λ^* (compare Propositions 2.5, 2.7, and 2.10). Furthermore *the second-order iteration has a faster convergence rate than the first-order iteration.* On the other hand, *the second-order iteration requires availability and computation of second derivatives as well as more overhead than the first-order iteration.* The first-order iteration has an additional advantage which will become apparent when we consider convex programming problems in Chapter 5. We shall see there that for such problems, the first-order iteration is guaranteed to converge even without differentiability assumptions and for an arbitrary starting multiplier λ_0. By contrast, the second-order iteration requires second derivatives for its implementation and in general convergence can be guaranteed only for a limited region of initial multipliers. This suggests that *for problems with inherently convex structure the first-order iteration is more robust than the second-order iteration.*

It is worth noting that from (32) of Section 2.2 and (5), we have $\nabla^2 d_c(\lambda^*) = -[\nabla^2 p(0) + cI]^{-1}$ and therefore

$$\lim_{c \to \infty} \frac{|\nabla^2 d_c(\lambda^*)^{-1} + cI|}{c} = 0.$$

Thus the first-order iteration approaches the second-order iteration as $c \to \infty$.

2.3.3 Quasi-Newton Versions of the Second-Order Iteration

It is possible to eliminate the need for availability and computation of second derivatives if a quasi-Newton method such as the DFP or BFGS method, described in Section 1.3.5, is used for minimization of the augmented Lagrangian. In these methods, one obtains usually (but not always) a good estimate D_k of $\{\nabla^2_{xx} L_{c_k}[x(\lambda_k, c_k), \lambda_k]\}^{-1}$ which, in view of (5), can be used in turn to generate an estimate of $\nabla^2 d_{c_k}(\lambda_k)$ by

$$\nabla^2 d_{c_k}(\lambda_k) \sim -\nabla h(x_k)' D_k \nabla h(x_k).$$

We thus can use the approximate version of the Newton iteration (24), (25) given by

$$(49) \qquad \lambda_{k+1} = \lambda_k + \{\nabla h[x(\lambda_k, c_k)]' D_k \nabla h[x(\lambda_k, c_k)]\}^{-1} h[x(\lambda_k, c_k)].$$

This type of method avoids computation of second derivatives at the expense of what is usually an insignificant degradation of rate of convergence over the second-order iteration (24), (25). For problems with a quadratic objective function and linear constraints, if the starting point x_0 in the first unconstrained minimization is such that a complete set of n iterations of the quasi-Newton method is required for termination, then the final inverse Hessian approximation D_k is exact (Proposition 1.21), and application of (49) will yield λ^* in a single iteration; i.e., $\lambda_1 = \lambda^*$. For this, it is necessary, of course, that the initial penalty parameter satisfies $c_0 > -\gamma$, where γ is the minimum eigenvalue of $\nabla^2 p(0)$, for, otherwise, $L_{c_0}(\cdot, \lambda_0)$ is unbounded below and has no local minimum.

Another advantage of this approach is that the matrix D_k obtained via the quasi-Newton method at the end of the kth minimization may be used to generate a good starting matrix for the quasi-Newton method at the next minimization. If $c_{k+1} \neq c_k$, we have

$$\nabla^2_{xx} L_{c_{k+1}}(x^*, \lambda^*) = \nabla^2_{xx} L_{c_k}(x^*, \lambda^*) + (c_{k+1} - c_k) \nabla h(x^*) \nabla h(x^*)',$$

so if D_k is a good approximation to $\nabla^2_{xx} L_{c_k}(x^*, \lambda^*)^{-1}$, then it is reasonable that

$$\bar{D}_k = \{D_k^{-1} + (c_{k+1} - c_k) \nabla h(x_k) \nabla h(x_k)'\}^{-1}$$

should be a good approximation to $\nabla^2_{xx} L_{c_{k+1}}(x^*, \lambda^*)^{-1}$. By using the matrix identity of Section 1.2, we also have

$$\bar{D}_k = D_k - D_k \nabla h(x_k) [\nabla h(x_k)' D_k \nabla h(x_k) + (c_{k+1} - c_k)^{-1} I]^{-1} \nabla h(x_k)' D_k.$$

Depending on whether one actually works with a Hessian or inverse Hessian approximation in the quasi-Newton scheme, one formula may be preferable to the other. If a separate penalty parameter is used for each constraint, then the corresponding formula is

$$\bar{D}_k = \left\{D_k^{-1} + \sum_{i=1}^{m} (c_{k+1}^i - c_k^i)\nabla h_i(x_k)\nabla h_i(x_k)'\right\}^{-1},$$

where c^i is the penalty parameter used for the ith constraint. An equivalent form is given by

$$\bar{D}_k = D_k - D_k N_k(N_k' D_k N_k + C_k^{-1})^{-1} N_k' D_k,$$

where N_k is the matrix having as columns the gradients $\nabla h_i(x_k)$ for which $c_{k+1}^i \neq c_k^i$ and C_k is the diagonal matrix having the nonzero penalty parameter differences $c_{k+1}^i - c_k^i$ along the diagonal.

Another idea for approximating the inverse Hessian $[\nabla^2 d_{c_k}(\lambda_k)]^{-1}$ is based on the formula [compare (31) of Section 2.2 and (5)]

$$-[\nabla^2 d_{c_k}(\lambda_k)]^{-1} = \nabla^2 p[h(x_k)] + c_k I.$$

Thus, the second-order iteration can be written as

$$\lambda_{k+1} = \lambda_k + \{\nabla^2 p[h(x_k)] + c_k I\} h(x_k).$$

Now during the course of the computation, we obtain the values and gradients of $p(u)$ at several points, since we have, for each $j \leq k$,

$$p(u_j) = f(x_j), \qquad \nabla p(u_j) = \lambda_j + c_j u_j, \qquad u_j = h(x_j).$$

From these function values and gradients, we can generate an approximation of $\nabla^2 p[h(x_k)]$ via a quasi-Newton iteration. This approximation can be used, in turn, in place of $\nabla^2 p[h(x_k)]$ in the second-order iteration above. Unfortunately, several points are necessary before a reasonable approximation to $\nabla^2 p$ can be obtained, so the idea can lead to substantial improvements only for problems where the number of constraints m is small. Note however that *this scheme is applicable regardless of whether a quasi-Newton method is used for unconstrained minimization of the augmented Lagrangian.*

2.3.4 *Geometric Interpretation of the Second-Order Multiplier Iteration*

We finally offer a geometric interpretation of the second-order iteration in terms of the primal functional p. Given λ_k, c_k, and x_k such that

$$\nabla_x L_{c_k}(x_k, \lambda_k) = 0,$$

the second-order iterate λ_{k+1} is given by

$$\lambda_{k+1} = \lambda_k + \{\nabla h(x_k)'[\nabla_{xx}^2 L_{c_k}(x_k, \lambda_k)]^{-1} \nabla h(x_k)\}^{-1} h(x_k).$$

This equation can be written in terms of the primal functional p as

(50) $$\lambda_{k+1} = \lambda_k + [\nabla^2 p(u_k) + c_k I]u_k = \tilde{\lambda}_k + \nabla^2 p(u_k)u_k,$$

where

(51) $$u_k = h(x_k), \qquad \tilde{\lambda}_k = \lambda_k + c_k u_k.$$

If we form the second-order Taylor series expansion of p around u_k

$$\tilde{p}_k(u) = p(u_k) + \nabla p(u_k)'(u - u_k) + \tfrac{1}{2}(u - u_k)'\nabla^2 p(u_k)(u - u_k),$$

we obtain

(52) $$\nabla \tilde{p}_k(0) = \nabla p(u_k) - \nabla^2 p(u_k)u_k.$$

Since (compare with Fig. 2.2) we have

(53) $$\tilde{\lambda}_k = -\nabla p(u_k),$$

it follows from (50)–(53) that

$$\nabla \tilde{p}_k(0) = -\lambda_{k+1}$$

as shown in Fig. 2.6. In other words *the second-order iteration yields the predicted value of* $-\nabla p(0)$ *based on a second-order Taylor series expansion*

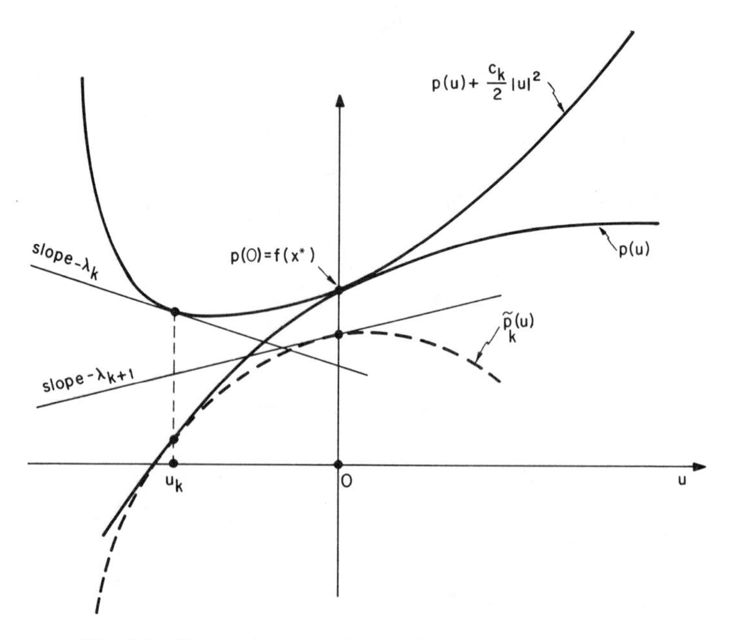

FIG. 2.6 Geometric interpretation of the second-order iteration

of p around $u_k = h(x_k)$. By contrast the first-order iteration yields the predicted value of $-\nabla p(0)$ based on a first-order Taylor series expansion of p around $u_k = h(x_k)$ (compare with Fig. 2.2).

2.4 Multiplier Methods with Partial Elimination of Constraints

In the algorithms of Sections 2.2 and 2.3, all the equality constraints were eliminated by means of a penalty. In some cases however, it is of interest to consider algorithms where only part of the constraints are eliminated by means of a penalty, while the remaining constraints are retained explicitly. A typical example is a problem of the form

$$\text{minimize} \quad f(x)$$
$$\text{subject to} \quad h(x) = 0, \qquad x \geq 0,$$

where the dimension n of the vector x is large and $h(x) = 0$ represents a small number of nonlinear constraints. While, in addition to the constraints $h(x) = 0$, it is possible to eliminate the simple inequality constraints $x \geq 0$ by means of a penalty function, it is probably desirable in most cases to handle these constraints directly by suitable modifications of unconstrained minimization methods (compare with Section 1.5). The corresponding method of multipliers consists of (simply) constrained minimizations of the form

$$\text{minimize} \quad f(x) + \lambda_k' h(x) + \tfrac{1}{2}c_k|h(x)|^2$$
$$\text{subject to} \quad x \geq 0$$

yielding vectors x_k, followed by multiplier updates of the form

$$\lambda_{k+1} = \lambda_k + c_k h(x_k).$$

In this section, we provide an analysis of multiplier methods of this type. We restrict attention to the case where the explicitly retained constraints are equalities. There is no loss of generality in doing so since, as will be seen in Section 3.1, a parallel analysis can be given for inequality constraints after they are converted to equalities by using additional variables.

Suppose that (ECP) can be written as

$$\text{minimize} \quad f(x) \tag{1}$$
$$\text{subject to} \quad h_1(x) = 0, \qquad h_2(x) = 0,$$

where $(h_1, h_2) = h$, $h_1: R^n \to R^{m_1}$, and $h_2: R^n \to R^{m_2}$. For $c \geq 0$, consider the *partial augmented Lagrangian* function $L_{1,c}: R^{n+m_1} \to R$ defined by

$$L_{1,c}(x, \lambda_1) = f(x) + \lambda_1' h_1(x) + \tfrac{1}{2}c|h_1(x)|^2. \tag{2}$$

The (first-order) multiplier method with partial elimination of constraints is defined, under Assumption (S), by the iteration

(3) $$\lambda_{1,k+1} = \lambda_{1,k} + c_k h_1(x_k),$$

where x_k solves locally in a neighborhood of x^* the problem

(4) $$\text{minimize} \quad L_{1,c}(x, \lambda_{1,k})$$
$$\text{subject to} \quad h_2(x) = 0.$$

As before, $\{c_k\}$ is a sequence of positive penalty parameters, and $\lambda_{1,0}$ is chosen a priori.

As mentioned earlier, we have in mind primarily cases where the constrained problem (4) can be solved quite easily and possibly even more easily than the problem of unconstrained minimization of the ordinary augmented Lagrangian.

The analysis of partial multiplier iterations of the form (3) is very similar to the one given in the previous two sections. In fact it is possible to argue that the method (3), (4) is really no different than the ordinary multiplier method. To see this, partition x as

$$x = \begin{bmatrix} x_1 \\ x_2 \end{bmatrix},$$

where $x_1 \in R^{n-m_2}$ and $x_2 \in R^{m_2}$, and assume without loss of generality that the $m_2 \times m_2$ gradient matrix $\nabla_{x_2} h_2(x_1^*, x_2^*)$ is nonsingular. Then, using the implicit function theorem, it is possible to solve near $x^* = (x_1^*, x_2^*)$ the system of equations

(5) $$h_2(x_1, x_2) = 0$$

and obtain x_2 in terms of x_1 as an implicit function $\phi(x_1)$. Then problem (1) becomes

(6) $$\text{minimize} \quad f[x_1, \phi(x_1)]$$
$$\text{subject to} \quad h_1[x_1, \phi(x_1)] = 0,$$

while problem (4) becomes

(7) $$\text{minimize} \quad L_{1,c}[x_1, \phi(x_1), \lambda_{1,k}]$$
$$\text{subject to} \quad x_1 \in R^{m_1}.$$

It is easy to see that the partial multiplier iteration (3) is nothing but the ordinary multiplier iteration for problem (6) and involves in effect unconstrained minimizations of the form (7). It is thus possible to extend all the results of Sections 2.2 and 2.3 to cover partial multiplier iterations of the form (3) by making use of the implicit solution of the system of equations

(5). It is probably more straightforward however to develop these results directly by simply paralleling the analysis of Sections 2.2 and 2.3. Some of the details will be left to the reader.

The following proposition parallels Proposition 2.4. Here and *in what follows in this section we assume that* (x^*, λ^*) *satisfy Assumption* (S) *and* λ_1^* (λ_2^*) *is the vector of the first* m_1 *(last* m_2*) coordinates of* λ^*.

Proposition 2.11: Assume (S) holds, and let $\bar{c}$ be a positive scalar† such that

$$z'\nabla_{xx}^2 L_{\bar{c}}(x^*, \lambda^*)z > 0 \qquad \forall\, z \neq 0, \quad \nabla h_2(x^*)'z = 0.$$

There exist positive scalars δ, ε, and M such that

(a) For all (λ_1, c) in the set $D \subset R^{m_1+1}$ defined by

$$D = \{(\lambda_1, c) \,|\, |\lambda_1 - \lambda_1^*| < \delta c, \bar{c} \leq c\},$$

the problem

$$\begin{aligned} &\text{minimize} && L_{1,c}(x, \lambda_1) \\ &\text{subject to} && h_2(x) = 0 \end{aligned}$$

has a unique solution denoted $x(\lambda_1, c)$. The function $x(\cdot, \cdot)$ is continuously differentiable in the interior of D, and for all $(\lambda_1, c) \in D$, we have

$$|x(\lambda_1, c) - x^*| \leq M|\lambda_1 - \lambda_1^*|/c.$$

(b) For all $(\lambda_1, c) \in D$, we have

$$|\tilde{\lambda}_1(\lambda_1, c) - \lambda_1^*| \leq M|\lambda_1 - \lambda_1^*|/c,$$

where

$$\tilde{\lambda}_1(\lambda_1, c) = \lambda_1 + c h_1[x(\lambda_1, c)].$$

(c) For all $(\lambda_1, c) \in D$, there exists a vector $\tilde{\lambda}_2(\lambda_1, c)$ such that

$$\nabla_x \bar{L}_c[x(\lambda_1, c), \lambda_1, \tilde{\lambda}_2(\lambda_1, c)] = 0,$$

$$z'\nabla_{xx}^2 \bar{L}_c[x(\lambda_1, c)), \lambda_1, \tilde{\lambda}_2(\lambda_1, c)]z > 0 \qquad \forall\, z \neq 0, \quad \nabla h_2[x(\lambda_1, c)]'z = 0,$$

where $\bar{L}_c(x, \lambda_1, \lambda_2) = L_{1.c}(x, \lambda_1) + \lambda_2' h_2(x)$. Furthermore, the matrix $\nabla h[x(\lambda_1, c)]$ has rank m, and we have $\tilde{\lambda}_2(\lambda_1^*, c) = \lambda_2^*$ for all $c \geq \bar{c}$.

Proof: For $c > 0$, consider the system of equations in $(x, \tilde{\lambda}_1, \lambda_1, \lambda_2, c)$,

$$\nabla f(x) + \nabla h_1(x)\tilde{\lambda}_1 + \nabla h_2(x)\lambda_2 = 0, \tag{8a}$$

$$h_1(x) + (\lambda_1 - \tilde{\lambda}_1)/c = 0, \tag{8b}$$

$$h_2(x) = 0. \tag{8c}$$

† We leave it to the reader to verify that Assumption (S) implies that such a scalar $\bar{c}$ exists.

By introducing the variables $t \in R^m$ and $\gamma \in R$ defined by

$$t = (\lambda_1 - \lambda_1^*)/c, \qquad \gamma = 1/c,$$

we write system (8) as

(9a) $$\nabla f(x) + \nabla h_1(x)\tilde{\lambda}_1 + \nabla h_2(x)\lambda_2 = 0$$

(9b) $$h_1(x) + t + \gamma\lambda_1^* - \gamma\tilde{\lambda}_1 = 0$$

(9c) $$h_2(x) = 0$$

For $t = 0$ and $\gamma \in [0, 1/\bar{c}]$, system (9) has the solution $x = x^*$, $\tilde{\lambda}_1 = \lambda_1^*$, $\lambda_2 = \lambda_2^*$. The Jacobian, with respect to $(x, \tilde{\lambda}_1, \lambda_2)$, at such a solution is

$$\begin{bmatrix} \nabla_{xx}^2 L_0(x^*, \lambda^*) & \nabla h_1(x^*) & \nabla h_2(x^*) \\ \nabla h_1(x^*)' & -\gamma I & 0 \\ \nabla h_2(x^*)' & 0 & 0 \end{bmatrix}.$$

Similarly as in the proof of Proposition 2.4, we show that this Jacobian is invertible for all $\gamma \in [0, 1/\bar{c}]$. We then apply the second implicit function theorem of Section 1.2 to system (9), and the remainder of the proof proceeds along the lines of the proof of Proposition 2.4, We leave the details to the reader. Q.E.D.

In Section 2.2, we saw that the eigenvalues of the primal functional play a significant role in the convergence analysis. Within the framework of the present section it is appropriate to define the *partial primal functional* $p_1 : S(0; \delta) \to R$ by

$$p_1(u_1) = \min\{f(x) \mid h_1(x) = u_1, h_2(x) = 0, x \in S(x^*; \varepsilon)\},$$

where ε and δ are sufficiently small scalars [compare with Section 2.2.3 and Eq. (26) in that section]. Clearly, we have

$$p_1(u_1) = p(u_1, 0) \qquad \forall\, u_1 \text{ with } |u_1| < \delta,$$

where $p(u) = p(u_1, u_2)$ is the primal functional of Section 2.2.3. Thus

$$\nabla p_1(u_1) = \nabla_{u_1} p(u_1, 0),$$

$$\nabla^2 p_1(u_1) = \nabla_{u_1 u_1}^2 p(u_1, 0).$$

We leave it to the reader to transfer the argument of the proof of Proposition 2.5 and show the following.

Proposition 2.12: Let (S) hold. For any scalar c, we have

$$z'\nabla_{xx}^2 L_{1,c}(x^*, \lambda_1^*)z > 0 \qquad \forall\, z \neq 0, \quad \nabla h_2(x^*)'z = 0,$$

if and only if

$$c > \max\{-e_1, \ldots, -e_{m_1}\},$$

where $e_1, \ldots, e_{m_1}$ are the eigenvalues of $\nabla^2 p_1(0)$.

We also have the following convergence result.

Proposition 2.13: Assume (S) holds, and let $\bar{c}$ and δ be as in Proposition 2.11. Denote by $e_1, \ldots, e_{m_1}$ the eigenvalues of the matrix $\nabla^2 p_1(0)$. Assume also that

$$\bar{c} > \max\{-2e_1, \ldots, -2e_{m_1}\}.$$

There exists a scalar δ_1 with $0 < \delta_1 \leq \delta$ such that if $\{c_k\}$ and $\lambda_{1,0}$ satisfy

$$|\lambda_{1,0} - \lambda_1^*|/c_0 < \delta_1, \qquad \bar{c} \leq c_k \leq c_{k+1} \qquad \forall\, k = 0, 1, \ldots,$$

then the sequence $\{\lambda_{1,k}\}$ generated by

$$\lambda_{1,k+1} = \lambda_{1,k} + c_k h[x(\lambda_{1,k}, c_k)]$$

is well defined, and we have $\lambda_{1,k} \to \lambda_1^*$ and $x(\lambda_{1,k}, c_k) \to x^*$. Furthermore if $\limsup_{k\to\infty} c_k = c^* < \infty$ and $\lambda_{1,k} \neq \lambda_1^*$ for all k, there holds

$$\limsup_{k\to\infty} \frac{|\lambda_{1,k+1} - \lambda_1^*|}{|\lambda_{1,k} - \lambda_1^*|} \leq \max_{i=1,\ldots,m_1} \left|\frac{e_i}{e_i + c^*}\right|,$$

while if $c_k \to \infty$ and $\lambda_{1,k} \neq \lambda_1^*$ for all k, there holds

$$\lim_{k\to\infty} \frac{|\lambda_{1,k+1} - \lambda_1^*|}{|\lambda_{1,k} - \lambda_1^*|} = 0.$$

The proof of this and other extensions of the results of Sections 2.2 and 2.3 are obtained by means of the following observation. For ε and δ as in the definition of the partial primal functional, the problem

$$\text{(10)} \qquad \begin{aligned} &\text{minimize} \quad L_{1,c}(x, \lambda_1) \\ &\text{subject to} \quad h_2(x) = 0, \qquad x \in S(x^*; \varepsilon), \end{aligned}$$

can also be written as a problem in the variables (x, u_1) of the form

$$\begin{aligned} &\text{minimize} \quad L_{1,c}(x, \lambda_1) \\ &\text{subject to} \quad h_1(x) = u_1, \qquad h_2(x) = 0, \qquad u_1 \in S(0; \delta), \qquad x \in S(x^*; \varepsilon). \end{aligned}$$

This problem in turn is equivalent to

$$\text{(11)} \qquad \begin{aligned} &\text{minimize} \quad p_1(u_1) + \lambda_1' u_1 + \tfrac{1}{2}c|u_1|^2 \\ &\text{subject to} \quad u_1 \in S(0; \delta), \end{aligned}$$

in the sense that if $u_1(\lambda_1, c)$ solves this problem and $x[u_1(\lambda_1, c)]$ solves the problem

$$\begin{aligned} &\text{minimize} \quad f(x) \\ &\text{subject to} \quad h_1(x) = u_1(\lambda_1, c), \qquad h_2(x) = 0, \qquad x \in S(x^*; \varepsilon), \end{aligned}$$

then the vector $x(\lambda_1, c) = x[u_1(\lambda_1, c)]$ solves problem (10), and we have $u_1(\lambda_1, c) = h_1[x(\lambda_1, c)]$. It can be seen now that the partial multiplier iteration $\lambda_{1,k+1} = \lambda_{1,k} + c_k h[x(\lambda_{1,k}, c_k)]$ is in effect the (ordinary) multiplier iteration for the problem

$$\begin{aligned} &\text{minimize} \quad p_1(u_1) \\ &\text{subject to} \quad u_1 = 0, \end{aligned} \tag{12}$$

which involves unconstrained minimizations of the form (11). By applying Proposition 2.7 to problem (12), we obtain a proof of Proposition 2.13.

Similarly by working with problem (12), we can define a *partial dual functional*

$$\begin{aligned} d_{1,c}(\lambda_1) &= \min_{u_1 \in S(0;\delta)} \{p_1(u_1) + \lambda_1' u_1 + \tfrac{1}{2}c|u_1|^2\} \\ &= \min_{\substack{x \in S(x^*;\varepsilon) \\ h_2(x)=0}} \{f(x) + \lambda_1' h_1(x) + \tfrac{1}{2}c|h_1(x)|^2\} \end{aligned}$$

for (λ_1, c) in a set of the form (compare with Proposition 2.11)

$$D = \{(\lambda_1, c) \,|\, |\lambda_1 - \lambda_1^*| < c\delta, \bar{c} \le c\}.$$

The gradient and Hessian matrix of d_c are given by [compare with (3), (5) of Section 2.3]

$$\nabla d_{1,c}(\lambda_1) = h_1[x(\lambda_1, c)],$$

$$\nabla^2 d_{1,c}(\lambda_1) = -\{\nabla^2 p_1[u_1(\lambda_1, c)] + cI\}^{-1} = -\{\nabla^2 p_1[h_1[x(\lambda_1, c)]] + cI\}^{-1}.$$

Using these expressions, we can define Newton's method for maximizing $d_{1,c}$, and we can also prove a convergence result similar to Proposition 2.9. As shown earlier, the Hessian of p_1 is given by

$$\nabla^2 p_1(u_1) = \nabla^2_{u_1 u_1} p(u_1, 0),$$

where p is the primal functional of Section 2.2.3. Thus, we can compute $\nabla^2 p_1(u_1)$ using Eq. (31) of Section 2.2.3 and the solution $x(\lambda_1, c)$ and corresponding Lagrange multiplier of the partially constrained problem

$$\begin{aligned} &\text{minimize} \quad L_{1,c}(x, \lambda_1) \\ &\text{subject to} \quad h_2(x) = 0. \end{aligned}$$

2.5 Asymptotically Exact Minimization in Methods of Multipliers

The multiplier methods considered in the previous sections have the drawback that unconstrained minimization of the augmented Lagrangian must be carried out exactly prior to updating the Lagrange multiplier. In practice, this can be achieved only approximately since unconstrained minimization of the augmented Lagrangian requires in general an infinite number of iterations. Furthermore computational experience has shown that insisting on accurate unconstrained minimization can be computationally wasteful. More efficient schemes result if the unconstrained minimization is terminated and the multiplier is updated as soon as some stopping criterion is satisfied. In this section, we consider the case where the stopping criterion becomes more stringent after every multiplier iteration so that minimization is asymptotically exact.

First-Order Iteration

Consider the first-order iteration

$$\lambda_{k+1} = \lambda_k + c_k h(x_k), \tag{1}$$

where x_k satisfies

$$|\nabla_x L_{c_k}(x_k, \lambda_k)| \leq \varepsilon_k, \tag{2}$$

and $\{\varepsilon_k\}$ is a sequence such that $\varepsilon_k \geq 0$ for all k and $\varepsilon_k \to 0$.

The following result relates to this method and extends Proposition 2.4.

Proposition 2.14: Assume (S) holds, and let $\bar{c}$ be a positive scalar such that

$$\nabla_{xx}^2 L_{\bar{c}}(x^*, \lambda^*) > 0. \tag{3}$$

There exist positive scalars δ, ε, and M such that:

(a) For all (λ, c, α) in the set $D \subset R^{m+1+n}$ defined by

$$D = \{(\lambda, c, \alpha) \,|\, (|\lambda - \lambda^*|^2/c^2 + |\alpha|^2)^{1/2} < \delta, \bar{c} \leq c\}, \tag{4}$$

there exists a unique vector $x_\alpha(\lambda, c)$ within $S(x^*; \varepsilon)$ satisfying

$$\nabla_x L_c[x_\alpha(\lambda, c), \lambda] = \alpha. \tag{5}$$

The function x_α is continuously differentiable in the interior of D, and, for all $(\lambda, c, \alpha) \in D$, we have

$$|x_\alpha(\lambda, c) - x^*| \leq M(|\lambda - \lambda^*|^2/c^2 + |\alpha|^2)^{1/2}. \tag{6}$$

(b) For all $(\lambda, c, \alpha) \in D$, we have

$$|\tilde{\lambda}_\alpha(\lambda, c) - \lambda^*| \le M(|\lambda - \lambda^*|^2/c^2 + |\alpha|^2)^{1/2}, \tag{7}$$

where

$$\tilde{\lambda}_\alpha(\lambda, c) = \lambda + ch[x_\alpha(\lambda, c)]. \tag{8}$$

(c) For all $(\lambda, c, \alpha) \in D$, the matrix $\nabla^2_{xx} L_c[x_\alpha(\lambda, c), \lambda]$ is positive definite, and the matrix $\nabla h[x_\alpha(\lambda, c)]$ has rank m.

Proof: The proof is very similar to that of Proposition 2.4. For $c > 0$, we consider the system of equations in $(x, \tilde{\lambda}, \lambda, c, \alpha)$

$$\nabla f(x) + \nabla h(x)\tilde{\lambda} = \alpha, \qquad h(x) + (\lambda - \tilde{\lambda})/c = 0. \tag{9}$$

We introduce the variables $t \in R^m$ and $\gamma \in R$, defined by

$$t = (\lambda - \lambda^*)/c \qquad \text{and} \qquad \gamma = 1/c, \tag{10}$$

and write system (9) as

$$\nabla f(x) + \nabla h(x)\tilde{\lambda} = \alpha, \qquad h(x) + t + \gamma\lambda^* - \gamma\tilde{\lambda} = 0. \tag{11}$$

For $t = 0$, $\gamma \in [0, 1/\bar{c}]$, and $\alpha = 0$, system (11) has the solution $x = x^*$, $\tilde{\lambda} = \lambda^*$. As in the proof of Proposition 2.4, we can apply the second implicit function theorem of Section 1.2 to system (11) and assert the existence of $\varepsilon > 0$ and $\delta > 0$ and unique continuously differentiable functions $\hat{x}_\alpha(t, \gamma)$, $\hat{\lambda}_\alpha(t, \gamma)$ defined on $S(K; \delta)$, where $K = \{(0, \gamma, 0) \mid \gamma \in [0, 1/\bar{c}]\}$ such that $(|\hat{x}_\alpha(t, \gamma) - x^*|^2 + |\hat{\lambda}_\alpha(t, \gamma) - \lambda^*|^2)^{1/2} < \varepsilon$ for all $(t, \gamma, \alpha) \in S(K; \delta)$ and satisfying

$$\nabla f[\hat{x}_\alpha(t, \gamma)] + \nabla h[\hat{x}_\alpha(t, \gamma)]\hat{\lambda}_\alpha(t, \gamma) = \alpha, \tag{12}$$

$$h[\hat{x}_\alpha(t, \gamma)] + t + \gamma\lambda^* - \gamma\hat{\lambda}_\alpha(t, \gamma) = 0. \tag{13}$$

Furthermore δ and ε can be chosen so that $\nabla h[\hat{x}_\alpha(t, \gamma)]$ has rank m, and

$$\nabla^2_{xx} L_0[\hat{x}_\alpha(t, \gamma), \hat{\lambda}_\alpha(t, \gamma)] + c\nabla h[\hat{x}_\alpha(t, \gamma)]\nabla h[\hat{x}_\alpha(t, \gamma)]' > 0 \tag{14}$$

for $(t, \gamma, \alpha) \in D$. This proves parts (a) and (c) except for (6) similarly as in the proof of Proposition 2.4.

To show (6) and (7), we differentiate (12) and (13) with respect to t, γ, and α. We obtain

$$\begin{bmatrix} \nabla_t \hat{x}_\alpha(t, \gamma)' & \nabla_\gamma \hat{x}_\alpha(t, \gamma)' & \nabla_\alpha \hat{x}_\alpha(t, \gamma)' \\ \nabla_t \hat{\lambda}_\alpha(t, \gamma)' & \nabla_\gamma \hat{\lambda}_\alpha(t, \gamma)' & \nabla_\alpha \hat{\lambda}_\alpha(t, \gamma)' \end{bmatrix} = A(t, \gamma, \alpha) \begin{bmatrix} 0 & 0 & I \\ -I & \hat{\lambda}_\alpha(t, \gamma) - \lambda^* & 0 \end{bmatrix},$$

where

$$A(t, \gamma, \alpha) = \begin{bmatrix} \nabla^2_{xx} L_0[\hat{x}_\alpha(t, \gamma), \hat{\lambda}_\alpha(t, \gamma)] & \nabla h[\hat{x}_\alpha(t, \gamma)] \\ \nabla h[\hat{x}_\alpha(t, \gamma)]' & -\gamma I \end{bmatrix}^{-1}. \tag{15}$$

We have, for all (t, γ, α) such that $|(t, \alpha)| < \delta$ and $\gamma \in [0, 1/\bar{c}]$,

(16)

$$\begin{bmatrix} \hat{x}_\alpha(t, \gamma) - x^* \\ \hat{\lambda}_\alpha(t, \gamma) - \lambda^* \end{bmatrix} = \int_0^1 A(\zeta t, \zeta\gamma, \zeta\alpha) \begin{bmatrix} 0 & 0 & I \\ -I & \hat{\lambda}_{\zeta\alpha}(\zeta t, \zeta\gamma) - \lambda^* & 0 \end{bmatrix} \begin{bmatrix} t \\ \gamma \\ \alpha \end{bmatrix} d\zeta.$$

Let μ be such that $|A(t, \gamma, \alpha)| \leq \mu$ for all $|(t, \alpha)| < \delta$ and $\gamma \in [0, 1/\bar{c}]$, and take δ sufficiently small to ensure that $\mu\delta < 1$. We have, from (16),

$$(|\hat{x}_\alpha(t, \gamma) - x^*|^2 + |\hat{\lambda}_\alpha(t, \gamma) - \lambda^*|^2)^{1/2} \leq \mu(|(t, \alpha)| + \max_{0 \leq \zeta \leq 1} |\hat{\lambda}_{\zeta\alpha}(\zeta t, \zeta\gamma) - \lambda^*|\gamma),$$

and, from this point on, the proof of (6) and (7) proceeds exactly as in the proof of Proposition 2.4 with $|(t, \alpha)|$ replacing $|t|$. Q.E.D.

We can obtain a convergence result from Proposition 2.14 as follows. Consider the iteration

$$\lambda_{k+1} = \lambda_k + c_k h(x_k), \tag{17}$$

where $0 < c_k \leq c_{k+1}$ for all k, and assume that for some sequence $\{\varepsilon_k\}$ with $\varepsilon_k \to 0$, $0 \leq \varepsilon_k$ for all k, we have

$$|\alpha_k| \leq \varepsilon_k, \qquad k = 0, 1, \ldots, \tag{18}$$

where

$$\alpha_k = \nabla_x L_{c_k}(x_k, \lambda_k). \tag{19}$$

Assume that for some $\bar{k}$ we have

$$\varepsilon_{\bar{k}} < \delta/\sqrt{2}, \qquad |\lambda_{\bar{k}} - \lambda^*| < c_{\bar{k}}\delta/\sqrt{2}, \qquad c_{\bar{k}} \geq \max\{\bar{c}, \sqrt{2}M\}, \tag{20}$$

where $\bar{c}$, M, and δ are as in Proposition 2.14. Then $(\lambda_{\bar{k}}, c_{\bar{k}}, \alpha_{\bar{k}}) \in D$, and assuming $x_{\bar{k}}$ is the unique point corresponding to $(\lambda_{\bar{k}}, c_{\bar{k}}, \alpha_{\bar{k}})$ as in Proposition 2.14, we have, in view of $c_{\bar{k}+1} \geq c_{\bar{k}} \geq \sqrt{2}M$,

$$|\lambda_{\bar{k}+1} - \lambda^*| \leq M(|\lambda_{\bar{k}} - \lambda^*|^2/c_{\bar{k}}^2 + |\alpha_{\bar{k}}|^2)^{1/2} < M\delta \leq c_{\bar{k}+1}\delta/\sqrt{2}.$$

Given also that $\varepsilon_{\bar{k}+1} \leq \varepsilon_{\bar{k}} < \delta/\sqrt{2}$, we obtain

$$(\lambda_{\bar{k}+1}, c_{\bar{k}+1}, \alpha_{\bar{k}+1}) \in D.$$

This argument can be repeated, and we are thus led to the conclusion that if, for some $\bar{k}$, (20) holds, then we have

$$(\lambda_k, c_k, \alpha_k) \in D \qquad \forall\, k \geq \bar{k}$$

provided that for each k the vector x_k generated by the algorithm is the unique point corresponding to $(\lambda_k, c_k, \alpha_k)$ as in Proposition 2.14. This means that the estimate (7) is applicable for $k \geq \bar{k}$, and we have

$$|\lambda_{k+1} - \lambda^*| \leq M(|\lambda_k - \lambda^*|^2/c_k^2 + \varepsilon_k^2)^{1/2} \qquad \forall\, k \geq \bar{k}.$$

Since $c_k \ge \sqrt{2}M$ for $k \ge \bar{k}$, we obtain

$$|\lambda_{k+1} - \lambda^*| \le (\tfrac{1}{2}|\lambda_k - \lambda^*|^2 + M^2\varepsilon_k^2)^{1/2} \qquad \forall\, k \ge \bar{k},$$

from which

$$|\lambda_{k+1} - \lambda^*|^2 \le \tfrac{1}{2}|\lambda_k - \lambda^*|^2 + M^2\varepsilon_k^2 \qquad \forall\, k \ge \bar{k}.$$

Hence, for $m \ge 1$,

$$|\lambda_{\bar{k}+m} - \lambda^*|^2 \le 2^{-m}|\lambda_{\bar{k}} - \lambda^*|^2 + \sum_{i=1}^{m} 2^{-(m-i)}M^2\varepsilon_{\bar{k}+i}^2.$$

It is easily seen that $\varepsilon_k \to 0$ implies $\lim_{m\to\infty} \sum_{i=1}^{m} 2^{-(m-i)}M^2\varepsilon_{\bar{k}+i}^2 = 0$, and it follows that $|\lambda_{\bar{k}+m} - \lambda^*| \to 0$ as $m \to \infty$; i.e.,

$$\lim_{k\to\infty} \lambda_k = \lambda^*.$$

Also from (6), we obtain

$$|x_k - x^*| \le M(|\lambda_k - \lambda^*|^2/c_k^2 + \varepsilon_k^2)^{1/2} \qquad \forall\, k \ge \bar{k},$$

and it follows that

$$\lim_{k\to\infty} x_k = x^*.$$

However, the rate of convergence of $\{|\lambda_k - \lambda^*|\}$ and $\{|x_k - x^*|\}$ need not be linear. In order to achieve a linear rate of convergence, it is necessary that the tolerance ε_k decreases to zero as fast as $|\lambda_k - \lambda^*|/c_k$. This can be achieved by replacing the stopping criterion (18), (19) by the stronger condition

$$|\nabla_x L_{c_k}(x_k, \lambda_k)| \le \min\{\varepsilon_k, \gamma|h(x_k)|\}, \tag{21}$$

where γ is some scalar. We have, from (7),

$$|\lambda_{k+1} - \lambda^*| \le M(|\lambda_k - \lambda^*|^2/c_k^2 + \gamma^2|h(x_k)|^2)^{1/2}. \tag{22}$$

Using (17), we obtain

$$|\lambda_k - \lambda^* + c_k h(x_k)| \le M(|\lambda_k - \lambda^*|^2/c_k^2 + \gamma^2|h(x_k)|^2)^{1/2}$$

from which

$$\begin{aligned} c_k|h(x_k)| &\le M(|\lambda_k - \lambda^*|^2/c_k^2 + \gamma^2|h(x_k)|^2)^{1/2} + |\lambda_k - \lambda^*| \\ &\le (M/c_k + 1)|\lambda_k - \lambda^*| + M\gamma|h(x_k)|. \end{aligned}$$

For $c_k > M\gamma$, the relation above yields

$$|h(x_k)| \le [(M + c_k)/c_k(c_k - M\gamma)]|\lambda_k - \lambda^*| \qquad \forall\, k \ge \bar{k}.$$

Substituting in (22), we obtain

$$|\lambda_{k+1} - \lambda^*| \le M\left[1 + \frac{\gamma^2(M + c_k)^2}{(c_k - M\gamma)^2}\right]^{1/2} \frac{|\lambda_k - \lambda^*|}{c_k}.$$

For $c_k > M(1 + 2\gamma)$, this relationship can be strengthened to yield finally

$$(23)\quad |\lambda_{k+1} - \lambda^*| \le M(1 + 4\gamma^2)^{1/2}|\lambda_k - \lambda^*|/c_k \le M(1 + 2\gamma)|\lambda_k - \lambda^*|/c_k.$$

Relation (23) holds if x_k satisfies the stopping criterion (21) and c_k exceeds the (unknown) threshold level $M(1 + 2\gamma)$. If (23) is effective and $c_k > M(1 + 2\gamma)$ for all k sufficiently large, then $\lambda_k \to \lambda^*$ and $x_k \to x^*$. The convergence rate is at least linear if $c_k \to c^* < \infty$ and superlinear if $c_k \to \infty$, similarly as in the method with exact minimization.

The preceding analysis shows that if the sequence $\{\varepsilon_k\}$ used in the stopping criterion (21) is bounded above by a sufficiently small positive number then, given any initial multiplier λ_0, there exists a generally unknown threshold level for the penalty parameter c_0. If c_0 exceeds this level convergence is obtained. It is actually possible to obtain a sharper convergence and rate-of-convergence result. To this end, we consider the following algorithmic model:

Two sequences $\{\varepsilon_k\}$ and $\{\gamma_k\}$, with $0 < \varepsilon_{k+1} < \varepsilon_k$, $\varepsilon_k \to 0$, $0 < \gamma_{k+1} < \gamma_k$, and $\gamma_k \to 0$, are given. An initial multiplier λ_0 and a penalty parameter sequence $\{c_k\}$ are also given and are assumed to satisfy

$$\bar{c} \le c_k \le c_{k+1}, \qquad k = 0, 1, \ldots,$$

$$(|\lambda_0 - \lambda^*|^2/c_0^2 + \varepsilon_0^2)^{1/2} \le \delta,$$

where $\bar{c}$ and δ are as in Proposition 2.14. For $k = 0, 1, \ldots$ and for any $(\lambda_k, c_k, \varepsilon_k)$ satisfying

$$(24)\qquad (|\lambda_k - \lambda^*|^2/c_k^2 + \varepsilon_k^2)^{1/2} \le \delta,$$

we consider the sets

$$(25)\qquad X(\lambda_k, c_k, \varepsilon_k) = \{x_\alpha(\lambda_k, c_k) \mid |\alpha| \le \min\{\varepsilon_k, \gamma_k |h[x_\alpha(\lambda_k, c_k)]|\}\},$$

$$(26)\qquad \Lambda(\lambda_k, c_k, \varepsilon_k) = \{\lambda_k + c_k h(x) \mid x \in X(\lambda_k, c_k, \varepsilon_k)\}.$$

We focus attention at the iteration given by

$$(27)\qquad x_k \in X(\lambda_k, c_k, \varepsilon_k),$$

$$(28)\qquad \lambda_{k+1} = \lambda_k + c_k h(x_k).$$

Iteration (27), (28) is equivalent to λ_{k+1} being any element of the set $\Lambda(\lambda_k, c_k, \varepsilon_k)$. We say that the iteration is *well defined* if λ_0, $\{\varepsilon_k\}$, $\{c_k\}$ are such that, for $k = 0, 1, \ldots$, if $(\lambda_k, c_k, \varepsilon_k)$ satisfies (24), then every $\lambda_{k+1} \in \Lambda(\lambda_k, c_k, \varepsilon_k)$ satisfies

$$(|\lambda_{k+1} - \lambda^*|^2/c_{k+1}^2 + \varepsilon_{k+1}^2)^{1/2} \leq \delta.$$

The idea underlying the elaborate construction given above is to ensure that multipliers generated by iteration (27), (28) lie within the region for which approximate local minima of the augmented Lagrangian can be guaranteed to exist in accordance with Proposition 2.14.

The following proposition provides a convergence and rate-of-convergence result for iteration (27), (28). Its proof utilizes the machinery developed in the proof of Proposition 2.14 similarly as the proof of the related Proposition 2.7 made use of the arguments of Proposition 2.4. We leave the straightforward but lengthy details to the reader.

Proposition 2.15: Assume (S) holds, and let $\bar{c}$ and δ be as in Proposition 2.14. Denote by $e_1, \ldots, e_m$ the eigenvalues of the matrix $\nabla^2 p(0)$ given by (32) of Section 2.2. Assume also that

$$\bar{c} > \max\{-2e_1, \ldots, -2e_m\}.$$

There exist positive scalars δ_1, γ such that if

$$|\lambda_0 - \lambda^*|/c_0 < \delta_1, \qquad \gamma_0 \leq \gamma, \qquad \bar{c} \leq c_k \leq c_{k+1} \qquad \forall\, k = 0, 1, \ldots,$$

then iteration (27), (28) is well defined and any generated sequences $\{\lambda_k\}$ and $\{x_k\}$ converge to λ^* and x^*, respectively. Furthermore, if $\lambda_k \neq \lambda^*$ for all k, we have

$$\limsup_{k\to\infty} \frac{|\lambda_{k+1} - \lambda^*|}{|\lambda_k - \lambda^*|} \leq \max_{i=1,\ldots,m} \left| \frac{e_i}{e_i + c^*} \right| \qquad \text{if} \quad \lim_{k\to\infty} c_k = c^* < \infty,$$

$$\limsup_{k\to\infty} \frac{|\lambda_{k+1} - \lambda^*|}{|\lambda_k - \lambda^*|} = 0 \qquad \text{if} \quad c_k \to \infty.$$

Second-Order Iteration

The key to deriving the proper form of the second-order iteration when unconstrained minimization is not exact lies with the result of Proposition 2.8 and Eqs. (30) and (34) of Section 2.3 in particular. The second-order iteration should consist of the first-order iteration followed by an iteration of Newton's method for solving the system of first-order necessary conditions. Based on equations (30) and (34) of Section 2.3, we obtain the iteration

$$\text{(29)} \qquad \lambda_{k+1} = \lambda_k + \{\nabla h(x_k)'[\nabla_{xx}^2 L_{c_k}(x_k, \lambda_k)]^{-1}\nabla h(x_k)\}^{-1} \times [h(x_k) - \nabla h(x_k)'[\nabla_{xx}^2 L_{c_k}(x_k, \lambda_k)]^{-1}\nabla_x L_{c_k}(x_k, \lambda_k)].$$

The same reasoning together with Proposition 2.14 and the argument of the proof of Proposition 2.9 yields the following result:

Proposition 2.16: Assume (S) holds and let $\bar{c}$ and δ be as in Proposition 2.14. Then given any scalar $\gamma > 0$, there exists a scalar δ_2 with $0 < \delta_2 \le \delta$ such that, for (λ, c, α) in the set D_2 defined by

$$D_2 = \{(\lambda, c, \alpha) | (|\lambda - \lambda^*|^2/c^2 + |\alpha|^2)^{1/2} < \delta_2 c, \bar{c} \le c\},$$

there holds

$$|\hat{\lambda}_\alpha(\lambda, c) - \lambda^*| \le \gamma(|\lambda - \lambda^*|^2/c^2 + |\alpha|^2)^{1/2},$$

where

$$\hat{\lambda}_\alpha(\lambda, c) = \lambda + B_c(\lambda, \alpha)^{-1}\{h[x_2(\lambda, c)] - l_c(\lambda, \alpha)\},$$

$$B_c(\lambda, \alpha) = \nabla h[x_\alpha(\lambda, c)]'\{\nabla_{xx}^2 L_c[x_\alpha(\lambda, c), \lambda]\}^{-1}\nabla h[x_\alpha(\lambda, c)],$$

$$l_c(\lambda, \alpha) = \nabla h[x_\alpha(\lambda, c)]'\{\nabla_{xx}^2 L_c[x_\alpha(\lambda, c), \lambda]\}^{-1}\nabla_x L_c[x_\alpha(\lambda, c), \lambda].$$

If in addition $\nabla^2 f$ and $\nabla^2 h_i$, $i = 1, \ldots, m$, are Lipschitz continuous in a neighborhood of x^*, there exists a scalar M_2 such that for all $(\lambda, c, \alpha) \in D_2$ there holds

$$|\hat{\lambda}_\alpha(\lambda, c) - \lambda^*| \le M_2(|\lambda - \lambda^*|^2/c^2 + |\alpha|^2).$$

There is also an analog of Proposition 2.10 that can be proved for iteration (29), assuming that $\bar{c} \le c_k \le c_{k+1}$ for all k, and x_k satisfies for all k the stopping criterion

$$|\nabla_x L_{c_k}(x_k, \lambda_k)| \le \min\{\varepsilon_k, \gamma_k |h(x_k)|\},$$

where $0 < \varepsilon_{k+1} < \varepsilon_k$, $0 < \gamma_{k+1} < \gamma_k$, $\varepsilon_k \to 0$, and $\gamma_k \to 0$.

2.6 Primal–Dual Methods Not Utilizing a Penalty Function

One of the first Lagrange multiplier methods proposed for solving the equality constrained problem

$$\text{(ECP)} \qquad \text{minimize } f(x) \quad \text{subject to } h(x) = 0$$

consists of sequential minimizations of the form

$$\text{(1)} \qquad \text{minimize } L_0(x, \lambda_k) \quad \text{subject to } x \in R^n$$

yielding vectors x_k, followed by multiplier iterations of the form

(2) $$\lambda_{k+1} = \lambda_k + \alpha h(x_k),$$

where α is a positive stepsize parameter.

A method of this type is particularly useful in separable problems having, for example, the form

(3) $$\text{minimize} \quad \sum_{i=1}^{n} f_i(\xi_i)$$
$$\text{subject to} \quad \sum_{i=1}^{n} h_i(\xi_i) = 0,$$

where $x = (\xi_1, \ldots, \xi_n)$. For such a problem, the minimization of the Lagrangian $L_0(\cdot, \lambda_k)$ can be decomposed into n one-dimensional minimizations

(4) $$\min_x L_0(x, \lambda_k) = \sum_{i=1}^{n} \left\{ \min_{\xi_i} [f_i(\xi_i) + \lambda_k' h_i(\xi_i)] \right\}$$

with considerable simplification resulting.

Essential for the validity of this method is the presence of some kind of convex structure in the problem. In local versions of the method (which are the only ones that will be examined in this section), one focuses on a local minimum x^* satisfying the sufficiency Assumption (S) of Section 2.2 *and the additional local convexity condition*

(5) $$\nabla_{xx}^2 L_0(x^*, \lambda^*) > 0.$$

There are also global versions of the theory where the underlying problem is a convex programming problem (see, e.g., Lasdon, 1970).

The analysis and indeed the motivation of method (1), (2) is based on local duality. In fact all the necessary analysis has already been carried out in Sections 2.2–2.5 and can be brought to bear by means of the following simple observation:

For a positive scalar parameter α, consider the problem

(6) $$\text{minimize} \quad f(x) - \tfrac{1}{2}\alpha |h(x)|^2$$
$$\text{subject to} \quad h(x) = 0.$$

It is a simple matter to verify that *this problem is equivalent to* (ECP) *in the sense that the two problems have the same local minimum–Lagrange multiplier pairs. If any such pair satisfies Assumption* (S), *for one problem, it also satisfies it for the other.* It is evident now that the *iteration*

$$\lambda_{k+1} = \lambda_k + \alpha h(x_k),$$

where x_k minimizes $L_0(\cdot, \lambda_k)$ locally around x^* is simply the first-order method of multipliers for problem (6) *with (constant) penalty parameter equal to* α.

Thus the analysis of the preceding sections can be used to infer convergence and rate-of-convergence results for the method. In particular, it is easy to verify the following facts assuming (S) and the local convexity condition (5).

(a) There exist $\varepsilon > 0$ and $\delta > 0$ such that for all $\lambda \in S(\lambda^*; \delta)$ the problem

$$\begin{aligned} &\text{minimize} \quad L_0(x, \lambda) \\ &\text{subject to} \quad x \in S(x^*; \varepsilon) \end{aligned}$$

has a unique solution denoted $x(\lambda)$ and such that $\nabla^2_{xx} L_0[x(\lambda), \lambda] > 0$ and $\nabla h[x(\lambda)]$ has rank m (compare with Proposition 2.4).

(b) A dual functional d is defined by

$$d(\lambda) = L_0[x(\lambda), \lambda] \qquad \forall\, \lambda \in S(\lambda^*; \delta)$$

and has gradient and Hessian given by

$$\begin{aligned} \nabla d(\lambda) &= h[x(\lambda)] && \forall\, \lambda \in S(\lambda^*; \delta) \\ \nabla^2 d(\lambda) &= -\nabla h[x(\lambda)]'\{\nabla^2_{xx} L_0[x(\lambda), \lambda]\}^{-1} \nabla h[x(\lambda)] && \forall\, \lambda \in S(\lambda^*; \delta) \end{aligned}$$

(compare with Section 2.3).

(c) If p is the primal functional of (ECP) (compare with Section 2.2.3), then the primal functional of problem (6) is given by

$$\bar{p}(u) = p(u) - \tfrac{1}{2}\alpha |u|^2,$$

and if e is the minimum eigenvalue of $\nabla^2 p(0)$ then $(e - \alpha)$ is the minimum eigenvalue of $\nabla^2 \bar{p}(0)$.

(d) There exists a $\delta_1 \in (0, \delta]$ such that the iteration

$$\lambda_{k+1} = \lambda_k + \alpha h[x(\lambda_k)]$$

is well defined and converges to λ^* if $\lambda_0 \in S(\lambda^*; \delta_1)$, $\alpha > 0$, and $\alpha > -2(e - \alpha)$ (compare with Proposition 2.7) or equivalently if

$$0 < \alpha < 2e, \tag{7}$$

where e is the minimum eigenvalue of the primal functional p of (ECP). Furthermore if $\lambda_k \neq \lambda^*$ for all k there holds

$$\limsup_{k \to \infty} \frac{|\lambda_{k+1} - \lambda^*|}{|\lambda_k - \lambda^*|} \le \max_{i=1,\dots,m} \left| \frac{e_i - \alpha}{e_i} \right|,$$

where $e_1, \dots, e_m$ are the eigenvalues of $\nabla^2 p(0)$.

(e) There exists a $\delta_2 \in (0, \delta]$ such that the second-order iteration

$$\lambda_{k+1} = \lambda_k + \{\nabla h[x(\lambda_k)]'\{\nabla_{xx}^2 L_0[x(\lambda_k), \lambda_k]\}^{-1}\nabla h[x(\lambda_k)]\}^{-1}h[x(\lambda_k)]$$

is well defined and converges to λ^* if $\lambda_0 \in S(\lambda^*; \delta_2)$ (compare with Proposition 2.10). Furthermore, the rate of convergence is superlinear (at least order 2 if $\nabla^2 f$ and $\nabla^2 h_i$, $i = 1, \ldots, m$, are Lipschitz continuous in an open sphere centered at x^*).

Notice from (7) that the region of stepsizes that guarantee convergence depends on the minimum eigenvalue e of $\nabla^2 p(0)$ and is generally unknown in practice. Furthermore a good initial choice λ_0 is necessary to guarantee convergence. These facts limit the usefulness of the simple primal–dual methods of this section to large-scale problems with special structure [for example, the separable problem (3)], which satisfy the local convexity condition (5). It is possible to construct primal–dual methods for large-scale separable problems satisfying Assumption (S) in place of the stronger local convexity condition (5), but this requires a more elaborate structure (see Bertsekas, 1979b).

2.7 Notes and Sources

Notes on Section 2.1: The basic idea of penalty function methods is quite old. An extensive work which had substantial influence on further developments is Fiacco and McCormick (1968). The rate of convergence of the quadratic penalty function method was analyzed by Poljak (1971).

Notes on Section 2.2: The quadratic method of multipliers was first proposed independently by Hestenes (1969) and Powell (1969). It was also proposed a year later by Haarhoff and Buys (1970). The thesis by Buys (1972) and the paper by Rupp (1972) provided the first local convergence results for a fixed value of the penalty parameter. Related results were also given by Wierzbicki (1971). Convergence results of a global nature and for a variable penalty parameter were first given independently in Bertsekas (1973, 1976a) and Poljak and Tretjakov (1973). A related result was given in Hestenes (1975). A sharp bound on the rate of convergence was first given in Bertsekas (1975c). The issue of convergence to a single limit point of the sequence $\{x_k\}$ generated by a method of multipliers is addressed in Polak and Tits (1979). The convergence analysis in this section follows Bertsekas (1979a) and sharpens the results of earlier works while weakening some of the assumptions.

Notes on Section 2.3: The local duality framework for the method of multipliers was developed independently by Buys (1972) and Luenberger (1973). The stepsize analysis of Section 2.3.1 extends the one of Bertsekas (1975c). The stepsize rule (20) is new. An alternative stepsize rule has been proposed by Jijtontrum (1980). The convergence and rate-of-convergence analysis for the second-order iteration improves on earlier results in Bertsekas (1976b, 1978). The first quasi-Newton version of the second-order iteration was suggested independently by Fletcher (1975) and Brusch (1973). The second quasi-Newton scheme is new.

Notes on Section 2.4: Multiplier methods with partial elimination of constraints were first considered in Bertsekas (1977). The convergence analysis given here is an improvement over the one in that reference.

Notes on Section 2.5: Multiplier methods with inexact minimization were proposed and analyzed by Buys (1972), Bertsekas (1973, 1975c, 1976a), and Poljak and Tretjakov (1973). Proposition 2.14 improves a result of Bertsekas (1973) and Poljak and Tretjakov (1973), while Proposition 2.16 improves a result of Bertsekas (1978).

Notes on Section 2.6: These methods were pioneered by Everett (1963). Additional relevant works are Poljak (1970), Luenberger (1973), and Lasdon (1970). Ideas related to methods of multipliers have been used for algorithmic solution of special types of large-scale separable problems for which the local convexity assumption is not satisfied – see Stephanopoulos and Westerberg (1975), Stoilow (1977), Watanabe *et al.* (1978). A different and more general approach has been proposed in Bertsekas (1979b).

Chapter 3

The Method of Multipliers for Inequality Constrained and Nondifferentiable Optimization Problems

3.1 One-Sided Inequality Constraints

Consider a nonlinear programming problem involving both equality and inequality constraints

$$\text{(NLP)} \qquad \begin{aligned} &\text{minimize} \quad f(x) \\ &\text{subject to} \quad h(x) = 0, \qquad g(x) \le 0, \end{aligned}$$

where $f: R^n \to R$, $h: R^n \to R^m$, and $g: R^n \to R^r$ are given functions and $m \le n$. The components of h and g are denoted by $h_1, \ldots, h_m$ and $g_1, \ldots, g_r$, respectively.

As discussed in Section 1.4, it is possible to convert (NLP) into an equality constrained problem by introducing a vector of additional variables $z = (z_1, \ldots, z_r)$. This problem is given by

$$(1) \qquad \begin{aligned} &\text{minimize} \quad f(x) \\ &\text{subject to} \quad h_1(x) = \cdots = h_m(x) = 0, \\ &\qquad\qquad g_1(x) + z_1^2 = \cdots = g_r(x) + z_r^2 = 0. \end{aligned}$$

We have that x^* is a local (global) minimum of (NLP) if and only if $(x^*, z_1^*, \ldots, z_r^*)$, where $z_j^* = \sqrt{-g_j(x^*)}$, $j = 1, \ldots, r$, is a local (global) minimum of problem (1).

Based on this conversion, we shall extend all the algorithms and results of Chapter 2 to (NLP). Essentially, no new analysis is required for this extension.

Consider first the augmented Lagrangian for problem (1) defined for $c > 0$ by

$$\bar{L}_c(x, z, \lambda, \mu) = f(x) + \lambda' h(x) + \tfrac{1}{2}c|h(x)|^2 + \sum_{j=1}^{r}\{\mu_j[g_j(x) + z_j^2] + \tfrac{1}{2}c|g_j(x) + z_j^2|^2\}. \quad (2)$$

In applying the methods of Chapter 2 to problem (1), we must minimize the augmented Lagrangian (2) with respect to (x, z) for various values of λ, μ, and c. An important point here is that *minimization of* $\bar{L}_c(x, z, \lambda, \mu)$ *with respect to* z *can be carried out explicitly for each fixed* x. To see this, note that

$$\min_z \bar{L}_c(x, z, \mu) = f(x) + \lambda' h(x) + \tfrac{1}{2}c|h(x)|^2 + \sum_{j=1}^{r} \min_{z_j}\{\mu_j[g_j(x) + z_j^2] + \tfrac{1}{2}c|g_j(x) + z_j^2|^2\}. \quad (3)$$

The minimization with respect to z_j is equivalent to

$$\min_{u_j \geq 0}\{\mu_j[g_j(x) + u_j] + \tfrac{1}{2}c|g_j(x) + u_j|^2\}. \quad (4)$$

The function in braces above is quadratic in u_j. Its unconstrained (global) minimum is the scalar $\hat{u}_j$ at which the derivative is zero. We have

$$\mu_j + c[g_j(x) + \hat{u}_j] = 0$$

from which

$$\hat{u}_j = -[(\mu_j/c) + g_j(x)].$$

There are two possibilities. Either $\hat{u}_j \geq 0$ in which case $\hat{u}_j$ solves problem (4), or else the solution of problem (4) is $u_j^* = 0$. Thus the solution of problem (4) is

$$u_j^* = \max\{0, -[(\mu_j/c) + g_j(x)]\}, \quad (5)$$

and we have

$$g_j(x) + u_j^* = \max\{g_j(x), -(\mu_j/c)\}. \quad (6)$$

Let us use the notation

$$g_j^+(x, \mu_j, c) = \max\{g_j(x), -(\mu_j/c)\}, \tag{7}$$

$$g^+(x, \mu, c) = \begin{bmatrix} g_1^+(x, \mu_1, c) \\ \vdots \\ g_r^+(x, \mu_r, c) \end{bmatrix}. \tag{8}$$

Then, from (3)–(8), we obtain

$$\min_z \bar{L}_c(x, z, \lambda, \mu) = f(x) + \lambda' h(x) + \tfrac{1}{2}c|h(x)|^2 + \mu' g^+(x, \mu, c) + \tfrac{1}{2}c|g^+(x, \mu, c)|^2.$$

We are thus led to the following definition of the *augmented Lagrangian for* (NLP)

$$L_c(x, \lambda, \mu) = f(x) + \lambda' h(x) + \mu' g^+(x, \mu, c) + \tfrac{1}{2}c\{|h(x)|^2 + |g^+(x, \mu, c)|^2\}. \tag{9}$$

An alternative expression for $L_c(x, \lambda, \mu)$ is given by

$$L_c(x, \lambda, \mu) = f(x) + \lambda' h(x) + \tfrac{1}{2}c|h(x)|^2 + \frac{1}{2c}\sum_{j=1}^{r}\{[\max\{0, \mu_j + cg_j(x)\}]^2 - \mu_j^2\}. \tag{10}$$

The equality of the expressions (9) and (10) can be verified by a straightforward calculation. The form of the last term in (10) is shown in Fig. 3.1.

The conclusion from the preceding discussion is that the problem

$$\text{minimize} \quad \bar{L}_c(x, z, \lambda, \mu, c) \qquad \text{subject to} \quad (x, z) \in R^{n+r} \tag{11}$$

is equivalent to the problem

$$\text{minimize} \quad L_c(x, \lambda, \mu) \qquad \text{subject to} \quad x \in R^n, \tag{12}$$

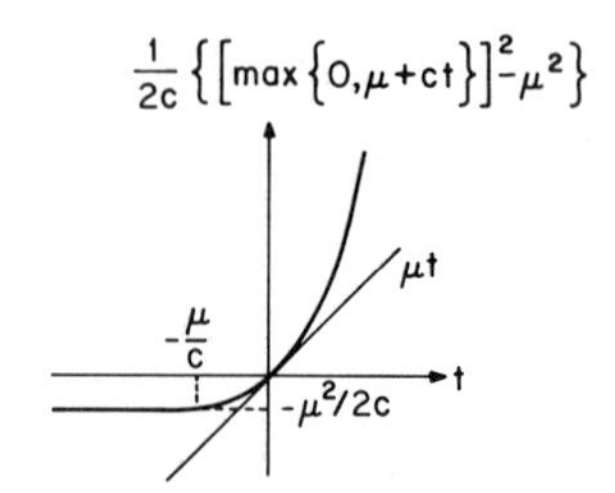

FIG. 3.1 Penalty function for one-sided inequality constraints

and $[x(\lambda, \mu, c), z(\lambda, \mu, c)]$ *is a solution of problem* (11) *if and only if* $x(\lambda, \mu, c)$ *is a solution or problem* (12) *and* [compare with (5)]

(13) $$z_j(\lambda, \mu, c)^2 = \max\{0, -[(\mu_j/c) + g_j[x(\lambda, \mu, c)]]\} \qquad \forall\, j = 1, \ldots, r.$$

As a result, the methods of Chapter 2 can be applied to problem (NLP) after it has been converted to the equality-constrained problem (1), but *the computation itself need not involve the additional variables* $z_1, \ldots, z_r$ *since we can solve in place of problem* (11) *the equivalent problem* (12).

We now develop some results, relating to problem (1) and the augmented Lagrangian (9), which allow an almost mechanical extension of all the algorithms and results of Chapter 2 to (NLP).

Proposition 3.1: (a) If f, h, and g are continuous on a subset S of R^n, then $L_c(\cdot, \lambda, \mu)$ is continuous on S for each λ, μ, and $c > 0$.

(b) If $f, h, g \in C^1$ on an open subset S of R^n, then $L_c(\cdot, \lambda, \mu) \in C^1$ on S for each λ, μ, and $c > 0$.

(c) If $f, h, g \in C^2$ on an open subset S of R^n, then $L_c(\cdot, \lambda, \mu) \in C^2$ on the set

(14) $$\hat{S}_{\mu, c} = S \cap \{x \,|\, g_j(x) \neq -\mu_j/c \;\; \forall\, j = 1, \ldots, r\}$$

for each λ, μ, and $c > 0$.

Proof: The proof follows from the expression (10) for $L_c(x, \lambda, \mu)$.

Q.E.D.

Much of the analysis of Chapter 2 focused on a local minimum x^* and rested on Assumption (S) of Section 2.2. Here again we focus attention at a local minimum for (NLP) satisfying an analogous assumption stated below.

Assumption (S$^+$): The vector x^* is a strict local minimum and a regular point of (NLP), and $f, h, g \in C^2$ on some open sphere centered at x^*. Furthermore x^* together with associated Lagrange multiplier vectors λ^*, μ^* satisfies

$$z'\left[\nabla^2 f(x^*) + \sum_{i=1}^{m} \lambda_i^* \nabla^2 h_i(x^*) + \sum_{j=1}^{r} \mu_j^* \nabla^2 g_j(x^*)\right] z > 0$$

for all $z \neq 0$ with $\nabla h(x^*)'z = 0$, and $\nabla g_j(x^*)'z = 0$ for all $j \in A(x^*) = \{j \,|\, g_j(x^*) = 0\}$. In addition, μ^* satisfies the strict complementarity assumption

$$g_j(x^*) = 0 \Rightarrow \mu_j^* > 0, \qquad j = 1, \ldots, r.$$

Restating Proposition 1.32, we have:

Proposition 3.2: If x^* satisfies Assumption (S$^+$) then the local minimum $(x^*, \sqrt{-g_1(x^*)}, \ldots, \sqrt{-g_r(x^*)})$ of the equality constrained problem (1) satisfies Assumption (S) of Section 2.2.

In view of Propositions 3.1 and 3.2, it is evident that we can extend the algorithms and analysis of Chapter 2 by applying them first to problem (1) and then transfering them to (NLP). Consider for example the first-order multiplier iteration. If $x(\lambda_k, \mu_k, c_k)$ is obtained by minimization of $L_{c_k}(\cdot, \lambda_k, \mu_k)$, the first-order multiplier iteration for problem (1) is given by [compare with (6)–(8)]

$$\lambda_{k+1} = \lambda_k + c_k h[x(\lambda_k, \mu_k, c_k)], \tag{15}$$

$$\mu_{k+1} = \mu_k + c_k g^+[x(\lambda_k, \mu_k, c_k), \mu_k, c_k]. \tag{16}$$

In view of (7), we can also write (16) as

$$\mu_{k+1}^j = \mu_k^j + c_k \max\{-(\mu_k^j/c_k), g_j[x(\lambda_k, \mu_k, c_k)]\}$$

and finally

$$\mu_{k+1}^j = \max\{0, \mu_k^j + c_k g_j[x(\lambda_k, \mu_k, c_k)]\}, \tag{17}$$

where μ_k^j denotes the jth coordinate of μ_k. Equation (17) gives the inequality constraint analog of the first-order iteration. Note that from this equation, it follows that *if $x(\lambda_k, \mu_k, c_k) \to x^*$, then the multipliers corresponding to constraints that are inactive at x^* converge to zero in a finite number of iterations.*

Duality and Second-Order Iterations

The duality theory of Section 2.3 can also be extended in a straightforward manner. Under (S^+) the dual functional is defined for all (λ, μ, c) in the set

$$D = \{(\lambda, \mu, c) \,|\, |(\lambda, \mu) - (\lambda^*, \mu^*)| \leq \delta c, \bar{c} \leq c\}. \tag{18}$$

via the equation

$$d_c(\lambda, \mu) = \min_{x \in S(x^*; \varepsilon)} L_c(x, \lambda, \mu), \tag{19}$$

where δ, ε, and $\bar{c}$ are as in Proposition 2.4 applied to problem (1). We have, for $(\lambda, \mu, c) \in D$,

$$\nabla_\lambda d_c(\lambda, \mu) = h[x(\lambda, \mu, c)], \tag{20}$$

$$\nabla_\mu d_c(\lambda, \mu) = g^+[x(\lambda, \mu, c), \mu, c], \tag{21}$$

where $x(\lambda, \mu, c)$ is the solution of the minimization problem in (19). The Hessian of d_c can be easily computed by writing (21) as [compare with (7), (8)]

$$\partial d_c(\lambda, \mu)/\partial \mu_j = \max\{g_j[x(\lambda, \mu, c)], -\mu_j/c\}, \qquad j = 1, \ldots, r. \tag{22}$$

If $x(\lambda, \mu, c)$ belongs to the set $\hat{S}_{\mu,c}$ of (14), we have that d_c is twice continuously differentiable in a neighborhood of (λ, μ, c). For all indices j such that $g_j[x(\lambda, \mu, c)] < -\mu_j/c$, we have

$$\frac{\partial^2 d_c(\lambda, \mu)}{\partial \mu_i \, \partial \mu_j} = \begin{cases} -1/c & \text{if} \quad i = j, \\ 0 & \text{if} \quad i \neq j, \end{cases} \tag{23a}$$

$$\frac{\partial^2 d_c(\lambda, \mu)}{\partial \lambda_i \, \partial \mu_j} = 0 \qquad \forall \, i = 1, \ldots, m. \tag{23b}$$

For indices j, such that $g_j[x(\lambda, \mu, c)] > -\mu_j/c$, the corresponding second derivatives are computed using the formulas of Section 2.3 by treating the inequality constraints $g_j(x) \leq 0$ as equalities. It can be seen that for (λ, μ) sufficiently close to (λ^*, μ^*), the vector $x(\lambda, \mu, c)$ belongs to $\hat{S}_{\mu,c}$ by the strict complementarity assumption, so for such (λ, μ) the Hessian $\nabla^2 d_c(\lambda, \mu)$ exists.

To give an explicit formula for $\nabla^2 d_c$ when $x(\lambda, \mu, c) \in \hat{S}_{\mu,c}$, assume without loss of generality that for some index p we have

$$g_j[x(\lambda, \mu, c)] > -\mu_j/c, \qquad j = 1, \ldots, p$$

$$g_j[x(\lambda, \mu, c)] < -\mu_j/c, \qquad j = p + 1, \ldots, r.$$

Then $\nabla^2 d_c$ has the form

$$\nabla^2 d_c(\lambda, \mu) = \begin{bmatrix} B_c(\lambda, \mu) & 0 \\ 0 & -c^{-1}I \end{bmatrix}, \tag{24}$$

where I is the $(r - p) \times (r - p)$ identity matrix and

$$B_c(\lambda, \mu) = -N'\{\nabla^2_{xx} L_c[x(\lambda, \mu, c), \lambda, \mu]\}^{-1} N,$$

where N is the matrix having as columns

$$\nabla h_1[x(\lambda, \mu, c)], \ldots, \nabla h_m[x(\lambda, \mu, c)], \quad \nabla g_1[x(\lambda, \mu, c)], \ldots, \nabla g_p[x(\lambda, \mu, c)].$$

The Newton iteration takes the form

$$\begin{bmatrix} \lambda_{k+1} \\ \mu_{k+1} \end{bmatrix} = \begin{bmatrix} \lambda_k \\ \mu_k \end{bmatrix} - [\nabla^2 d_{c_k}(\lambda_k, \mu_k)]^{-1} \nabla d_{c_k}(\lambda_k, \mu_k). \tag{25}$$

In view of (22) and (24), we have

$$\mu^j_{k+1} = 0 \qquad \forall \, j \notin A_k,$$

where

$$A_k = \{j \,|\, g_j[x(\lambda_k, \mu_k, c_k)] > -\mu^j_k/c_k\}.$$

The set A_k may be viewed as *the set of indices of inequality constraints estimated to be active at* x^*. The Newton iteration can perhaps be described better with words than with equations. *We set equal to zero the multipliers of inequality constraints estimated to be inactive at* x^* $(j \notin A_k)$, *and we treat the remaining multipliers as if they correspond to equality constraints.*

When employing the Newton iteration (25), it is quite possible that some of the multipliers μ_{k+1}^j will turn out to be negative. On the other hand, we know that $\mu^* \geq 0$, so it appears sensible to set the negative multipliers to zero, i.e., to use in place of (25) the iteration

$$\begin{bmatrix} \lambda_{k+1} \\ \mu_{k+1} \end{bmatrix} = \begin{bmatrix} \bar{\lambda}_k \\ \bar{\mu}_k^+ \end{bmatrix}, \tag{26}$$

$$\begin{bmatrix} \bar{\lambda}_k \\ \bar{\mu}_k \end{bmatrix} = \begin{bmatrix} \lambda_k \\ \mu_k \end{bmatrix} - [\nabla^2 d_{c_k}(\lambda_k, \mu_k)]^{-1} \nabla d_{c_k}(\lambda_k, \mu_k), \tag{27}$$

where, for every $\mu \in R^r$, we denote

$$\mu^+ = \begin{bmatrix} \max\{0, \mu^1\} \\ \vdots \\ \max\{0, \mu^r\} \end{bmatrix}.$$

This can be justified in two ways. First, we clearly have

$$|\mu_{k+1} - \mu^*| \leq |\bar{\mu}_k - \mu^*|,$$

so that μ_{k+1} is closer to the solution μ^* than $\bar{\mu}_k$. Second, it can be seen from (10) that for all x, λ, μ, we have

$$L_c(x, \lambda, \mu) \leq L_c(x, \lambda, \mu^+).$$

It follows that

$$d_{c_k}(\lambda_{k+1}, \mu_{k+1}) \geq d_{c_k}(\bar{\lambda}_k, \bar{\mu}_k),$$

so the value of the dual functional cannot be decreased by replacing $\bar{\mu}_k$ by μ_{k+1}. These facts are sufficient to establish that every convergence and rate of convergence result that can be shown for iteration (25) can also be shown for iteration (26), (27). At the same time, they suggest that iteration (26), (27) may provide some computational savings over iteration (25).

3.2 Two-Sided Inequality Constraints

Many problems encountered in practice involve two-sided constraints of the form

$$\alpha_j \leq g_j(x) \leq \beta_j,$$

where α_j and β_j are some scalars. Each two-sided constraint could of course be separated into two one-sided constraints which could be treated as discussed in the previous section. This would require, however, the assignment of two multipliers per two-sided constraint. We describe a more efficient approach which requires only *one multiplier per two-sided constraint.*

We consider for simplicity the following problem involving exclusively two-sided constraints. The reader can make appropriate adjustments for the case where there are additional equality or one-sided inequality constraints.

$$\text{(1)} \qquad \text{minimize} \quad f(x)$$
$$\text{subject to} \quad \alpha_j \le g_j(x) \le \beta_j, \qquad j = 1, \ldots, r,$$

where $f: R^n \to R$, $g_j: R^n \to R$, and α_j and β_j, $j = 1, \ldots, r$, are given scalars with $\alpha_j < \beta_j$. Problem (1) is equivalent to the problem

$$\text{(2)} \qquad \text{minimize} \quad f(x)$$
$$\text{subject to} \quad \alpha_j \le g_j(x) - u_j \le \beta_j, \qquad u_j = 0, \quad j = 1, \ldots, r.$$

Now consider a multiplier method for problem (2) where only the constraints $u_j = 0$ are eliminated by means of a quadratic penalty function. This corresponds to partial elimination of constraints discussed in Section 2.4. The method consists of sequential minimizations over x and $u_1, \ldots, u_r$ of the form

$$\text{(3)} \qquad \text{minimize} \quad f(x) + \sum_{j=1}^{r} \{\mu_k^j u_j + \tfrac{1}{2} c_k |u_j|^2\}$$
$$\text{subject to} \quad \alpha_j \le g_j(x) - u_j \le \beta_j, \qquad j = 1, \ldots, r.$$

The multipliers μ_k^j are updated by means of the iteration

$$\text{(4)} \qquad \mu_{k+1}^j = \mu_k^j + c_k u_k^j, \qquad j = 1, \ldots, r,$$

where $u_k^1, \ldots, u_k^r$ together with a vector x_k solve problem (3). Now similarly as in the previous section, minimization in problem (3) can be carried out first with respect to u_j yielding the equivalent problem

$$\text{(5)} \qquad \text{minimize} \quad f(x) + \sum_{j=1}^{r} p_j[g_j(x), \mu_k^j, c_k]$$
$$\text{subject to} \quad x \in R^n,$$

where

$$p_j[g_j(x), \mu_k^j, c_k] = \min_{\alpha_j \le g_j(x) - u_j \le \beta_j} \{\mu_k^j u_j + \tfrac{1}{2} c_k |u_j|^2\}.$$

A straightforward calculation shows that the minimum above is attained at the point u_k^j given by

$$(6) \qquad u_k^j = \begin{cases} g_j(x) - \beta_j & \text{if} \quad \mu_k^j + c_k[g_j(x) - \beta_j] > 0, \\ g_j(x) - \alpha_j & \text{if} \quad \mu_k^j + c_k[g_j(x) - \alpha_j] < 0, \\ -\mu_k^j/c_k & \text{otherwise,} \end{cases}$$

and p_j is given by

$$(7) \quad p_j[g_j(x), \mu_k^j, c_k]$$
$$= \begin{cases} \mu_k^j[g_j(x) - \beta_j] + \frac{1}{2}c|g_j(x) - \beta_j|^2 & \text{if } \mu_k^j + c_k[g_j(x) - \beta_j] > 0, \\ \mu_k^j[g_j(x) - \alpha_j] + \frac{1}{2}c|g_j(x) - \alpha_j|^2 & \text{if } \mu_k^j + c_k[g_j(x) - \alpha_j] < 0, \\ -(\mu_j^k)^2/2c_k & \text{otherwise.} \end{cases}$$

It is easily seen that if $g_j \in C^1$, then $p_j[g_j(x), \mu_k^j, c_k]$ is continuously differentiable in x. If $g_j \in C^2$, then $p_j[g_j(x), \mu_k^j, c_k]$ is twice continuously differentiable on the set $\{x \mid \mu_k^j + c_k[g_j(x) - \beta_j] \neq 0, \; \mu_k^j + c_k[g_j(x) - \alpha_j] \neq 0\}$. The form of the function p_j is shown in Fig. 3.2.

The conclusion from the preceding analysis is that a method of multipliers for problem (1) consists of sequential minimizations of the form (5), (7), which do not involve the variables $u_1, \ldots, u_r$. The (first-order) multiplier iteration is given by [compare with (4), (6)]

$$\mu_{k+1}^j = \begin{cases} \mu_k^j + c_k[g_j(x_k) - \beta_j] & \text{if} \quad \mu_k^j + c_k[g_j(x_k) - \beta_j] > 0, \\ \mu_k^j + c_k[g_j(x_k) - \alpha_j] & \text{if} \quad \mu_k^j + c_k[g_j(x_k) - \alpha_j] < 0, \\ 0 & \text{otherwise,} \end{cases}$$

where x_k solves problem (5). It is also possible to develop a second-order iteration which is best described verbally as follows (compare with the procedure for one-sided constraints described in the previous section).

For every index j such that

$$g_j(x_k) - \beta_j \leq -\mu_k^j/c_k \leq g_j(x_k) - \alpha_j,$$

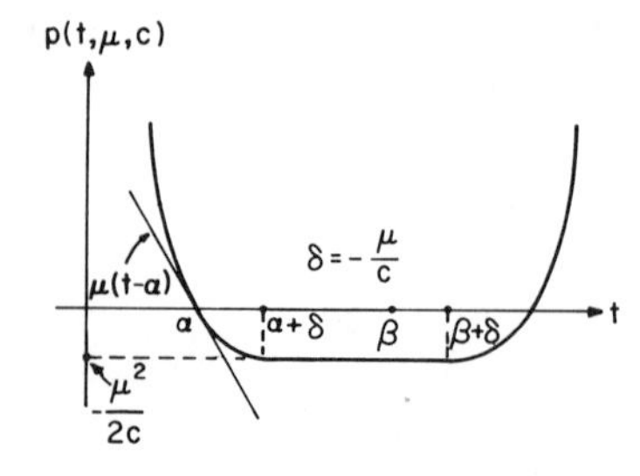

FIG. 3.2 Penalty function for two-sided inequality constraints

set $\mu_{k+1}^j = 0$. For every other index j, treat the constraint $g_j(x) \geq \alpha_j$ or the constraint $g_j(x) \leq \beta_j$ as an equality constraint depending on whether

$$-\mu_k^j/c_k > g_j(x_k) - \alpha_j \qquad \text{or} \qquad -\mu_k^j/c_k < g_j(x_k) - \beta_j.$$

3.3 Approximation Procedures for Nondifferentiable and Ill-Conditioned Optimization Problems

Many optimization problems of interest can be written as

$$\text{minimize} \quad f[x, \gamma_1[g_1(x)], \ldots, \gamma_m[g_m(x)]]$$

$$\text{subject to} \quad h[x, \gamma_1[g_1(x)], \ldots, \gamma_m[g_m(x)]] = 0,$$

where $f: R^n \to R$, $g_i: R^n \to R^{r_i}$, and h, γ_i are given functions.

We are primarily interested in the case where $f, h, g_i \in C^1$, but the presence of the functions γ_i introduces difficulties in the numerical solution of the problem in the sense that, if the functions γ_i were replaced by some real-valued continuously differentiable functions $\tilde{\gamma}_i$, then the problem could be solved in a relatively easy manner. For example, the functions γ_i may induce constraints, nondifferentiabilities, or ill-conditioning.

We shall initially focus on the simpler case where the problem is of the form

$$\text{minimize} \quad f(x) + \sum_{i=1}^{m} \gamma_i[g_i(x)] \tag{1}$$

$$\text{subject to} \quad x \in X \subset R^n,$$

and we shall subsequently discuss the more general case. In connection with problem (1), we shall assume that for each i, $\gamma_i: R^{r_i} \to (-\infty, +\infty]$ is an extended real-valued, lower semicontinuous, convex function with $\gamma_i(t) < \infty$ for at least one $t \in R^{r_i}$. (Such functions will be referred to as closed proper convex functions in Chapter 5.)

We provide some examples of functions γ_i that are of interest in connection with problem (1). In the first five examples, γ_i is a function defined on the real line R ($r_i = 1$).

Example 1: *Equality Constraints*:

$$\gamma_i(t) = \begin{cases} 0 & \text{if} \quad t = 0, \\ +\infty & \text{otherwise.} \end{cases} \tag{2}$$

Here, the presence of $\gamma_i[g_i(x)]$ in problem (1) is equivalent to an additional equality constraint

$$g_i(x) = 0.$$

Example 2: *One-Sided Inequality Constraints*:

$$\gamma_i(t) = \begin{cases} 0 & \text{if} \quad t \le 0, \\ +\infty & \text{otherwise.} \end{cases} \tag{3}$$

Here, the presence of $\gamma_i[g_i(x)]$ induces the constraint

$$g_i(x) \le 0.$$

Example 3: *Two-Sided Inequality Constraints*:

$$\gamma_i(t) = \begin{cases} 0 & \text{if} \quad \alpha_i \le t \le \beta_i, \\ +\infty & \text{otherwise,} \end{cases} \tag{4}$$

where α_i and β_i are scalars with $\alpha_i < \beta_i$.

Example 4: *Polyhedral Functions*:

$$\gamma_i(t) = \max\{0, t\}, \tag{5}$$

$$\gamma_i(t) = |t|, \tag{6}$$

$$\gamma_i(t) = \begin{cases} |t| & \text{if} \quad |t| \le \alpha, \\ +\infty & \text{otherwise,} \end{cases} \tag{7}$$

$$\gamma_i(t) = \begin{cases} \max\limits_{j=1,\ldots,r} \{\gamma_j t + \delta_j\} & \text{if} \quad \alpha \le t \le \beta, \\ +\infty & \text{otherwise,} \end{cases} \tag{8}$$

where α, β, γ_j, and δ_j are given scalars.

Example 5: *Ill-Conditioning Terms*:

$$\gamma_i(t) = \tfrac{1}{2}st^2, \tag{9}$$

$$\gamma_i(t) = \alpha e^{\beta t}, \tag{10}$$

where s, α, and β are given scalars with $s > 0$ and $\alpha > 0$. The term (9) may induce ill-conditioning in problem (1) if s is very large, while the term (10) may induce ill-conditioning in problem (1) if β is very large. More generally, if the second derivatives or third derivatives of γ_i are very large, relative to other terms in the cost functional, the numerical solution of problem (1) may run into serious difficulties.

Example 6: *Minimax Problems*: For $t = (t_1, t_2, \ldots, t_{r_i}) \in R^{r_i}$, consider

$$\gamma_i(t) = \max\{t_1, t_2, \ldots, t_{r_i}\}, \tag{11}$$

$$\gamma_i(t) = \max\{|t_1|, |t_2|, \ldots, |t_{r_i}|\}, \tag{12}$$

$$\gamma_i(t) = \max_{|z-\alpha| \le 1} t'z, \tag{13}$$

where α is a given vector in R^{r_i}.

This section presents an approach for solving numerical problems of the type described above. The approach consists of the approximation of problem (1) by a sequence of optimization problems which involve relatively well-behaved objective functions. The approximation is effected by introducing additional variables and constraints in problem (1), thus forming an equivalent constrained minimization problem. This problem is subsequently handled by the method of multipliers. For the case of Examples 1, 2, and 3, our approach turns out to be identical to multiplier methods introduced earlier. Thus the present section in effect extends the range of applicability of the methods already discussed.

Throughout this section, we shall restrict ourselves to first-order multiplier iterations. Second-order iterations are also possible and can be developed along lines used earlier.

It is clear that problem (1) is equivalent to the following problem:

$$\text{(14)} \qquad \text{minimize} \quad f(x) + \sum_{i=1}^{m} \gamma_i[g_i(x) - u_i],$$

$$\text{subject to} \quad x \in X, \qquad u_i = 0, \qquad i = 1, \ldots, m,$$

where we have introduced the additional vectors

$$u_i \in R^{r_i}, \qquad i = 1, \ldots, m.$$

A method of multipliers for the problem above is based on sequential minimization over $x, u_1, \ldots, u_m$ of the form

$$\text{(15)} \qquad \text{minimize} \quad f(x) + \sum_{i=1}^{m} \{\gamma_i[g_i(x) - u_i] + y_k^{i\prime} u_i + \tfrac{1}{2}c_k|u_i|^2\}$$

$$\text{subject to} \quad x \in X,$$

where y_k^i are multiplier vectors in R^{r_i}, c_k is a positive scalar penalty parameter and prime denotes transposition. Equivalently, problem (15) is written as

$$\text{(16)} \qquad \text{minimize} \quad f(x) + \sum_{i=1}^{m} p_{c_k}^i[g_i(x), y_k^i]$$

$$\text{subject to} \quad x \in X,$$

where

$$\text{(17)} \qquad p_{c_k}^i[g_i(x), y_k^i] = \min_{u_i} \{\gamma_i[g_i(x) - u_i] + y_k^{i\prime} u_i + \tfrac{1}{2}c_k|u_i|^2\}.$$

The initial multiplier vectors y_0^i, $i = 1, \ldots, m$, are arbitrary, and after each minimization (16), the multiplier vectors y_k^i are updated by means of

$$\text{(18)} \qquad y_{k+1}^i = y_k^i + c_k u_i^k, \qquad i = 1, \ldots, m,$$

where u_i^k, $i = 1, \ldots, m$, solve (15), together with some vector x_k. Alternate methods could be obtained by using a nonquadratic penalty function in (15); in fact, in some cases, the use of such nonquadratic penalty functions is essential. We shall restrict ourselves for the moment to quadratic penalty functions and discuss methods based on nonquadratic penalty functions in Section 5.1.3.

It is important to note that the function $p_{c_k}^i$ of (17) is *both real-valued and continuously differentiable* in x, provided the function g_i is continuously differentiable. Hence, problem (16) can be solved by the powerful methods available for differentiable functions whenever f and g_i are differentiable. These properties of the function $p_{c_k}^i$ can be inferred from the following result.

Proposition 3.3: Let $\gamma\colon R^r \to (-\infty, +\infty]$ be a lower semicontinuous convex function, and assume that $\gamma(t) < +\infty$ for at least one vector $t \in R^r$. Also let λ be any vector in R^r and $c > 0$ be a scalar. Then, the function $p_c(\cdot, \lambda)$ defined by

$$p_c(t, \lambda) = \inf_u \{\gamma(t - u) + \lambda' u + \tfrac{1}{2} c |u|^2\} \tag{19}$$

is real-valued, convex, and continuously differentiable in t. Furthermore, the infimum with respect to u in (19) is attained at a unique point for every $t \in R^r$.

Proof: The function $p_c(\cdot, t)$ is the infimal convolution (Rockafellar, 1970) of the convex function γ and the quadratic convex function $h\colon R^r \to R$ defined by

$$h(u) = \lambda' u + \tfrac{1}{2} c |u|^2. \tag{20}$$

Since

$$h(u) \to \infty \qquad \text{as} \quad |u| \to \infty,$$

it follows from Corollary 9.2.2 of Rockafellar (1970) that $p_c(\cdot, \lambda)$ is convex and the infimum is attained for each λ by some u. Since h is strictly convex and real-valued, it follows that $p_c(\cdot, \lambda)$ is also real-valued and the infimum is attained at a single point. Also, h is a smooth function and from Corollary 26.3.2 of Rockafellar (1970), it follows that $p_c(\cdot, \lambda)$ is an essentially smooth convex function. Since it is also real-valued, it is continuously differentiable. Q.E.D.

The interpretation of $p_c(\cdot, \lambda)$ in the proof above as the infimal convolution of γ and h defined by (20) is useful in visualizing the form of $p_c(\cdot, \lambda)$. The epigraph of $p_c(\cdot, \lambda)$ is obtained as the vector sum of the epigraph of the functions $\gamma(\cdot)$ and $h(\cdot)$ (see Rockafellar, 1970, Theorem 5.4).

In some cases, it is useful to work with a dual expression for the function $p_c(\cdot, \lambda)$ of (19). This expression is given in the following lemma, the proof of which follows by straightforward application of Fenchel's duality theorem (Rockafellar, 1970, Theorem 31.1).

Proposition 3.4: The function $p_c(\cdot, \lambda)$ of (19) is also given by

$$p_c(t, \lambda) = \sup_{u^*}\left\{t'u^* - \gamma^*(u^*) - \frac{1}{2c}|u^* - \lambda|^2\right\}, \tag{21}$$

where

$$\gamma^*(u^*) = \sup_u\{u'u^* - \gamma(u)\}$$

is the convex conjugate function of γ. Furthermore, the supremum in (21) is attained at a unique point $u^*(t, \lambda, c)$, and we have

$$u^*(t, \lambda, c) = \lambda + cu(t, \lambda, c) = \nabla_t p_c(t, \lambda), \tag{22}$$

where $u(t, \lambda, c)$ is the unique point attaining the infimum in (19) and $\nabla_t p_c$ is the gradient of p_c with respect to t.

The correspondence between Eqs. (22) and (18) is often convenient in the analysis of specific cases.

It is to be noted that, even though we employed the additional vectors $u_1, \ldots, u_m$ in order to introduce the algorithm, the numerical computation itself need not involve these vectors, since, in the cases of interest to us, the functions $p_{c_k}^i$ of (16)–(17) can be obtained in explicit form. Furthermore, the minimizing vectors u_i^k of (18) can be expressed directly in terms of minimizing vectors x_k in problem (16), since u_i^k is uniquely defined in terms of x_k, c_k, and y_k^i as the minimizing vector in (17). We provide the corresponding analysis for the examples given earlier.

Example 1: For the case where $\gamma_i(t)$ is given by (2), we obtain from (17)–(18):

$$p_{c_k}^i[g_i(x), y_k^i] = y_k^i g_i(x) + \tfrac{1}{2}c_k[g_i(x)]^2,$$
$$y_{k+1}^i = y_k^i + c_k g_i(x_k), \qquad i = 1, \ldots, m.$$

In this case, the iteration reduces to the ordinary first-order multiplier iteration for equality constraints.

Example 2: For $\gamma_i(t)$ given by (3), we obtain from (17)–(18) by straightforward calculation:

$$p_{c_k}^i[g_i(x), y_k^i] = \frac{1}{2c_k}\{(\max\{0, y_k^i + c_k g_i(x)\})^2 - (y_k^i)^2\},$$
$$y_{k+1}^i = \max\{0, y_k^i + c_k g_i(x_k)\}, \qquad i = 1, \ldots, m.$$

The algorithm reduces to the first-order multiplier method for inequality constraints.

Example 3: For the case of two-sided inequality constraints, where $\gamma_i(t)$ is given by (4), we obtain

$$p_{c_k}^i[g_i(x), y_k^i] = \begin{cases} y_k^i[g_i(x) - \beta_i] + \frac{1}{2}c_k[g_i(x) - \beta_i]^2 & \text{if } \beta_i - y_k^i/c_k \le g_i(x), \\ y_k^i[g_i(x) - \alpha_i] + \frac{1}{2}c_k[g_i(x) - \alpha_i]^2 & \text{if } g_i(x) \le \alpha_i - y_k^i/c_k, \\ -(y_k^i)^2/2c_k, & \text{otherwise;} \end{cases}$$

$$y_{k+1}^i = \begin{cases} y_k^i + c_k[g_i(x_k) - \beta_i] & \text{if } \beta_i - y_k^i/c_k \le g_i(x_k), \\ y_k^i + c_k[g_i(x_k) - \alpha_i] & \text{if } g_i(x_k) \le \alpha_i - y_k^i/c_k, \\ 0 & \text{otherwise.} \end{cases}$$

The iteration reduces to the first-order iteration for two-sided inequality constraints given in the previous section.

Example 4: Consider the case where $\gamma_i(t) = \max\{0, t\}$. Then, by straightforward calculation, we obtain

$$p_{c_k}^i[g_i(x), y_k^i] = \begin{cases} g_i(x) - (1 - y_k^i)^2/2c_k & \text{if } g_i(x) \ge (1 - y_k^i)/c_k, \\ -(y_k^i)^2/2c_k & \text{if } g_i(x) \le -y_k^i/c_k, \\ y_k^i g_i(x) + \frac{1}{2}c_k[g_i(x)]^2 & \text{if } -y_k^i/c_k \le g_i(x) \le (1 - y_k^i)/c_k; \end{cases}$$

$$y_{k+1}^i = \begin{cases} 1 & \text{if } g_i(x_k) \ge (1 - y_k^i)/c_k, \\ 0 & \text{if } g_i(x_k) \le -y_k^i/c_k, \\ y_k^i + c_k g_i(x_k) & \text{if } -y_k^i/c_k \le g_i(x_k) \le (1 - y_k^i)/c_k. \end{cases}$$

Notice that a single multiplier per term $\gamma_i[g_i(x)]$ is utilized. If one were to convert the problem to a nonlinear programming problem of the form

$$\text{minimize} \quad f(x) + \sum_{i=1}^{m} z_i$$

$$\text{subject to} \quad g_i(x) \le z_i, \qquad 0 \le z_i, \qquad i = 1, \ldots, m,$$

where z_i are additional variables, then two multipliers per term $\gamma_i[g_i(x)]$ would be required in order for the problem to be solved by the method of multipliers.

The case where $\gamma_i(t)$ is given by (6) can be converted to the earlier case by writing

$$|t| = -t + \max\{0, 2t\}.$$

Let $\gamma_i(t)$ be given by (7), where α is some positive number. Such terms appear, for example, in the cost functional of minimum-fuel problems in optimal control. We have by straightforward calculation

$$p_{c_k}^i[g_i(x), y_k^i] = \begin{cases} \alpha + y_k^i[g_i(x) + \alpha] + \frac{1}{2}c_k[g_i(x) + \alpha]^2 & \text{if } g_i(x) \le -\alpha - (1 + y_k^i)/c_k, \\ -g_i(x) - (1 + y_k^i)^2/2c_k & \text{if } -\alpha - (1 + y_k^i)/c_k \le g_i(x) \le -(1 + y_k^i)/c_k, \\ y_k^i g_i(x) + \frac{1}{2}c_k[g_i(x)]^2 & \text{if } -(1 + y_k^i)/c_k \le g_i(x) \le (1 - y_k^i)/c_k, \\ g_i(x) - (1 - y_k^i)^2/2c_k & \text{if } (1 - y_k^i)/c_k \le g_i(x) \le \alpha + (1 - y_k^i)/c_k, \\ \alpha + y_k^i[g_i(x) - \alpha] + \frac{1}{2}c_k[g_i(x) - \alpha]^2 & \text{if } \alpha + (1 - y_k^i)/c_k \le g_i(x), \end{cases}$$

and iteration (18) takes the form

$$y_{k+1}^i = \begin{cases} y_k^i + c_k[g_i(x_k) + \alpha], & \text{if } g_i(x_k) \le -\alpha - (1 + y_k^i)/c_k, \\ -1 & \text{if } -\alpha - (1 + y_k^i)/c_k \le g_i(x_k) \le -(1 + y_k^i)/c_k, \\ y_k^i + c_k g_i(x_k), & \text{if } -(1 + y_k^i)/c_k \le g_i(x_k) \le (1 - y_k^i)/c_k, \\ 1, & \text{if } (1 - y_k^i)/c_k \le g_i(x_k) \le \alpha + (1 - y_k^i)/c_k. \\ y_k^i + c_k[g_i(x_k) - \alpha], & \text{if } \alpha + (1 - y_k^i)/c_k \le g_i(x_k). \end{cases}$$

Notice that a single multiplier per term γ_i is utilized in place of four multipliers per term γ_i for the ordinary method of multipliers.

Similarly, one may obtain the function $p_{c_k}^i$ and iteration (18) in explicit form for the function $\gamma_i(t)$ given by (8). Again, only one multiplier per term is required in place of $r + 2$ multipliers for the ordinary multiplier method.

Example 5: Let $\gamma_i(t) = \frac{1}{2}s_i t^2$ [compare with (9)]. Then, we have by straightforward calculation:

$$p_{c_k}^i[g_i(x), y_k^i] = [s_i/(s_i + c_k)]\{\tfrac{1}{2}c_k[g_i(x)]^2 + y_k^i g_i(x) - (y_k^i)^2/2s_i\},$$

and iteration (18) takes the form

$$y_{k+1}^i = y_k^i + c_k[s_i g_i(x_k) - y_k^i]/(s_i + c_k).$$

Notice that the second derivative of $p_{c_k}^i(\cdot, y_k^i)$ given above is $s_i c_k/(s_i + c_k)$ and can be made arbitrarily small by choosing c_k sufficiently small.

The case where $\gamma_i(t)$ is given by (10) requires a slightly different approximation method and a nonquadratic penalty function. It will be examined in Chapter 5.

Example 6: Let $\gamma(t) = \max\{t_1, \ldots, t_r\}$ [compare with (11)]. From Eq. (21), we have

$$p_c(t, \lambda) = \sup_{u^*}\left\{t'u^* - \gamma^*(u^*) - \frac{1}{2c}|u^* - \lambda|^2\right\},$$

where the convex conjugate function of γ can be easily calculated as

$$\gamma^*(u^*) = \begin{cases} 0 & \text{if } \sum_{i=1}^r u_i^* = 1, \quad u_i^* \geq 0, \quad i = 1, \ldots, r, \\ +\infty & \text{otherwise.} \end{cases}$$

Hence

$$p_c(t, \lambda) = \max_{\substack{\sum_{i=1}^r u_i^* = 1 \\ u_i^* \geq 0, i=1,\ldots,r}} \left\{t'u^* - \frac{1}{2c}|u^* - \lambda|^2\right\}. \tag{23}$$

By introducing a Lagrange multiplier $\mu(t, \lambda, c)$ corresponding to the constraint

$$\sum_{i=1}^r u_i^* = 1$$

and carrying out the straightforward optimization in (23), we obtain

$$p_c(t, \lambda) = \frac{1}{2c}\sum_{i=1}^r \{(\max\{0, \lambda_i + c[t_i - \mu(t, \lambda, c)]\})^2 - \lambda_i^2\} + \mu(t, \lambda, c).$$

The maximizing vector $\bar{u}^*$ in (23) has coordinates given by

$$\bar{u}_i^* = \max\{0, \lambda_i + c[t_i - \mu(t, \lambda, c)]\}, \qquad i = 1, \ldots, r, \tag{24}$$

and the Lagrange multiplier $\mu(t, \lambda, c)$ is determined from

$$\sum_{i=1}^r \max\{0, \lambda_i + c[t_i - \mu(t, \lambda, c)]\} = 1.$$

In the context of problem (1), a term of the form

$$\gamma[g(x)] = \max\{g^1(x), g^2(x), \ldots, g^r(x)\}$$

is approximated by

$$p_{c_k}[g(x), y_k] = \frac{1}{2c_k} \sum_{i=1}^{r} \{(\max\{0, y_k^i + c_k[g_i(x) - \mu[g(x), y_k, c_k]]\})^2 - (y_k^i)^2\}$$
$$+ \mu[g(x), y_k, c_k].$$

The gradient with respect to x of the expression above is obtained from (22) and (24):

$$\nabla p_{c_k}[g(x), y_k] = \sum_{i=1}^{r} \nabla g_i(x) \max\{0, y_k^i + c_k[g_i(x) - \mu[g(x), y_k, c_k]]\}.$$

The scalar $\mu[g(x), y_k, c_k]$ is determined from

$$\sum_{i=1}^{r} \max\{0, y_k^i + c_k[g_i(x) - \mu[g(x), y_k, c_k]]\} = 1.$$

It is easy to see that the value of $\mu[g(x), y_k, c_k]$ can be computed from the relation above with very little effort.

Regarding the multiplier iteration, we have [see (18), (22), and (24)]

$$\text{(25)} \quad y_{k+1}^i = \max\{0, y_k^i + c_k[g_i(x_k) - \mu[g(x_k), y_k, c_k]]\}, \qquad i = 1, \ldots, r.$$

For the case where

$$\gamma[g(x)] = \max\{|g_1(x)|, |g_2(x)|, \ldots, |g_r(x)|\},$$

a very similar calculation as the one for the previous case yields the following:

$$p_{c_k}[g(x), y_k] = \sum_{i=1}^{r} \tilde{p}_{c_k}^i[g(x), y_k] + \mu[g(x), y_k, c_k],$$

where

$$\tilde{p}_{c_k}^i[g(x), y_k] = \begin{cases} y_k^i[g_i(x) - \mu[g(x), y_k, c_k]] + \frac{1}{2}c_k[g_i(x) - \mu[g(x), y_k, c_k]]^2 \\ \qquad \text{if} \quad y_k^i + c_k[g_i(x) - \mu[g(x), y_k, c_k]] \geq 0, \\ y_k^i[g_i(x) + \mu[g(x), y_k, c_k]] + \frac{1}{2}c_k[g_i(x) + \mu[g(x), y_k, c_k]]^2 \\ \qquad \text{if} \quad y_k^i + c_k[g_i(x) + \mu[g(x), y_k, c_k]] \leq 0, \\ -(y_k^i)^2/2c_k \qquad \text{otherwise.} \end{cases}$$

The gradient of $p_{c_k}[g(x), y_k]$ with respect to x is given by

$$\nabla p_{c_k}[g(x), y_k] = \sum_{i=1}^{r} \nabla g_i(x) \bar{u}_i^*(x, y_k, c_k),$$

where

$$\bar{u}_i^*(x, y_k, c_k) = \begin{cases} y_k^i + c_k[g_i(x) - \mu[g(x), y_k, c_k]] & \text{if } y_k^i + c_k[g_i(x) \\ & \quad -\mu[g(x), y_k, c_k]] \geq 0, \\ y_k^i + c_k[g_i(x) + \mu[g(x), y_k, c_k]] & \text{if } y_k^i + c_k[g_i(x) \\ & \quad +\mu[g(x), y_k, c_k]] \leq 0, \\ 0 & \text{otherwise.} \end{cases}$$

The scalar $\mu[g(x), y_k, c_k]$ is determined from

$$\mu[g(x), y_k, c_k] = 0 \qquad \text{if} \quad \sum_{i=1}^{r} |y_k^i + c_k g_i(x)| \leq 1,$$

$$\sum_{i=1}^{r} \max\{0, |y_k^i + c_k g_i(x)| - c_k \mu[g(x), y_k, c_k]\} = 1 \qquad \text{otherwise.}$$

The multiplier iteration is given by

$$y_{k+1}^i = \bar{u}_i^*(x_k, y_k, c_k), \qquad i = 1, \ldots, r,$$

where $\bar{u}_i^*$ is defined above.

The case where $\gamma_i(t)$ is given by (13) and other related cases where γ is the support function of a relatively simple set can be handled in a similar manner. We note that an alternative approximation procedure for minimax problems, based on an exponential penalty function, is given in Section 5.1.3. This procedure has the advantage that it leads to twice differentiable approximating functions and for many problems should be preferable over the one described above.

Generalized Minimax Problems

Similar approximation procedures can be employed for solution of generalized versions of problem (1) such as the problem

$$\text{(26)} \qquad \begin{aligned} &\text{minimize} \quad f[x, \gamma_1[g_1(x)], \ldots, \gamma_m[g_m(x)]] \\ &\text{subject to} \quad h[x, \gamma_1[g_1(x)], \ldots, \gamma_m[g_m(x)]] = 0, \end{aligned}$$

where $\gamma_i: R^{r_i} \to R$, $i = 1, \ldots, m$, are convex real-valued functions and $f: R^{n+m} \to R$, $g_i: R^n \to R^{r_i}$, and $h: R^{n+m} \to R^s$, are continuously differentiable functions. A special case of particular interest is when γ_i in problem (26) is of the form (11)

$$\gamma_i(t) = \max\{t_1, t_2, \ldots, t_r\}, \qquad i = 1, \ldots, m.$$

The corresponding method of multipliers consists of sequential unconstrained minimizations of the form

$$\text{(27)} \qquad \begin{aligned} \text{minimize} \quad & f[x, p_{c_k}[g_1(x), y_{1,k}], \ldots, p_{c_k}[g_m(x), y_{m,k}]] \\ & + \lambda_k' h[x, p_{c_k}[g_1(x), y_{1,k}], \ldots, p_{c_k}[g_m(x), y_{m,k}]] \\ & + \tfrac{1}{2} c_k |h[x, p_{c_k}[g_1(x), y_{1,k}], \ldots, p_{c_k}[g_m(x), y_{m,k}]]|^2, \end{aligned}$$

where c_k is the penalty parameter sequence, $y_{i,k}$ is the multiplier corresponding to γ_i, and λ_k is the multiplier corresponding to h. If x_k solves (perhaps approximately) problem (27), the multipliers $y_{i,k}$ are updated by means of [compare with (25)]

(28) $$y_{i,k+1}^j = \max\{0, y_{i,k}^j + c_k[g_i^j(x_k) - \mu[g_i(x_k), y_{i,k}, c_k]]\},$$
$$i = 1, \ldots, m, \quad j = 1, \ldots, r,$$

and the multiplier λ_k is updated by means of the usual iteration

(29) $$\lambda_{k+1} = \lambda_k + c_k h[x_k, p_{c_k}[g_1(x_k), y_{1,k}], \ldots, p_{c_k}[g_m(x_k), y_{m,k}]].$$

Under assumptions paralleling the second-order sufficiency assumption (S) *of Chapter* 2, *it is possible to show convergence of this algorithm without the need to increase* c_k *to infinity*. It is also possible to construct a (local) duality theory similar to the one of Section 2.3 and interpret iterations (28) and (29) as steepest ascent iterations for maximizing a related dual functional. Second-order iterations are also possible. The corresponding analysis closely parallels the one in Chapter 2, but is tedious and will not be given here. We refer to Papavassilopoulos (1977) for an account.

Finally we mention that the approach of this section applies to situations where the functions γ_i appear in the objective function and constraints in forms different than in problem (26). For example, the approach is applicable to a problem of the form

$$\text{minimize} \quad f[x, \gamma_1[g_1(x, \gamma_2(x))]]$$
$$\text{subject to} \quad h(x) = 0.$$

In other words it is possible that the functions γ_i may contain as arguments other such functions. It is interesting to note in this connection that the function $\max\{t_1, t_2, \ldots, t_r\}$ can be expressed in terms of the simpler function $\gamma: R \to R$ given by

$$\gamma(t) = \max\{0, t\}$$

by means of the equation

$$\max\{t_1, t_2, \ldots, t_r\} = t_1 + \gamma(t_2 - t_1 + \gamma(t_3 - t_2 + \cdots + \gamma(t_r - t_{r-1}) \cdots)).$$

Another interesting situation of this type is when the objective function can be expressed as a concatenation of operators of the type $\max\{\cdot, \cdot, \ldots, \cdot\}$ as for example in dynamic programming. For an application of this type in a problem of power system scheduling see Bertsekas, Lauer, Sandell, and Posbergh (1981). The general approach in such problems is to replace the functions γ_i as they appear in the cost function and the constraints by suitable approximating functions and sequentially solve the resulting approximate minimization problems. Each minimization is followed by multiplier updates using the appropriate formulas.

3.4 Notes and Sources

Notes on Section 3.1: The proper form of the quadratic augmented Lagrangian function for inequality constraints was first given and analyzed by Rockafellar (1971, 1973b).

Notes on Section 3.2: The treatment of two-sided inequality constraints by using a single multiplier per constraint was first given in Bertsekas (1976b, 1977).

Notes on Section 3.3: Approximation procedures based on the method of multipliers for nondifferentiable optimization problems were introduced in Bertsekas (1974b, 1977). When the multiplier updating formulas (28) and (29) are used, the performance of the method is very similar to that of the first-order method of multipliers, and indeed, if the functions $\max\{t_1, t_2, \ldots, t_r\}$ enter linearly in the objective function and do not appear in the constraints, the two methods are mathematically equivalent. The corresponding local duality theory and second-order algorithms may be found in the M.S. thesis by Papavassilopoulos (1977). His results can be strengthened by using an analysis that parallels the one given in Chapter 2. Relations with the proximal point algorithm are explored in Poljak (1979).

Chapter 4

Exact Penalty Methods and Lagrangian Methods

The methods presented in Chapters 2 and 3 require the solution of several unconstrained or partially constrained minimization problems. It is thus quite interesting that it is possible to construct methods which require the solution of only a single unconstrained problem. We call such methods *exact penalty methods* and consider them in the first three sections of this chapter.

In the fourth section, we consider a class of seemingly unrelated methods which attempt to solve the system of equations and inequalities that constitute the necessary optimality conditions for the constrained optimization problem. The methods here are similar to those used for solving systems of nonlinear equations. We term these methods *Lagrangian methods* in view of the prominent role played by the Lagrangian function and Lagrange multiplier iterations.

An important disadvantage of Lagrangian methods is that they require a good starting point in order to converge to an optimal solution, i.e., they converge only locally. In order to enlarge their region of convergence, it is necessary to combine them with other methods that have satisfactory global convergence properties. Such combinations are discussed in the last section of this chapter and here we find that the method of multipliers of Chapters 2 and 3 and the exact penalty methods of this chapter are well suited for this purpose.

Throughout this chapter we shall consider the problem

(NLP) $$\text{minimize} \quad f(x)$$
$$\text{subject to} \quad h(x) = 0, \qquad g(x) \leq 0,$$

where $f\colon R^n \to R$, $h\colon R^n \to R^m$, $g\colon R^n \to R^r$, and $m \leq n$.

We shall also consider the special cases of (NLP)

(ECP) $$\text{minimize} \quad f(x)$$
$$\text{subject to} \quad h(x) = 0$$

and

(ICP) $$\text{minimize} \quad f(x)$$
$$\text{subject to} \quad g(x) \leq 0.$$

The components of h and g are denoted $h_1, \ldots, h_m$ and $g_1, \ldots, g_r$, respectively. *A standing assumption will be that $f, h, g \in C^1$ on R^n.*

TERMINOLOGY: In what follows we shall encounter several constrained optimization problems involving differentiable functions, equality constraints, inequality constraints, or a mixture of both. We shall say that a pair (triple) of vectors is a *Kuhn–Tucker (K–T for short) pair (triple)* if it satisfies the first-order necessary optimality conditions of Proposition 1.29, referred to as the *K–T conditions*. For example, (x^*, λ^*, μ^*) is a K–T triple for (NLP) if

$$\nabla f(x^*) + \nabla h(x^*)\lambda^* + \nabla g(x^*)\mu^* = 0,$$

$$h(x^*) = 0, \qquad g(x^*) \leq 0, \qquad \mu^* \geq 0, \qquad \mu_j^* g_j(x^*) = 0, \qquad j = 1, \ldots, r.$$

Much of the analysis of this chapter focuses on K–T pairs satisfying the second-order sufficiency assumptions (S) or (S^+) of Sections 2.2 and 3.1, respectively.

4.1 Nondifferentiable Exact Penalty Functions

We shall show that solutions of (NLP) are related to solutions of the (nondifferentiable) unconstrained problem

$(NDP)_c$ $$\text{minimize} \quad f(x) + cP(x)$$
$$\text{subject to} \quad x \in R^n,$$

where $c > 0$ and the function P is defined by

$$P(x) = \max\{0, g_1(x), \ldots, g_r(x), |h_1(x)|, \ldots, |h_m(x)|\}. \tag{1}$$

To see why something like this should be true, consider the special case

$$\text{(ECP)} \qquad \begin{aligned} &\text{minimize} \quad f(x) \\ &\text{subject to} \quad h(x) = 0, \end{aligned}$$

and let x^* be a strict local minimum satisfying, together with a corresponding Lagrange multiplier vector λ^*, Assumption (S) of Section 2.2. Consider also the primal functional $p: S(0; \delta) \to R$ defined in Section 2.2.3 and given by

$$p(u) = \min\{f(x) \mid h(x) = u, x \in S(x^*; \varepsilon)\}$$

[compare with Section 2.2.3, Eq. (26)]. Then

(2)

$$\inf_{x \in S(x^*; \varepsilon)} [f(x) + c \max\{|h_1(x)|, \ldots, |h_m(x)|\}]$$

$$= \inf_{u \in \{z \mid h(x) = z, x \in S(x^*; \varepsilon)\}} \inf_{h(x) = u, x \in S(x^*; \varepsilon)} [f(x) + c \max\{|h_1(x)|, \ldots, |h_m(x)|\}]$$

$$= \inf_{u \in \{z \mid h(x) = z, x \in S(x^*; \varepsilon)\}} p_c(u),$$

where

$$p_c(u) = p(u) + c \max\{|u_1|, \ldots, |u_m|\}.$$

Since $\nabla p(0) = -\lambda^*$, by the mean value theorem, we have, for each u and some $\bar{\alpha} \in [0, 1]$,

$$p(u) = p(0) - \lambda^{*\prime} u + \tfrac{1}{2} u' \nabla^2 p(\bar{\alpha} u) u.$$

Thus

$$\text{(3)} \qquad p_c(u) = p(0) - \sum_{i=1}^{m} \lambda_i^* u_i + c \max\{|u_1|, \ldots, |u_m|\} + \tfrac{1}{2} u' \nabla^2 p(\bar{\alpha} u) u.$$

Assume that, for some $\gamma > 0$,

$$c \geq \sum_{i=1}^{m} |\lambda_i^*| + \gamma.$$

Then we have

$$c \max\{|u_1|, \ldots, |u_m|\} \geq \left(\sum_{i=1}^{m} |\lambda_i^*| + \gamma \right) \max\{|u_1|, \ldots, |u_m|\}$$

$$\geq \sum_{i=1}^{m} \lambda_i^* u_i + \gamma \max\{|u_1|, \ldots, |u_m|\}$$

Using this relation in (3), we obtain

$$p_c(u) \geq p(0) + \gamma \max\{|u_1|, \ldots, |u_m|\} + \tfrac{1}{2} u' \nabla^2 p(\bar{\alpha} u) u.$$

For $|u|$ sufficiently small, the last term is dominated by the next to last term, so we have

$$p_c(u) > p(0) = p_c(0)$$

for all $u \neq 0$ in a neighborhood of the origin. Hence $u = 0$ is a strict local minimum of p_c as shown in Fig. 4.1. By using (2) and the fact that x^* is a strict local minimum of (ECP), it follows that if $c > \sum_{i=1}^m |\lambda_i^*|$, then x^* is a strict local minimum of $f + cP$.

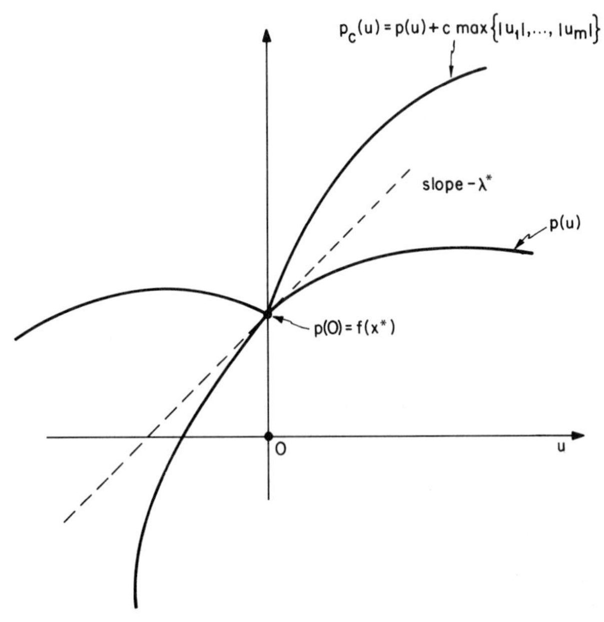

FIG. 4.1

The preceding analysis can be extended to the general case of (NLP) simply by converting the inequality constraints to equalities as in Section 3.1. We have thus proved by means of this abbreviated but simple argument the following proposition.

Proposition 4.1: Let x^* be a strict local minimum of (NLP) satisfying, together with corresponding Lagrange multiplier vectors λ^* and μ^*, Assumption (S^+) of Section 3.1. Then, if

$$c > \sum_{i=1}^{m} |\lambda_i^*| + \sum_{j=1}^{r} \mu_j^*,$$

the vector x^* is a strict unconstrained local minimum of $f + cP$.

Proposition 4.1 indicates that solution of (NLP) may be attempted by solving the unconstrained problem $(\mathrm{NDP})_c$. We would like to prove other similar results that require less restrictive assumptions than (S^+). In the process we shall develop several results that are useful for the construction of algorithms. We first consider the case where there are no equality constraints. We then extend the analysis to the general case by converting each constraint $h_i(x) = 0$ to the two inequality constraints $h_i(x) \leq 0$ and $-h_i(x) \leq 0$.

Inequality Constrained Problems

Consider the problem

(ICP)
$$\text{minimize} \quad f(x)$$
$$\text{subject to} \quad g(x) \leq 0,$$

and for $c > 0$, the corresponding problem

$(\mathrm{NDP})_c$
$$\text{minimize} \quad f(x) + cP(x)$$
$$\text{subject to} \quad x \in R^n$$

For notational convenience, we denote by g_0 the function which is identically zero

$$g_0(x) = 0 \qquad \forall\, x \in R^n, \tag{4}$$

and thus

$$P(x) = \max\{g_0(x), g_1(x), \ldots, g_r(x)\} \tag{5}$$

For $x \in R^n$, $d \in R^n$, and $c > 0$, we use the notation

$$J(x) = \{j \,|\, g_j(x) = P(x), j = 0, 1, \ldots, r\}. \tag{6}$$

$$\theta_c(x; d) = \max\{[\nabla f(x) + c\nabla g_j(x)]'d \,|\, j \in J(x)\} \tag{7}$$

Definition: We say that $x^* \in R^n$ is a *critical point* of $f + cP$ if for all $d \in R^n$ there holds

$$\theta_c(x^*; d) \geq 0.$$

We note that $\theta_c(x^*; d)$ in the above definition can be shown [compare with (9)] to be the Gateaux differential of $f + cP$ at x^* in the direction d (Ortega and Rheinboldt, 1970, p. 65, Luenberger, 1969, p. 171). Our definition of critical point is consistent with analogous definitions for nondifferentiable functions that are Gateaux differentiable. The following two propositions show that descent directions of $f + cP$ can be found only at noncritical points. Furthermore, such directions can be obtained from the following (convex) quadratic program, in $(d, \xi) \in R^{n+1}$,

$$(\mathrm{QP})_c(x, H, J) \quad \text{minimize} \quad \nabla f(x)'d + \tfrac{1}{2}d'Hd + c\xi$$
$$\text{subject to} \quad g_j(x) + \nabla g_j(x)'d \le \xi, \qquad j \in J,$$

where $c > 0$, H is a positive definite matrix and J is an index set containing $J(x)$; i.e.,

$$0 < c, \qquad 0 < H, \qquad J(x) \subset J \subset \{0, 1, \ldots, r\}. \tag{8}$$

It is easily seen that $(\mathrm{QP})_c(x, H, J)$ *has a unique optimal solution (in view of* $H > 0, c > 0$), *and at least one Lagrange multiplier vector* (Proposition 1.33).

Proposition 4.2: (a) For all $x \in R^n$, $d \in R^n$, and $\alpha > 0$,

$$f(x + \alpha d) + cP(x + \alpha d) - f(x) - cP(x) = \alpha\theta_c(x; d) + o(\alpha), \tag{9}$$

where $\lim_{\alpha \to 0^+} o(\alpha)/\alpha = 0$. As a result, if $\theta_c(x; d) < 0$, then there exists $\bar{\alpha} > 0$ such that

$$f(x + \alpha d) + cP(x + \alpha d) < f(x) + cP(x) \qquad \forall\, \alpha \in (0, \bar{\alpha}].$$

(b) For any $x \in R^n$, $H > 0$, and J with $J(x) \subset J \subset \{0, 1, \ldots, r\}$, if (d, ξ) is the optimal solution of $(\mathrm{QP})_c(x, H, J)$ and $d \neq 0$, then

$$\theta_c(x; d) \le -d'Hd < 0. \tag{10}$$

Proof: (a) We have, for all $\alpha > 0$ and $j \in J(x)$,

$$\begin{aligned} &f(x + \alpha d) + cg_j(x + \alpha d) \\ &\qquad = f(x) + \alpha\nabla f(x)'d + c[g_j(x) + \alpha\nabla g_j(x)'d] + o_j(\alpha), \end{aligned}$$

where $\lim_{\alpha \to 0^+} o_j(\alpha)/\alpha = 0$. Hence

$$\begin{aligned} &f(x + \alpha d) + c\max\{g_j(x + \alpha d) \mid j \in J(x)\} \\ &\quad = f(x) + \alpha\nabla f(x)'d + c\max\{g_j(x) + \alpha\nabla g_j(x)'d \mid j \in J(x)\} + o(\alpha), \\ &\quad = f(x) + cP(x) + \alpha\theta_c(x; d) + o(\alpha) \end{aligned}$$

where $\lim_{\alpha \to 0^+} o(\alpha)/\alpha = 0$. We have, for all α that are sufficiently small,

$$\max\{g_j(x + \alpha d) \mid j \in J(x)\} = \max\{g_j(x + \alpha d) \mid j = 0, 1, \ldots, r\} = P(x + \alpha d).$$

Combining the two above relations we obtain (9).

(b) We have $g_j(x) + \nabla g_j(x)'d \le \xi$ for all $j \in J$. Since $g_j(x) = P(x)$ for all $j \in J(x)$, it follows that $\nabla g_j(x)'d \le \xi - P(x)$ for all $j \in J(x)$ and therefore using the definition of θ_c we have

$$\theta_c(x; d) \le \nabla f(x)'d + c[\xi - P(x)]. \tag{11}$$

Let $\{\mu_j \mid j \in J\}$ be a set of Lagrange multipliers for $(\text{QP})_c(x, H, J)$. The K–T conditions are written

(12) $$\nabla f(x) + Hd + \sum_{j \in J} \mu_j \nabla g_j(x) = 0,$$

(13) $$c - \sum_{j \in J} \mu_j = 0,$$

(14) $$g_j(x) + \nabla g_j(x)'d \leq \xi, \qquad \mu_j \geq 0 \qquad \forall\, j \in J,$$

(15) $$\mu_j[g_j(x) + \nabla g_j(x)'d - \xi] = 0 \qquad \forall\, j \in J.$$

From (12) we obtain

(16) $$\nabla f(x)'d + d'Hd + \sum_{j \in J} \mu_j \nabla g_j(x)'d = 0$$

while from Eqs. (13), (15) and the fact that $g_j(x) \leq P(x)$ for all $j \in J$ we have

(17) $$\begin{aligned} \sum_{j \in J} \mu_j \nabla g_j(x)'d &= \sum_{j \in J} \mu_j \xi - \sum_{j \in J} \mu_j g_j(x) \\ &\geq \sum_{j \in J} \mu_j [\xi - P(x)] \\ &= c[\xi - P(x)]. \end{aligned}$$

Combining (16) and (17) we obtain

(18) $$\nabla f(x)'d + d'Hd + c[\xi - P(x)] \leq 0$$

Adding (11) and (18) we obtain finally

$$\theta_c(x; d) + d'Hd \leq 0$$

which is the desired result. Q.E.D.

Proposition 4.3. (a) If x^* is a critical point of $f + cP$, then the quadratic program $(\text{QP})_c(x^*, H, J)$ has $\{d = 0, \xi = P(x^*)\}$ as its optimal solution for every J and H with

(19) $$0 < H, \qquad J(x^*) \subset J \subset \{0, 1, \ldots, r\}.$$

(b) If $\{d = 0, \xi = P(x^*)\}$ is the optimal solution of some quadratic program $(\text{QP})_c(x^*, H, J)$ where H and J satisfy (19), then x^* is a critical point of $f + cP$.

Proof: (a) If x^* is critical then Proposition 4.2b shows that $\{d = 0, \xi = P(x^*)\}$ is the optimal solution of $(\text{QP})_c(x^*, H, J)$.

(b) Suppose $\{d = 0, \xi = P(x^*)\}$ is the optimal solution of $(\text{QP})_c(x^*, H, J)$. Then the K–T conditions [compare with Eqs. (12) and (13)] yield for some set of Lagrange multipliers $\{\mu_j \mid j \in J\}$

(20) $$\nabla f(x^*) + \sum_{j \in J} \mu_j \nabla g_j(x^*) = 0, \qquad \sum_{j \in J} \mu_j = c.$$

If x^* is not a critical point of $f + cP$ then, by definition, there must exist $d \in R^n$ such that

$$\nabla f(x^*)'d + c\nabla g_j(x^*)'d < 0 \qquad \forall\, j \in J(x^*). \tag{21}$$

The inequality constraints of $(\text{QP})_c(x^*, H, J)$ corresponding to indices $j \notin J(x^*)$ must be inactive, so $\mu_j = 0$ for all $j \notin J(x^*)$. Therefore, in view of (20), at least one of the multipliers $\mu_j, j \in J(x^*)$ must be positive. By multiplying (21) by μ_j and adding over $J(x^*)$ we obtain

$$\sum_{j \in J(x^*)} \mu_j \nabla f(x^*)'d + c \sum_{j \in J(x^*)} \mu_j \nabla g_j(x^*)'d < 0$$

or

$$c\left[\nabla f(x^*) + \sum_{j \in J(x^*)} \mu_j \nabla g_j(x^*)\right]'d < 0$$

which contradicts (20). Q.E.D.

Just as the quadratic program $(\text{QP})_c(x, H, J)$ is related to $(\text{NDP})_c$, there is a quadratic program associated with the nonlinear program (ICP). This program is

$$(\text{QP})_0(x, H, J) \quad \text{minimize} \quad \nabla f(x)'d + \tfrac{1}{2}d'Hd$$
$$\text{subject to} \quad g_j(x) + \nabla g_j(x)'d \le 0 \qquad \forall\, j \in J,$$

where

$$0 < H, \qquad J(x) \subset J \subset \{0, 1, \ldots, r\}. \tag{22}$$

For notational convenience, we allow the possibility $0 \in J$ which corresponds to the inequality $g_0(x) + \nabla g_0(x)'d \le 0$ or $0 \le 0$. This inequality is superfluous and can be assigned an arbitrary nonnegative Lagrange multiplier. Note that the program $(\text{QP})_0(x, H, J)$ may not be feasible for some x and J. If it is feasible, it has a unique optimal solution and at least one Lagrange multiplier vector which are related to K–T pairs of (ICP) as in the following proposition.

Proposition 4.4: If a pair $\{x^*, (\mu_1^*, \ldots, \mu_r^*)\}$ is a K–T pair for (ICP), there exists a $\mu_0^* \ge 0$ such that $\{d^* = 0, \{\mu_j^* \mid j \in J\}\}$ is a K–T pair for $(\text{QP})_0(x^*, H, J)$ for all H and J satisfying (22). Conversely, if $\{d^* = 0, \{\mu_j^* \mid j \in J\}\}$ is a K–T pair for $(\text{QP})_0(x^*, H, J)$ for some H and J satisfying (22), then $\{x^*, (\mu_1^*, \ldots, \mu_r^*)\}$ is a K–T pair for (ICP), where we define $\mu_j^* = 0$ for all $j \notin J$.

Proof: The K–T conditions for (ICP) are

$$\nabla f(x^*) + \sum_{j=1}^{r} \mu_j^* \nabla g_j(x^*) = 0, \tag{23}$$

$$g(x^*) \le 0, \qquad \mu_j^* \ge 0, \qquad \mu_j^* g_j(x^*) = 0 \qquad \forall\, j = 1, \ldots, r. \tag{24}$$

By taking $\mu_0^* = 0$ and using the fact that $g_0(x) \triangleq 0$, we see that these conditions imply that $\{0, \{\mu_j^* \mid j \in J\}\}$ satisfy the K–T conditions for $(\text{QP})_0(x^*, H, J)$. Conversely, if we write the conditions for $\{0, \{\mu_j^* \mid j \in J\}\}$ to be a K–T pair for $(\text{QP})_0(x^*, H, J)$, we find that they imply (23) and (24). Q.E.D.

The next proposition shows that if $(\text{QP})_0(x, H, J)$ is feasible, then its optimal solution can also be obtained by solving $(\text{QP})_c(x, H, J \cup \{0\})$ for c sufficiently large.

Proposition 4.5: If $\{d, \{\mu_j \mid j \in J\}\}$ is a K–T pair of $(\text{QP})_0(x, H, J)$ and

$$c \geq \sum_{\substack{j \in J \\ j \neq 0}} \mu_j,$$

then $\{d, \xi = 0, \{\bar{\mu}_j \mid j \in \bar{J}\}\}$ is a K–T pair of $(\text{QP})_c(x, H, \bar{J})$ where

$$\bar{J} = J \cup \{0\}, \qquad \bar{\mu}_j = \mu_j \quad \forall\, j \in \bar{J},\ j \neq 0, \qquad \bar{\mu}_0 = c - \sum_{\substack{j \in J \\ j \neq 0}} \mu_j.$$

Proof: The hypothesis implies that

$$\nabla f(x) + \sum_{j \in J} \mu_j \nabla g_j(x) + Hd = 0, \qquad g_j(x) + \nabla g_j(x)'d \leq 0 \qquad \forall\, j \in J,$$

$$\mu_j \geq 0, \qquad \mu_j[g_j(x) + \nabla g_j(x)'d] = 0 \qquad \forall\, j \in J.$$

Using the definition of $\bar{J}$, $\bar{\mu}_j$, and the fact that $g_0(x) \triangleq 0$, we see that these relations imply

$$\nabla f(x) + \sum_{j \in \bar{J}} \bar{\mu}_j \nabla g_j(x) + Hd = 0, \qquad c = \sum_{j \in \bar{J}} \bar{\mu}_j,$$

$$g_j(x) + \nabla g_j(x)'d \leq 0 \qquad \forall\, j \in \bar{J},$$

$$\bar{\mu}_j \geq 0, \qquad \bar{\mu}_j[g_j(x) + \nabla g_j(x)'d] = 0 \qquad \forall\, j \in \bar{J}.$$

These are precisely the K–T conditions for $\{d, \xi = 0, \{\bar{\mu}_j \mid j \in \bar{J}\}\}$ in connection with $(\text{QP})_c(x, H, \bar{J})$. Q.E.D.

An immediate consequence of the preceding proposition is the following result showing that K–T pairs of the nonlinear program (ICP) give rise to critical points of $f + cP$ provided c is sufficiently large.

Proposition 4.6: If $\{x^*, (\mu_1^*, \ldots, \mu_r^*)\}$ is a K–T pair of (ICP), then x^* is a critical point of $f + cP$ for all c with

$$c \geq \sum_{j=1}^{r} \mu_j^*.$$

Proof: By Proposition 4.4, there exists $\mu_0^* \geq 0$, such that $\{d^* = 0, (\mu_0^*, \mu_1^*, \ldots, \mu_r^*)\}$ is a K–T pair of $(\mathrm{QP})_0(x^*, H, \{0, 1, \ldots, r\})$. It follows, from Proposition 4.5, that if $c \geq \sum_{j=1}^r \mu_j^*$, then $\{d^* = 0, \xi^* = 0\}$ is an optimal solution of $(\mathrm{QP})_c(x^*, H, \{0, 1, \ldots, r\})$. By Proposition 4.3, x^* is a critical point of $f + cP$. Q.E.D.

While each K–T pair of (ICP) gives rise to a critical point of $f + cP$, the reverse is not true. It is possible in general that critical points of $f + cP$ do not correspond to K–T pairs of (ICP), which is somewhat unfortunate since we are contemplating solution of (ICP) by unconstrained minimization of $f + cP$. The following three propositions, among other things, delineate situations where this difficulty does not arise.

Proposition 4.7: Let $X \subset R^n$ be a compact set such that, for all $x \in X$, the set of gradients

$$\{\nabla g_j(x) \mid j \in J(x), j \neq 0\}$$

is linearly independent. There exists a $c^* \geq 0$ such that for every $c > c^*$:

(a) If x^* is a critical point of $f + cP$ and $x^* \in X$, there exists a $\mu^* \in R^r$ such that (x^*, μ^*) is a K–T pair for (ICP).

(b) If (x^*, μ^*) is a K–T pair of (ICP) and $x^* \in X$, then x^* is a critical point of $f + cP$.

For the proof of Proposition 4.7, we shall need the following lemma:

Lemma 4.8: If X is a compact set satisfying the assumption of Proposition 4.7, then for each $x \in X$ there exists a unique vector $\bar{\mu}(x) = [\bar{\mu}_1(x), \ldots, \bar{\mu}_r(x)]$ minimizing over $\mu = (\mu_1, \ldots, \mu_r)$ the function

$$q_x(\mu) = |\nabla f(x) + \sum_{j=1}^r \mu_j \nabla g_j(x)|^2 + \sum_{j=1}^r [P(x) - g_j(x)]^2 \mu_j^2. \tag{25}$$

The function $\bar{\mu}(\cdot)$ is continuous over X, and if (x^*, μ^*) is a K–T pair of (ICP) with $x^* \in X$, then

$$\bar{\mu}(x^*) = \mu^*.$$

Proof: To show uniqueness of the minimizing vector of (25), it will suffice to show that the second-order term of $q_x(\mu)$

$$|\sum_{j=1}^r \mu_j \nabla g_j(x)|^2 + \sum_{j=1}^r [P(x) - g_j(x)]^2 \mu_j^2$$

cannot be zero unless $\mu = 0$. Indeed if this term is zero, then $\mu_j = 0$ for all $j = 1, \ldots, r$ with $P(x) > g_j(x)$ while at the same time $\sum_{j=1}^r \mu_j \nabla g_j(x) = 0$.

Hence

$$\sum_{\substack{1 \le j \le r \\ j \in J(x)}} \mu_j \nabla g_j(x) = 0.$$

Since $\{\nabla g_j(x) \mid j \in J(x), j \neq 0\}$ is a linearly independent set by hypothesis it follows that $\mu_j = 0$ for all j with $g_j(x) = P(x)$. Hence $\mu = 0$.

Continuity of $\bar{\mu}$ follows from continuity of ∇f, ∇g_j, and P. If (x^*, μ^*) is a K–T pair for (ICP), then $q_{x^*}(\mu^*) = 0$. Hence μ^* minimizes $q_{x^*}(\cdot)$, and it follows that $\mu^* = \bar{\mu}(x^*)$. Q.E.D.

Proof of Proposition 4.7: Let

$$c^* = \max_{x \in X} \sum_{j=1}^{r} \bar{\mu}_j(x),$$

where $\bar{\mu}_j(\cdot)$ is as in Lemma 4.8. The maximum in the above equation is attained since X is compact by hypothesis and $\bar{\mu}_j(\cdot)$ is continuous by Lemma 4.8.

(a) If $x^* \in X$ is a critical point of $f + cP$, then, by Proposition 4.3, $\{d = 0, \xi = P(x^*)\}$ is the optimal solution of $(\mathrm{QP})_c(x^*, H, \{0, 1, \ldots, r\})$. Hence, there exist $\mu_0^*, \mu_1^*, \ldots, \mu_r^*$ such that

$$\nabla f(x^*) + \sum_{j=0}^{r} \mu_j^* \nabla g_j(x^*) = 0, \qquad c = \sum_{j=0}^{r} \mu_j^*, \tag{26}$$

$$\mu_j^* \ge 0, \qquad \mu_j^*[g_j(x^*) - P(x^*)] = 0, \qquad j = 0, 1, \ldots, r. \tag{27}$$

Since $g_0(x) \triangleq 0$, we obtain

$$\nabla f(x^*) + \sum_{j=1}^{r} \mu_j^* \nabla g_j(x^*) = 0, \qquad \mu_j^*[g_j(x^*) - P(x^*)] = 0 \qquad \forall\, j = 1, \ldots, r.$$

Using the above equations and Lemma 4.8, it follows that $\mu_j^* = \bar{\mu}_j(x^*)$ for all $j = 1, \ldots, r$. If $c > c^*$, then we obtain

$$\mu_0^* = c - \sum_{j=1}^{r} \mu_j^* = c - \sum_{j=1}^{r} \bar{\mu}_j(x^*) \ge c - c^* > 0.$$

Since $0 = \mu_0^*[g_0(x^*) - P(x^*)] = -\mu_0^* P(x^*)$, it follows that $P(x^*) = 0$ and x^* is feasible for (ICP). It follows from (26) and (27) that $\{x^*, (\mu_1^*, \ldots, \mu_r^*)\}$ is a K–T pair for (ICP).

(b) If (x^*, μ^*) is a K–T pair for (ICP) and $x^* \in X$, then, by Lemma 4.8, we have $\mu^* = \bar{\mu}(x^*)$. If $c > c^*$, then

$$c > \sum_{j=1}^{r} \mu_j^*,$$

and using Proposition 4.6, we obtain that x^* is a critical point of $f + cP$. Q.E.D.

The next two propositions are similar to Proposition 4.7 but employ convexity assumptions in place of the linear independence assumption.

Proposition 4.9: Assume that $g_1, \ldots, g_r$ are convex over R^n and that there exists a vector $\bar{x}$ such that

$$g_j(\bar{x}) < 0 \qquad \forall j = 1, \ldots, r.$$

Then for every compact set X, there exists a $c^* \geq 0$ such that for all $c > c^*$:

(a) If x^* is a critical point of $f + cP$ and $x^* \in X$, there exists a $\mu^* \in R^r$ such that (x^*, μ^*) is a K–T pair for (ICP).

(b) If (x^*, μ^*) is a K–T pair for (ICP) and $x^* \in X$, then x^* is a critical point of $f + cP$.

It is convenient to state the main argument needed for the proof of Proposition 4.9 as a lemma.

Lemma 4.10: Let $X \subset R^n$ be a set such that for each $x \in X$ the system of inequalities in d

$$g_j(x) + \nabla g_j(x)'d \leq 0, \qquad j \in J(x),$$

has at least one solution. Fix $H > 0$, and assume that there exists a $c^* \geq 0$ with the following property:

For each $x \in X$, $(\mathrm{QP})_0(x, H, J(x))$ has a set of Lagrange multipliers

$$\{\mu_j(x) \mid j \in J(x)\}$$

satisfying

$$c^* \geq \sum_{j \in J(x)} \mu_j(x)$$

Then for all $c > c^*$:

(a) If $x^* \in X$ is a critical point of $f + cP$ and $x^* \in X$, there exists a $\mu^* \in R^r$ such that (x^*, μ^*) is a K–T pair for (ICP).

(b) If (x^*, μ^*) is a K–T pair for (ICP) and $x^* \in X$, then x^* is a critical point of $f + cP$.

Proof: (a) Assume that $x^* \in X$ is critical. Let $\{d^*, \{\mu_j(x^*) \mid j \in J(x^*)\}\}$ be the corresponding K–T pair of $(\mathrm{QP})_0(x^*, H, J(x^*))$. Let $c > c^*$. Since $c > \sum_{j \in J(x^*)} \mu_j(x^*)$, it follows from Proposition 4.5 that $\{d^*, \xi = 0\}$ is the optimal solution of $(\mathrm{QP})_c(x^*, H, J(x^*))$. Since x^* is critical, Proposition 4.3 shows that

$$d^* = 0, \qquad P(x^*) = 0.$$

It follows from Proposition 4.4 that $\{x^*, (\mu_1^*, \ldots, \mu_r^*)\}$, where

$$\mu_j^* = \begin{cases} \mu_j(x^*) & \text{for } j \in J(x^*), \quad j \neq 0, \\ 0 & \text{for } j \notin J(x^*), \quad j \neq 0. \end{cases}$$

is a K–T pair for (ICP).

(b) Assume that $\{x^*, (\mu_1^*, \ldots, \mu_r^*)\}$ is a K–T pair for (ICP) and $x^* \in X$. Then, by Proposition 4.4, $d^* = 0$ is the optimal solution of $(\mathrm{QP})_0(x^*, H, J(x^*))$. Let $\mu_j(x^*)$ be the Lagrange multipliers satisfying $c^* \geq \sum_{j \in J(x^*)} \mu_j(x^*)$ according to the hypothesis. It follows from Proposition 4.5 that $\{d^* = 0, \xi^* = 0\}$ is an optimal solution of $(\mathrm{QP})_c(x^*, H, J(x^*))$ for all $c \geq c^*$. Using Proposition 4.3, we obtain that x^* is a critical point of $f + cP$ for all $c \geq c^*$. Q.E.D.

Proof of Proposition 4.9: Fix $H > 0$. By convexity of g_j, we have

$$g_j(x) + \nabla g_j(x)'(\bar{x} - x) \leq g_j(\bar{x}) < 0 \qquad \forall\, x \in R^n, \quad j = 1, \ldots, r.$$

Hence for every $x \in R^n$, the program $(\mathrm{QP})_0(x, H, J(x))$ has $\bar{d} = (\bar{x} - x)$ as a feasible solution. Let $d(x)$ be its optimal solution and $\{\mu_j(x) \mid j \in J(x)\}$ be a corresponding set of Lagrange multipliers. We have that $d(x)$ minimizes

$$\nabla f(x)'d + \tfrac{1}{2}d'Hd + \sum_{j \in J(x)} \mu_j(x)[g_j(x) + \nabla g_j(x)'d]$$

over all d while

$$\mu_j(x)[g_j(x) + \nabla g_j(x)'d(x)] = 0 \qquad \forall\, j \in J(x).$$

Hence

$$\begin{aligned} (28) \qquad & \nabla f(x)'d(x) + \tfrac{1}{2}d(x)'Hd(x) \\ & \leq \nabla f(x)'(\bar{x} - x) + \tfrac{1}{2}(\bar{x} - x)'H(\bar{x} - x) \\ & \quad + \sum_{j \in J(x)} \mu_j(x)[g_j(x) + \nabla g_j(x)'(\bar{x} - x)] \\ & \leq \nabla f(x)'(\bar{x} - x) + \tfrac{1}{2}(\bar{x} - x)'H(\bar{x} - x) + \sum_{j \in J(x)} \mu_j(x)g_j(\bar{x}) \\ & \leq \nabla f(x)'(\bar{x} - x) + \tfrac{1}{2}(\bar{x} - x)'H(\bar{x} - x) - b \sum_{j \in J(x)} \mu_j(x), \end{aligned}$$

where

$$b = \min\{-g_j(\bar{x}) \mid j = 1, \ldots, r\} > 0.$$

We also have

$$\begin{aligned} (29) \qquad 0 \leq \tfrac{1}{2}|H^{-1/2}\nabla f(x) + H^{1/2}d(x)|^2 = \tfrac{1}{2}\nabla f(x)'H^{-1}\nabla f(x) & \\ + \nabla f(x)'d(x) & \\ + \tfrac{1}{2}d(x)'Hd(x). & \end{aligned}$$

By combining (28) and (29), we obtain

$$\sum_{j \in J(x)} \mu_j(x) \le c(x) \qquad \forall\, x \in R^n,$$

where

$$c(x) = [\tfrac{1}{2}\nabla f(x)'H^{-1}\nabla f(x) + \nabla f(x)'(\bar{x} - x) + \tfrac{1}{2}(\bar{x} - x)'H(\bar{x} - x)]/b.$$

Given a compact set X and a fixed $H > 0$, define

$$c^* = \max_{x \in X} c(x),$$

and note that

$$c^* \ge c(x) \ge \sum_{j \in J(x)} \mu_j(x) \qquad \forall\, x \in X.$$

The result now follows from Lemma 4.10. Q.E.D.

Proposition 4.11: Assume that $f, g_1, \ldots, g_r$ are convex over R^n and that (ICP) has at least one Lagrange multiplier vector $\mu^* = (\mu_1^*, \ldots, \mu_r^*)$, in the sense that $\mu_j^* \ge 0, j = 1, \ldots, r$, and

$$\inf_{x \in R^n} \{f(x) + \mu^{*\prime} g(x)\} = \inf_{g(x) \le 0} f(x).$$

Then, for every $c > \sum_{j=1}^{r} \mu_j^*$, a vector x^* is a global minimum of $f + cP$ if and only if x^* is a global minimum of (ICP).

We postpone the proof of Proposition 4.11 until Chapter 5, where we shall show a stronger version (Proposition 5.25.)

The following two examples illustrate the limitations of the preceding results.

Example 1: Let $n = 2$, $r = 1$, and for all $x = (x_1, x_2)$,

$$f(x) = (x_1 - 1)^2 + x_2^2, \qquad g_1(x) = x_1^2.$$

Here f and g_1 are convex and (ICP) has a unique optimal solution $\{x_1^* = 0, x_2^* = 0\}$. Consider the function

$$f(x) + cP(x) = (x_1 - 1)^2 + x_2^2 + c \max\{0, x_1^2\}$$
$$= (x_1 - 1)^2 + x_2^2 + cx_1^2.$$

For every $c > 0$, it has a unique critical point $\{x_1(c), x_2(c)\}$ (in fact a global minimum) given by

$$x_1(c) = 1/(1 + c), \qquad x_2(c) = 0.$$

Thus the optimal solution $\{x_1^* = 0, x_2^* = 0\}$ of (ICP) is not a critical point of $f + cP$ for any $c > 0$. Conversely, none of the critical points $\{x_1(c), x_2(c)\}$,

$c > 0$, is an optimal solution of (ICP). Here $\{x_1^* = 0, x_2^* = 0\}$ is not a regular point $[\nabla g_1(x^*) = 0]$, and it can be verified that there is no corresponding Lagrange multiplier μ_1^*. Thus Proposition 4.7 cannot be applied to a compact set containing $\{x_1^* = 0, x_2^* = 0\}$, and the assumption of Proposition 4.11 is violated. Because there is no $\bar{x}$ such that $g_1(\bar{x}) < 0$, the assumption of Proposition 4.9 is also violated.

Example 2: Let $n = 1$, $r = 2$, and for all x,

$$f(x) = 0, \qquad g_1(x) = -x, \qquad g_2(x) = 1 - x^2.$$

The function

$$P(x) = \max\{0, -x, 1 - x^2\}$$

is shown in Fig. 4.2.

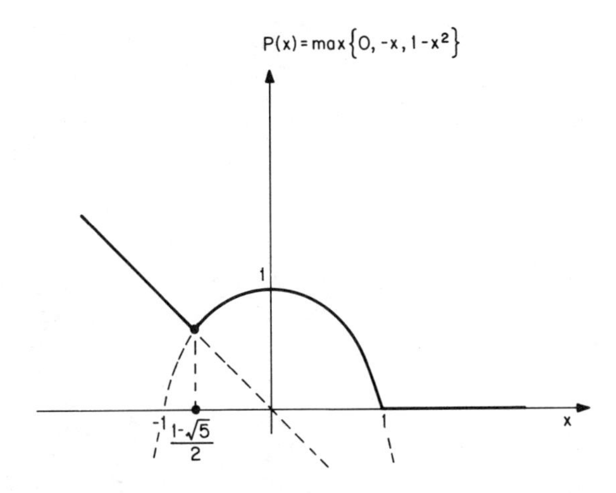

FIG. 4.2 Function $P(x)$ for Example 2

Since $f(x) \triangleq 0$, the critical points of $f + cP$ do not depend on c. They are

$$x = \tfrac{1}{2}(1 - \sqrt{5}), \qquad x = 0, \qquad 1 \le x.$$

Of these, only the ones with $x \ge 1$ correspond to K–T pairs of (ICP) (each with Lagrange multipliers $\mu_1^* = 0$ and $\mu_2^* = 0$). Proposition 4.7 applies to these points with $c^* = 0$. The critical points $\frac{1}{2}(1 - \sqrt{5})$ and 0 are not covered by Proposition 4.7 since the corresponding sets of gradients $\{\nabla g_j(x) \,|\, g_j(x) = P(x), j = 1, 2\}$ are linearly dependent. Propositions 4.9 and 4.11 are inapplicable, since g_2 is not convex.

Example 2 illustrates the type of difficulties that are unavoidable if we attempt to solve (ICP) by minimizing $f + cP$. The minimization method will be attracted to the infeasible local minimum $\frac{1}{2}(1 - \sqrt{5})$ if started near it independently of the value of c (compare with Fig. 4.2). A very similar situation occurs in connection with the quadratic penalty method (compare with the example following Proposition 2.3).

Extension to Mixed Equality and Inequality Constraints

Consider now the general problem

$$\text{(NLP)} \qquad \begin{aligned} &\text{minimize} \quad f(x) \\ &\text{subject to} \quad h(x) = 0, \qquad g(x) \le 0. \end{aligned}$$

We can convert this problem into one of the form (ICP) by converting each equality constraint into two inequality constraints.

By denoting

$$h_i(x) = g_{r+2i-1}(x), \qquad -h_i(x) = g_{r+2i}(x),$$

we obtain the problem

$$\begin{aligned} &\text{minimize} \quad f(x) \\ &\text{subject to} \quad g_j(x) \le 0, \qquad j = 1, \ldots, r + 2m. \end{aligned}$$

We can then transfer in a straightforward manner the analysis for (ICP) to (NLP). We summarize these results, leaving many of the details to the reader.

Denote, for all $x \in R^n$, $d \in R^n$, and $c > 0$,

$$\begin{aligned} g_0(x) &= 0, \\ P(x) &= \max\{g_0(x), g_1(x), \ldots, g_r(x), |h_1(x)|, \ldots, |h_m(x)|\}, \\ J(x) &= \{j \mid g_j(x) = P(x), j = 1, \ldots, r\}, \\ I(x) &= \{i \mid |h_i(x)| = P(x), i = 1, \ldots, m\}, \\ \theta_c(x; d) &= \nabla f(x)'d + c \max\{\nabla g_j(x)'d, \zeta_i(x; d) \mid j \in J(x), i \in I(x)\}, \end{aligned}$$

where

$$\zeta_i(x; d) = \begin{cases} \nabla h_i(x)'d & \text{if } h_i(x) > 0 \\ -\nabla h_i(x)'d & \text{if } h_i(x) < 0 \\ |\nabla h_i(x)'d| & \text{if } h_i(x) = 0 \end{cases}$$

A vector x is said to be a *critical point* of $f + cP$ if

$$\theta_c(x; d) \ge 0 \qquad \forall\, d \in R^n.$$

We have that if x is a critical point of $f + cP$, then (compare with Proposition 4.3) the quadratic program

$$\begin{aligned} (\mathrm{QP})_c(x, H, J, I) \quad \text{minimize} \quad & \nabla f(x)'d + \tfrac{1}{2}d'Hd + c\xi \\ \text{subject to} \quad & g_j(x) + \nabla g_j(x)'d \le \xi, \qquad j \in J, \\ & |h_i(x) + \nabla h_i(x)'d| \le \xi, \qquad i \in I, \end{aligned}$$

has $\{d = 0, \xi = P(x)\}$ as its optimal solution for each H, J, and I with

$$(30) \qquad 0 < H, \qquad J(x) \subset J \subset \{0, 1, \ldots, r\}, \qquad I(x) \subset I \subset \{1, \ldots, m\}.$$

Conversely, if $\{d = 0, \xi = P(x)\}$ is the optimal solution of $(\mathrm{QP})_c(x, H, J, I)$ for some H, J, and I satisfying (30), then x is a critical point of $f + cP$.

If x is not a critical point of $f + cP$, then there is a $d \in R^n$ such that $\theta_c(x; d) < 0$. Such a d is a descent direction for $f + cP$ and can be obtained by solving $(\mathrm{QP})_c(x, H, J, I)$ (compare with Proposition 4.2).

Consider the quadratic program

$$\begin{aligned} (\mathrm{QP})_0(x, H, J, I) \quad \text{minimize} \quad & \nabla f(x)'d + \tfrac{1}{2}d'Hd \\ \text{subject to} \quad & g_j(x) + \nabla g_j(x)'d \le 0, \qquad j \in J, \\ & h_i(x) + \nabla h_i(x)'d = 0, \qquad i \in I, \end{aligned}$$

where H, J, and I satisfy (30). If $\{x^*, (\mu_1^*, \ldots, \mu_r^*), (\lambda_1^*, \ldots, \lambda_m^*)\}$ is a K–T triple of (NLP), then $\{d^* = 0, \{\mu_j^* \,|\, j \in J\}, \{\lambda_i^* \,|\, i \in I\}\}$, where $\mu_0^* = 0$, is a K–T triple of $(\mathrm{QP})_0(x^*, H, J, I)$. Conversely if $\{d^* = 0, \{\mu_j^* \,|\, j \in J\}, \{\lambda_i^* \,|\, i \in I\}\}$ is a K–T triple of $(\mathrm{QP})_0(x^*, H, J, I)$ then $\{x^*, (\mu_1^*, \ldots, \mu_r^*), (\lambda_1^*, \ldots, \lambda_m^*)\}$ is a K–T triple of (NLP), where $\mu_j^* = \lambda_i^* = 0$ for all $j \notin J$ and $i \notin I$ (compare with Proposition 4.4).

If d is the optimal solution of $(\mathrm{QP})_0(x, H, J, I)$, with corresponding Lagrange multipliers $\{\mu_j \,|\, j \in J\}$, $\{\lambda_i \,|\, i \in I\}$, and

$$c \ge \sum_{\substack{j \in J \\ j \ne 0}} \mu_j + \sum_{i \in I} |\lambda_i|,$$

then $\{d, \xi = 0\}$ is the optimal solution of $(\mathrm{QP})_c(x, H, J, I)$ (compare with Proposition 4.5).

If $\{x^*, (\mu_1^*, \ldots, \mu_r^*), (\lambda_1^*, \ldots, \lambda_m^*)\}$ is a K–T triple of (ICP), then x^* is a critical point of $f + cP$ for all c with

$$c \ge \sum_{j=1}^{r} \mu_j^* + \sum_{i=1}^{m} |\lambda_i^*|$$

(compare with Proposition 4.6).

We state formally the analog of Proposition 4.7:

Proposition 4.12: Let $X \subset R^n$ be a compact set such that for all $x \in X$ the set of gradients

$$\{\nabla g_j(x), \nabla h_i(x) \mid j \in J(x), j \neq 0, i \in I(x)\}$$

is linearly independent. There exists a $c^* \geq 0$ such that for every $c > c^*$:

(a) If x^* is a critical point of $f + cP$ and $x^* \in X$, there exist $\mu^* \in R^r$ and $\lambda^* \in R^m$ such that (x^*, μ^*, λ^*) is a K–T triple for (NLP).

(b) If (x^*, μ^*, λ^*) is a K–T triple of (NLP) and $x^* \in X$, then x^* is a critical point of $f + cP$.

The proof of Proposition 4.12 is nearly identical to the proof of Proposition 4.7. As a threshold value c^*, one may take

$$c^* = \max_{x \in X} \left\{ \sum_{j=1}^{r} \bar{\mu}_j(x) + \sum_{i=1}^{m} |\bar{\lambda}_i(x)| \right\},$$

where $\bar{\mu}(x)$ and $\bar{\lambda}(x)$ are the unique minimizing vectors of

$$q_x(\mu, \lambda) = |\nabla f(x) + \nabla g(x)\mu + \nabla h(x)\lambda|^2 + \sum_{j=1}^{r} [P(x) - g_j(x)]^2 \mu_j^2$$

$$+ \sum_{i=1}^{m} [P(x) - |h_i(x)|]^2 \lambda_i^2.$$

4.2 Linearization Algorithms Based on Nondifferentiable Exact Penalty Functions

4.2.1 Algorithms for Minimax Problems

We first consider an algorithm for finding critical points of $f + cP$, where $c > 0$,

$$P(x) = \max\{g_0(x), g_1(x), \ldots, g_r(x)\} \qquad \forall\, x \in R^n,$$

$$g_0(x) = 0 \qquad \forall\, x \in R^n,$$

and $f, g_j \in C^1, j = 1, \ldots, r$. We subsequently specialize the algorithm and the corresponding convergence analysis to (ICP).

Linearization Algorithm: A vector $x_0 \in R^n$ is chosen and the kth iteration of the algorithm is given by

(1) $$x_{k+1} = x_k + \alpha_k d_k,$$

where α_k is a nonnegative scalar stepsize, and d_k is a direction obtained by solving the quadratic program in (d, ξ)

$$(\mathrm{QP})_c(x_k, H_k, J_k) \quad \text{minimize } \nabla f(x_k)'d + \tfrac{1}{2}d'H_k d + c\xi$$
$$\text{subject to } \quad g_j(x_k) + \nabla g_j(x_k)'d \le \xi, \qquad j \in J_k.$$

We require that H_k and J_k satisfy

$$0 < H_k, \qquad J_\delta(x_k) \subset J_k \subset \{0, 1, \ldots, r\},$$

where

$$J_\delta(x_k) = \{j \mid g_j(x_k) \ge P(x_k) - \delta, j = 0, 1, \ldots, r\},$$

and δ is some positive scalar which is fixed throughout the algorithm. The stepsize α_k is chosen by any one of the stepsize rules listed below:

(a) *Minimization rule:* Here α_k is chosen so that

$$f(x_k + \alpha_k d_k) + cP(x_k + \alpha_k d_k) = \min_{\alpha \ge 0}\{f(x_k + \alpha d_k) + cP(x_k + \alpha d_k)\}.$$

(b) *Limited minimization rule:* A fixed scalar $s > 0$ is selected and α_k is chosen so that

$$f(x_k + \alpha_k d_k) + cP(x_k + \alpha_k d_k) = \min_{\alpha \in [0, s]}\{f(x_k + \alpha d_k) + cP(x_k + \alpha d_k)\}.$$

(c) *Armijo rule:* Fixed scalars s, β, and σ, with $s > 0$, $\beta \in (0, 1)$, and $\sigma \in (0, \frac{1}{2})$, are selected and we set $\alpha_k = \beta^{m_k}s$, where m_k is the first nonnegative integer m for which

$$(2) \quad f(x_k) + cP(x_k) - f(x_k + \beta^m s d_k) - cP(x_k + \beta^m s d_k) \ge \sigma\beta^m s d_k' H_k d_k.$$

We do not discuss the rather complex question of practical (approximate) implementation of the minimization rules. On the other hand, it is easy to show that if $d_k \ne 0$, the Armijo rule will yield a stepsize after a finite number of arithmetic operations. To see this, note that by Proposition 4.2, we have for all $\alpha > 0$,

$$(3) \quad f(x_k) + cP(x_k) - f(x_k + \alpha d_k) - cP(x_k + \alpha d_k) = -\alpha\theta_c(x_k; d_k) + o(\alpha)$$
$$\ge \alpha d_k' H_k d_k + o(\alpha).$$

Hence if $\bar{\alpha} > 0$ is such that for $\alpha \in (0, \bar{\alpha}]$ we have $(1 - \sigma)\alpha d_k' H_k d_k + o(\alpha) \ge 0$, then it follows using (3) that

$$f(x_k) + cP(x_k) - f(x_k + \alpha d_k) - cP(x_k + \alpha d_k) \ge \sigma\alpha d_k' H_k d_k \qquad \forall\, \alpha \in (0, \bar{\alpha}].$$

Therefore there is an integer m such that (2) is satisfied. Note also that if $d_k = 0$ then, by Proposition 4.3, x_k is a critical point of $f + cP$.

Note that, in implementing the algorithm instead of solving $(\mathrm{QP})_c(x_k, H_k, J_k)$, it is possible to solve a dual problem. The reader who is familiar with duality theory can verify that one such dual problem involving maximization with respect to the Lagrange multipliers $\mu_j, j \in J_k$, is given by

$$\text{maximize} \quad -\tfrac{1}{2}\left[\nabla f(x_k) + \sum_{j \in J_k} \mu_j \nabla g_j(x_k)\right]' H_k^{-1}\left[\nabla f(x_k) + \sum_{j \in J_k} \mu_j \nabla g_j(x_k)\right] + \sum_{j \in J_k} \mu_j g_j(x_k)$$

$$\text{subject to} \quad \sum_{j \in J_k} \mu_j = c, \qquad \mu_j \geq 0 \qquad \forall\, j \in J_k.$$

It may be advantageous to solve the dual problem above, since it has a simpler constraint set than $(\text{QP})_c(x_k, H_k, J_k)$ and possibly a smaller number of variables.

We have the following convergence result:

Proposition 4.13: Let $\{x_k\}$ be a sequence generated by the linearization algorithm where the stepsize α_k is chosen by either the minimization rule or the limited minimization rule or the Armijo rule. Assume that there exist positive scalars γ and Γ such that

$$\gamma |z|^2 \leq z' H_k z \leq \Gamma |z|^2 \qquad \forall\, z \in R^n, \quad k = 0, 1, \ldots. \tag{4}$$

Then every limit point of $\{x_k\}$ is a critical point of $f + cP$.

Proof: We provide a proof by contradiction. Assume that a subsequence $\{x_k\}_K$ generated by the algorithm using the Armijo rule converges to a vector x which is not a critical point of $f + cP$. We may assume without loss of generality that for some index set J, we have

$$J_\delta(x) \subset J \subset \{0, 1, \ldots, r\}, \qquad J_k = J \qquad \forall\, k \in K.$$

Since $f(x_k) + cP(x_k)$ is monotonically decreasing, we have $f(x_k) + cP(x_k) \to f(x) + cP(x)$ and hence also $\{f(x_k) + cP(x_k) - f(x_{k+1}) - cP(x_{k+1})\} \to 0$. By the definition of the Armijo rule, we have

$$f(x_k) + cP(x_k) - f(x_{k+1}) - cP(x_{k+1}) \geq \sigma \alpha_k d_k' H_k d_k.$$

Hence

$$\alpha_k d_k' H_k d_k \to 0. \tag{5}$$

Since for $k \in K$, d_k is the optimal solution of $(\text{QP})_c(x_k, H_k, J)$, it follows that for some set of Langrange multipliers $\{\mu_j^k \mid j \in J\}$ and all $k \in K$, we have

$$\nabla f(x_k) + \sum_{j \in J} \mu_j^k \nabla g_j(x_k) + H_k d_k = 0, \qquad c = \sum_{j \in J} \mu_j^k, \tag{6}$$

$$\mu_j^k \geq 0, \qquad \mu_j^k[g_j(x_k) + \nabla g_j(x_k)' d_k - \xi_k] = 0 \qquad \forall\, j \in J, \tag{7}$$

where

$$\xi_k = \max_{j \in J}\{g_j(x_k) + \nabla g_j(x_k)' d_k\}.$$

The relations $c = \sum_{j \in J} \mu_j^k$ and $\mu_j^k \geq 0$ imply that the subsequences $\{\mu_j^k\}_K$ are bounded. Hence, without loss of generality, we may assume that for some $\mu_j, j \in J$, we have

$$\{\mu_j^k\}_K \to \mu_j \qquad \forall\, j \in J. \tag{8}$$

Using assumption (4), we can also assume without loss of generality that

$$\{H_k\}_K \to H \tag{9}$$

for some positive definite matrix H.

Now from (5), it follows that there are two possibilities. Either

$$\liminf_{\substack{k \to \infty \\ k \in K}} \{|d_k|\} = 0 \tag{10}$$

or else

$$\liminf_{\substack{k \to \infty \\ k \in K}} \alpha_k = 0, \qquad \liminf_{\substack{k \to \infty \\ k \in K}} \{|d_k|\} > 0. \tag{11}$$

If (10) holds, then we may assume without loss of generality that $\{d_k\}_K \to 0$ and from (6)–(9), we have

$$\nabla f(x) + \sum_{j \in J} \mu_j \nabla g_j(x) = 0, \qquad c = \sum_{j \in J} \mu_j,$$

$$\mu_j \geq 0, \qquad \mu_j[g_j(x) - \xi] = 0 \qquad \forall\, j \in J,$$

where $\xi = \max_{j \in J} g_j(x)$. Hence the quadratic program $(\mathrm{QP})_c(x, H, J)$ has $\{d = 0, P(x)\}$ as its optimal solution, while we have $J(x) \subset J_\delta(x) \subset J \subset \{0, 1, \ldots, r\}$. From Proposition 4.3, it follows that x is a critical point of $f + cP$ thus contradicting the hypothesis made earlier.

If (11) holds, we may assume without loss of generality that

$$\{\alpha_k\}_K \to 0. \tag{12}$$

Since (6), (8), and (9) show that $\{d_k\}_K$ is a bounded sequence, we may assume without loss of generality that

$$\{d_k\}_K \to d \neq 0, \tag{13}$$

where d is some vector which cannot be zero in view of (11). Since $\{\alpha_k\}_K \to 0$, it follows, in view of the definition of the Armijo rule, that the initial stepsize s

will be reduced at least once for all $k \in K$ after some index $\bar{k}$. This means that for all $k \in K$, $k \geq \bar{k}$,

$$f(x_k) + cP(x_k) - f(x_k + \bar{\alpha}_k d_k) - cP(x_k + \bar{\alpha}_k d_k) < \sigma \bar{\alpha}_k d_k' H_k d_k, \tag{14}$$

where

$$\bar{\alpha}_k = \alpha_k / \beta. \tag{15}$$

From Proposition 4.2a we have

(16)

$$f(x_k) + cP(x_k) - f(x_k + \bar{\alpha}_k d_k) - cP(x_k + \bar{\alpha}_k d_k) = -\theta_c(x_k; \bar{\alpha}_k d_k) + o(\bar{\alpha}_k),$$

where

$$\lim_{\substack{k \to \infty \\ k \in K}} \frac{o(\bar{\alpha}_k)}{\bar{\alpha}_k} = 0, \tag{17}$$

while from Proposition 4.2b, we have

$$-\theta_c(x_k; d_k) \geq d_k' H_k d_k.$$

Combining this relation with (14) and (16), we obtain

$$(1 - \sigma) d_k' H_k d_k + o(\alpha_k)/\alpha_k < 0.$$

In view of (9), (13), and (17) this leads to a contradiction. This completes the proof of the proposition for the case of the Armijo rule.

Consider now the minimization rule and let $\{x_k\}_K$ converge to a vector x which is not a critical point of $f + cP$. Let $\tilde{x}_{k+1}$ be the point that would be generated from x_k via the Armijo rule and let $\tilde{\alpha}_k$ be the corresponding stepsize. We have

$$f(x_k) + cP(x_k) - f(x_{k+1}) - cP(x_{k+1}) \geq f(x_k) + cP(x_k) -$$
$$f(\tilde{x}_{k+1}) - cP(\tilde{x}_{k+1}) \geq \sigma \tilde{\alpha}_k d_k' H_k d_k.$$

By simply replacing α_k by $\tilde{\alpha}_k$ in the arguments of the earlier proof, we obtain a contradiction. In fact, this line of argument establishes that any stepsize rule that gives a larger reduction in objective function value at each iteration than the Armijo rule inherits its convergence properties. This proves also the proposition for the limited minimization rule. Q.E.D.

It is possible to relax somewhat condition (4) and still be able to prove the result of Proposition 4.13 similarly as in Section 1.3.1. For example, the reader may wish to verify that (4) can be replaced by the condition

$$\gamma |w(x_k)|^{q_1} |z|^2 \leq z' H_k z \leq \Gamma |w(x_k)|^{q_2} |z|^2 \qquad \forall\, z \in R^n, \quad k = 0, 1, \ldots,$$

where q_1 and q_2 are some nonnegative scalars and $w(\cdot)$ is a continuous function such that $w(x) \neq 0$ if x is not a critical point of $f + cP$ [for example, $w(x) = \min\{\theta_c(x; d) \,|\, |d| \leq 1\}$].

The method of proof of Proposition 4.13 can be used to show that if an alternative form of the Armijo rule given by

$$f(x_k) + cP(x_k) - f(x_k + \beta^{m_k} s d_k) - cP(x_k + \beta^{m_k} s d_k) \geq -\sigma \xi_c(x_k; \beta^{m_k} s d_k),$$

where ξ_c is defined by

$$\xi_c(x; d) = \nabla f(x)'d + c \max\{g_j(x) + \nabla g_j(x)'d \mid j = 0, 1, \ldots, r\} - cP(x), \tag{18}$$

then the result of Proposition 4.13 also holds. The verification of this fact is left for the reader (see also Section 4.5.3). This form of the Armijo rule has been suggested by Mayne and Polak (1978). In contrast with (2), it requires the evaluation of the gradients of all the constraint functions.

4.2.2. Algorithms for Constrained Optimization Problems

Consider the inequality constrained problem

$$\text{(ICP)} \qquad \begin{aligned} &\text{minimize} \quad f(x) \\ &\text{subject to} \quad g_j(x) \leq 0, \qquad j = 1, \ldots, r. \end{aligned}$$

We know that each K–T pair (x^*, μ^*) of (ICP) gives rise to a critical point of $f + cP$ provided $c \geq \sum_{j=1}^{r} \mu_j^*$. Thus we can apply the linearization algorithm for finding critical points of $f + cP$. The difficulty with this is that we may not know a suitable threshold value for c. Under these circumstances, a possible approach is to choose an initial value c_0 for c and increase it as necessary at each iteration k if the algorithm indicates that the current value c_k is inadequate. An underestimate for a suitable value of c_k is $\sum_{j \in J_k, j \neq 0} \mu_j^k$, where $\{\mu_j^k \mid j \in J_k\}$ are Lagrange multipliers obtained by solving $(\text{QP})_0(x_k, H_k, J_k)$ (compare with Proposition 4.5). At the same time, we know that if

$$c_k \geq \sum_{\substack{j \in J_k \\ j \neq 0}} \mu_j^k,$$

then the problem $(\text{QP})_0(x_k, H_k, J_k)$ is equivalent to $(\text{QP})_{c_k}(x_k, H_k, J_k \cup \{0\})$ in the sense that d_k is the optimal solution of the former if and only if $(d_k, 0)$ is the optimal solution of the latter (Proposition 4.5). So by solving $(\text{QP})_0(x_k, H_k, J_k)$, we not only solve $(\text{QP})_{c_k}(x_k, H_k, J_k \cup \{0\})$, as needed in the linearization algorithm, but we also simultaneously obtain an underestimate for a suitable value of c_k. These considerations lead to the following algorithm:

Modified Linearization Algorithm: A vector $x_0 \in R^n$ and a penalty parameter $c_0 > 0$ are chosen. The kth iteration of the algorithm is given by

$$x_{k+1} = x_k + \alpha_k d_k, \qquad c_{k+1} = \bar{c}_k,$$

where α_k is chosen by any one of the stepsize rules given in Section 4.2.1 with c replaced by $\bar{c}_k$. [For example, the minimization rule takes the form

$$f(x_k + \alpha_k d_k) + \bar{c}_k P(x_k + \alpha_k d_k) = \min_{\alpha \geq 0}\{f(x_k + \alpha d_k) + \bar{c}_k P(x_k + \alpha d_k)\}.$$

The vector d_k and the scalar $\bar{c}_k$ depend on x_k, c_k, a matrix H_k, and an index set J_k satisfying

$$0 < H_k, \qquad J_\delta(x_k) \subset J_k \subset \{0, 1, \ldots, r\},$$

where

$$J_\delta(x_k) = \{j \mid g_j(x_k) \geq P(x_k) - \delta, j = 0, 1, \ldots, r\}$$

and $\delta > 0$ is a scalar fixed throughout the algorithm. They are obtained depending on which of the following cases holds true as follows:

CASE 1: There exists $d \in R^n$ satisfying

$$g_j(x_k) + \nabla g_j(x_k)'d \leq 0 \qquad \forall\, j \in J_k. \tag{19}$$

In this case d_k is the unique solution of

$$(\mathrm{QP})_0(x_k, H_k, J_k) \quad \begin{array}{ll} \text{minimize} & \nabla f(x_k)'d + \tfrac{1}{2}d'H_k d \\ \text{subject to} & g_j(x_k) + \nabla g_j(x_k)'d \leq 0 \qquad \forall\, j \in J_k \end{array}$$

and $\bar{c}_k$ is defined by

$$\bar{c}_k = \begin{cases} \sum_{j \in J_k, j \neq 0} \mu_j^k + \varepsilon & \text{if } \sum_{j \in J_k, j \neq 0} \mu_j^k \geq c_k, \\ c_k & \text{otherwise,} \end{cases}$$

where $\{\mu_j^k \mid j \in J_k\}$ is a set of Lagrange multipliers for $(\mathrm{QP})_0(x_k, H_k, J_k)$, and $\varepsilon > 0$ is a scalar that is fixed throughout the algorithm.

NOTES: (1) When there are equality constraints of the form $h_i(x) = 0$, they can be treated by conversion to the inequalities $h_i(x) \leq 0$ and $-h_i(x) \leq 0$. In that case, the corresponding quadratic program is

$$\begin{array}{lll} \text{minimize} & \nabla f(x_k)'d + \tfrac{1}{2}d'H_k d & \\ \text{subject to} & g_j(x_k) + \nabla g_j(x_k)'d \leq 0 & \forall\, j \in J_k, \\ & h_i(x_k) + \nabla h_i(x_k)'d = 0 & \forall\, i \in I_k, \end{array}$$

where I_k is an index set containing $\{i \mid |h_i(x_k)| \geq P(x_k) - \delta\}$. The definition of $\bar{c}_k$ becomes

$$\bar{c}_k = \begin{cases} \sum_{j \in J_k, j \neq 0} \mu_j^k + \sum_{i \in I_k} |\lambda_i^k| + \varepsilon & \text{if } \sum_{j \in J_k, j \neq 0} \mu_j^k + \sum_{i \in I_k} |\lambda_i^k| \geq c_k, \\ c_k & \text{otherwise,} \end{cases}$$

where $\{\mu_j^k, \lambda_i^k \mid j \in J_k, i \in I_k\}$ is a set of Lagrange multipliers for the quadratic program.

(2) In place of $(QP)_0(x_k, H_k, J_k)$, it may be easier to solve the dual problem in $\mu_j, j \in J_k$,

$$\text{maximize} \quad -\tfrac{1}{2}[\nabla f(x_k) + \sum_{j \in J_k} \mu_j \nabla g_j(x_k)]' H_k^{-1} [\nabla f(x_k) + \sum_{j \in J_k} \mu_j \nabla g_j(x_k)] + \sum_{j \in J_k} \mu_j g_j(x_k)$$

$$\text{subject to} \quad \mu_j \geq 0, \qquad j \in J_k.$$

CASE 2: There does not exist $d \in R^n$ satisfying (19). In this case, d_k together with some $\xi_k > 0$ are the unique solution of

$$(QP)_{c_k}(x_k, H_k, J_k) \quad \text{minimize} \quad \nabla f(x_k)'d + \tfrac{1}{2} d' H_k d_k + c_k \xi$$
$$\text{subject to} \quad g_j(x_k) + \nabla g_j(x_k)'d \leq \xi, \qquad j \in J_k,$$

and

$$\bar{c}_k = c_k.$$

We observe that the sequence $\{c_k\}$ generated by the algorithm is unbounded only if the sequence $\{x_k\}$ is such that the system (19) is feasible for an infinite number of indices k with $\sum_{j \in J_k, j \neq 0} \mu_j^k \geq c_k$. Otherwise, for some $\bar{c} > 0$, we have $c_k = \bar{c}$ for all k sufficiently large, and Proposition 4.5 implies that the algorithm is equivalent to the earlier linearization algorithm for which Proposition 4.13 applies. In this way, we obtain the following convergence result.

Proposition 4.14: Let $\{x_k\}$ be a sequence generated by the modified linearization algorithm where the stepsize α_k is chosen by either the minimization rule or the limited minimization rule or the Armijo rule. Assume that there exist positive scalars γ and Γ such that

$$\gamma |z|^2 \leq z' H_k z \leq \Gamma |z|^2 \qquad \forall\, z \in R^n, \quad k = 0, 1, \ldots.$$

(a) If there exist $\bar{k}$ and $\bar{c}$ such that

$$c_k = \bar{c} \qquad \forall\, k \geq \bar{k}, \tag{20}$$

then every limit point of $\{x_k\}$ is a critical point of $f + \bar{c}P$. If, in addition, the system of inequalities

$$g_j(x_k) + \nabla g_j(x_k)'d \leq 0 \qquad \forall\, j \in J_k \tag{21}$$

is feasible for an infinite set of indices K, then every limit point of $\{x_k, \mu^k\}_K$ is a K–T pair of (ICP), where for $k \in K$, we have that

$$\mu^k = (\mu_1^k, \ldots, \mu_r^k),$$

$\{\mu_j^k \mid j \in J_k\}$ is a set of Lagrange multipliers of $(QP)_0(x_k, H_k, J_k)$, and $\mu_j^k = 0$ for $j \notin J_k$.

(b) If the functions $g_1, \ldots, g_r$ are convex, there exists a vector $\bar{x} \in R^n$ such that

$$g_j(\bar{x}) < 0 \qquad \forall\, j = 1, \ldots, r,$$

and the sequence $\{x_k\}$ is bounded, then every limit point of $\{x_k, \mu^k\}$ is a K–T pair of (ICP).

Proof: (a) The proof follows from the remarks preceding the statement of the proposition.

(b) Under the assumptions of this part, the system (21) is feasible for all k. An argument which is very similar to the one used in the proof of Proposition 4.9 shows that

$$\sum_{\substack{j \in J_k \\ j \neq 0}} \mu_j^k \leq \frac{\frac{1}{2}\nabla f(x_k)' H_k^{-1} \nabla f(x_k) + \nabla f(x_k)'(\bar{x} - x_k) + \frac{1}{2}(\bar{x} - x_k)' H_k (\bar{x} - x_k)}{\min\{-g_j(\bar{x}) \,|\, j = 1, 2, \ldots, r\}}$$

It follows that if $\{x_k\}$ is bounded then $\{\sum_{j \in J_k, j \neq 0} \mu_j^k\}$ is also bounded which implies that (20) holds for some $\bar{k}$. The result follows from part (a). Q.E.D.

It is interesting to note that the proof of Proposition 4.14b hinges on the fact that its assumptions guarantee that the system (21) is feasible and the sequence of multipliers $\{\mu_j^k \,|\, j \in J_k, j \neq 0\}$ of $(\mathrm{QP})_0(x_k, H_k, J_k)$ is bounded if $\{x_k\}$ and $\{H_k\}$ are bounded. There are assumptions other than the ones of Proposition 4.14b that guarantee boundedness of $\{\mu_j^k \,|\, j \in J_k,\ j \neq 0\}$. For example, the reader may wish to verify a convergence result similar to Proposition 4.14b for the problem

$$\text{(ECP)} \qquad \text{minimize} \quad f(x)$$
$$\text{subject to} \quad h_i(x) = 0, \qquad i = 1, \ldots, m,$$

under the assumption that the set $\{\nabla h_i(x) \,|\, i = 1, \ldots, m\}$ is linearly independent for all $x \in R^n$.

Implementation Aspects

One of the drawbacks of the modified linearization algorithm is that the value of the penalty parameter c_k may increase rapidly during the early stages of the algorithm, while during the final stage of the algorithm a much smaller value of c_k may be adequate to enforce convergence to a K–T pair of (ICP). A large value of c_k results in very sharp corners of the surfaces of equal cost of the penalized objective $f + c_k P$ along the boundary of the constraint set, and can have a substantial adverse effect on algorithmic progress. In this connection, it is interesting to note that if the system $g_j(x_k) + \nabla g_j(x_k)'d \leq 0$, $j \in J_k$ is feasible, then *the direction d_k is independent of c_k while the stepsize α_k*

depends strongly on c_k. For this reason, it may be important to provide schemes that allow for reduction of c_k if circumstances appear to be favorable. One possibility is to monitor the progress of the generated sequence $\{(x_k, \mu^k)\}$ towards satisfying the K–T conditions. If at some iteration k, a pair (x_k, μ^k) is obtained for which a measure of violation of the K–T conditions [for example, $|H_k d_k| + P(x_k)$ or something similar] is reduced by a certain factor over the last time c was reduced, then we set $c_{k+1} = \sum_{j \in J_k,\, j \neq 0} \mu_j^k + \varepsilon$ even if $\sum_{j \in J_k,\, j \neq 0} \mu_j^k < c_k$. This guarantees that even if c_k is reduced for an infinite set of indices K then every limit point of $\{x_k, \mu^k\}_K$ will be a K–T pair. If c_k is reduced only a finite number of times, the resulting scheme is essentially the same as the one involving no reduction for the purposes of convergence analysis.

A most important question relates to the choice of the matrices H_k. In unconstrained minimization, one tries to employ a stepsize $\alpha_k = 1$ together with matrices H_k which approximate the Hessian of the objective function at a solution. A natural analog for the constrained case would be to choose H_k close to the Hessian of the Lagrangian function

$$L(x, \mu) = f(x) + \mu' g(x)$$

evaluated at a K–T pair (x^*, μ^*). Indeed if the objective function is positive definite quadratic, the constraints are linear, and $J_0 = \{0, 1, \ldots, r\}$, then the corresponding algorithm will find the optimal solution in a single iteration.

There are two difficulties relating to such an approach. The first is that $\nabla_{xx}^2 L(x^*, \mu^*)$ may not be positive definite. Actually this is not as serious as might appear. As we discuss more fully in Sections 4.4.2 and 4.5.2, what is important is that H_k approximate closely $\nabla_{xx}^2 L(x^*, \mu^*)$ only on the subspace tangent to the active constraints. Under second-order sufficiency assumptions on (x^*, μ^*), this can be done with positive definite H_k, since then $\nabla_{xx}^2 L(x^*, \mu^*)$ is positive definite on this subspace.

The second difficulty relates to the fact that even if we were to choose H_k equal to the (generally unknown) matrix $\nabla_{xx}^2 L(x^*, \mu^*)$ and even if this matrix is positive definite, it may happen that arbitrarily close to x^* a stepsize $\alpha_k = 1$ is not acceptable by the algorithm because it does not decrease the value of the objective fucntion $f + cP$. An example illustrating this fact is given in Section 4.5.3. This example shows that unless modifications are introduced in the linearization method, we cannot expect to prove a superlinear rate of convergence for broad classes of problems even with a favorable choice of the matrices H_k. We shall consider such modifications in Section 4.5.3, where we shall discuss the possibility of combining the linearization method with superlinearly convergent Lagrangian methods.

When the linearization method converges to a local minimum of (NLP) satisfying the sufficiency Assumption (S^+) of Section 3.1, it can be normally expected to converge at least at a linear rate. As an indication of this, we note that when there are no constraints the method is equivalent to the (scaled) steepest descent method. The proof of a linear convergence rate result is sketched in Section 4.4.1 (see also Pschenichny and Danilin, 1975).

It is interesting to note one important special case where the linearization method with H_k equal to the identity can be expected to converge superlinearly under reasonable assumptions. This is the case where $f(x) \triangleq 0$, and (ICP) is equivalent to the problem of solving the system of nonlinear inequalities

$$g_j(x) \leq 0, \qquad j = 1, 2, \ldots, r.$$

We refer again to Pschenichny and Danilin (1975) for related analysis.

4.3 Differentiable Exact Penalty Functions

4.3.1 Exact Penalty Functions Depending on x and λ

In this section, we show that it is possible to construct a differentiable unconstrained optimization problem involving *joint minimization in x and λ* and having optimal solutions that are related to K–T pairs of the problem

$$\text{(ECP)} \qquad \begin{aligned} &\text{minimize} \quad f(x) \\ &\text{subject to} \quad h(x) = 0, \end{aligned}$$

where *we assume that* $f, h \in C^2$ *on* R^n. To see that something like this is possible consider the Lagrangian function

$$L(x, \lambda) = f(x) + \lambda' h(x), \tag{1}$$

the necessary conditions for optimality

$$\nabla_x L(x, \lambda) = 0, \qquad \nabla_\lambda L(x, \lambda) = h(x) = 0, \tag{2}$$

and the unconstrained optimization problem

$$\begin{aligned} &\text{minimize} \quad \tfrac{1}{2}|h(x)|^2 + \tfrac{1}{2}|\nabla_x L(x, \lambda)|^2 \\ &\text{subject to} \quad (x, \lambda) \in R^n \times R^m. \end{aligned} \tag{3}$$

It is clear that (x^*, λ^*) is a K–T pair for (ECP) if and only if (x^*, λ^*) is a global minimum of (3). It is thus possible to attempt the solution of (ECP) by solving instead the unconstrained problem (3). A drawback of this approach is, however, that the distinction between local minima and local maxima of (ECP) is completely lost when passing to problem (3).

As an alternative to problem (3), we may consider the problem

$$\text{(4)} \qquad \begin{aligned} &\text{minimize} \quad P(x, \lambda; c, \alpha) \\ &\text{subject to} \quad (x, \lambda) \in R^n \times R^m, \end{aligned}$$

where P is defined by

$$\text{(5)} \qquad P(x, \lambda; c, \alpha) = L(x, \lambda) + \tfrac{1}{2}c|h(x)|^2 + \tfrac{1}{2}\alpha|\nabla_x L(x, \lambda)|^2$$

and c and α are some positive scalar parameters. As initial motivation for this, we mention the fact that, for all $c > 0$ and $\alpha > 0$, *if* (x^*, λ^*) *is any* K–T *pair of* (ECP) *then* (x^*, λ^*) *is a critical point of* $P(\cdot, \cdot; c, \alpha)$ [compare with (2) and (5)]. Our main hope, however, is that by introducing $L(x, \lambda)$ in the objective function and by appropriately choosing c and α, we can build into the unconstrained problem (4) a preference towards local minima versus local maxima of (ECP). Before going into this, we examine the relation of critical points of P with K–T pairs of (ECP).

Proposition 4.15: Let $X \times \Lambda$ be a compact subset of $R^n \times R^m$. Assume that $\nabla h(x)$ has rank m for all $x \in X$. There exists a scalar $\bar{\alpha} > 0$ and, for each $\alpha \in (0, \bar{\alpha}]$, a scalar $\bar{c}(\alpha) > 0$ such that for all c and α with

$$\alpha \in (0, \bar{\alpha}], \qquad c \geq \bar{c}(\alpha),$$

every critical point of $P(\cdot, \cdot; c, \alpha)$ belonging to $X \times \Lambda$ is a K–T pair of (ECP). If $\nabla^2_{xx} L(x, \lambda)$ is positive semidefinite for all $(x, \lambda) \in X \times \Lambda$, then $\bar{\alpha}$ can be taken to be any positive scalar.

Proof: The gradient of P is given by

$$\text{(6)} \qquad \nabla P = \begin{bmatrix} \nabla_x P \\ \nabla_\lambda P \end{bmatrix} = \begin{bmatrix} \nabla_x L + c\nabla h h + \alpha \nabla^2_{xx} L \nabla_x L \\ h + \alpha \nabla h' \nabla_x L \end{bmatrix},$$

where all gradients are evaluated at the same point $(x, \lambda) \in X \times \Lambda$. At any critical point of P in $X \times \Lambda$, we have $\nabla P = 0$ which can be written as

$$\text{(7)} \qquad \begin{bmatrix} I + \alpha \nabla^2_{xx} L & c\nabla h \\ \alpha \nabla h' & I \end{bmatrix} \begin{bmatrix} \nabla_x L \\ h \end{bmatrix} = 0.$$

Let $\bar{\alpha} > 0$ be such that for all $\alpha \in (0, \bar{\alpha}]$ the matrix $I + \alpha \nabla^2_{xx} L$ is positive definite on $X \times \Lambda$. (If $\nabla^2_{xx} L$ is positive semidefinite on $X \times \Lambda$, then $\bar{\alpha}$ can be taken to be any positive scalar.) Then from the first equation of system (7), we obtain

$$\text{(8)} \qquad \nabla_x L = -c(I + \alpha \nabla^2_{xx} L)^{-1} \nabla h h,$$

and substitution in the second equation yields

$$\text{(9)} \qquad [\alpha c \nabla h'(I + \alpha \nabla^2_{xx} L)^{-1} \nabla h - I]h = 0.$$

For any $\alpha \in (0, \bar{\alpha}]$, we can choose $\bar{c}(\alpha) > 0$ such that, for all $c \geq \bar{c}(\alpha)$, the matrix on the left above is positive definite on $X \times \Lambda$. For such c and α, we obtain from (9) $h = 0$ and from (8) $\nabla_x L = 0$. Q.E.D.

The proof of the proposition can be adapted to show that there exist $\bar{\alpha} > 0$ and $\bar{c} > 0$ such that, for all $\alpha \in (0, \bar{\alpha}]$ and $c \in (0, \bar{c}]$, every critical point of P in $X \times \Lambda$ is a K–T pair of (ECP). However, as will be seen in Proposition 4.16, there are other reasons that make us prefer a large rather than a small value of c.

The result of Proposition 4.15 might lead one to hypothesize that if h has rank m on the entire space R^n then all the critical points of $P(\cdot, \cdot; c, \alpha)$ are K–T pairs of (ECP). This is not true however. Even under quite favorable circumstances P can have, for every $c > 0$ and $\alpha > 0$, critical points that are unrelated to K–T pairs of (ECP). According to Proposition 4.15, these spurious critical points move towards "infinity" as $c \to \infty$ and $\alpha \to 0$. We illustrate this situation by an example.

Example 1: Consider the scalar problem where

$$f(x) = \tfrac{1}{6}x^3, \qquad h(x) = x, \qquad P(x, \lambda; c, \alpha) = \tfrac{1}{6}x^3 + \lambda x + \tfrac{1}{2}cx^2 + \tfrac{1}{2}\alpha|\tfrac{1}{2}x^2 + \lambda|^2.$$

Here $\{x^* = 0, \lambda^* = 0\}$ is the unique K–T pair. Critical points of P are obtained by solving the equations

$$\nabla_x P = \tfrac{1}{2}x^2 + \lambda + cx + \alpha x(\tfrac{1}{2}x^2 + \lambda) = 0, \qquad \nabla_\lambda P = x + \alpha(\tfrac{1}{2}x^2 + \lambda) = 0.$$

From the second equation, we obtain

$$\lambda = -x/\alpha - \tfrac{1}{2}x^2,$$

and substitution in the first equation yields, after a straightforward calculation,

$$x[x - c + (1/\alpha)] = 0.$$

By solving these equations, we obtain that the critical points of P are $\{x^* = 0, \lambda^* = 0\}$ and $\{x(c, \alpha) = c - 1/\alpha, \lambda(c, \alpha) = (1 - c^2\alpha^2)/2\alpha^2\}$. It can be seen that, for every $c > 0$ and $\alpha > 0$ with $c\alpha \neq 1$, the critical point $[x(c, \alpha), \lambda(c, \alpha)]$ is not a K–T pair of (ECP). On the other hand, for any fixed $\alpha > 0$, we have $\lim_{c \to \infty} x(c, \alpha) = \infty$ and $\lim_{c \to \infty} \lambda(c, \alpha) = -\infty$ which is consistent with the conclusion of Proposition 4.15.

The next example shows that if $\nabla_{xx}^2 L$ is not positive semidefinite on $X \times \Lambda$ then the upper bound $\bar{\alpha}$ cannot be chosen arbitrarily.

Example 2: Let $n = 2$, $m = 1$, and

$$f(x_1, x_2) = -\tfrac{1}{2}x_1^2, \qquad h(x_1, x_2) = x_2.$$

Here $\{x_1^* = 0, x_2^* = 0, \lambda^* = 0\}$ is the unique K–T pair (a global maximum). Also ∇h is constant and equal to $(0, 1)$ so that the rank assumption in Proposition 4.15 is satisfied. Take $\alpha = 1$. We have, for every $c > 0$,

$$P(x, \lambda; c, 1) = -\tfrac{1}{2}x_1^2 + \lambda x_2 + \tfrac{1}{2}cx_2^2 + \tfrac{1}{2}x_1^2 + \tfrac{1}{2}\lambda^2 = \lambda x_2 + \tfrac{1}{2}cx_2^2 + \tfrac{1}{2}\lambda^2.$$

Since P is independent of x_1, any vector of the form $\{x_1 = y, x_2 = 0, \lambda = 0\}$ with $y \in R$ is a critical point of P and of these only the vector $\{x_1^* = 0, x_2^* = 0, \lambda^* = 0\}$ is a K–T pair of (ECP).

The next proposition indicates how local minima of (ECP) relate to unconstrained local minima of P.

Proposition 4.16: Assume $f, h \in C^3$ on R^n.

(a) If x^* is a strict local minimum of (ECP) satisfying, together with a corresponding Lagrange multiplier vector λ^*, the second-order sufficiency assumption (S) of Section 2.2, then for every $\alpha > 0$ there exists a $\bar{c}(\alpha) > 0$ such that, for all $c \geq \bar{c}(\alpha)$, (x^*, λ^*) is a strict unconstrained local minimum of P, and the matrix $\nabla^2 P(x^*, \lambda^*; c, \alpha)$ is positive definite.

(b) Let (x^*, λ^*) be a K–T pair of (ECP). Assume there exists $z \in R^n$ such that $\nabla h(x^*)'z = 0$ and $z'\nabla_{xx}^2 L(x^*, \lambda^*)z < 0$. Then there exists $\bar{\alpha} > 0$ such that for each $\alpha \in (0, \bar{\alpha}]$ and $c > 0$, (x^*, λ^*) is not an unconstrained local minimum of P.

Proof: (a) By differentiating ∇P as given by (6) and taking into account the fact that $\nabla_x L(x^*, \lambda^*) = 0$ and $h(x^*) = 0$, we obtain, via a straightforward calculation,

$$\nabla^2 P(x^*, \lambda^*; c, \alpha) = \begin{bmatrix} H & N \\ N' & 0 \end{bmatrix} + \begin{bmatrix} cNN' & 0 \\ 0 & 0 \end{bmatrix} + \alpha \begin{bmatrix} H \\ N' \end{bmatrix} [H \quad N],$$

where

$$H = \nabla_{xx}^2 L(x^*, \lambda^*), \qquad N = \nabla h(x^*).$$

We can write, for any $(z, w) \in R^{n+m}$,

$$[z' \quad w'] \nabla^2 P(x^*, \lambda^*; c, \alpha) \begin{bmatrix} z \\ w \end{bmatrix} = Q(z, w) + cR(z, w), \tag{10}$$

where Q and R are the quadratic forms

$$Q(z, w) = z'Hz + 2w'N'z + \alpha|Hz + Nw|^2, \tag{11}$$

$$R(z, w) = z'NN'z. \tag{12}$$

If $(z, w) \neq (0, 0)$ and $R(z, w) = 0$, then $N'z = 0$ which implies

$$Q(z, w) = z'Hz + \alpha|Hz + Nw|^2.$$

By the second-order sufficiency assumption we have $z'Hz > 0$ if $z \neq 0$ while if $z = 0$ then $w \neq 0$ and the full rank assumption on N implies $|Hz + Nw|^2 = |Nw|^2 > 0$. In either case, we obtain $Q(z, w) > 0$ for all $(z, w) \neq 0$ with $R(z, w) = 0$. Since R is positive semidefinite, by Lemma 1.25, there exists a $\bar{c}(\alpha) > 0$ such that for all $c \geq \bar{c}(\alpha)$ the quadratic form $Q + cR$, or equivalently the matrix $\nabla^2 P(x^*, \lambda^*; c, \alpha)$, is positive definite. Hence, (x^*, λ^*) is a strict local minimum of P.

(b) Let z be such that $N'z = 0$ and $z'Hz < 0$. For $\alpha < -z'Hz/|Hz|^2$ we obtain, from (10)–(12),

$$[z' \quad 0]\nabla^2 P(x^*, \lambda^*; c, \alpha)\begin{bmatrix} z \\ 0 \end{bmatrix} = z'Hz + \alpha|Hz|^2 < 0,$$

which implies that $\nabla^2 P(x^*, \lambda^*; c, \alpha)$ is not positive semidefinite. Hence, (x^*, λ^*) cannot be a local minimum of P. Q.E.D.

If in Example 2 we take $\alpha > 1$, then we see that the global maximum $\{x_1^* = 0, x_2^* = 0\}$ gives rise to a global minimum of P, and this shows that the upper bound on α is necessary for the conclusion of Proposition 4.16b. Also in Example 1 by computing $\nabla^2 P$ at the global minimum $\{x^* = 0, \lambda^* = 0\}$, we find that it is positive definite if and only if $\alpha c > 1$, so the lower bound $c(\alpha)$ on c is necessary for the conclusion of Proposition 4.16a.

Proposition 4.16b shows in particular that local maxima of (ECP) satisfying the second-order sufficiency conditions for optimality cannot give rise to unconstrained local minima of P provided α is chosen small enough, while under sufficiency assumptions local minima of (ECP) give rise to local minima of P provided c is chosen large enough. This supports our contention that by employing the exact penalty function P in place of $\frac{1}{2}|h(x)|^2 + \frac{1}{2}|\nabla_x L(x, \lambda)|^2$ we provide a preference towards local minima rather than local maxima of (ECP).

We are still not completely satisfied, however, in view of the fact that this property depends on proper choice of both parameters c and α. It turns out that we can eliminate the effect of the parameter α and simultaneously gain some additional flexibility by considering the function

$$P(x, \lambda; c, M) = L(x, \lambda) + \tfrac{1}{2}c|h(x)|^2 + \tfrac{1}{2}|M(x)\nabla_x L(x, \lambda)|^2, \tag{13}$$

where, for each x, $M(x)$ is a $p \times n$ matrix where $m \leq p \leq n$. *We assume that* $M \in C^1$ *on the open set*

$$X^* = \{x \,|\, \nabla h(x) \text{ has rank } m\}. \tag{14}$$

Note that, for $p = n$ and

$$M(x) = \sqrt{\alpha} I,$$

we obtain the function $P(x, \lambda; c, \alpha)$ considered earlier.

We calculate the gradient of P. Let $m_1, \dots, m_p$ denote the columns of M' so that

$$M(x) = \begin{bmatrix} m_1(x)' \\ \vdots \\ m_p(x)' \end{bmatrix},$$

and let $e_1, \dots, e_p$ denote the columns of the $p \times p$ identity matrix, so that

$$m_i(x) = M(x)'e_i, \qquad i = 1, \dots, p,$$

$$m_i(x)' = e_i'M(x), \qquad i = 1, \dots, p.$$

Using this notation and suppressing the argument of all functions, we have

$$\begin{aligned} \nabla_x(\tfrac{1}{2}|M\nabla_x L|^2) &= \nabla_x\left[\tfrac{1}{2}\sum_{i=1}^{p}(m_i'\nabla_x L)^2\right] \\ &= \sum_{i=1}^{p}(\nabla_{xx}^2 L m_i + \nabla m_i \nabla_x L)m_i'\nabla_x L \\ &= \nabla_{xx}^2 L M'\left(\sum_{i=1}^{p} e_i e_i'\right)M\nabla_x L + \left(\sum_{i=1}^{p}\nabla m_i \nabla_x L e_i'\right)M\nabla_x L. \end{aligned}$$

Since $\sum_{i=1}^{p} e_i e_i' = I$, we finally obtain

$$\nabla_x(\tfrac{1}{2}|M\nabla_x L|^2) = \left(\nabla_{xx}^2 L M' + \sum_{i=1}^{p}\nabla m_i \nabla_x L e_i'\right)M\nabla_x L.$$

Using this expression and (13), we obtain

(15a) $$\nabla_x P = \nabla_x L + c\nabla h h + \left(\nabla_{xx}^2 L M' + \sum_{i=1}^{p}\nabla m_i \nabla_x L e_i'\right)M\nabla_x L,$$

(15b) $$\nabla_\lambda P = h + \nabla h' M' M \nabla_x L,$$

where the argument of all functions in the expressions above is the same (typical) vector $(x, \lambda) \in X^* \times R^m$.

The following result is an immediate consequence of the form of P and ∇P given by (13) and (15).

Proposition 4.17: If (x^*, λ^*) is a K–T pair of (ECP) and $x^* \in X^*$, then (x^*, λ^*) is a critical point of $P(\cdot, \cdot; c, M)$ and

$$P(x^*, \lambda^*; c, M) = f(x^*).$$

The following three propositions apply to the case where $p = m$, and the $m \times m$ matrix $M(x)\nabla h(x)$ is nonsingular in some subset of R^n. This can be true only if $\nabla h(x)$ has rank m in which case any choice of M of the form

(16) $$M(x) = A(x)\nabla h(x)',$$

where $A(x)$ is an $m \times m$ nonsingular continuously differentiable matrix on X^*, makes $M(x)\nabla h(x)$ nonsingular. For example, we may choose

$$M(x) = \eta \nabla h(x)',$$

where η is any positive scalar.

Proposition 4.18: Let $X \times \Lambda$ be a compact subset of $X^* \times R^m$, where X^* is given by (14) and assume that $M(x)\nabla h(x)$ is an $m \times m$ nonsingular matrix for all $x \in X$. Then there exists a $\bar{c} > 0$ such that, for all $c \geq \bar{c}$, every critical point of $P(\cdot, \cdot\,; c, M)$ belonging to $X \times \Lambda$ is a K–T pair of (ECP).

Proof: By (15), the condition $\nabla_\lambda P = 0$ at some point of $X \times \Lambda$ implies

$$M \nabla_x L = -(\nabla h' M')^{-1} h, \tag{17}$$

so if at this point $\nabla_x P = 0$ also holds, we obtain, using (15) and (17),

$$\begin{aligned}
0 &= M \nabla_x P \\
&= M \nabla_x L + cM\nabla h h + M\left(\nabla_{xx}^2 L M' + \sum_{i=1}^m \nabla m_i \nabla_x L e_i'\right) M \nabla_x L \\
&= -(\nabla h' M')^{-1} h + cM\nabla h h \\
&\quad - M\left(\nabla_{xx}^2 L M' + \sum_{i=1}^m \nabla m_i \nabla_x L e_i'\right)(\nabla h' M')^{-1} h \\
&= \left\{ cM\nabla h - \left[I + M\left(\nabla_{xx}^2 L M' + \sum_{i=1}^m \nabla m_i \nabla_x L e_i'\right)\right](\nabla h' M')^{-1} \right\} h.
\end{aligned}$$

Since $M\nabla h$ is invertible on X, $M \in C^1$ on X^*, and $X \times \Lambda$ is compact, there exists a $\bar{c} > 0$ such that, for all $c \geq \bar{c}$, the matrix on the right in the above expression is nonsingular. Thus, if $c \geq \bar{c}$, then for every point in $X \times \Lambda$, with $\nabla_x P = 0$ and $\nabla_\lambda P = 0$, we obtain $h = 0$, and from (17), $M\nabla_x L = 0$. Using (15a), we also obtain $\nabla_x L = 0$. So, for $c \geq \bar{c}$, every critical point of $P(\cdot, \cdot\,; c, M)$ in $X \times \Lambda$ is a K–T pair of (ECP). Q.E.D.

We note that Example 1, given earlier, satisfies the assumption of Proposition 4.18 on every compact set with $M(x) = \sqrt{\alpha}\nabla h(x)'$. We saw in that example that, for every $c > 0$ and $\alpha > 0$, P has a spurious critical point that is not a K–T pair. This critical point moves towards "infinity" as c increases, which is consistent with the conclusion of the proposition.

The next proposition and corollary show that isolated local minima of (ECP) on compact sets, which are also regular points, give rise to isolated local minima of P for c sufficiently large.

Proposition 4.19: Let (x^*, λ^*) be a K–T pair of (ECP) and X be a compact subset of X^*. Assume that x^* is the unique global minimum of f

over $X \cap \{x \mid h(x) = 0\}$ and that x^* lies in the interior of X. Assume further that $M(x^*)\nabla h(x^*)$ is an $m \times m$ nonsingular matrix. Then, for every compact set $\Lambda \subset R^m$ containing λ^* in its interior, there exists a $\bar{c} > 0$ such that, for all $c \geq \bar{c}$, (x^*, λ^*) is the unique global minimum of $P(\cdot, \cdot; c, M)$ over $X \times \Lambda$.

Proof: Let $\Lambda \subset R^m$ be a compact set such that λ^* belongs to the interior of Λ. Assume that the conclusion of the proposition is false. Then, for any integer k, there exists $c_k \geq k$ and a global minimum (x_k, λ_k) of $P(\cdot, \cdot; c_k, M)$ over $X \times \Lambda$ such that $(x_k, \lambda_k) \neq (x^*, \lambda^*)$. Therefore

$$P(x_k, \lambda_k; c_k, M) \leq P(x^*, \lambda^*; c_k, M) = f(x^*), \tag{18}$$

where the last equality follows from Proposition 4.17. Hence we have

$$\limsup_{k \to \infty} P(x_k, \lambda_k; c_k, M) \leq f(x^*). \tag{19}$$

We shall show that $\{(x_k, \lambda_k)\}$ converges to (x^*, λ^*). Indeed if $(\bar{x}, \bar{\lambda})$ is a limit point of $\{(x_k, \lambda_k)\}$, then since $c_k \to \infty$ we must have $h(\bar{x}) = 0$ and

$$f(\bar{x}) + \tfrac{1}{2}|M(\bar{x})\nabla_x L(\bar{x}, \bar{\lambda})|^2 \leq f(x^*) \tag{20}$$

for otherwise (19) would be violated. Since also $\bar{x} \in X$ and x^* is the unique global minimum of f over $X \cap \{x \mid h(x) = 0\}$, it follows from (20) that $\bar{x} = x^*$ and $M(\bar{x})\nabla_x L(\bar{x}, \bar{\lambda}) = M(x^*)\nabla_x L(x^*, \bar{\lambda}) = 0$. Taking into account the fact that $\nabla_x L(x^*, \lambda^*) = 0$, we obtain

$$M(x^*)\nabla h(x^*)\bar{\lambda} = M(x^*)\nabla h(x^*)\lambda^*.$$

Since $M(x^*)\nabla h(x^*)$ is invertible we have $\bar{\lambda} = \lambda^*$.

Since $\{(x_k, \lambda_k)\}$ converges to (x^*, λ^*), it follows that there are open spheres S_{x^*} and S_{λ^*} contained in the interior of X and Λ, and centered at x^* and λ^*, respectively, such that $(x_k, \lambda_k) \in S_{x^*} \times S_{\lambda^*}$ for all k sufficiently large. Furthermore we can choose S_{x^*} so that $M(x)\nabla h(x)$ is invertible in the closure of S_{x^*}. By Proposition 4.18, there exists a $\bar{c} > 0$ such that, for all $c \geq \bar{c}$, every critical point of $P(\cdot, \cdot; c, M)$ in the closure of $S_{x^*} \times S_{\lambda^*}$ is a K–T pair. Hence for k sufficiently large, (x_k, λ_k) is a K–T pair, implying $h(x_k) = 0$, and from (18), $f(x_k) \leq f(x^*)$. Since x^* is the unique global minimum of f over $X \cap \{x \mid h(x) = 0\}$, it follows that $x_k = x^*$ for all k sufficiently large. Since $\nabla_x L(x_k, \lambda_k) = 0$ and $\nabla h(x_k)(= \nabla h(x^*))$ has rank m, it follows that $\lambda_k = \lambda^*$ for all k sufficiently large. This contradicts the hypothesis $(x_k, \lambda_k) \neq (x^*, \lambda^*)$ for all k. Q.E.D.

Corollary 4.20: Let (x^*, λ^*) be a K–T pair of (ECP) such that x^* is the unique local minimum of (ECP) over an open sphere S_{x^*}, centered at x^* with $S_{x^*} \subset X^*$, and $\nabla h(x^*)$ has rank m. Then there exists a $\bar{c} > 0$ and an open sphere S_{λ^*} centered at λ^* such that, for all $c \geq \bar{c}$, (x^*, λ^*) is the unique local minimum of $P(\cdot, \cdot; c, M)$ over $S_{x^*} \times S_{\lambda^*}$.

The next proposition shows that local minima of P over any bounded set give rise to local minima of (ECP) provided c is sufficiently large.

Proposition 4.21: Let $X \times \Lambda$ be a compact subset of $X^* \times R^m$ and assume that $M(x)\nabla h(x)$ is an $m \times m$ nonsingular matrix for all $x \in X$. Then there exists a $\bar{c} > 0$ such that, for all $c \geq \bar{c}$, if (x^*, λ^*) is an unconstrained local minimum of $P(\cdot, \cdot; c, M)$ belonging to $X \times \Lambda$, then x^* is a local minimum of (ECP).

Proof: By Proposition 4.18, there exists $\bar{c} > 0$ such that, for all $c \geq \bar{c}$, if (x^*, λ^*) is an unconstrained local minimum of P, then (x^*, λ^*) is a K–T pair of (ECP). This implies

$$P(x^*, \lambda^*; c, M) = f(x^*) \qquad \forall\, c \geq \bar{c},$$

and that there exist open spheres S_{x^*}, S_{λ^*} centered at x^* and λ^*, respectively, such that

$$P(x^*, \lambda^*; c, M) \leq P(x, \lambda; c, M) \qquad \forall\, x \in S_{x^*} \cap \{x \mid h(x) = 0\}, \quad \lambda \in S_{\lambda^*}.$$

The last two relations yield

$$\text{(21)}\quad f(x^*) \leq f(x) + \tfrac{1}{2}|M(x)\nabla f(x) + M(x)\nabla h(x)\lambda|^2 \quad \forall\, x \in S_{x^*} \cap \{x \mid h(x) = 0\}, \quad \lambda \in S_{\lambda^*}.$$

By the continuity and rank assumptions, there exists an open sphere $\hat{S}_{x^*}$ centered at x^* such that

$$\text{(22)}\qquad \lambda = -[M(x)\nabla h(x)]^{-1} M(x)\nabla f(x) \in S_{\lambda^*} \qquad \forall\, x \in \hat{S}_{x^*}.$$

By combining (21) and (22), we obtain

$$f(x^*) \leq f(x) \qquad \forall\, x \in \hat{S}_{x^*} \cap \{x \mid h(x) = 0\},$$

which implies that x^* is a local minimum of (ECP). Q.E.D.

Proposition 4.21 illustrates the advantage gained by using the $m \times n$ matrix M in the formulation of the exact penalty function. Under a nonsingularity condition on $M\nabla h$ we can, by proper choice of the *single* parameter c, guarantee that, within a bounded set, local minima of $P(\cdot, \cdot; c, M)$ can arise only from local minima of (ECP). By contrast, it was necessary to choose appropriately both c and α in the penalty function $P(\cdot, \cdot; c, \alpha)$ in order to achieve the same effect. The price for this is a more complex expression for both P and its derivatives.

We finally mention the more general penalty functions

$$P_\tau(x, \lambda; c, \alpha) = L(x, \lambda) + \tfrac{1}{2}(c + \tau|\lambda|^2)|h(x)|^2 + \tfrac{1}{2}\alpha|\nabla_x L(x, \lambda)|^2$$

and

$$P_\tau(x, \lambda; c, M) = L(x, \lambda) + \tfrac{1}{2}(c + \tau|\lambda|^2)|h(x)|^2 + \tfrac{1}{2}|M(x)\nabla_x L(x, \lambda)|^2,$$

where $\tau \geq 0$ is some fixed scalar. When $\tau = 0$, we obtain the penalty functions examined earlier. A possible advantage in using a positive scalar τ is that if $f + \frac{1}{2}c|h|^2$ is bounded below, then $P_\tau(\cdot, \cdot; c, M)$ is also bounded below for $\tau > 0$ but this is not necessarily true for $\tau = 0$. This can have a beneficial effect in the performance of unconstrained methods for minimizing P. It is possible to show that the results of this section generalize to the penalty functions P_τ. As an aid in this, note that the extra term

$$\tfrac{1}{2}\tau|\lambda|^2|h(x)|^2$$

contributes to $\nabla^2_{xx}P_\tau$ at a K–T pair (x^*, λ^*) the term $\tau|\lambda^*|^2\nabla h(x^*)\nabla h(x^*)'$ and does not otherwise affect the Hessian $\nabla^2 P_\tau$ at (x^*, λ^*). Thus, as far as $\nabla^2 P_\tau$ is concerned, the effect of the added term $\frac{1}{2}\tau|\lambda|^2|h(x)|^2$ at (x^*, λ^*) is the same as adding $\tau|\lambda^*|^2$ to the penalty parameter c.

4.3.2 *Exact Penalty Functions Depending Only on x*

If our ultimate objective is to solve (ECP) by minimizing with respect to (x, λ) the exact penalty function P, we can take advantage of the fact that P is quadratic in λ and minimize explicitly P with respect to λ. Consider the set

$$X^* = \{x \mid \nabla h(x) \text{ has rank } m\}. \tag{23}$$

Let us choose, for $x \in X^*$,

$$M(x) = [\nabla h(x)'\nabla h(x)]^{-1}\nabla h(x)'. \tag{24}$$

Then $M(x)\nabla h(x)$ equals the identity, and we have

$$P(x, \lambda; c, M) = f(x) + \lambda' h(x) + \tfrac{1}{2}c|h(x)|^2 + \tfrac{1}{2}|M(x)\nabla f(x) + \lambda|^2. \tag{25}$$

By setting $\nabla_\lambda P = 0$, we obtain

$$h(x) + M(x)\nabla f(x) + \lambda = 0, \tag{26}$$

so the minimum of P with respect to λ is attained at

$$\hat{\lambda}(x) = -h(x) - [\nabla h(x)'\nabla h(x)]^{-1}\nabla h(x)'\nabla f(x). \tag{27}$$

Substituting in (25) and using (26), we obtain

$$\text{(28)}\quad \min_{\lambda} P(x, \lambda; c, M) = f(x) - h(x)'[\nabla h(x)'\nabla h(x)]^{-1}\nabla h(x)'\nabla f(x) + \tfrac{1}{2}(c-1)|h(x)|^2.$$

We are thus led to consideration of the function

$$\text{(29)}\qquad \hat{P}(x; c) = f(x) + \lambda(x)'h(x) + \tfrac{1}{2}c|h(x)|^2,$$

where

$$\text{(30)}\qquad \lambda(x) = -[\nabla h(x)'\nabla h(x)]^{-1}\nabla h(x)'\nabla f(x) \qquad \forall\, x \in X^*.$$

Note that from (28) we have, for all $x \in X^*$,

$$\text{(31)}\qquad \hat{P}(x; c) = \min_{\lambda} P(x, \lambda; c + 1, M),$$

where

$$\text{(32)}\qquad M(x) = [\nabla h(x)'\nabla h(x)]^{-1}\nabla h(x)' \qquad \forall\, x \in X^*.$$

From (31), we obtain

$$\text{(33)}\qquad \hat{P}(x; c) = P[x, \hat{\lambda}(x); c + 1, M] \qquad \forall\, x \in X^*,$$

where $\hat{\lambda}(x)$ is given by (27). Differentiation with respect to x yields

$$\text{(34)}\quad \nabla\hat{P}(x; c) = \nabla_x P[x, \hat{\lambda}(x); c + 1, M] + \nabla\hat{\lambda}(x)\nabla_\lambda P[x, \hat{\lambda}(x); c + 1, M].$$

Since

$$\text{(35)}\qquad \nabla_\lambda P[x, \hat{\lambda}(x); c + 1, M] = 0 \qquad \forall\, x \in X^*,$$

by definition of $\hat{\lambda}(x)$, we obtain, using (34) and (15a),

$$\begin{aligned}\text{(36)}\quad \nabla\hat{P}(x; c) &= \nabla_x P[x, \hat{\lambda}(x); c + 1, M] \\ &= \nabla_x L[x, \hat{\lambda}(x)] + (c + 1)\nabla h(x)h(x) \\ &\quad + \left\{\nabla^2_{xx} L[x, \hat{\lambda}(x)]M(x)' + \sum_{i=1}^{m} \nabla m_i(x)\nabla_x L[x, \hat{\lambda}(x)]e_i'\right\} \\ &\quad \times M(x)\nabla_x L[x, \hat{\lambda}(x)],\end{aligned}$$

where $\hat{\lambda}$ is given by (27), M is given by (32), m_i is the ith column of M', and e_i is the ith column of the $m \times m$ identity matrix. We have the following proposition.

Proposition 4.22: The following hold true for the set X^* and the function $\hat{P}$ defined by (23) and (29), (30).

(a) If (x^*, λ^*) is a K–T pair of (ECP) and $x \in X^*$, then x^* is a critical point of $\hat{P}(\cdot; c)$ for all $c > 0$.

(b) Let X be a compact subset of X^*. There exists a $\bar{c} > 0$ such that, for all $c \geq \bar{c}$, if x^* is a critical point of $\hat{P}(\cdot; c)$ and $x^* \in X$, then $[x^*, \lambda(x^*)]$ is a K–T pair of (ECP).

(c) Let (x^*, λ^*) be a K–T pair of (ECP) and X be a compact subset of X^*. Assume that x^* is the unique global minimum of f over $X \cap \{x \mid h(x) = 0\}$ and that x^* lies in the interior of X. Then there exists a $\bar{c} > 0$ such that, for all $c \geq \bar{c}$, x^* is the unique global minimum of $\hat{P}(\cdot\,; c)$ over X.

(d) Let X be a compact subset of X^*. There exists a $\bar{c} > 0$ such that, for all $c \geq \bar{c}$, if x^* is a local unconstrained minimum of $\hat{P}(\cdot\,; c)$ belonging to X, then x^* is a local minimum of (ECP).

Proof: (a) We have $\nabla f(x^*) + \nabla h(x^*)\lambda^* = 0$ from which $\nabla h(x^*)'\nabla f(x^*) + \nabla h(x^*)'\nabla h(x^*)\lambda^* = 0$. Since $\nabla h(x^*)$ has rank m and $h(x^*) = 0$, it follows that $\lambda^* = \hat{\lambda}(x^*)$ [compare with (27)]. Since $\nabla_x L[x^*, \hat{\lambda}(x^*)] = 0$ and $h(x^*) = 0$, it follows from (36) that $\nabla \hat{P}(x^*; c) = 0$ for all $c > 0$.

(b) From (34) and (35), it follows that $x^* \in X$ is a critical point of $\hat{P}(\cdot\,; c)$ if and only if $[x^*, \hat{\lambda}(x^*)]$ is a critical point of $P(\cdot, \cdot\,; c + 1, M)$. Consider the compact set $\Lambda = \{\lambda \mid \lambda = \hat{\lambda}(x), x \in X\}$. By Proposition 4.18, there exists a $\bar{c} > 0$ such that if $c \geq \bar{c}$ and $[x^*, \hat{\lambda}(x^*)] \in X \times \Lambda$ is a critical point of $P(\cdot, \cdot\,; c + 1, M)$, then $[x^*, \hat{\lambda}(x^*)]$ is a K–T pair of (ECP). Since for $h(x^*) = 0$, we have $\hat{\lambda}(x^*) = \lambda(x^*)$, the result follows.

(c) By Proposition 4.19, for any compact set Λ containing λ^* in its interior, there exists a $\bar{c} > 0$ such that, for all $c \geq \bar{c}$, (x^*, λ^*) is a global minimum of $P(\cdot, \cdot\,; c + 1, M)$ over $X \times \Lambda$. The result follows using (31).

(d) From (31) and (33), it follows that $x^* \in X$ is a local unconstrained minimum of $\hat{P}(\cdot\,; c)$ if and only if $[x^*, \hat{\lambda}(x^*)]$ is a local unconstrained minimum of $P(\cdot, \cdot\,; c + 1, M)$. The result follows from Proposition 4.21. Q.E.D.

The reader may wish to verify that, for Example 1 of Section 4.3.1, the function $\hat{P}(\cdot\,; c)$ has, for every $c > 0$, a critical point $x(c)$ that does not correspond to a K–T pair of (ECP). For this critical point, we have $\lim_{c \to \infty} x(c) = \infty$ which is consistent with the conclusion of Proposition 4.22b.

We note that the form of the function $\hat{P}$ of (29) depends on the particular choice for M given in (25). Different choices for M yield different exact penalty functions. Other functions can also be obtained by minimization of $P_\tau(\cdot, \cdot\,; c, M)$ over λ for positive values of τ.

4.3.3 Algorithms Based on Differentiable Exact Penalty Functions

We have examined so far in this section the following three basic types of exact penalty functions of varying degrees of complexity

$$P(x, \lambda; c, \alpha) = L(x, \lambda) + \tfrac{1}{2}c|h(x)|^2 + \tfrac{1}{2}\alpha|\nabla_x L(x, \lambda)|^2,$$
$$P(x, \lambda; c, M) = L(x, \lambda) + \tfrac{1}{2}c|h(x)|^2 + \tfrac{1}{2}|M(x)\nabla_x L(x, \lambda)|^2,$$
$$\hat{P}(x; c) = L[x, \lambda(x)] + \tfrac{1}{2}c|h(x)|^2,$$

where

$$\lambda(x) = -[\nabla h(x)'\nabla h(x)]^{-1}\nabla h(x)'\nabla f(x).$$

The preceding analysis suggests that unconstrained minimization of any one of these penalty functions is a valid approach for solving (ECP). Any unconstrained minimization algorithm based on derivatives can be used for minimization of P or $\hat{P}$. However it is important to use an algorithm that takes into account the special structure of these functions. The salient feature of this structure is that *the gradients of P and $\hat{P}$ involve the second derivatives of f and h.* If these second derivatives are unavailable or are difficult to compute, they can be suitably approximated by using first derivatives. As an example, take the gradient of $P(x, \lambda; c, \alpha)$. From (6), we have

$$\nabla_x P(x, \lambda; c, \alpha) = \nabla_x L(x, \lambda) + c\nabla h(x)h(x) + \alpha\nabla_{xx}^2 L(x, \lambda)\nabla_x L(x, \lambda),$$

$$\nabla_\lambda P(x, \lambda; c, \alpha) = h(x) + \alpha\nabla h(x)'\nabla_x L(x, \lambda).$$

At any point (x, λ), the troublesome term

$$\nabla_{xx}^2 L(x, \lambda)\nabla_x L(x, \lambda)$$

can be approximated by

$$t^{-1}\{\nabla_x L[x + t\nabla_x L(x, \lambda), \lambda] - \nabla_x L(x, \lambda)\},$$

where t is a small positive scalar. Thus we can bypass the need of computing $\nabla_{xx}^2 L(x, \lambda)$ by means of a single additional evaluation of ∇f and ∇h. A similar approach can be used for the other penalty functions.

If second derivatives can be computed relatively easily then there arises the possibility of using a Newton-like scheme for unconstrained minimization. The difficulty with this is that the Hessian matrix of P or $\hat{P}$ involves third derivatives of f and h. It turns out, however, that *at* K–T *pairs of* (ECP), *the third derivative terms vanish,* so they can be neglected in a Newton-like algorithm without loss of the superlinear rate of convergence property.

Consider first the function $P(x, \lambda; c, \alpha)$. We can write

$$P(x, \lambda; c, \alpha) = L(x, \lambda) + \tfrac{1}{2}\nabla L(x, \lambda)'K\nabla L(x, \lambda),$$

where K is the matrix

$$K = \begin{bmatrix} \alpha I & 0 \\ 0 & cI \end{bmatrix}.$$

We have

$$\nabla P(x, \lambda; c, \alpha) = [I + \nabla^2 L(x, \lambda)K]\nabla L(x, \lambda), \tag{37}$$

while at a K–T pair (x^*, λ^*), we have

$$\nabla^2 P(x^*, \lambda^*; c, \alpha) = \nabla^2 L(x^*, \lambda^*) + \nabla^2 L(x^*, \lambda^*) K \nabla^2 L(x^*, \lambda^*). \tag{38}$$

Thus $\nabla^2 P(x^*, \lambda^*; c, \alpha)$ involves only first and second derivatives of f and h. Consider the Newton-like method

$$\begin{bmatrix} x_{k+1} \\ \lambda_{k+1} \end{bmatrix} = \begin{bmatrix} x_k \\ \lambda_k \end{bmatrix} + \alpha_k \begin{bmatrix} \Delta x_k \\ \Delta \lambda_k \end{bmatrix}, \tag{39}$$

where α_k is a positive stepsize parameter and

$$\begin{bmatrix} \Delta x_k \\ \Delta \lambda_k \end{bmatrix} \triangleq d_k = -B_k^{-1} \nabla P(x_k, \lambda_k; c, \alpha), \tag{40}$$

$$B_k = \nabla^2 L(x_k, \lambda_k) + \nabla^2 L(x_k, \lambda_k) K \nabla^2 L(x_k, \lambda_k). \tag{41}$$

Since B_k approaches $\nabla^2 P(x^*, \lambda^*; c, \alpha)$ as (x_k, λ_k) approaches a K–T pair (x^*, λ^*), we conclude (compare with Proposition 1.15) that the method is well defined and converges superlinearly if $\nabla^2 P(x^*, \lambda^*; c, \alpha) > 0$ and α_k is chosen by the Armijo rule with unity initial stepsize. Now from (37), (40), and (41), we have

$$\begin{aligned} d_k &= -(\nabla^2 L + \nabla^2 L K \nabla^2 L)^{-1}(I + \nabla^2 L K)\nabla L \\ &= -(\nabla^2 L + \nabla^2 L K \nabla^2 L)^{-1}(\nabla^2 L + \nabla^2 L K \nabla^2 L)\nabla^2 L^{-1} \nabla L, \end{aligned}$$

and finally

$$d_k = -\nabla^2 L(x_k, \lambda_k)^{-1} \nabla L(x_k, \lambda_k).$$

An important observation is that d_k *is the Newton direction for solving the system of equations* $\nabla L = 0$. Thus *iteration* (39) *coupled with a stepsize procedure based on descent of the exact penalty function* $P(x, \lambda; c, \alpha)$ *can be alternately viewed as a means for enlarging the region of convergence of Newton's method for solving the system* $\nabla L = 0$. We shall discuss more specific methods of this type in Section 4.5.2.

Consider next the function $P(x, \lambda; c, M)$. We can write

$$P(x, \lambda; c, M) = L(x, \lambda) + \tfrac{1}{2} \nabla L(x, \lambda)' K(x) \nabla L(x, \lambda),$$

where

$$K(x) = \begin{bmatrix} M(x)'M(x) & 0 \\ 0 & cI \end{bmatrix}. \tag{42}$$

We have

$$\nabla P(x, \lambda; c, M) = [I + \tfrac{1}{2}\nabla^2 L K + \tfrac{1}{2}\nabla(K \nabla L)]\nabla L, \tag{43}$$

where $\nabla(K\nabla L)$ denotes the gradient matrix of the function $K\nabla L$ with respect to (x, λ). At any K–T pair (x^*, λ^*), we have

$$\text{(44)} \quad \nabla^2 P(x^*, \lambda^*; c, M) = \nabla^2 L(x^*, \lambda^*) + \tfrac{1}{2}\nabla^2 L(x^*, \lambda^*)K(x^*)\nabla^2 L(x^*, \lambda^*) + \tfrac{1}{2}\nabla(K(x^*)\nabla L(x^*, \lambda^*))\nabla^2 L(x^*, \lambda^*).$$

We are thus led to consideration of the Newton-like method

$$\text{(45)} \qquad \begin{bmatrix} x_{k+1} \\ \lambda_{k+1} \end{bmatrix} = \begin{bmatrix} x_k \\ \lambda_k \end{bmatrix} + \alpha_k \begin{bmatrix} \Delta x_k \\ \Delta \lambda_k \end{bmatrix},$$

where α_k is a positive stepsize parameter and

$$\text{(46)} \qquad \begin{bmatrix} \Delta x_k \\ \Delta \lambda_k \end{bmatrix} \triangleq d_k = -B_k^{-1}\nabla P(x_k, \lambda_k; c, \alpha),$$

$$\text{(47)} \qquad B_k = \nabla^2 L(x_k, \lambda_k) + \tfrac{1}{2}\nabla^2 L(x_k, \lambda_k)K(x_k)\nabla^2 L(x_k, \lambda_k) + \tfrac{1}{2}\nabla(K(x_k)\nabla L(x_k, \lambda_k))\nabla^2 L(x_k, \lambda_k).$$

In view of (43), (46), and (47), we can also write

$$d_k = -\nabla^2 L(x_k, \lambda_k)^{-1}\nabla L(x_k, \lambda_k).$$

Thus the direction of descent for the Newton-like method is again the Newton direction for solving the system $\nabla L = 0$.

In both Newton-like methods presented for minimizing $P(x, \lambda; c, \alpha)$ and $P(x, \lambda; c, M)$, it may be necessary to introduce modifications in order to improve their global convergence properties. Such modifications together with quasi-Newton versions are given in Section 4.5.2.

Finally consider the function $\hat{P}(x; c)$. We have, from (29),

$$\text{(48)} \qquad \nabla \hat{P}(x, c) = \nabla_x L[x, \lambda(x)] + \nabla\lambda(x)h(x) + c\nabla h(x)h(x).$$

If (x^*, λ^*) is a K–T pair and $\nabla h(x^*)$ has rank m, then $\lambda^* = \lambda(x^*)$ and, by differentiating $\nabla \hat{P}$ at x^* and using the facts $\nabla_x L(x^*, \lambda^*) = 0$ and $h(x^*) = 0$, we obtain

$$\text{(49)} \quad \nabla^2 \hat{P}(x^*; c) = \nabla_{xx}^2 L[x^*, \lambda(x^*)] + \nabla\lambda(x^*)\nabla h(x^*)' + \nabla h(x^*)\nabla\lambda(x^*)' + c\nabla h(x^*)\nabla h(x^*)'.$$

Thus we may consider a Newton-like method of the form

$$x_{k+1} = x_k + \alpha_k d_k,$$

where α_k is a stepsize parameter and d_k is obtained by solving the system of equations

$$H_k d_k = -\nabla \hat{P}(x_k; c),$$

where

$$H_k = \nabla^2_{xx} L[x_k, \lambda(x_k)] + \nabla\lambda(x_k)\nabla h(x_k)' + \nabla h(x_k)\nabla\lambda(x_k)' + c\nabla h(x_k)\nabla h(x_k)'.$$

Again it may be necessary to introduce modifications similar to those for Newton's method (compare with Section 1.3.3) in order to improve the global convergence properties of the method. Additional Newton-like methods and quasi-Newton versions for minimizing the penalty function $\hat{P}$ will be analyzed in Section 4.5.2.

Choice of the Penalty Parameter

For each of the exact penalty methods examined so far, the penalty parameter c must be chosen sufficiently high, for otherwise the method breaks down. We can gain some insight regarding the proper range of values for c by considering a problem with quadratic objective function and linear constraints

$$\text{(50)} \qquad \begin{aligned} &\text{minimize} \quad f(x) = \tfrac{1}{2}x'Hx \\ &\text{subject to} \quad N'x = 0, \end{aligned}$$

where we assume that $f(x) > 0$ for all $x \neq 0$ with $N'x = 0$, and that N has rank m. (This corresponds to the case where the K–T pair $\{x^* = 0, \lambda^* = 0\}$ satisfies the second-order sufficiency conditions for optimality.) Consider first the function

$$\text{(51)} \qquad \hat{P}(x; c) = \tfrac{1}{2}x'Hx + \lambda(x)'N'x + \tfrac{1}{2}c|N'x|^2,$$

where

$$\lambda(x) = -(N'N)^{-1}N'Hx.$$

Appropriate values of c are those for which the Hessian $\nabla^2\hat{P}$ is positive definite. We have

$$\text{(52)} \qquad \nabla^2\hat{P} = \nabla^2\psi(x) + cNN',$$

where

$$\psi(x) = L[x, \lambda(x)] = \tfrac{1}{2}x'Hx + \lambda(x)'N'x.$$

By differentiation of ψ, we obtain

$$\nabla^2\psi = H - HN(N'N)^{-1}N' - N(N'N)^{-1}N'H.$$

Denote

$$\text{(53)} \qquad E = N(N'N)^{-1}N', \qquad \hat{E} = I - E.$$

Then a straightforward calculation yields

(54) $$\nabla^2 \psi = H - HE - EH = \hat{E}H\hat{E} - EHE.$$

The matrices $\hat{E}$ and E are projection matrices for the subspaces

$$\mathscr{C} = \{x \mid N'x = 0\}$$

and its orthogonal complement

$$\mathscr{C}^\perp = \{N\xi \mid \xi \in R^m\},$$

respectively. By this, we mean that any vector $x \in R^n$ can be written as

$$x = \hat{E}x + Ex,$$

and $\hat{E}x$ is the orthogonal projection of x on $\mathscr{C}$, while Ex is the orthogonal projection of x on $\mathscr{C}^\perp$. Furthermore, we have

(55) $$\hat{E}x = x, \qquad Ex = 0 \qquad \forall\, x \in \mathscr{C},$$

(56) $$Ex = x, \qquad \hat{E}x = 0 \qquad \forall\, x \in \mathscr{C}^\perp.$$

From (54), (55), and (56), we obtain

$$x'\nabla^2 \psi x = x'Hx \qquad \forall\, x \in \mathscr{C},$$

$$x'\nabla^2 \psi x = -x'Hx \qquad \forall\, x \in \mathscr{C}^\perp.$$

Thus we find that $\nabla^2 \psi$ *has the same curvature as* H *on the subspace* $\mathscr{C}$ *(the constraint set), and the opposite curvature of* H *on the subspace* $\mathscr{C}^\perp$ *(the subspace orthogonal to the constraint set).*

Returning to (52), we see that the term cNN' cannot influence the curvature of $\nabla^2 \hat{P}$ along $\mathscr{C}$. Its purpose is to counteract the possibly negative curvature of $\nabla^2 \psi$ along $\mathscr{C}^\perp$ or equivalently the possibly positive (!) curvature of H along $\mathscr{C}^\perp$. More precisely, from (53), we have $EN = N$, so using (52) and (54), we can write

$$\nabla^2 \hat{P} = \hat{E}H\hat{E} + E(cNN' - H)E,$$

$$x'\nabla^2 \hat{P} x = (\hat{E}x)'H(\hat{E}x) + (Ex)'(cNN' - H)(Ex) \qquad \forall\, x \in R^n.$$

It follows that $\nabla^2 \hat{P}$ is positive definite if and only if

(57) $$z'Hz > 0 \qquad \forall\, z \neq 0, \quad z \in \mathscr{C}$$

$$cz'NN'z > z'Hz \qquad \forall\, z \neq 0, \quad z \in \mathscr{C}^\perp.$$

By assumption, we have that (57) holds. Thus we obtain

$$\nabla^2 \hat{P} > 0 \Leftrightarrow c > \max\left\{ \frac{z'Hz}{z'NN'z} \,\middle|\, |z| = 1, \quad z \in \mathscr{C}^\perp \right\}.$$

It follows that *if H is negative semidefinite on the subspace $\mathscr{C}^{\perp}$ any positive value of c is suitable for use in the exact penalty function. Otherwise there is a lower bound $\bar{c} > 0$ that suitable values of c must exceed.* This implies in particular that *for convex programming problems we can expect that high values of c may be necessary.* The preceding observations show that *there is a striking difference between the roles of the penalty parameter c in the method of multipliers and in the exact penalty methods of this section. In the method of multipliers, the threshold value for c increases as the curvature of the objective along $\mathscr{C}^{\perp}$ becomes smaller, while the exact opposite is true for the exact penalty methods of this section.* This may be viewed as a fundamental difference between the two types of methods.

Consider next the penalty function

$$P(x, \lambda; c, M) = \tfrac{1}{2}x'Hx + \lambda'N'x + \tfrac{1}{2}c|N'x|^2 + \tfrac{1}{2}|M(Hx + N\lambda)|^2,$$

where M is a $p \times n$ matrix with $m \le p \le n$ and such that MN has rank m. We are interested in conditions on c and M that guarantee that $\nabla^2 P$ is positive definite. Consider the function

$$\tilde{P}(x; c, M) = \min_{\lambda} P(x, \lambda; c, M).$$

Since P is positive definite quadratic in λ for every x, the minimization above can be carried out explicitly and the minimizing vector is given by

$$\hat{\lambda}(x) = -(N'M'MN)^{-1}(N' + N'M'MH)x.$$

Substitution in the expression for P yields

$$\begin{aligned}\tilde{P}(x; c, M) = \tfrac{1}{2}x'[&H + HM'MH + cNN' \\ &- (N + HM'MN)(N'M'MN)^{-1}(N' + N'M'MH)]x.\end{aligned}$$

It can be easily verified that $\nabla^2 P$ is positive definite if and only if

$$\nabla^2 \tilde{P}(x; c, M) > 0.$$

Consider the matrices

$$E_M = M'MN(N'M'MN)^{-1}N', \qquad \hat{E}_M = I - E_M. \tag{58}$$

A straightforward calculation shows that $\tilde{P}$ may also be written as

$$\begin{aligned}\tilde{P}(x; c, M) = \tfrac{1}{2}x'[&\hat{E}_M'H\hat{E}_M - E_M'HE_M + cNN' - N(N'M'MN)^{-1}N']x \\ &+ \tfrac{1}{2}(MHx)'[I - MN(N'M'MN)^{-1}N'M'](MHx).\end{aligned}$$

The matrix $[I - MN(N'M'MN)^{-1}N'M']$ is a projection matrix and is therefore positive semidefinite. Hence the second term in the right-hand side in the previous equation is nonnegative, and it follows that in order that $\nabla^2 \tilde{P}(x; c, M) > 0$ it is sufficient that

(59)
$$x'[\hat{E}_M' H \hat{E}_M - E_M' H E_M + cNN' - N(N'M'MN)^{-1}N']x > 0 \qquad \forall\, x \neq 0.$$

Consider the subspace

$$\mathscr{C} = \{x \mid N'x = 0\}.$$

For any $x \in R^n$, we have, using (58),

$$N'E_M x = N'x, \qquad N'\hat{E}_M x = N'(I - E_M)x = 0.$$

Hence

(60)
$$\hat{E}_M x \in \mathscr{C} \qquad \forall\, x \in R^n.$$

We have also

$$N'x = N'Ex,$$

where E is given by (53). In view of (58), this implies

$$E_M x = E_M Ex.$$

By using the above two equations, we can write (59) as

$$(\hat{E}_M x)' H (\hat{E}_M x) + (Ex)'[cNN' - E_M' H E_M - N(N'M'MN)^{-1}N'](Ex) > 0 \qquad \forall\, x \neq 0.$$

In view of the fact that $\hat{E}_M x \in \mathscr{C}$ [compare with (60)] and the hypothesis $z'Hz > 0\ \forall\, z \neq 0$ with $z \in \mathscr{C}$, the first term in the previous relation is nonnegative. Hence, the relation will hold if and only if

(61)
$$c > \max\left\{ \frac{z'[E_M' H E_M + N(N'M'MN)^{-1}N']z}{z'NN'z} \,\middle|\, |z| = 1, z \in \mathscr{C}^\perp \right\}.$$

This in turn implies that the matrix $\nabla^2 P$ will be positive definite if c satisfies (61).

Consider now the case $M = \sqrt{\alpha} I$, for which we have $P(x, \lambda; c, M) = P(x, \lambda; c, \alpha)$. Since every vector $z \in \mathscr{C}^\perp$ can be represented as $z = N\xi$, where $\xi \in R^m$, and $E_M = N(N'N)^{-1}N'$, relation (61) is easily shown to be equivalent to

$$c\alpha(N'N)^2 - \alpha N'HN - N'N > 0,$$

or by right and left multiplication with $(N'N)^{-1}$,

(62)
$$c\alpha I - \alpha(N'N)^{-1}(N'HN)(N'N)^{-1} - (N'N)^{-1} > 0.$$

This relation suggests rules for selection of the parameters c and α. Given α, one should select c sufficiently large so that (62) holds. If the value of α is not sufficiently small to the extent that unconstrained minimization yields critical points of P which are not local minima of (ECP), then α must be reduced but *this reduction must be accompanied by a corresponding increase of c so that* (62) *holds. A good rule of thumb is therefore to increase c so as to keep the product $c\alpha$ roughly constant.*

Automatic Adjustment of the Penalty Parameter

Since a proper range of values of the penalty parameter may be unknown in a practical situation, it may be useful to provide for a scheme that automatically increases c if the results of the computation indicate that the value currently used is inadequate. We provide an informal discussion of such schemes.

Consider first the penalty function $P(x, \lambda; c, M)$. One possibility for automatic penalty adjustment is based on the idea that if the penalty parameter were increased only finitely often, say to a maximum value $\bar{c}$, then the unconstrained minimization algorithm will normally converge to a critical point $(\bar{x}, \bar{\lambda})$ of $P(\cdot, \cdot; \bar{c}, M)$. Thus we would have

$$\lim_{k \to \infty} \nabla P(x_k, \lambda_k; c_k, M) = \nabla P(\bar{x}, \bar{\lambda}; \bar{c}, M) = 0. \tag{63}$$

Now if we had $h(\bar{x}) = 0$, it can be seen from (15) that (63) implies $\nabla_x L(\bar{x}, \bar{\lambda}) = 0$ so that $(\bar{x}, \bar{\lambda})$ is a K–T pair. If $h(\bar{x}) \neq 0$, then, for k sufficiently large and any positive scalar γ, we would have

$$|F(x_{k+1}, \lambda_{k+1}; c_k, M)| < \gamma |h(x_{k+1})|, \tag{64}$$

where F is any continuous function such that $F = 0$ when $\nabla P = 0$. Thus when (64) holds, it provides us with an indication that the current value c_k is inadequate and should be increased.

Thus we are led to a scheme whereby *at each iteration k we perform an iteration of an unconstrained algorithm for minimizing $P(\cdot, \cdot; c_k, M)$ to obtain (x_{k+1}, λ_{k+1}). We then check to see if* (64) *is satisfied. If so we increase c_k to $c_{k+1} = \beta c_k$, where $\beta > 1$ is a fixed scalar factor. Otherwise we set $c_{k+1} = c_k$ and continue.*

In order for such a scheme to have a good chance of success it is necessary to show that, under normal conditions, if c_k becomes large enough then (64) will not hold so that c_k will normally be increased finitely often. This can be guaranteed *if $M(x)\nabla h(x)$ is an $m \times m$ nonsingular matrix in the region of interest $X^* = \{x \,|\, \nabla h(x)$ has rank $m\}$*. We choose

$$F = M\nabla_x P - \left[I + M\left(\nabla^2_{xx} LM' + \sum_{i=1}^{p} \nabla m_i \nabla_x L e_i'\right)\right](\nabla h' M')^{-1} \nabla_\lambda P. \tag{65}$$

The motivation for this complicated formula will become apparent later where it will be seen that in special cases it leads to a simple test [compare with (66)]. We have, using (15a) and (15b),

$$M\nabla_x P = cM\nabla hh + \left[I + M\left(\nabla_{xx}^2 LM' + \sum_{i=1}^{p} \nabla m_i \nabla_x Le_i'\right)\right] M\nabla_x L,$$

$$M\nabla_x L = (\nabla h' M')^{-1}(\nabla_\lambda P - h),$$

from which we obtain

$$F = \left\{cM\nabla h - \left[I + M\left(\nabla_{xx}^2 LM' + \sum_{i=1}^{m} \nabla m_i \nabla_x Le_i'\right)\right](\nabla h' M')^{-1}\right\}h.$$

It is clear that, given any $\gamma > 0$ and any compact subset $X \times \Lambda \subset X^* \times R^m$, there exists a $\bar{c} > 0$ such that

$$|F(x, \lambda; c, M)| \geq \gamma |h(x)| \qquad \forall\, c \geq \bar{c}, \quad (x, \lambda) \in X \times \Lambda.$$

So if $\{(x_k, \lambda_k)\}$ remains within a bounded subset of $X^* \times R^m$, inequality (64) will never be satisfied if c_k increases beyond a certain level.

To summarize, if the algorithm with the automatic penalty adjustment scheme just described is used to generate a sequence $\{(x_k, \lambda_k, c_k)\}$ there are only two possibilities:

(a) There is no compact subset of $X^* \times R^m$ containing the sequence $\{(x_k, \lambda_k)\}$.

(b) The sequence $\{(x_k, \lambda_k)\}$ belongs to a compact subset of $X^* \times R^m$ in which case c_k is constant for k sufficiently large. If the unconstrained algorithm used to minimize $P(\cdot, \cdot; c, M)$ has the property that, for every $c > 0$, all limit points of sequences it generates are critical points of $P(\cdot, \cdot; c, M)$, then all limit points of $\{(x_k, \lambda_k)\}$ are K–T pairs of (ECP).

We can similarly construct a penalty adjustment scheme for the penalty function $\hat{P}$. Since [compare with (24), (27), (28), and (31)]

$$\hat{P}(x; c) = \min_{\lambda} P(x, \lambda; c + 1, M) = P[x, \hat{\lambda}(x); c + 1, M],$$

where

$$\hat{\lambda}(x) = -h(x) - [\nabla h(x)'\nabla h(x)]^{-1}\nabla h(x)'\nabla f(x),$$

$$M(x) = [\nabla h(x)'\nabla h(x)]^{-1}\nabla h(x)',$$

any unconstrained algorithm for minimizing $\hat{P}(\cdot; c)$ may be viewed as an unconstrained algorithm for minimizing $P(\cdot, \cdot; c + 1, M)$. We are thus reduced to the case examined earlier. This leads to the test [compare with (64)]

$$|F(x_{k+1}, \lambda_{k+1}; c_k + 1, M)| < \gamma |h(x_{k+1})|,$$

where F is given by (65) and

$$\lambda_{k+1} = \hat{\lambda}(x_{k+1}).$$

Since [compare with (35) and (36)]

$$\nabla_\lambda P(x_{k+1}, \lambda_{k+1}; c_k + 1, M) = 0,$$
$$\nabla_x P(x_{k+1}, \lambda_{k+1}; c_k + 1, M) = \nabla \hat{P}(x_{k+1}; c_k),$$

the test above can also be written as

$$|M(x_{k+1}) \nabla \hat{P}(x_{k+1}; c_k)| < \gamma |h(x_{k+1})|. \tag{66}$$

Whenever the relation above holds, we increase c_k by multiplying it with a scalar $\beta > 1$. Similar statements regarding convergence can be made as for the scheme given earlier in connection with the exact penalty function $P(x, \lambda; c, M)$.

Extensions to Inequality Constraints

Some of the preceding results and algorithms admit straightforward extensions for problems involving inequality constraints. This can be done by converting inequality constraints to equality constraints. Consider the equality constrained problem

$$\text{(67)} \qquad \begin{aligned} &\text{minimize} \quad f(x) \\ &\text{subject to} \quad g_j(x) + z_j^2 = 0, \qquad j = 1, \ldots, r, \end{aligned}$$

obtained from (ICP) by introducing the additional variables z_j, $j = 1, \ldots, r$. Consider also the corresponding exact penalty function

$$\text{(68)} \quad P(x, z, \mu; c, \alpha) = f(x) + \sum_{j=1}^{r} \{\mu_j[g_j(x) + z_j^2] + \tfrac{1}{2}c[g_j(x) + z_j^2]^2\}$$
$$+ \tfrac{1}{2}\alpha |\nabla_x L(x, \mu)|^2 + 2\alpha \sum_{j=1}^{r} z_j^2 \mu_j^2,$$

where

$$L(x, \mu) = f(x) + \mu' g(x).$$

Minimization of P with respect to (x, z, μ) can be carried out by minimizing first with respect to z and by subsequently minimizing the resulting function with respect to (x, μ). A straightforward calculation yields

$$\text{(69)} \quad P^+(x, \mu; c, \alpha) \triangleq \min_z P(x, z, \mu; c, \alpha) = f(x) + \tfrac{1}{2}\alpha |\nabla_x L(x, \mu)|^2$$
$$+ \frac{1}{2c} \sum_{j=1}^{r} \{[\max\{0, \mu_j + 2\alpha\mu_j^2 + cg_j(x)\}]^2$$
$$- (\mu_j + 2\alpha\mu_j^2)^2 - 4\alpha c \mu_j^2 g_j(x)\}$$

with the minimum attained at

$$z_j^2(x, \mu; c, \alpha) = \max\{0, -[(\mu_j + 2\alpha\mu_j^2)/c] - g_j(x)\}.$$

Thus minimization of P can be carried out by minimizing instead the function P^+ of (69) which does not involve the additional variables z_j.

If instead of the penalty function (68), we use the penalty function

$$P_\tau(x, z, \mu; c, \alpha) = f(x) + \sum_{j=1}^{r} \left\{\mu_j[g_j(x) + z_j^2] + \frac{c + \tau|\mu|^2}{2}[g_j(x) + z_j^2]^2\right\}$$
$$+ \tfrac{1}{2}\alpha|\nabla_x L(x, \mu)|^2 + 2\alpha \sum_{j=1}^{r} z_j^2 \mu_j^2,$$

where $\tau > 0$, then we can similarly eliminate z and obtain the penalty function

$$(70) \quad P_\tau^+(x, \mu; c, \alpha) \triangleq \min_z P_\tau(x, z, \mu; c, \alpha) = f(x) + \tfrac{1}{2}\alpha|\nabla_x L(x, \mu)|^2$$
$$+ \frac{1}{2(c + \tau|\mu|^2)}$$
$$\times \sum_{j=1}^{r} \{[\max\{0, \mu_j + 2\alpha\mu_j^2 + (c + \tau|\mu|^2)g_j(x)\}]^2$$
$$- (\mu_j + 2\alpha\mu_j^2)^2 - 4\alpha(c + \tau|\mu|^2)\mu_j^2 g_j(x)\}.$$

The minimum is attained at

$$z_j^2(x, \mu; c, \alpha) = \max\left\{0, -\frac{\mu_j + 2\alpha\mu_j^2}{c + \tau|\mu|^2} - g_j(x)\right\}.$$

A similar procedure can be used for the penalty function $P_\tau(x, z, \mu; c, M)$. In connection with problem (67), let us choose

$$M(x, z) = \sqrt{\eta}[\overline{M}(x)\ 2Z],$$

where η is a positive scalar, $\overline{M}(x)$ is a continuous $r \times n$ matrix, and Z is the diagonal matrix

$$Z = \begin{bmatrix} z_1 & & 0 \\ & \ddots & \\ 0 & & z_r \end{bmatrix}.$$

we have

$$P_\tau(x, z, \mu; c, \overline{M}) = f(x) + \sum_{j=1}^{r} \left\{\mu_j[g_j(x) + z_j^2] + \frac{c + \tau|\mu|^2}{2}[g_j(x) + z_j^2]^2\right\}$$
$$+ \tfrac{1}{2}\eta|\overline{M}(x)\nabla_x L(x, \mu) + 4Z^2\mu|^2.$$

Again minimization of P with respect to (x, z, μ) can be carried out by minimizing first with respect to z and by subsequently minimizing the resulting function with respect to (x, μ). It is straightforward to verify that

$$(71)\quad P_\tau^+(x, \mu; c, \overline{M}) \triangleq \min_z P_\tau(x, z, \mu; c, \overline{M})$$

$$= f(x) + \mu' g(x) + \tfrac{1}{2}(c + \tau|\mu|^2)|g(x)|^2 + \tfrac{1}{2}\eta|\overline{M}(x)\nabla_x L(x, \mu)|^2$$

$$- \sum_{j=1}^{r} \frac{[\min\{0, (c + \tau|\mu|^2)g_j(x) + \mu_j + 4\eta\mu_j \overline{m}_j(x)'\nabla_x L(x, \mu)\}]^2}{2(c + \tau|\mu|^2 + 16\eta\mu_j^2)}$$

where $m_j(x)'$ is the jth row of the matrix $\overline{M}(x)$ (see also DiPillo and Grippo, 1979b).

Unfortunately, when the penalty function $\hat{P}(x, z; c)$ is used in conjunction with problem (67), it does not seem possible to eliminate the additional variables z_j. However, Glad and Polak (1979) have been able to construct an exact penalty function analogous to $\hat{P}$ for problem (ICP) that does not employ additional variables. The same reference gives a corresponding superlinearly convergent algorithm under an assumption that is somewhat stronger than Assumption (S^+).

Newton-like algorithms for minimizing the penalty functions $P_\tau^+(\cdot, \cdot; c, \alpha)$ and $P_\tau^+(\cdot, \cdot; c, \overline{M})$ will be given in Section 4.5.2.

Extensions to Nonnegativity Constraints

When the only inequality constraints are nonnegativity constraints on the variables, it may be worthwhile to consider an alternative approach. Consider the problem

$$(\text{ECP})^+ \qquad \text{minimize} \quad f(x)$$
$$\text{subject to} \quad h(x) = 0, \qquad x \geq 0,$$

where $f: R^n \to R$ and $h: R^n \to R^m$. An equivalent problem is obtained by making the change of variables

$$x_i = z_i^2, \qquad i = 1, \ldots, n,$$

where x_i, $i = 1, \ldots, n$ are the coordinates of x. The problem is then transformed into

$$(\text{ECP})_T^+ \qquad \text{minimize} \quad \hat{f}(z)$$
$$\text{subject to} \quad \hat{h}(z) = 0,$$

where

$$z = (z_1, \ldots, z_n)$$

and

$$\hat{f}(z) = f(z_1^2, \ldots, z_n^2), \qquad \hat{h}(z) = h(z_1^2, \ldots, z_n^2).$$

It is easily seen that

$$\nabla\hat{f}(z) + \nabla\hat{h}(z)\lambda = Z[\nabla f(z_1^2, \ldots, z_n^2) + \nabla h(z_1^2, \ldots, z_n^2)\lambda],$$

where

$$Z = \begin{bmatrix} 2z_1 & & 0 \\ & \ddots & \\ 0 & & 2z_n \end{bmatrix}.$$

Consider the expression for the penalty function $P_\tau(z, \lambda; c, \alpha)$ for problem $(\text{ECP})_\text{T}^+$. Based on the relation above we find that the variables $z_1, \ldots, z_n$ enter in this expression exclusively in squared form so that by using the substitution $x_i = z_i^2$ it is possible to write this expression in terms of the variables x_i. It takes the form

$$\tilde{P}_\tau(x, \lambda; c, \alpha) = L(x, \lambda) + \tfrac{1}{2}\nabla L(x, \lambda)' K(x, c, \alpha)\nabla L(x, \lambda),$$

where

$$L(x, \lambda) = f(x) + \lambda' h(x),$$

$$K(x, c, \alpha) = \left[\begin{array}{ccc|c} 4\alpha x_1 & & 0 & \\ & \ddots & & 0 \\ 0 & & 4\alpha x_n & \\ \hline & 0 & & (c + \tau|\lambda|^2)I \end{array}\right].$$

Thus the unconstrained minimization problem

$$\begin{aligned} &\text{minimize} \quad P_\tau(z, \lambda; c, \alpha) \\ &\text{subject to} \quad z \in R^n, \qquad \lambda \in R^m, \end{aligned}$$

is equivalent to the (simply) constrained problem

$$\begin{aligned} &\text{minimize} \quad \tilde{P}_\tau(x, \lambda; c, \alpha) \\ &\text{subject to} \quad x \geq 0, \qquad \lambda \in R^m. \end{aligned}$$

By solving this latter problem for suitable values of c and α, we can, based on the theory of this section, reasonably hope to obtain a solution of $(\text{ECP})_\text{T}^+$ and hence also of $(\text{ECP})^+$. A similar approach can be developed by using in connection with $(\text{ECP})^+$ the penalty function $P_\tau(z, \lambda; M, c)$ with

$$M(z) = \nabla\hat{h}(z)'$$

or

$$M(z) = [\nabla\hat{h}(z)'\nabla\hat{h}(z)]^{-1}\nabla\hat{h}(z)'.$$

4.4 Lagrangian Methods—Local Convergence

The methods to be examined in this section may be viewed as methods for solving the system of nonlinear equations (and possibly inequalities) that represent the necessary conditions for optimality of the constrained minimization problem. Thus the necessary conditions for optimality of (ECP)

$$\nabla f(x) + \nabla h(x)\lambda = 0, \qquad h(x) = 0, \tag{1}$$

are viewed as a system of $(n + m)$ nonlinear equations with $(n + m)$ unknowns—the vectors x and λ.

We can view system (1) as a special case of the general system

$$F(z) = 0, \tag{2}$$

where $F: R^p \to R^p$ and p is a positive integer. A general class of methods for solving system (2) is given by

$$z_{k+1} = G(z_k), \qquad k = 0, 1, \ldots, \tag{3}$$

where $G: R^p \to R^p$ is some continuous function. If $\{z_k\}$ generated by (3) converges to a vector z^*, then by continuity of G, we must have $z^* = G(z^*)$, so G must be chosen so that its fixed points are solutions of (2). General tools for showing convergence of iteration (3) are various fixed point theorems of the contraction mapping type. We give one such theorem that is often quite useful. Its proof may be easily deduced from the analysis in Ortega and Rheinboldt (1970, p. 300). We first introduce the following definition:

Definition: A vector $z^* \in R^p$ is said to be a *point of attraction* of iteration (3) if there exists an open set S such that if $z_0 \in S$ then the sequence $\{z_k\}$ generated by (3) belongs to S and converges to z^*.

Ostrowski's Theorem: Assume that $G: R^p \to R^p$ has a fixed point z^*, and that $G \in C^1$ on an open set containing z^*. Assume further that all eigenvalues of $\nabla G(z^*)'$ lie strictly within the unit circle of the complex plane. Then z^* is a point of attraction of iteration (3), and if the sequence $\{z_k\}$ generated by (3) converges to z^*, the rate of convergence of $\{|z_k - z^*|\}$ is at least linear.

In what follows in this section, we consider various Lagrangian methods starting with a first-order method which does not require second derivatives. We then examine Newton-like methods and their quasi-Newton versions. Throughout this section, we focus on local convergence properties, i.e., questions of convergence and rate of convergence from a starting point that is sufficiently close to a solution. Our presentation, however, is geared towards preparing the ground for the developments of Section 4.5 where modifications of Lagrangian methods will be introduced with the purpose of improving their global convergence properties.

4.4.1 First-Order Methods

The simplest of all Lagrangian methods for the equality constrained problem

(ECP) $$\text{minimize} \quad f(x)$$
$$\text{subject to} \quad h(x) = 0$$

is given by

$$x_{k+1} = x_k - \alpha \nabla_x L(x_k, \lambda_k), \tag{4}$$

$$\lambda_{k+1} = \lambda_k + \alpha \nabla_\lambda L(x_k, \lambda_k), \tag{5}$$

where L is the Lagrangian function

$$L(x, \lambda) = f(x) + \lambda' h(x)$$

and $\alpha > 0$ is a scalar stepsize. We have the following result:

Proposition 4.23: Let (x^*, λ^*) be a K–T pair of (ECP) such that $f, h \in C^2$ in an open set containing x^*. Assume that the matrix $\nabla h(x^*)$ has rank m and the matrix $\nabla_{xx}^2 L(x^*, \lambda^*)$ is positive definite. There exists $\bar{\alpha} > 0$, such that for all $\alpha \in (0, \bar{\alpha}]$, (x^*, λ^*) is a point of attraction of iteration (4), (5), and if the sequence $\{(x_k, \lambda_k)\}$ generated by (4), (5) converges to (x^*, λ^*), then the rate of convergence of $\{|(x_k, \lambda_k) - (x^*, \lambda^*)|\}$ is at least linear.

Proof: The proof consists of showing that, for α sufficiently small, the hypothesis of Ostrowski's theorem is satisfied. Indeed for $\alpha > 0$, consider the mapping $G_\alpha: R^{n+m} \to R^{n+m}$ defined by

$$G_\alpha(x, \lambda) = \begin{bmatrix} x - \alpha \nabla_x L(x, \lambda) \\ \lambda + \alpha \nabla_\lambda L(x, \lambda) \end{bmatrix}.$$

Clearly (x^*, λ^*) is a fixed point of G_α, and we have

$$\nabla G_\alpha(x^*, \lambda^*)' = I - \alpha B, \tag{6}$$

where

$$B = \begin{bmatrix} \nabla_{xx}^2 L(x^*, \lambda^*) & \nabla h(x^*) \\ -\nabla h(x^*)' & 0 \end{bmatrix}. \tag{7}$$

We shall show that the real part of each eigenvalue of B is strictly positive, and then the result will follow from (6) by using Ostrowski's theorem. For any complex vector y, denote by $\hat{y}$ its complex conjugate, and for any complex number γ, denote by $\text{Re}(\gamma)$ its real part. Let β be an eigenvalue of B, and let

$(z, w) \neq (0, 0)$ be a corresponding eigenvector where z and w are complex vectors of dimension n and m, respectively. We have

$$\text{(8)} \quad \operatorname{Re}\left\{[\hat{z}' \quad \hat{w}']B\begin{bmatrix} z \\ w \end{bmatrix}\right\} = \operatorname{Re}\left\{\beta[\hat{z}' \quad \hat{w}']\begin{bmatrix} z \\ w \end{bmatrix}\right\} = \operatorname{Re}(\beta)(|z|^2 + |w|^2),$$

while at the same time, using (7),

$$\text{(9)} \quad \operatorname{Re}\left\{[\hat{z}' \quad \hat{w}']B\begin{bmatrix} z \\ w \end{bmatrix}\right\} = \operatorname{Re}\{\hat{z}'\nabla^2_{xx}L(x^*, \lambda^*)z + \hat{z}'\nabla h(x^*)w - \hat{w}'\nabla h(x^*)'z\}.$$

Since we have for any real $n \times m$ matrix M

$$\operatorname{Re}\{\hat{z}'M'w\} = \operatorname{Re}\{\hat{w}'Mz\}$$

it follows from (8) and (9) that

$$\text{(10)} \quad \operatorname{Re}\{\hat{z}'\nabla^2_{xx}L(x^*, \lambda^*)z\} = \operatorname{Re}\left\{[\hat{z}' \quad \hat{w}']B\begin{bmatrix} z \\ w \end{bmatrix}\right\} = \operatorname{Re}(\beta)(|z|^2 + |w|^2).$$

Since for any positive definite matrix A we have

$$\operatorname{Re}\{\hat{z}'Az\} > 0 \qquad \forall\, z \neq 0,$$

it follows from (10) and the positive definiteness assumption on $\nabla^2_{xx}L(x^*, \lambda^*)$ that either $\operatorname{Re}(\beta) > 0$ or else $z = 0$. But if $z = 0$ the equation $B\left[\begin{smallmatrix} z \\ w \end{smallmatrix}\right] = \beta\left[\begin{smallmatrix} z \\ w \end{smallmatrix}\right]$ yields

$$\nabla h(x^*)w = 0.$$

Since $\nabla h(x^*)$ has rank m it follows that $w = 0$. This contradicts our earlier assumption that $(z, w) \neq (0, 0)$. Consequently we must have $\operatorname{Re}(\beta) > 0$.

Q.E.Q.

By appropriately scaling the vectors x and λ, we can show that the result of Proposition 4.23 holds also for the more general iteration

$$x_{k+1} = x_k - \alpha D\nabla_x L(x_k, \lambda_k), \qquad \lambda_{k+1} = \lambda_k + \alpha M\nabla_\lambda L(x_k, \lambda_k),$$

where D and M are any positive definite symmetric matrices of appropriate dimension (compare with Section 1.3.2). However the restrictive positive definiteness assumption on $\nabla^2_{xx}L(x^*, \lambda^*)$ is essential for the conclusion to hold.

There are other first-order Lagrangian methods available in the literature. As an example, we mention the linearization method of Section 4.2 with a constant stepsize and constant matrix H_k which can be shown to converge locally with a linear rate to a K–T pair satisfying Assumption (S^+) of Section 3.1 provided the stepsize is sufficiently small. We sketch a proof of this fact for the case of problem (ECP) and the choice $H_k \equiv I$. The method takes the form

$$x_{k+1} = x_k + \alpha d(x_k),$$

where $d(x_k)$ is the solution of the quadratic program

$$\begin{aligned} &\text{minimize} \quad \nabla f(x_k)'d + \tfrac{1}{2}|d|^2 \\ &\text{subject to} \quad h(x_k) + \nabla h(x_k)'d = 0, \end{aligned}$$

and $\alpha > 0$ is a constant stepsize parameter. If $\nabla h(x_k)$ has rank m, the Lagrange multiplier for this program can be calculated to be

$$\hat{\lambda}(x_k) = [\nabla h(x_k)'\nabla h(x_k)]^{-1}[h(x_k) - \nabla h(x_k)'\nabla f(x_k)],$$

and it follows from the condition $\nabla f(x_k) + \nabla h(x_k)\hat{\lambda}(x_k) + d(x_k) = 0$ that

$$d(x_k) = -\nabla_x L[x_k, \hat{\lambda}(x_k)].$$

So the method takes the form

$$x_{k+1} = x_k - \alpha \nabla_x L[x_k, \hat{\lambda}(x_k)].$$

The result of Proposition 4.26 in Section 4.4.2, together with Ostrowski's theorem, can be used to show that if α is sufficiently small, this iteration converges locally with a linear rate to a local minimum x^* satisfying Assumption (S). For a detailed analysis together with an extension of this result to the case of inequality constraints we refer the reader to Pschenichny and Danilin (1975).

4.4.2 Newton-like Methods for Equality Constraints

Consider the system of necessary optimality conditions for (ECP)

$$\nabla f(x) + \nabla h(x)\lambda = 0, \qquad h(x) = 0, \tag{11}$$

or equivalently

$$\nabla L(x, \lambda) = 0. \tag{12}$$

Newton's method for solving this system is given by

$$x_{k+1} = x_k + \Delta x_k, \qquad \lambda_{k+1} = \lambda_k + \Delta\lambda_k, \tag{13}$$

where $(\Delta x_k, \Delta\lambda_k) \in R^{n+m}$ is obtained by solving the system of equations

$$\nabla^2 L(x_k, \lambda_k)\begin{bmatrix}\Delta x_k \\ \Delta\lambda_k\end{bmatrix} = -\nabla L(x_k, \lambda_k). \tag{14}$$

We have

$$\nabla^2 L(x_k, \lambda_k) = \begin{bmatrix} H_k & N_k \\ N_k' & 0 \end{bmatrix}, \qquad \nabla L(x_k, \lambda_k) = \begin{bmatrix} \nabla_x L(x_k, \lambda_k) \\ h(x_k) \end{bmatrix}, \tag{15}$$

where

$$H_k = \nabla_{xx}^2 L(x_k, \lambda_k), \qquad N_k = \nabla h(x_k). \tag{16}$$

Thus, the system (14) takes the form

$$\begin{bmatrix} H_k & N_k \\ N_k' & 0 \end{bmatrix} \begin{bmatrix} \Delta x_k \\ \Delta \lambda_k \end{bmatrix} = -\begin{bmatrix} \nabla_x L(x_k, \lambda_k) \\ h(x_k) \end{bmatrix}. \tag{17}$$

We say that (x_{k+1}, λ_{k+1}) is *well defined* by the Newton iteration (13), (14) if the matrix $\nabla^2 L(x_k, \lambda_k)$ is invertible. Note that at a K–T pair (x^*, λ^*) satisfying the sufficiency Assumption (S) of Section 2.2, we have that $\nabla^2 L(x^*, \lambda^*)$ is invertible (Lemma 1.27). As a result, $\nabla^2 L(x, \lambda)$ is invertible in a neighborhood of (x^*, λ^*), and within this neighborhood points generated by the Newton iteration are well defined. In the subsequent discussion, when stating various local convergence properties of the Newton iteration in connection with such a K–T pair, we implicitly restrict the iteration within a neighborhood where it is well defined.

The local convergence properties of the method can be inferred from Proposition 1.17, and in fact we have already made use of these properties in Section 2.3.2 (compare with Proposition 2.8 and the subsequent analysis). For purposes of convenient reference, we provide the corresponding result in the following proposition.

Proposition 4.24: Let x^* be a strict local minimum and a regular point of (ECP) satisfying together with a corresponding Lagrange multiplier vector λ^* the sufficiency Assumption (S) of Section 2.2. Then (x^*, λ^*) is a point of attraction of the Newton iteration (13), (14). Furthermore if $\{(x_k, \lambda_k)\}$ generated by (13), (14) converges to (x^*, λ^*) the rate of convergence of $\{|(x_k, \lambda_k) - (x^*, \lambda^*)|\}$ is superlinear (at least order two if $\nabla^2 f$ and $\nabla^2 h_i$, $i = 1, \ldots, m$, are Lipschitz continuous in an open set containing x^*).

Alternative Implementations of Newton's Method

We first observe that *if H_k is invertible and N_k has rank m* we can provide a more explicit expression for the Newton iteration. Indeed the system (17) can be written

$$H_k \Delta x_k + N_k \Delta \lambda_k = -\nabla_x L(x_k, \lambda_k). \tag{18}$$

$$N_k' \Delta x_k = -h(x_k). \tag{19}$$

By multiplying the first equation with $N_k' H_k^{-1}$ and using the second equation, it follows that

$$-h(x_k) + N_k' H_k^{-1} N_k \Delta\lambda_k = -N_k' H_k^{-1} \nabla_x L(x_k, \lambda_k).$$

Since N_k has rank m, the matrix $N_k' H_k^{-1} N_k$ is nonsingular, and we obtain

$$\lambda_{k+1} - \lambda_k = \Delta\lambda_k = (N_k' H_k^{-1} N_k)^{-1}[h(x_k) - N_k' H_k^{-1} \nabla_x L(x_k, \lambda_k)]. \tag{20}$$

Since

$$\begin{aligned}\nabla_x L(x_k, \lambda_k) &= \nabla f(x_k) + N_k \lambda_k = \nabla f(x_k) + N_k \lambda_{k+1} - N_k \Delta\lambda_k \\ &= \nabla_x L(x_k, \lambda_{k+1}) - N_k \Delta\lambda_k,\end{aligned}$$

we also have

$$(N_k' H_k^{-1} N_k)^{-1} N_k' H_k^{-1} \nabla_x L(x_k, \lambda_k) = \lambda_k + (N_k' H_k^{-1} N_k)^{-1} N_k' H_k^{-1} \nabla f(x_k),$$
$$\nabla_x L(x_k, \lambda_k) + N_k \Delta\lambda_k = \nabla_x L(x_k, \lambda_{k+1}).$$

Using these equations in (20) and (18), we finally obtain

$$\lambda_{k+1} = (N_k' H_k^{-1} N_k)^{-1}[h(x_k) - N_k' H_k^{-1} \nabla f(x_k)], \tag{21}$$

$$x_{k+1} = x_k - H_k^{-1} \nabla_x L(x_k, \lambda_{k+1}). \tag{22}$$

Another way to write the same equations is based on the observation that for every scalar c we have, from (19),

$$cN_k N_k' \Delta x = -cN_k h(x_k),$$

and substitution in (18) yields

$$(H_k + cN_k N_k')\Delta x_k + N_k \Delta\lambda_k = -\nabla_x L[x_k, \lambda_k + ch(x_k)].$$

Thus, if $(H_k + cN_k N_k)^{-1}$ exists, then we obtain by the same type of calculation used to obtain (21) and (22):

$$\begin{aligned}\hat{\lambda}_{k+1} &= [N_k'(H_k + cN_k N_k')^{-1} N_k]^{-1} \\ &\quad \times [h(x_k) - N_k'(H_k + cN_k N_k')^{-1} \nabla f(x_k)],\end{aligned} \tag{23}$$

$$\lambda_{k+1} = \hat{\lambda}_{k+1} - ch(x_k), \tag{24}$$

$$x_{k+1} = x_k - (H_k + cN_k N_k')^{-1} \nabla_x L(x_k, \hat{\lambda}_{k+1}). \tag{25}$$

Note that for $c = 0$, Eqs. (23)–(25) reduce to (21) and (22). An advantage that (23)–(25) may offer is that the matrix H_k may not be invertible while $(H_k + cN_k N_k')$ may be invertible for some values of c. For example, if (x^*, λ^*) satisfy Assumption (S) then H_k need not be invertible, while for sufficiently large c and (x_k, λ_k) sufficiently close to (x^*, λ^*), we have that $(H_k + cN_k N_k')$ *is not only invertible but also positive definite.* An additional advantage offered by this property is that it allows us to differentiate between local minima and local maxima, for if (x_k, λ_k) is near a *local maximum–Lagrange multiplier pair satisfying the sufficiency conditions for optimality, then* $(H_k + cN_kN_k')$ *will not be positive definite for any value of c.* Note that positive definiteness of $(H_k + cN_kN_k')$ can be easily detected if the Cholesky factorization method is used for solving the various linear systems of equations in (23) and (25).

A third implementation of the Newton iteration is based on the observation that Eqs. (22) and (19) can be written as

$$\nabla f(x_k) + H_k \Delta x_k + N_k \lambda_{k+1} = 0, \qquad h(x_k) + N_k' \Delta x_k = 0,$$

and are therefore the necessary optimality conditions for $(\Delta x_k, \lambda_{k+1})$ to be a K–T pair of the quadratic program

$$\text{(26)} \qquad \begin{aligned} &\text{minimize} \quad \nabla f(x_k)'\Delta x + \tfrac{1}{2}\Delta x' H_k \Delta x \\ &\text{subject to} \quad h(x_k) + N_k'\Delta x = 0. \end{aligned}$$

Thus we can obtain $(\Delta x_k, \lambda_{k+1})$ by solving this problem. This implementation is not particularly useful for practical purposes but provides an interesting connection with linearization methods. This relation can be made more explicit by noting that the solution Δx_k of (26) is unaffected if H_k is replaced by any matrix of the form $(H_k + \bar{c}N_kN_k')$, where $\bar{c} \in R$, thereby obtaining the program

$$\text{(27)} \qquad \begin{aligned} &\text{minimize} \quad \nabla f(x_k)'\Delta x + \tfrac{1}{2}\Delta x'(H_k + \bar{c}N_kN_k')\Delta x \\ &\text{subject to} \quad h(x_k) + N_k'\Delta x = 0. \end{aligned}$$

To see that problems (26) and (27) have the same solution Δx_k, simply note that they have the same constraints while their objective functions differ by the constant term $(1/2)\bar{c}\Delta x' N_k N_k' \Delta x = (1/2)\bar{c}|h(x_k)|^2$. Near a local minimum–Lagrange multiplier pair (x^*, λ^*) satisfying Assumption (S), we have that $(H_k + \bar{c}N_kN_k')$ is positive definite if $\bar{c}$ is sufficiently large and the quadratic program (27) is positive definite. We see therefore that, under these circumstances, *the Newton iteration can be viewed in effect as a special case of the linearization method of Section* 4.2 *with a constant unity stepsize, and scaling matrix* $\bar{H}_k = H_k + \bar{c}N_kN_k'$ *where* $\bar{c}$ *is any scalar for which* $\bar{H}_k$ *is positive definite.*

Still another implementation of Newton's method which offers computational advantages in certain situations will be given in Section 4.5.2.

Descent Properties of Newton's Method

Since we would like to improve the global convergence properties of Newton's method, it is of interest to search for functions for which $(x_{k+1} - x_k)$ is a descent direction at x_k. By this, we mean functions $F: R^n \to R$ such that

$$F[x_k + \alpha(x_{k+1} - x_k)] < F(x_k) \qquad \forall\, \alpha \in (0, \bar{\alpha}],$$

if $x_k \neq x^*$ and $\bar{\alpha}$ is a sufficiently small positive scalar. We have already developed the necessary machinery for proving the following proposition.

Proposition 4.25: Let x^* be a strict local minimum of (ECP), satisfying together with a corresponding Lagrange multiplier vector λ^* the sufficiency assumption (S) of Section 2.2. There exists a neighborhood S of (x^*, λ^*) such that if $(x_k, \lambda_k) \in S$ and $x_k \neq x^*$, then (x_{k+1}, λ_{k+1}) is well defined by the Newton iteration (13), (14) and the following hold true:

(a) For every $c > 0$, the vector $(x_{k+1} - x_k)$ is a descent direction at x_k for the exact penalty function

$$f(x) + c \max\{|h_1(x)|, \ldots, |h_m(x)|\}. \tag{28}$$

(b) The vector $\{(x_{k+1} - x_k), (\lambda_{k+1} - \lambda_k)\}$ is a descent direction at (x_k, λ_k) for the exact penalty function

$$F(x, \lambda) = \tfrac{1}{2}|\nabla L(x, \lambda)|^2. \tag{29}$$

Furthermore given any scalar $r > 0$, there exists a $\delta > 0$ such that if

$$|(x_k - x^*, \lambda_k - \lambda^*)| < \delta,$$

we have

$$F(x_{k+1}, \lambda_{k+1}) \leq rF(x_k, \lambda_k). \tag{30}$$

(c) Let $M(x)$ be a continuous $p \times n$ matrix with $m \leq p \leq n$ and such that $M(x^*)\nabla h(x^*)$ has rank m. For every x_k, λ_k and $c > 0$ for which the matrix

$$\nabla^2 L(x_k, \lambda_k) + \nabla^2 L(x_k, \lambda_k) \begin{bmatrix} M(x_k)'M(x_k) & 0 \\ 0 & cI \end{bmatrix} \nabla^2 L(x_k, \lambda_k) \tag{31}$$

is positive definite, the vector $\{(x_{k+1} - x_k), (\lambda_{k+1} - \lambda_k)\}$ is a descent direction at (x_k, λ_k) of the exact penalty function

$$P(x, \lambda; c, M) = L(x, \lambda) + \tfrac{1}{2}c|h(x)|^2 + \tfrac{1}{2}|M(x)\nabla_x L(x, \lambda)|^2. \tag{32}$$

(d) For every $c \in R$ for which $(H_k + cN_k N_k')$ is positive definite, the vector $(x_{k+1} - x_k)$ is a descent direction at x_k of the augmented Lagrangian function $L_c(\cdot, \lambda_{k+1})$.

Proof: (a) Take $\bar{c} > 0$ sufficiently large and a neighborhood S of (x^*, λ^*) which is sufficiently small, so that for $(x_k, \lambda_k) \in S$, the matrix $(H_k + \bar{c}N_k N_k)$ is positive definite. Since Δx_k is the solution of the quadratic program (27), it follows from Proposition 4.2 that if $x_k \neq x^*$, then Δx_k is a descent direction of (28).

(b) We have

$$\begin{bmatrix} x_{k+1} - x_k \\ \lambda_{k+1} - \lambda_k \end{bmatrix} = -\nabla^2 L(x_k, \lambda_k)^{-1} \nabla L(x_k, \lambda_k)$$

and

$$\nabla F(x_k, \lambda_k) = \nabla^2 L(x_k, \lambda_k) \nabla L(x_k, \lambda_k).$$

So

$$[(x_{k+1} - x_k)', (\lambda_{k+1} - \lambda_k)'] \nabla F(x_k, \lambda_k) = -|\nabla L(x_k, \lambda_k)|^2 < 0,$$

and the descent property follows.

From Proposition 4.24, we have that, given any $\bar{r} > 0$, there exists a $\bar{\delta} > 0$ such that for $|(x_k - x^*, \lambda_k - \lambda^*)| < \bar{\delta}$ we have

$$|(x_{k+1} - x^*, \lambda_{k+1} - \lambda^*)| \le \bar{r} |(x_k - x^*, \lambda_k - \lambda^*)|. \tag{33}$$

For every (x, λ), we have, by the mean value theorem,

$$\nabla L(x, \lambda) = B \begin{bmatrix} x - x^* \\ \lambda - \lambda^* \end{bmatrix},$$

where each row of B is the corresponding row of $\nabla^2 L$ at a point between (x, λ) and (x^*, λ^*). Since $\nabla^2 L(x^*, \lambda^*)$ is invertible, it follows that there is an $\varepsilon > 0$ and scalars $\mu > 0$ and $M > 0$ such that for $|(x - x^*, \lambda - \lambda^*)| < \varepsilon$, we have

$$\mu |(x - x^*, \lambda - \lambda^*)| \le |\nabla L(x, \lambda)| \le M |(x - x^*, \lambda - \lambda^*)|. \tag{34}$$

From (33) and (34), it follows that for each $\bar{r} > 0$ there exists $\delta > 0$ such that, for $|(x_k - x^*, \lambda_k - \lambda^*)| < \delta$,

$$|\nabla L(x_{k+1}, \lambda_{k+1})| \le (M\bar{r}/\mu) |\nabla L(x_k, \lambda_k)|.$$

or equivalently

$$F(x_{k+1}, \lambda_{k+1}) \le (M^2 \bar{r}^2/\mu^2) F(x_k, \lambda_k).$$

Given $r > 0$, we take $\bar{r} = (\mu/M)\sqrt{r}$ in the relation above, and (30) follows.

(c) This part was shown in effect in Section 4.3.3.

(d) From (24) and (25), we have

$$\begin{aligned} x_{k+1} - x_k &= -(H_k + cN_k N_k')^{-1} \nabla_x L[x_k, \lambda_{k+1} + ch(x_k)] \\ &= -(H_k + cN_k N_k')^{-1} \nabla_x L_c(x_k, \lambda_{k+1}), \end{aligned}$$

and the result follows. Q.E.D.

Variations of Newton's Method

A variation of Newton's method is obtained by introducing a positive parameter c_k in the second equation so that Δx_k and $\Delta\lambda_k$ are obtained by solving the system

$$H_k \Delta x_k + N_k \Delta\lambda_k = -\nabla_x L(x_k, \lambda_k), \tag{35}$$

$$N_k' \Delta x_k - c_k^{-1} \Delta\lambda_k = -h(x_k). \tag{36}$$

As $c_k \to \infty$, the system becomes in the limit the one corresponding to Newton's method. We can show that the system (35), (36) has a unique solution if either H_k^{-1} or $(H_k + c_k N_k N_k')^{-1}$ exists. Indeed when H_k^{-1} exists, we can write explicitly the solution. By multiplying (35) by $N_k' H_k^{-1}$ and by using (36), we obtain

$$c_k^{-1}\Delta\lambda_k - h(x_k) + N_k' H_k^{-1} N_k \Delta\lambda_k = -N_k' H_k^{-1} \nabla_x L(x_k, \lambda_k)$$

from which

$$\Delta\lambda_k = [c_k^{-1} I + N_k' H_k^{-1} N_k]^{-1} [h(x_k) - N_k' H_k^{-1} \nabla_x L(x_k, \lambda_k)]$$

and

$$\lambda_{k+1} = \lambda_k + [c_k^{-1} I + N_k' H_k^{-1} N_k]^{-1} [h(x_k) - N_k' H_k^{-1} \nabla_x L(x_k, \lambda_k)]. \tag{37}$$

From (35), we obtain

$$x_{k+1} = x_k - H_k^{-1} \nabla_x L(x_k, \lambda_{k+1}). \tag{38}$$

Also if $(H_k + c_k N_k N_k')^{-1}$ exists, by multiplying (36) with $c_k N_k$ and adding the resulting equation to (35), we obtain

$$(H_k + c_k N_k N_k')\Delta x_k = -\nabla_x L(x_k, \lambda_k) - c_k N_k h(x_k),$$

and finally

$$x_{k+1} = x_k - (H_k + c_k N_k N_k')^{-1} \nabla_x L_{c_k}(x_k, \lambda_k), \tag{39}$$

where L_{c_k} is the augmented Lagrangian function. Also from (36), we obtain

$$\lambda_{k+1} = \lambda_k + c_k[h(x_k) + N_k'(x_{k+1} - x_k)]. \tag{40}$$

Note that the preceding analysis shows that N_k *need not have rank* m *in order for the system* (35), (36) *to have a unique solution, while this is not true for the Newton iteration.* Another interesting fact that follows from (39) is that *if* $(H_k + c_k N_k N_k')$ *is positive definite, then* $(x_{k+1} - x_k)$ *is a descent direction for the augmented Lagrangian function* $L_{c_k}(\cdot, \lambda_k)$. Furthermore, if the constraints are linear, then (40) can be written as

$$\lambda_{k+1} = \lambda_k + c_k h(x_{k+1}),$$

while if in addition the objective function is quadratic and $(H_k + c_k N_k N_k')$ is positive definite, then from (39), x_{k+1} is the unique minimizing point of the augmented Lagrangian function $L_{c_k}(\cdot, \lambda_k)$. Hence, it follows that *if the constraints are linear,* $[h(x) = N'x - b]$, *the objective function is quadratic* $[f(x) = \frac{1}{2}x'Qx]$, *and for all k,* c_k *is such that* $(Q + c_k NN')$ *is positive definite, then the iteration* (39), (40) *is equivalent to the first-order method of multipliers of Section* 2.2. This suggests that if c_k is taken sufficiently large, then iteration (39), (40) should converge locally to a local minimum–Lagrange multiplier pair (x^*, λ^*) satisfying Assumption (S). Furthermore the rate of convergence should be superlinear if $c_k \to \infty$. Indeed this can be shown either directly or by appealing to the theory of consistent approximations (see Ortega and Rheinboldt, 1970, Theorems 11.2.2 and 11.2.3). The proof is routine and is left to the reader.

Another variation of Newton's method is given by

$$x_{k+1} = x_k - (H_k + c_k N_k N_k')^{-1} \nabla L_{c_k}(x_k, \lambda_k), \tag{41}$$

$$\lambda_{k+1} = \lambda_k + c_k h(x_{k+1}). \tag{42}$$

This iteration is the same as (39), (40) except that the term $h(x_{k+1})$ in (42) replaces its first-order linear approximation $[h(x_k) + N_k'(x_{k+1} - x_k)]$ in (40). When the constraints are linear, the two iterations are identical. It is possible to show that, if c_k is constant but sufficiently large, iteration (41), (42) converges locally to a K–T pair satisfying Assumption (S) at a linear rate. Since this iteration seems less interesting than (39), (40), as well as the iterations (43), (44) and (45), (46) that follow, we omit the proof.

Two more variations of Newton's method are obtained by replacing $(H_k + c_k N_k N_k')$ in (39) or (41) by $\nabla_{xx}^2 L_{c_k}(x_k, \lambda_k)$, thereby obtaining the iterations

$$x_{k+1} = x_k - [\nabla_{xx}^2 L_{c_k}(x_k, \lambda_k)]^{-1} \nabla_x L_{c_k}(x_k, \lambda_k), \tag{43}$$

$$\lambda_{k+1} = \lambda_k + c_k[h(x_k) + N_k'(x_{k+1} - x_k)] \tag{44}$$

and

$$x_{k+1} = x_k - [\nabla_{xx}^2 L_{c_k}(x_k, \lambda_k)]^{-1} \nabla_x L_{c_k}(x_k, \lambda_k), \tag{45}$$

$$\lambda_{k+1} = \lambda_k + c_k h(x_{k+1}). \tag{46}$$

Since

$$\nabla_{xx}^2 L_{c_k}(x_k, \lambda_k) - (H_k + c_k N_k N_k') = c_k \sum_{i=1}^{m} h_i(x_k) \nabla^2 h_i(x_k),$$

we see that if c_k is chosen in a way that $c_k h(x_k) \to 0$, then iteration (43), (44) becomes asymptotically identical with (39), (40) while (45), (46) becomes asymptotically identical with (41), (42). The condition $c_k h(x_k) \to 0$ can be enforced in a practical algorithm by means of a test on the magnitude of $|h(x_k)|$ which allows c_k to be increased by a factor $\beta > 1$ only if $|h(x_k)|$ has been decreased by a factor $\gamma > \beta$ over the previous time c_k was changed. Another simple way to enforce the condition $c_k h(x_k) \to 0$ is to keep c_k constant

$$c_k = c \qquad \forall\, k = 0, 1, \ldots .$$

Under these circumstances, if c is chosen sufficiently large, both iterations (43), (44) and (45), (46) can be shown to converge locally to a K–T pair (x^*, λ^*) satisfying Assumption (S) at a linear rate. We show this fact for iteration (45), (46). The proof for iteration (43), (44) is similar and will be omitted.

Let $c > 0$ be such that

$$\nabla_{xx}^2 L_c(x^*, \lambda^*) > 0.$$

For (x_k, λ_k) sufficiently near a K–T pair (x^*, λ^*) satisfying Assumption (S) so that $\nabla_{xx}^2 L_c(x_k, \lambda_k)$ is nonsingular and $\nabla h(x_k)$ has rank m, consider the iteration

$$x_{k+1} = x_k - [\nabla_{xx}^2 L_c(x_k, \lambda_k)]^{-1} \nabla_x L_c(x_k, \lambda_k),$$
$$\lambda_{k+1} = \lambda_k + c h(x_{k+1}).$$

We have, by the mean value theorem,

$$\nabla_x L_c(x_k, \lambda_k) = R_k(x_k - x^*) + N_k(\lambda_k - \lambda^*),$$

where each row of $R_k(N_k)$ equals the corresponding row of $\nabla_{xx}^2 L_c\,(\nabla h)$ evaluated at a point lying between (x_k, λ_k) and (x^*, λ^*). Similarly, we have

$$h(x_{k+1}) = \bar{N}_k'(x_{k+1} - x^*),$$

where each row of $\bar{N}_k'$ equals the corresponding row of $\nabla h'$ evaluated at a point between x_{k+1} and x^*. By combining the relations above we obtain

$$\begin{bmatrix} x_{k+1} - x^* \\ \lambda_{k+1} - \lambda^* \end{bmatrix} = \begin{bmatrix} A_k & B_k \\ C_k & D_k \end{bmatrix} \begin{bmatrix} x_k - x^* \\ \lambda_k - \lambda^* \end{bmatrix},$$

where the matrices A_k, B_k, C_k, and D_k are given by

$$A_k = I - \nabla_{xx}^2 L_c(x_k, \lambda_k)^{-1} R_k,$$
$$B_k = -\nabla_{xx}^2 L_c(x_k, \lambda_k)^{-1} N_k,$$
$$C_k = c\bar{N}_k' A_k,$$
$$D_k = I - c\bar{N}_k' \nabla_{xx}^2 L_c(x_k, \lambda_k)^{-1} N_k.$$

For any $\tilde{c} > 0$ such that $\nabla_{xx}^2 L_{\tilde{c}}(x^*, \lambda^*) > 0$, we have, by using the matrix identity of Section 1.2,

$$\begin{aligned} &\nabla_{xx}^2 L_c(x^*, \lambda^*)^{-1} \nabla h(x^*) \\ &\quad = [\nabla_{xx}^2 L_{\tilde{c}}(x^*, \lambda^*) + (c - \tilde{c}) \nabla h(x^*) \nabla h(x^*)']^{-1} \nabla h(x^*) \\ &\quad = \nabla_{xx}^2 L_{\tilde{c}}(x^*, \lambda^*)^{-1} \nabla h(x^*) \\ &\qquad \times \{I - [I/(c - \tilde{c}) + \nabla h(x^*)' \nabla_{xx}^2 L_{\tilde{c}}(x^*, \lambda^*)^{-1} \nabla h(x^*)]^{-1} \\ &\qquad \times \nabla h(x^*)' \nabla_{xx}^2 L_{\tilde{c}}(x^*, \lambda^*)^{-1} \nabla h(x^*)\}. \end{aligned}$$

Hence

$$\lim_{c \to \infty} \nabla^2_{xx} L_c(x^*, \lambda^*)^{-1} \nabla h(x^*) = 0.$$

We also have, from Eq. (32) of Section 2.2.3,

$$\lim_{c \to \infty} [I - c\nabla h(x^*)' \nabla^2_{xx} L_c(x^*, \lambda^*)^{-1} \nabla h(x^*)] = 0.$$

By using these relations, it is easy to see that given any $\varepsilon > 0$ there exists a $\bar{c}(\varepsilon) > 0$ such that for every $c \geq \bar{c}(\varepsilon)$ there is a neighborhood $N(c, \varepsilon)$ of (x^*, λ^*) within which $|A_k| < \varepsilon$, $|B_k| < \varepsilon$, $|C_k| < \varepsilon$, and $|D_k| < \varepsilon$. Hence, given any $r > 0$, there exists a $\bar{c}(r) > 0$ such that for every $c \geq \bar{c}(r)$ there is a neighborhood of (x^*, λ^*) within which there holds

$$|(x_{k+1} - x^*, \lambda_{k+1} - \lambda^*)| \leq r|(x_k - x^*, \lambda_k - \lambda^*)|.$$

It follows that if c is sufficiently large then (x^*, λ^*) is a point of attraction of iteration (45), (46). The convergence rate is at least linear with a convergence ratio that can be made arbitrarily small by choosing c sufficiently large.

Newton's Method in the Space of Primal Variables

As indicated by Proposition 4.24, it is necessary to have a good initial choice for both x and λ in order to ensure convergence of the Newton iteration. If however a good initial choice x_0 is available, then it is possible to obtain a good initial choice λ_0 from

$$\lambda_0 = \hat{\lambda}(x_0),$$

where the function $\hat{\lambda}$ is given for all x in the set

$$X^* = \{x \mid \nabla h(x) \text{ has rank } m\} \tag{47}$$

by

$$\hat{\lambda}(x) = [\nabla h(x)' \nabla h(x)]^{-1} [h(x) - \nabla h(x)' \nabla f(x)] \qquad \forall\, x \in X^*. \tag{48}$$

Indeed for any K–T pair (x^*, λ^*) such that $x^* \in X^*$, we have shown (Proposition 4.22) that $\hat{\lambda}(x^*) = \lambda^*$. Since $\hat{\lambda}(\cdot)$ is a continuous function on X^*, it follows that $\hat{\lambda}(x_0)$ is near λ^* if x_0 is near x^*. This leads to a Newton-like iteration whereby (x_{k+1}, λ_{k+1}) are obtained by solving the system

$$\begin{bmatrix} \nabla^2_{xx} L[x_k, \hat{\lambda}(x_k)] & \nabla h(x_k) \\ \nabla h(x_k)' & 0 \end{bmatrix} \begin{bmatrix} x_{k+1} - x_k \\ \lambda_{k+1} - \lambda_k \end{bmatrix} = -\begin{bmatrix} \nabla_x L(x_k, \lambda_k) \\ h(x_k) \end{bmatrix}. \tag{49}$$

This system can also be written as

$$\nabla^2_{xx} L[x_k, \hat{\lambda}(x_k)](x_{k+1} - x_k) + \nabla h(x_k)\lambda_{k+1} = -\nabla f(x_k), \tag{50}$$

$$\nabla h(x_k)'(x_{k+1} - x_k) = -h(x_k), \tag{51}$$

so x_{k+1} *is independent of* λ_k.

We derive now an explicit formula for x_{k+1}. From (51), we have

(52) $$\nabla h(x_k)[\nabla h(x_k)'\nabla h(x_k)]^{-1}\nabla h(x_k)'(x_{k+1} - x_k) = -\nabla h(x_k)[\nabla h(x_k)'\nabla h(x_k)]^{-1}h(x_k),$$

while from (50), we obtain

(53)
$$-\nabla h(x_k)[\nabla h(x_k)'\nabla h(x_k)]^{-1}\nabla h(x_k)'\nabla^2_{xx}L[x_k, \hat{\lambda}(x_k)](x_{k+1} - x_k) - \nabla h(x_k)\lambda_{k+1} = \nabla h(x_k)[\nabla h(x_k)'\nabla h(x_k)]^{-1}\nabla h(x_k)'\nabla f(x_k).$$

By adding (50), (52), and (53) and by making use of (48), we obtain

(54) $$\{E(x_k) + [I - E(x_k)]\nabla^2_{xx}L[x_k, \hat{\lambda}(x_k)]\}(x_{k+1} - x_k) = -\nabla_x L[x_k, \hat{\lambda}(x_k)],$$

where E is defined by

(55) $$E(x) = \nabla h(x)[\nabla h(x)'\nabla h(x)]^{-1}\nabla h(x)'.$$

If the matrix within braces in (54) is invertible, we can write

(56) $$x_{k+1} = x_k - \{E(x_k) + [I - E(x_k)]\nabla^2_{xx}L[x_k, \hat{\lambda}(x_k)]\}^{-1}\nabla_x L[x_k, \hat{\lambda}(x_k)].$$

We shall demonstrate shortly that the inverse above indeed exists for x_k sufficiently close to a local minimum x^* satisfying Assumption (S) (Proposition 4.26c). *We have thus obtained a Newton-like method which can be carried out in the space of primal variables x without any reference to the dual variables λ.*

We can develop iteration (56) by starting from a different viewpoint. Consider for $x \in X^*$ the equation

(57) $$\nabla_x L[x, \hat{\lambda}(x)] = 0.$$

The following proposition shows that K–T pairs of (ECP) can be obtained by solving this equation.

Proposition 4.26: Let $x^* \in X^*$ and assume $f, h \in C^2$ in a neighborhood of x^*. Then:

(a) (x^*, λ^*) is a K–T pair of (ECP) if and only if x^* is a solution of Eq. (57) and $\lambda^* = \hat{\lambda}(x^*)$.

(b) If x^* is a solution of Eq. (57), then the Jacobian matrix (with respect to x) of $\nabla_x L[x, \hat{\lambda}(x)]$ evaluated at x^* is given by

(58) $$\nabla(\nabla_x L[x^*, \hat{\lambda}(x^*)])' = E(x^*) + [I - E(x^*)]\nabla^2_{xx}L[x^*, \hat{\lambda}(x^*)].$$

(c) If x^* is a local minimum of (ECP) which together with $\lambda^* = \hat{\lambda}(x^*)$ satisfies Assumption (S) of Section 2.2, then the matrix (58) is nonsingular.

More specifically this matrix has m eigenvalues equal to one and its remaining $(n - m)$ eigenvalues are equal to the $(n - m)$ positive eigenvalues $\gamma_1, \ldots, \gamma_{n-m}$ of the matrix

$$[I - E(x^*)]\nabla^2_{xx}L(x^*, \lambda^*)[I - E(x^*)]. \tag{59}$$

(*Note:* It will be shown as part of the proof that the matrix (59) has exactly $(n - m)$ positive eigenvalues and a zero eigenvalue of multiplicity m.)

Proof: (a) If (x^*, λ^*) is a K–T pair then the equation $\nabla f(x^*) + \nabla h(x^*)\lambda^* = 0$ yields

$$\nabla h(x^*)'\nabla f(x^*) + \nabla h(x^*)'\nabla h(x^*)\lambda^* = 0$$

from which

$$\lambda^* = -[\nabla h(x^*)'\nabla h(x^*)]^{-1}\nabla h(x^*)'\nabla f(x^*).$$

Using (48) and the fact that $h(x^*) = 0$, we obtain $\lambda^* = \hat{\lambda}(x^*)$. Hence, $0 = \nabla f(x^*) + \nabla h(x^*)\hat{\lambda}(x^*) = \nabla_x L[x^*, \hat{\lambda}(x^*)]$, and it follows that x^* is a solution of Eq. (57).

Conversely let x^* be a solution of Eq. (57); i.e.,

$$\nabla_x L[x^*, \hat{\lambda}(x^*)] = 0. \tag{60}$$

From (48), we obtain

$$\begin{aligned} h(x^*) &= \nabla h(x^*)'\nabla f(x^*) + \nabla h(x^*)'\nabla h(x^*)\hat{\lambda}(x^*) \\ &= \nabla h(x^*)'\nabla_x L[x^*, \hat{\lambda}(x^*)]. \end{aligned} \tag{61}$$

By combining (60) and (61) and writing $\lambda^* = \hat{\lambda}(x^*)$, we obtain

$$\nabla_x L(x^*, \lambda^*) = 0, \qquad h(x^*) = 0,$$

showing that (x^*, λ^*) is a K–T pair for (ECP).

(b) Denote, for $x \in X^*$,

$$p(x) = \nabla_x L[x, \hat{\lambda}(x)]. \tag{62}$$

By differentiation, we obtain

$$\nabla p(x)' = \nabla^2_{xx}L[x, \hat{\lambda}(x)] + \nabla h(x)\nabla\hat{\lambda}(x)'. \tag{63}$$

From (55), it follows that, for $x \in X^*$,

$$[I - E(x)]\nabla h(x) = 0. \tag{64}$$

By applying $[I - E(x)]$ to both sides of (63) and using (64), we obtain

$$[I - E(x)]\nabla p(x)' = [I - E(x)]\nabla^2_{xx}L[x, \hat{\lambda}(x)] \qquad \forall\, x \in X^*. \tag{65}$$

Also from (48), we have $\nabla h(x)'\nabla_x L[x, \hat{\lambda}(x)] = h(x)$ or, equivalently,

$$h(x) - \nabla h(x)'p(x) = 0 \qquad \forall\, x \in X^*.$$

By differentiating and by taking into account the fact that $p(x^*) = 0$, we obtain

$$\nabla h(x^*)' - \nabla h(x^*)' \nabla p(x^*)' = 0.$$

Multiplying with $\nabla h(x^*)[\nabla h(x^*)' \nabla h(x^*)]^{-1}$ and using (55), we obtain

$$E(x^*) \nabla p(x^*)' = E(x^*). \tag{66}$$

By combining (65) and (66), it follows that

$$\nabla p(x^*)' = E(x^*) + [I - E(x^*)] \nabla_{xx}^2 L[x^*, \hat{\lambda}(x^*)],$$

which, in view of (62), is identical to (58).

(c) Let γ be an eigenvalue of the matrix (58) and let $y \neq 0$ be a corresponding eigenvector. We have

$$[E(x^*) + [I - E(x^*)] \nabla_{xx}^2 L(x^*, \lambda^*)] y = \gamma y. \tag{67}$$

By using the relation

$$E(x^*)[I - E(x^*)] = 0 \tag{68}$$

and by multiplying (67) in turn by $E(x^*)$ and $[I - E(x^*)]$, we obtain

$$E(x^*) y = \gamma E(x^*) y, \tag{69}$$

$$[I - E(x^*)] \nabla_{xx}^2 L(x^*, \lambda^*) y = \gamma [I - E(x^*)] y. \tag{70}$$

There are two possibilities:

(i) $E(x^*) y \neq 0$. Then it follows from (69) that $\gamma = 1$.
(ii) $E(x^*) y = 0$. In this case, $[I - E(x^*)] y = y$, and (70) yields

$$[I - E(x^*)] \nabla_{xx}^2 L(x^*, \lambda^*) [I - E(x^*)] y = \gamma y; \tag{71}$$

i.e., γ is an eigenvalue of the matrix (59) and y is a corresponding eigenvector. Since matrix (59) is symmetric, both γ and y are real. Hence from (71), we also obtain

$$y'[I - E(x^*)] \nabla_{xx}^2 L(x^*, \lambda^*) [I - E(x^*)] y = \gamma |y|^2,$$

from which, using the fact that $E(x^*) y = 0$, it follows that

$$y' \nabla_{xx}^2 L(x^*, \lambda^*) y = \gamma |y|^2. \tag{72}$$

By using (55), the equation $E(x^*) y = 0$ is written

$$\nabla h(x^*)[\nabla h(x^*)' \nabla h(x^*)]^{-1} \nabla h(x^*)' y = 0,$$

and by multiplying with $\nabla h(x^*)'$, we obtain

$$\nabla h(x^*)' y = 0.$$

By using Assumption (S) and the fact that $y \neq 0$, we obtain

$$y'\nabla^2_{xx} L(x^*, \lambda^*) y > 0.$$

In view of (72), this implies that

$$\gamma > 0.$$

Conversely, if $\bar{\gamma} \neq 0$ is an eigenvalue of matrix (59) and $\bar{y} \neq 0$ is the corresponding eigenvector, then both $\bar{\gamma}$ and $\bar{y}$ are real since this matrix is symmetric. We have

$$[I - E(x^*)]\nabla^2_{xx} L(x^*, \lambda^*)[I - E(x^*)]\bar{y} = \bar{\gamma}\bar{y} \tag{73}$$

and by multiplying with $E(x^*)$ and using the fact that $E(x^*)[I - E(x^*)] = 0$, we obtain $0 = \bar{\gamma} E(x^*)\bar{y}$ or

$$E(x^*)\bar{y} = 0. \tag{74}$$

Combining (73) and (74), we obtain

$$[E(x^*) + [I - E(x^*)]\nabla^2_{xx} L(x^*, \lambda^*)]\bar{y} = \bar{\gamma}\bar{y}.$$

Hence, $\bar{\gamma}$ is also an eigenvalue of matrix (58), and $\bar{y}$ is a corresponding eigenvector. This together with (74) and the facts already proved also imply that $\bar{\gamma} > 0$.

Summarizing, we have shown up to this point that each nonzero eigenvalue of matrix (59) is positive and is also an eigenvalue of matrix (58), and all the remaining eigenvalues of matrix (58) equal unity. The proposition will be proved if we can show that matrix (59) has a zero eigenvalue of multiplicity exactly m. It can be easily seen, using Assumption (S), that the nullspace of matrix (59) is the m-dimensional subspace $\{z \mid [I - E(x^*)]z = 0\}$. For symmetric matrices the multiplicity of the zero eigenvalue is equal to the dimension of the nullspace and the result follows. Q.E.D.

It can now be seen that iteration (56) is a Newton-like method for solving Eq. (57), where the Jacobian of $\nabla_x L[x, \hat{\lambda}(x)]$ is replaced by the matrix

$$E(x) + [I - E(x)]\nabla^2_{xx} L[x, \hat{\lambda}(x)].$$

Since at a solution these two matrices are equal by essentially repeating the proof of Proposition 1.17 (compare also with the proof of Proposition 4.25b), we obtain the following result:

Proposition 4.27: Let x^* be a local minimum of (ECP) satisfying, together with $\lambda^* = \hat{\lambda}(x^*)$, Assumption (S) of Section 2.2. Then:

(a) x^* is a point of attraction of iteration (56), and if a sequence $\{x_k\}$ generated by (56) converges to x^*, the rate of convergence of $\{|x_k - x^*|\}$ is superlinear.

(b) Given any scalar $r > 0$, there exists a $\delta > 0$ such that if $|x_k - x^*| < \delta$ then

$$|\nabla_x L[x_{k+1}, \hat{\lambda}(x_{k+1})]| \le r|\nabla_x L[x_k, \hat{\lambda}(x_k)]|. \tag{75}$$

It is worth mentioning that for $x_k \in X^*$ we can obtain both $\hat{\lambda}(x_k)$ and $\nabla_x L[x_k, \hat{\lambda}(x_k)]$ by solving the quadratic program

$$\begin{aligned} &\text{minimize} \quad \nabla f(x_k)'d + \tfrac{1}{2}|d|^2 \\ &\text{subject to} \quad h(x_k) + \nabla h(x_k)'d = 0. \end{aligned}$$

Indeed the K–T conditions for this program are $\nabla f(x_k) + \nabla h(x_k)\lambda + d = 0$ and $h(x_k) + \nabla h(x_k)'d = 0$, and it can be easily seen that the (unique) Lagrange multiplier vector of this program is $\hat{\lambda}(x)$, while the unique optimal solution is

$$d(x) = -\nabla_x L[x, \hat{\lambda}(x)].$$

4.4.3 Newton-like Methods for Inequality Constraints

There are two main approaches for developing Newton-like methods for problems with inequality constraints. In the first approach, inequality constraints are treated by separating them explicitly or implicitly into two groups. In the first group are those that are predicted to be active at a solution and these are treated essentially as equality constraints. In the second group are those that are predicted to be inactive at a solution, and these are essentially ignored. This will be referred to as the *active set approach.* In the second approach inequality constraints are treated directly. Since all the methods of this type that we shall consider involve the solution of quadratic programming subproblems, we refer to this approach as the *quadratic programming approach.*

For simplicity we restrict attention to the problem

$$\text{(ICP)} \qquad \begin{aligned} &\text{minimize} \quad f(x) \\ &\text{subject to} \quad g(x) \le 0. \end{aligned}$$

The methods to be described can be extended to handle additional equality constraints in a manner that should be obvious to the reader in light of the developments so far in this chapter.

Active Set Approaches

The first active set approach to be examined is based on a transformation by means of which the K–T conditions for (ICP) are converted into a system of nonlinear equations. For a fixed scalar $c > 0$, consider the open set $S_c^* \subset R^n \times R^r$ defined by

$$S_c^* = \{(x, \mu) \mid \mu_j + cg_j(x) \neq 0, \ j = 1, \ldots, r\} \tag{76}$$

and the system of equations on S_c^*

$$\nabla f(x) + \nabla_x g^+(x, \mu, c)\mu = 0, \tag{77}$$

$$g^+(x, \mu, c) = 0, \tag{78}$$

where the function g^+ is given by

$$g^+(x, \mu, c) = \begin{bmatrix} g_1^+(x, \mu_1, c) \\ \vdots \\ g_r^+(x, \mu_r, c) \end{bmatrix}, \tag{79}$$

$$g_j^+(x, \mu_j, c) = \max\{g_j(x), -\mu_j/c\}, \qquad j = 1, \ldots, r. \tag{80}$$

Note that g^+ is differentiable on S_c^* as many times as g, so the system (77), (78) is well defined. We remind the reader that g^+ appears in the definition of the augmented Lagrangian function for (ICP) which takes the form

$$L_c(x, \mu) = f(x) + \mu' g^+(x, \mu, c) + \tfrac{1}{2}c|g^+(x, \mu, c)|^2$$

[compare with Section 3.1, Eq. (9)].

The following proposition establishes the validity and relevance of Newton's method for solving the system (77), (78).

Proposition 4.28: Let $c > 0$ be a scalar.

(a) A pair (x^*, μ^*) belongs to S_c^* and is a solution of the system (77), (78) if and only if (x^*, μ^*) is a K–T pair of (ICP) satisfying the strict complementarity condition

$$\mu_j^* > 0 \Leftrightarrow g_j(x^*) = 0 \qquad \forall\, j = 1, \ldots, r. \tag{81}$$

(b) If (x^*, μ^*) is a K–T pair of (ICP) satisfying Assumption (S^+) of Section 3.1, then (x^*, μ^*) is a point of attraction of Newton's method for solving the system (77), (78). If $\{(x_k, \mu_k)\}$ generated by Newton's method converges to (x^*, μ^*), then the rate of convergence of $\{|(x_k, \mu_k) - (x^*, \mu^*)|\}$ is superlinear (of order at least two if $\nabla^2 f$ and $\nabla^2 g_j, j = 1, \ldots, r$, are Lipschitz continuous in a neighborhood of x^*).

Proof: (a) Assume that (x^*, μ^*) belongs to S_c^* and is a solution of the system (77), (78). Since $g^+(x^*, \mu^*, c) = 0$, it follows in view of (79), (80) that

$$g_j(x^*) \leq 0, \quad \mu_j^* \geq 0 \qquad \forall\, j = 1, \ldots, r,$$

$$g_j(x^*) = 0 \quad \text{if} \quad \mu_j^* > 0, \qquad g_j(x^*) < 0 \quad \text{if} \quad \mu_j^* = 0.$$

In view of these relations, the equation $\nabla f(x^*) + \nabla_x g^+(x^*, \mu^*, c)\mu^* = 0$ can be written as

$$\nabla f(x^*) + \nabla g(x^*)\mu^* = 0.$$

Hence all the K–T conditions, as well as the strict complementarity condition (81), are satisfied by (x^*, μ^*). The proof of the converse is straightforward and is left for the reader.

(b) There is a neighborhood of (x^*, μ^*) such that $g_j(x) > -\mu_j/c$ if $g_j(x^*) = 0$ and $g_j(x) < -\mu_j/c$ if $g_j(x^*) < 0$ for all (x, μ) in this neighborhood. Within this neighborhood, the functions appearing in the system (77), (78) are continuously differentiable and Proposition 4.24 applies. Q.E.D.

Consider now the implementation of Newton's method. Define, for $(x, \mu) \in S_c^*$,

$$L^+(x, \mu, c) = f(x) + \mu' g^+(x, \mu, c), \tag{82}$$

$$A_c(x, \mu) = \{j \mid g_j(x) > -\mu_j/c, \quad j = 1, \ldots, r\}, \tag{83}$$

and assume without loss of generality that $A_c(x, \mu) = \{1, \ldots, p\}$ for some integer p (which depends on x and μ). We may view $A_c(x, \mu)$ as the *active index set*, in the sense that indices in $A_c(x, \mu)$ are "predicted" by the algorithm to be active at the solution. By differentiation in the system (77), (78) we find that Newton's method consists of the iteration

$$\bar{x} = x + \Delta x, \qquad \bar{\mu} = \mu + \Delta\mu, \tag{84}$$

where $(\Delta x, \Delta\mu)$ is the solution of the system

$$\begin{bmatrix} \nabla^2_{xx} L^+(x, \mu, c) & N(x, \mu, c) & 0 \\ N(x, \mu, c)' & 0 & 0 \\ 0 & 0 & -(1/c)I \end{bmatrix} \begin{bmatrix} \Delta x \\ \Delta\mu_1 \\ \vdots \\ \Delta\mu_p \\ \Delta\mu_{p+1} \\ \vdots \\ \Delta\mu_r \end{bmatrix} = - \begin{bmatrix} \nabla_x L^+(x, \mu, c) \\ g_1^+(x, \mu, c) \\ \vdots \\ g_p^+(x, \mu, c) \\ g_{p+1}^+(x, \mu, c) \\ \vdots \\ g_r^+(x, \mu, c) \end{bmatrix}. \tag{85}$$

In the equation above, $N(x, \mu, c)$ is the $n \times p$ matrix having as columns the gradients $\nabla g_j(x)$, $j \in A_c(x, \mu)$, I is the $(r - p) \times (r - p)$ identity matrix, and the zero matrices have appropriate dimension. Since we have

$$g_j^+(x, \mu, c) = -\mu_j/c \qquad \forall\, j \notin A_c(x, \mu),$$

it follows, from (84) and (85), that

$$\bar{\mu}_j = 0 \qquad \forall\, j \notin A_c(x, \mu). \tag{86}$$

It also follows from (85) that the remaining variables Δx and $\Delta\mu_1, \ldots, \Delta\mu_p$ are obtained by solving the reduced system

$$\begin{bmatrix} \nabla^2_{xx} L^+(x, \mu, c) & N(x, \mu, c) \\ N(x, \mu, c)' & 0 \end{bmatrix} \begin{bmatrix} \Delta x \\ \Delta\mu_1 \\ \vdots \\ \Delta\mu_p \end{bmatrix} = - \begin{bmatrix} \nabla_x L^+(x, \mu, c) \\ g_1(x) \\ \vdots \\ g_p(x) \end{bmatrix}. \tag{87}$$

where we have made use of the fact that

$$g_j^+(x, \mu, c) = g_j(x) \qquad \forall\, j \in A_c(x, \mu). \tag{88}$$

If we note the fact that

$$\nabla_x L^+(x, \mu, c) = \nabla f(x) + \sum_{j \in A_c(x, \mu)} \nabla g_j(x)\mu_j, \tag{89}$$

we can see from (86), (87), and (89) that the Newton iteration can be described in a simple manner. *We set the Lagrange multipliers of constraints that are not in the active set $A_c(x, \mu)$ to zero, and treat the remaining constraints as if they are equalities.*

The second active set approach to be examined is based on the last Newton-like method described in the previous section. We consider the quadratic program

$$\begin{aligned} &\text{minimize} \quad \nabla f(x)'d + \tfrac{1}{2}|d|^2 \\ &\text{subject to} \quad g_j(x) + \nabla g_j(x)'d \le 0, \qquad j \in J_\delta(x), \end{aligned} \tag{90}$$

where

$$J_\delta(x) = \{j \mid g_j(x) \ge \max\{0, g_1(x), \ldots, g_r(x)\} - \delta\}$$

and $\delta > 0$ is a fixed scalar. For x such that the program (90) has a feasible solution, let $\hat{\mu}_j(x)$, $j \in J_\delta(x)$, be corresponding Lagrange multipliers and set $\hat{\mu}_j(x) = 0$ for $j \notin J_\delta(x)$. Let the *active index set* be

$$A(x) = \{j \mid \hat{\mu}_j(x) > 0\},$$

and assume, without loss of generality, that $A(x)$ contains the first p indices where $p \le r$. Define the $n \times p$ matrix $N(x)$ by

$$N(x) = [\nabla g_1(x) \cdots \nabla g_p(x)],$$

and let

$$E(x) = N(x)[N(x)'N(x)]^{-1}N(x)'.$$

A Newton-like method can now be defined by

$$\bar{x} = x + \Delta x,$$

where Δx is the solution of the system [compare with (56)]

$$\{E(x) + [I - E(x)]\nabla_{xx}^2 L[x, \hat{\mu}(x)]\}\Delta x = -\nabla_x L[x, \hat{\mu}(x)].$$

Again we see that *this method consists of treating the constraints in the active set as equalities and ignoring the remaining constraints.* It is relatively easy to show that if x^* is a local minimum of (ICP) satisfying the sufficiency Assumption (S^+) of Section 3.1, then x^* is a point of attraction of the method just described, and that the rate of convergence of $\{|x_k - x^*|\}$ is superlinear. We leave the verification of this fact as an exercise for the reader.

As a precautionary note, we finally mention that active set approaches depend strongly for their success on the choice of a starting point which is sufficiently favorable to enable accurate identification of the constraints that are active at the solution. For many problems such a choice is unavailable, so active set approaches are typically effective only when combined with methods that incorporate a mechanism for enforcing convergence from poor initial starting points. Such combinations will be considered in Section 4.5.

Quadratic Programming Approach

This approach is based on a direct extension of Newton's method to inequality constrained problems. Given (x_k, μ_k), we obtain (x_{k+1}, μ_{k+1}) as a K–T pair of the quadratic program

$$\text{(91)} \qquad \begin{aligned} &\text{minimize} \quad \nabla f(x_k)'(x - x_k) + \tfrac{1}{2}(x - x_k)'\nabla_{xx}^2 L(x_k, \mu_k)(x - x_k) \\ &\text{subject to} \quad g(x_k) + \nabla g(x_k)'(x - x_k) \le 0. \end{aligned}$$

Note that $\nabla_{xx}^2 L(x_k, \mu_k)$ need not be positive definite even near a K–T pair (x^*, μ^*) satisfying Assumption (S^+) of Section 3.1. For this reason it is necessary to show that, at least for (x_k, μ_k) near (x^*, μ^*), the program (91) has at least one K–T pair and to further specify which of its possibly multiple K–T pairs will be the next iterate (x_{k+1}, μ_{k+1}) of Newton's method. This can be done by making use of the implicit function theorem as we now show.

Consider the following system of $(n + r)$ equations, with unknowns the vectors $x \in R^n$, $\mu \in R^r$, $\bar{x} \in R^n$, and $\bar{\mu} \in R^r$,

$$\text{(92)} \qquad \nabla f(x) + \nabla g(x)\bar{\mu} + \nabla_{xx}^2 L(x, \mu)(\bar{x} - x) = 0,$$

$$\text{(93)} \qquad \bar{\mu}_j[g_j(x) + \nabla g_j(x)'(\bar{x} - x)] = 0, \qquad j = 1, \ldots, r.$$

Note that (92) and (93) are necessary conditions for $\{\bar{d} = (\bar{x} - x), \bar{\mu}\}$ to be a K–T pair of the quadratic program

$$\text{(94)} \qquad \begin{aligned} &\text{minimize} \quad \nabla f(x)'d + \tfrac{1}{2}d'\nabla_{xx}^2 L(x, \mu)d \\ &\text{subject to} \quad g(x) + \nabla g(x)'d \le 0, \end{aligned}$$

the remaining K–T conditions being

$$g(x) + \nabla g(x)'(\bar{x} - x) \leq 0, \qquad \bar{\mu} \geq 0. \tag{95}$$

Let (x^*, μ^*) be a K–T pair of (ICP) satisfying the sufficiency Assumption (S^+) of Section 3.1. Then $x = x^*$, $\mu = \mu^*$, $\bar{x} = x^*$, and $\bar{\mu} = \mu^*$ is a solution of the system (92), (93). The Jacobian matrix of this system, with respect to $(\bar{x}, \bar{\mu})$ evaluated at the solution (x^*, μ^*, x^*, μ^*), can be calculated to be

$$G^* = \left[\begin{array}{c|cccc} \nabla^2_{xx} L(x^*, \mu^*) & & \nabla g(x^*) & & \\ \hline \mu_1^* \nabla g_1(x^*)' & g_1(x^*) & 0 & \cdots & 0 \\ \mu_2^* \nabla g_2(x^*)' & 0 & g_2(x^*) & \cdots & 0 \\ \vdots & \vdots & & \ddots & \vdots \\ \mu_r^* \nabla g_r(x^*)' & 0 & \cdots & & g_r(x^*) \end{array}\right]. \tag{96}$$

In order to apply the implicit function theorem, we must show that G^* is nonsingular. Indeed if $(z, w_1, \ldots, w_r)$ is a vector in the nullspace of G^*, then we have

$$\nabla^2_{xx} L(x^*, \mu^*)z + \sum_{j=1}^{r} w_j \nabla g_j(x^*) = 0, \tag{97}$$

$$\mu_j^* \nabla g_j(x^*)'z + g_j(x^*)w_j = 0, \qquad j = 1, \ldots, r. \tag{98}$$

Let $A(x^*)$ denote the set of indices of active constraints at x^*

$$A(x^*) = \{j \mid g_j(x^*) = 0, j = 1, \ldots, r\}.$$

Since (x^*, μ^*) satisfy Assumption (S^+), we have the strict complementarity condition

$$\mu_j^* > 0 \Leftrightarrow j \in A(x^*), \qquad \mu_j^* = 0 \Leftrightarrow j \notin A(x^*).$$

These relations together with (98) imply that

$$w_j = 0 \qquad \forall\, j \notin A(x^*), \tag{99a}$$

$$\nabla g_j(x^*)'z = 0 \qquad \forall\, j \in A(x^*). \tag{99b}$$

Multiplying (97) with z', we obtain

$$z'\nabla^2_{xx} L(x^*, \mu^*)z = -\sum_{j=1}^{r} w_j z' \nabla g_j(x^*). \tag{100}$$

From the last three relations, it follows that, for all $j \in A(x^*)$,

$$z'\nabla^2_{xx} L(x^*, \mu^*)z = 0, \qquad \nabla g_j(x^*)'z = 0.$$

Using Assumption (S^+), we obtain

$$z = 0. \tag{101}$$

Therefore, using (97) and (99a) we have

$$\sum_{j \in A(x^*)} w_j \nabla g_j(x^*) = 0.$$

Since by (S^+) the gradients $\nabla g_j(x^*)$, $j \in A(x^*)$, are linearly independent, we obtain

$$w_j = 0 \qquad \forall\, j \in A(x^*). \tag{102}$$

From (99a), (101), and (102) it follows that the only vector in the nullspace of G^* is the zero vector. Hence G^* is nonsingular.

Now by applying the implicit function theorem to the system (92), (93), it follows that there exist open spheres S_1 and S_2 centered at (x^*, μ^*) and a continuous function $\phi(\cdot, \cdot): S_1 \to S_2$ with

$$\phi(x, \mu) = \begin{bmatrix} \bar{x}(x, \mu) \\ \bar{\mu}(x, \mu) \end{bmatrix},$$

such that

$$\bar{x}(x^*, \mu^*) = x^*, \qquad \bar{\mu}(x^*, \mu^*) = \mu^*,$$

and for all $(x, \mu) \in S_1$, there holds

$$\nabla f(x) + \nabla g(x)\bar{\mu}(x, \mu) + \nabla_{xx}^2 L(x, \mu)[\bar{x}(x, \mu) - x] = 0 \tag{103}$$

$$\bar{\mu}_j(x, \mu)[g_j(x) + \nabla g_j(x)'[\bar{x}(x, \mu) - x]] = 0 \qquad \forall\, j = 1, \ldots, r. \tag{104}$$

We can take S_1 sufficiently small, so that for all $(x, \mu) \in S_1$

$$g_j(x) + \nabla g_j(x)'[\bar{x}(x, \mu) - x] < 0 \qquad \forall\, j \notin A(x^*), \tag{105}$$

$$\bar{\mu}_j(x, \mu) > 0 \qquad \forall\, j \in A(x^*), \tag{106}$$

and $\bar{x}(x, \mu)$, $\bar{\mu}(x, \mu)$ is the solution of (92), (93) closest to (x^*, μ^*) in terms of Euclidean distance. Observe that (103)–(106) are the K–T conditions for $\{\bar{d} = \bar{x}(x, \mu) - x, \bar{\mu}(x, \mu)\}$ to be a K–T pair of the quadratic program (94). Furthermore $\{\bar{d}, \bar{\mu}(x, \mu)\}$ is the K–T pair of (94) which is closest to (x^*, μ^*) in terms of Euclidean distance.

We are now in a position to define the iteration of Newton's method for (ICP). For (x_k, μ_k) in the open sphere S_1 specified above via the implicit function theorem, *the iteration consists of*

$$x_{k+1} = \bar{x}(x_k, \mu_k), \qquad \bar{\mu}_{k+1} = \bar{\mu}(x_k, \mu_k), \tag{107}$$

where $[\bar{x}(x_k, \mu_k), \bar{\mu}(x_k, \mu_k)]$ *is the* K–T *pair of the quadratic program* (91) *which is closest to* (x^*, μ^*) *in terms of Euclidean distance.*

Note that from (104)–(106), we have

$$\bar{\mu}_j(x_k, \mu_k) > 0 \qquad \forall\, j \in A(x^*), \qquad \bar{\mu}_j(x_k, \mu_k) = 0 \qquad \forall\, j \notin A(x^*).$$

It follows that, for $(x_k, \mu_k) \in S_1$, the Newton iteration can be alternately described as follows:

We set

$$\mu_j^{k+1} = 0 \qquad \forall\, j \notin A(x^*),$$

and we obtain $[x_{k+1}, \{\mu_j^{k+1} \mid j \in A(x^*)\}]$ as the K–T pair of the quadratic program

$$\begin{aligned} &\text{minimize} \quad \nabla f(x_k)'(x - x_k) + \tfrac{1}{2}(x - x_k)'\nabla_{xx}^2 L(x_k, \mu_k)(x - x_k) \\ &\text{subject to} \quad g_j(x_k) + \nabla g_j(x_k)'(x - x_k) = 0, \qquad j \in A(x^*), \end{aligned}$$

which is closest to $[x^*, \{\mu_j^* \mid j \in A(x^*)\}]$. Equivalently,

$$[x_{k+1}, \{\mu_j^{k+1} \mid j \in A(x^*)\}]$$

are obtained by solving the system of equations

$$\nabla_{xx}^2 L(x_k, \mu_k)(x - x_k) + \sum_{j \in A(x^*)} \mu_j \nabla g_j(x_k) = -\nabla f(x_k), \tag{108}$$

$$\nabla g_j(x_k)'(x - x_k) = -g_j(x_k) \qquad j \in A(x^*). \tag{109}$$

Except for the additional term

$$\left[\sum_{j \notin A(x^*)} \mu_j^k \nabla^2 g_j(x_k)\right](x - x_k)$$

in (108), this system is the same as the one solved in Newton's method of the previous section applied to the equality constrained problem

$$\begin{aligned} &\text{minimize} \quad f(x) \\ &\text{subject to} \quad g_j(x) = 0, \qquad j \in A(x^*). \end{aligned}$$

Now, since $\mu_j^* = 0$, for $j \notin A(x^*)$, the term $[\sum_{j \notin A(x^*)} \mu_j^k \nabla^2 g_j(x_k)]$ can be made arbitrarily small by taking μ_k sufficiently close to μ^*. Based on this fact, it can be verified, by essentially repeating the proof of Proposition 1.17, that given any scalar $r > 0$, there exists a $\delta_r > 0$ such that if $|(x_k, \mu_k) - (x^*, \mu^*)| < \delta$ then

$$|x_{k+1} - x^*|^2 + \sum_{j \in A(x^*)} |\mu_j^{k+1} - \mu_j^*|^2 \le r^2\Big(|x_k - x^*|^2 + \sum_{j \in A(x^*)} |\mu_j^k - \mu_j^*|^2\Big).$$

Since $\mu_j^{k+1} = \mu_j^* = 0$ for $j \notin A(x^*)$, we also obtain

$$|x_{k+1} - x^*|^2 + |\mu_{k+1} - \mu^*|^2 \le r^2(|x_k - x^*|^2 + |\mu_k - \mu^*|^2)$$

or equivalently

$$|(x_{k+1} - x^*, \mu_{k+1} - \mu^*)| \le r|(x_k - x^*, \mu_k - \mu^*)|.$$

This implies that (x^*, μ^*) *is a point of attraction of iteration* (107) *and the rate of convergence is superlinear.* If $\nabla^2 f$ and $\nabla^2 g_j, j \in A(x^*)$, are Lipschitz continuous in a neighborhood of x^*, then the rate of convergence is superlinear of order at least two.

4.4.4 Quasi-Newton Versions

We can develop quasi-Newton versions of the Newton-like methods of Section 4.4.2 simply by replacing various Hessian or inverse Hessian matrices wherever they appear in Newton-like iterations by approximations obtained via quasi-Newton updating formulas such as the BFGS, DFP, and others (see Section 1.3.5).

Thus a quasi-Newton version of the Newton iteration (21), (22) is given by

$$\lambda_{k+1} = (N_k' H_k^{-1} N_k)^{-1}[h(x_k) - N_k' H_k^{-1} \nabla f(x_k)], \tag{110}$$

$$x_{k+1} = x_k - H_k^{-1} \nabla_x L(x_k, \lambda_{k+1}), \tag{111}$$

where $N_k = \nabla h(x_k)$ and H_k is an approximation to $\nabla_{xx}^2 L(x_k, \lambda_k)$. Another quasi-Newton version of the same iteration is given by

$$\lambda_{k+1} = (N_k' B_k N_k)^{-1}[h(x_k) - N_k' B_k \nabla f(x_k)], \tag{112}$$

$$x_{k+1} = x_k - B_k \nabla_x L(x_k, \lambda_{k+1}), \tag{113}$$

where B_k is an approximation to $[\nabla_{xx}^2 L(x_k, \lambda_k)]^{-1}$.

In a similar manner, one can provide quasi-Newton versions of variations of Newton's method (compare with (37)–(38), (43)–(44), and (45)–(46)] and of Newton's method for inequality constraints [compare with (91)].

There are a number of formulas for updating the approximating matrices H_k and B_k of (110)–(113). Some examples follow.

$$H_{k+1} = H_k + \frac{(y_k - H_k s_k)s_k' + s_k(y_k - H_k s_k)'}{s_k' s_k} - \frac{s_k'(y_k - H_k s_k)s_k s_k'}{(s_k' s_k)^2}, \tag{114}$$

$$H_{k+1} = H_k + \frac{(y_k - H_k s_k)y_k' + y_k(y_k - H_k s_k)'}{y_k' s_k} - \frac{s_k'(y_k - H_k s_k)y_k y_k'}{(y_k' s_k)^2}, \tag{115}$$

$$B_{k+1} = B_k + \frac{(s_k - B_k y_k)y_k' + y_k(s_k - B_k y_k)'}{y_k' y_k} - \frac{y_k'(s_k - B_k y_k)y_k y_k'}{(y_k' y_k)^2}, \tag{116}$$

$$B_{k+1} = B_k + \frac{(s_k - B_k y_k)s_k' + s_k(s_k - B_k y_k)'}{s_k' y_k} - \frac{y_k'(s_k - B_k y_k)s_k s_k'}{(s_k' y_k)^2}, \tag{117}$$

where

$$s_k = x_{k+1} - x_k, \tag{118}$$

$$y_k = \nabla_x L(x_{k+1}, \lambda_{k+1}) - \nabla_x L(x_k, \lambda_{k+1}). \tag{119}$$

The formula (114) stems from Powell (1970), while (115) is an analog of the Davidon–Fletcher–Powell formula considered in Section 1.3.5. The formula (116) stems from Greenstadt (1970), while (117) is an analog of the Broyden–Fletcher–Goldfarb–Shanno formula (Section 1.3.5).

The convergence analysis of the iterations just described follows a pattern established in papers by Broyden *et al.* (1973) and Dennis and Moré (1974). The main assumptions are that the starting matrices H_0 and B_0 are close to $\nabla_{xx}^2 L(x_0, \lambda_0)$ and $[\nabla_{xx}^2 L(x_0, \lambda_0)]^{-1}$, respectively, and that (x_0, λ_0) is close to a K–T pair (x^*, λ^*) satisfying Assumption (S). For the case of the formulas (115) and (117), it is also necessary to assume that $\nabla_{xx}^2 L(x^*, \lambda^*)$ is positive definite. These assumptions are of course quite restrictive, but it should be recalled that the analysis of this section is purely local in nature. The principal idea of the analysis is that the updating formulas are such that the differences $[H_k - \nabla_{xx}^2 L(x^*, \lambda^*)]$ and $[B_k - [\nabla_{xx}^2 L(x^*, \lambda^*)]^{-1}]$ remain small as $k \to \infty$ and tend to zero along the directions of interest. This in turn implies superlinear convergence of $\{|(x_k - x^*, \lambda_k - \lambda^*)|\}$. For a detailed analysis we refer the reader to Glad (1979), Han (1977a), Tapia (1977), and Gabay (1979). An alternative quasi-Newton approach, due to Powell (1978a), will be described in Section 4.5.3.

4.5 Lagrangian Methods—Global Convergence

In order to enlarge the region of convergence of Lagrangian methods, it is necessary to combine them with some other method that has satisfactory global convergence properties. We refer to such a method as a *global method*. The main ideas here are very similar to those underlying modifications of Newton's method for unconstrained minimization (compare with Section 1.3.3), although the resulting implementations tend to be somewhat more complex. Basically, we would like to have a combined method that when sufficiently close to a local minimum of (NLP) satisfying the sufficiency conditions for optimality switches automatically to a superlinearly convergent Lagrangian method, while when far away from such a point it switches automatically to the global method which is designed to make steady progress towards approaching the set of K–T pairs of (NLP). Prime candidates for use as global methods are various penalty and multiplier methods, such as those examined in Chapters 2 and 3, and exact penalty methods, such as those considered in this chapter.

There are many possibilities for combining global and Lagrangian methods, and the suitability of any one of these depends strongly on the problem at hand. For this reason, our main purpose in this section is not to develop and recommend specific algorithms, but rather to focus on the main guidelines for harmoniously interfacing global and Lagrangian methods while retaining the advantages of both. Our emphasis thus is placed on explaining ideas rather than proving specific convergence and rate of convergence theorems.

Once a global and a Lagrangian method have been selected, the main issue to be settled is the choice of what we shall call the *switching rule* and the *acceptance rule*. The switching rule determines on the basis of certain tests at each iteration whether a switch should be made to the Lagrangian method. The tests to be used depend on the information currently available, and their purpose is to determine whether an iteration of the Lagrangian method has a reasonable chance of success. As an example, for (ECP) such tests might include verification that ∇h has rank m and that $\nabla_{xx}^2 L$ is positive definite on the subspace $\{z \mid \nabla h' z = 0\}$. We hasten to add here that these tests should not require excessive computational overhead. In some cases a switch might be made without any test at all, subject only to the condition that the Lagrangian iteration is well defined.

The acceptance rule determines whether the results of the Lagrangian iteration will be accepted as they are, whether they will be modified, or whether they will be rejected completely and a switch will be made back to the global method. Typically, acceptance of the results of the Lagrangian iteration is based on improvement of some criterion of merit such as reduction of the value of some exact penalty function.

Nearly all the combined methods to be considered are motivated by the descent properties of Newton's method and its modifications discussed in Section 4.4.2 (compare with Proposition 4.25).

4.5.1 Combinations with Penalty and Multiplier Methods

One possibility for enlarging the region of convergence of Lagrangian methods is to combine them with methods of multipliers discussed in Chapters 2 and 3. The resulting combined methods tend to be very reliable, since they inherit the robustness of the method of multipliers. At the same time they typically require fewer iterations to converge within the same accuracy than pure methods of multipliers.

The simplest possibility is to switch to a Lagrangian method at the beginning (or the end) of each (perhaps approximate) unconstrained minimization of a method of multipliers and continue using the Lagrangian method as long as the value of the exact penalty function $|\nabla L|^2$ is being decreased by a certain factor at each iteration. If satisfactory progress in decreasing $|\nabla L|^2$ is not observed, a switch back to the method of multipliers is made. Another possibility is to attempt a switch to a Lagrangian method at each iteration. As an example, consider the following method for solving (ECP) which combines Newton's method for unconstrained minimization of the augmented Lagrangian together with the Lagrangian iteration (43), (44) of Section 4.4.2.

At iteration k, we have x_k, λ_k, and a penalty parameter c_k. We also have a positive scalar w_k, which represents a target value of the exact penalty function $|\nabla L|^2$ that must be attained in order to accept the Lagrangian iteration, and a positive scalar ε_k that controls the accuracy of the unconstrained minimization of the method of multipliers. At the kth iteration, we determine x_{k+1}, λ_{k+1}, w_{k+1}, and ε_{k+1} as follows:

We first form the modified Cholesky factorization $L_k L_k'$ of the matrix $\nabla_{xx}^2 L_{c_k}(x_k, \lambda_k)$ as in Section 1.3.3. In the process, we modify $\nabla_{xx}^2 L_{c_k}(x_k, \lambda_k)$ if it is not "sufficiently positive definite" (compare with Section 1.3.3). We then find the Newton direction

$$d_k = -(L_k L_k')^{-1} \nabla_x L_{c_k}(x_k, \lambda_k), \tag{1}$$

and if $\nabla_{xx}^2 L_{c_k}(x_k, \lambda_k)$ was found "sufficiently positive definite" during the factorization process, we also carry out the Lagrangian iteration [compare with (43) and (44) in Section 4.4.2)

$$\bar{x}_k = x_k + d_k, \tag{2}$$

$$\bar{\lambda}_k = \lambda_k + c_k[h(x_k) + \nabla h(x_k)'(\bar{x}_k - x_k)]. \tag{3}$$

If

$$|\nabla L(\bar{x}_k, \bar{\lambda}_k)|^2 \le w_k,$$

then we accept the Lagrangian iteration and we set

$$x_{k+1} = \bar{x}_k, \qquad \lambda_{k+1} = \bar{\lambda}_k, \qquad c_{k+1} = c_k, \qquad \varepsilon_{k+1} = \varepsilon_k,$$
$$w_{k+1} = \gamma_1 |\nabla L(\bar{x}_k, \bar{\lambda}_k)|^2,$$

where γ_1 is a fixed scalar with $0 < \gamma_1 < 1$. Otherwise we set

$$x_{k+1} = x_k + \alpha_k d_k,$$

where the stepsize is obtained from the Armijo rule (compare with Section 1.3.1)

$$\alpha_k = \beta^{m_k},$$

where m_k is the first nonnegative integer m such that

$$L_{c_k}(x_k, \lambda_k) - L_{c_k}(x_k + \beta^m d_k, \lambda_k) \ge -\sigma \beta^m d_k' \nabla_x L_{c_k}(x_k, \lambda_k)$$

and β and σ are fixed scalars with $\beta \in (0, 1)$ and $\sigma \in (0, \frac{1}{2})$. If

$$|\nabla_x L_{c_k}(x_{k+1}, \lambda_k)| \le \varepsilon_k,$$

implying termination of the current unconstrained minimization, we set

$$\lambda_{k+1} = \lambda_k + c_k h(x_k), \tag{4}$$

$$\varepsilon_{k+1} = \gamma_2 \varepsilon_k, \quad c_{k+1} = r c_k, \quad w_{k+1} = \gamma_2 |\nabla L(x_{k+1}, \lambda_{k+1})|^2,$$

where γ_2 and r are fixed scalars with $0 < \gamma_2 < 1$ and $r > 1$. If

$$|\nabla_x L_{c_k}(x_{k+1}, \lambda_k)| > \varepsilon_k,$$

we set

$$\lambda_{k+1} = \lambda_k, \qquad \varepsilon_{k+1} = \varepsilon_k, \qquad c_{k+1} = c_k, \qquad w_{k+1} = w_k,$$

and proceed with the next iteration.

The preceding algorithm is only an example of a large variety of methods that one can construct based on combinations of multiplier methods and Lagrangian iterations. For example, a quasi-Newton approximation could be used in place of the Newton direction (1); a different Lagrangian iteration could be used in place of (2), (3); a second-order multiplier iteration could be used in place of (4); etc. Finally, one can handle inequality constraints via Lagrangian iterations employing an active set strategy (compare with Section 4.4.3). We refer to the paper by Glad (1979) for some specific algorithms and computational results.

4.5.2 Combinations with Differentiable Exact Penalty Methods —Newton and Quasi-Newton Versions

We have already observed, in Section 4.3.3, that the Newton direction for solving the system of equations $\nabla L(x, \lambda) = 0$ of (ECP) approaches asymptotically the Newton direction for minimizing the exact penalty functions (compare with Section 4.3.1)

$$P_\tau(x, \lambda; c, \alpha) = L(x, \lambda) + \tfrac{1}{2}(c + \tau|\lambda|^2)|h(x)|^2 + \tfrac{1}{2}\alpha|\nabla_x L(x, \lambda)|^2 \tag{5}$$

and

$$P_\tau(x, \lambda; c, M) = L(x, \lambda) + \tfrac{1}{2}(c + \tau|\lambda|^2)|h(x)|^2 + \tfrac{1}{2}|M(x)\nabla_x L(x, \lambda)|^2, \tag{6}$$

where $c > 0$, $\alpha > 0$, $\tau \ge 0$, $M(\cdot)$ is continuous, and $M(x)\nabla h(x)$ is invertible for all x in the set X^* defined by

$$X^* = \{x \,|\, \nabla h(x) \text{ has rank } m\}.$$

More precisely for (x_k, λ_k) sufficiently close to a K–T pair (x^*, λ^*) satisfying the sufficiency assumption (S), the Lagrangian iteration

$$x_{k+1} = x_k + \Delta x_k, \qquad \lambda_{k+1} = \lambda_k + \Delta\lambda_k, \tag{7}$$

where

$$\begin{bmatrix} \Delta x_k \\ \Delta\lambda_k \end{bmatrix} = -\nabla^2 L(x_k, \lambda_k)^{-1} \nabla L(x_k, \lambda_k), \tag{8}$$

is well defined. Furthermore iteration (8) can be expressed as

$$\begin{aligned} \begin{bmatrix} \Delta x_k \\ \Delta\lambda_k \end{bmatrix} &= -B(x_k, \lambda_k; c, \alpha)\nabla P_\tau(x_k, \lambda_k; c, \alpha) \\ &= -B(x_k, \lambda_k; c, M)\nabla P_\tau(x_k, \lambda_k; c, M), \end{aligned}$$

where $B(\cdot, \cdot; c, \alpha)$, $B(\cdot, \cdot; c, M)$ are continuous matrices satisfying

$$B(x^*, \lambda^*; c, \alpha) = \nabla^2 P_\tau(x^*, \lambda^*; c, \alpha)^{-1},$$

$$B(x^*, \lambda^*; c, M) = \nabla^2 P_\tau(x^*, \lambda^*; c, M)^{-1}.$$

Based on this fact, we can introduce in the Lagrangian iteration (7), (8) a stepsize procedure based on descent of the penalty function (5) or (6) and combine the iteration with modifications such as those considered in connection with Newton's method for unconstrained minimization to enforce convergence from poor starting points. We consider two types of modifications. The first is based on a combination with the steepest descent method, while the second is based on modification of the Hessian $\nabla_{xx}^2 L(x_k, \lambda_k)$ to make it positive definite along the subspace which is tangent to the constraint surface.

It appears that the algorithms of this section that are based on descent of the penalty functions (5) and (6) are primarily useful in the case where second derivatives of the objective and constraint functions are available. Quasi-Newton versions of these algorithms are possible [see Eqs. (58)–(61) in this section], but then it seems preferable to use for descent purposes an exact penalty function depending only on x (compare with Section 4.3.2) as will be described in what follows [see Eqs. (74), (75) in this section]. *We assume throughout this section that* $f, h \in C^3$.

Combination with the Steepest Descent Method

Let us consider an algorithm that combines the Newton iteration (7), (8), a scaled steepest descent method with a positive definite scaling matrix D, and the Armijo stepsize rule with parameters $\sigma \in (0, \frac{1}{2})$, $\beta \in (0, 1)$, and unity initial stepsize. The algorithm consists of the iteration

$$x_{k+1} = x_k + \beta^{m_k}\Delta x_k, \qquad \lambda_{k+1} = \lambda_k + \beta^{m_k}\Delta\lambda_k, \tag{9}$$

where m_k is the first nonnegative integer m for which

$$\begin{aligned} P_\tau(x_k, \lambda_k; c, M) &- P_\tau(x_k + \beta^m\Delta x_k, \lambda_k + \beta^m\Delta\lambda_k; c, M) \\ &\geq -\sigma\beta^m[\Delta x_k'\nabla_x P_\tau(x_k, \lambda_k; c, M) + \Delta\lambda_k'\nabla_\lambda P_\tau(x_k, \lambda_k; c, M)]. \end{aligned} \tag{10}$$

The direction $(\Delta x_k, \Delta\lambda_k)$ is the Newton direction (8), if $\nabla^2 L(x_k, \lambda_k)$ is invertible and if

$$\begin{aligned} -[\Delta x_k'\nabla_x P_\tau(x_k, \lambda_k; c, M) &+ \Delta\lambda_k'\nabla_\lambda P_\tau(x_k, \lambda_k; c, M)] \\ &\geq \gamma|\nabla P_\tau(x_k, \lambda_k; c, M)|^q, \end{aligned} \tag{11}$$

where γ is a positive scalar with typically very small value, and q is a scalar with $q > 2$. (These tests represent the switching rule to the Lagrangian iteration.) Otherwise $(\Delta x_k, \Delta\lambda_k)$ is the scaled steepest descent direction

$$\begin{bmatrix} \Delta x_k \\ \Delta\lambda_k \end{bmatrix} = -D\nabla P_\tau(x_k, \lambda_k; c, M). \tag{12}$$

The preceding algorithm is not necessarily the most efficient for any given problem, but rather represents an example of how one can enlarge the region of convergence of the Lagrangian iteration (7), (8). It is straightforward (compare with the analysis in Sections 1.3.1 and 1.3.3) to verify the following facts:

(a) Every limit point of a sequence $\{(x_k, \lambda_k)\}$ generated by iteration (9)–(12) is a critical point of $P_\tau(\cdot, \cdot; c, M)$.

(b) Suppose (x^*, λ^*) is a K–T pair of (ECP), satisfying Assumption (S), and c is such that $\nabla^2 P_\tau(x^*, \lambda^*; c, M)$ is positive definite. If (x^*, λ^*) is a limit point of $\{(x_k, \lambda_k)\}$ generated by (9)–(12) then $\{(x_k, \lambda_k)\}$ actually converges to (x^*, λ^*). Furthermore, the rate of convergence is superlinear. In addition there exists an integer $\bar{k}$ such that, for all $k \geq \bar{k}$, $(\Delta x_k, \Delta\lambda_k)$ is given by the Newton direction (8) and the stepsize equals unity [$m_k = 0$ in (9)]. If (x_0, λ_0) is sufficiently close to (x^*, λ^*) then the same is true for all k.

A similar algorithm can be constructed in connection with the penalty function $P_\tau(\cdot, \cdot; c, \alpha)$, and convergence results analogous to the one stated above can be shown. In fact for this penalty function, it is possible to characterize somewhat more precisely the region of pairs (x_k, λ_k) for which the Newton direction (8) is a direction of descent. This result is given in the following exercise, the proof of which can be obtained by straightforward adaptation of the proof of Proposition 4.29 that follows.

Exercise: Consider the penalty function $P_\tau(\cdot, \cdot; c, \alpha)$ of (6) where $\tau \geq 0$ and $\alpha > 0$. Let X be a compact subset of X^* and Λ a compact subset of R^m such that, for some positive scalars γ and Γ, we have

$$\gamma|z|^2 \leq z'\nabla_{xx}^2 L(x, \lambda)z \leq \Gamma|z|^2$$

for all $(x, \lambda) \in X \times \Lambda$, and $z \in R^n$ with $\nabla h(x)'z = 0$. Then there exist scalars $\bar{c} > 0$ and $\beta > 0$ (depending on X and Λ) such that the solution $(\Delta x, \Delta\lambda)$ of the system

$$\begin{bmatrix} \nabla_{xx}^2 L(x, \lambda) & \nabla h(x) \\ \nabla h(x)' & 0 \end{bmatrix} \begin{bmatrix} \Delta x \\ \Delta\lambda \end{bmatrix} = -\begin{bmatrix} \nabla_x L(x, \lambda) \\ h(x) \end{bmatrix}$$

satisfies, for all $(x, \lambda) \in X \times \Lambda$ and $c \geq \bar{c}$.

$$\nabla P_\tau(x, \lambda; c, \alpha)' \begin{bmatrix} \Delta x \\ \Delta\lambda \end{bmatrix} \leq -\beta|\nabla P_\tau(x, \lambda; c, \alpha)|^2.$$

Positive Definitenesss Modification on the Tangent Plane and Quasi-Newton Versions

We remind the reader that one of the modified versions of Newton's method for the unconstrained problem

$$\begin{aligned} &\text{minimize} \quad f(x) \\ &\text{subject to} \quad x \in R^n \end{aligned}$$

consists of the iteration (compare with Section 1.3.3)

$$x_{k+1} = x_k - \alpha_k D_k \nabla f(x_k),$$

where D_k is a positive definite matrix of the form

$$D_k = (\nabla^2 f(x_k) + E_k)^{-1}$$

where E_k is a diagonal matrix which is either zero if $\nabla^2 f(x_k)$ is "sufficiently positive definite" or else is a positive definite matrix obtained via the Cholesky factorization process.

The natural constrained analog of this procedure is to *modify the Hessian of the Lagrangian* $\nabla_{xx}^2 L(x_k, \lambda_k)$ *by adding a matrix so as to make it positive definite along the tangent plane*

$$\mathscr{C}_k = \{z \mid \nabla h(x_k)'z = 0\}. \tag{13}$$

By this, we mean replacing $\nabla_{xx}^2 L(x_k, \lambda_k)$ by

$$H_k = \nabla_{xx}^2 L(x_k, \lambda_k) + E_k, \tag{14}$$

where E_k is an $n \times n$ matrix such that

$$z'H_k z > 0 \qquad \forall\, z \in \mathscr{C}_k, \quad z \neq 0.$$

Similarly as in unconstrained minimization, this modification can be embedded within a factorization process used for solving the system of equations

$$\begin{bmatrix} \nabla^2_{xx} L(x_k, \lambda_k) & \nabla h(x_k) \\ \nabla h(x_k)' & 0 \end{bmatrix} \begin{bmatrix} \Delta x \\ \Delta \lambda \end{bmatrix} = -\begin{bmatrix} \nabla_x L(x_k, \lambda_k) \\ h(x_k) \end{bmatrix} \tag{15}$$

that yields the Newton direction, as we now show.

Assume that x_k belongs to the set

$$X^* = \{x \,|\, \nabla h(x) \text{ has rank } m\}, \tag{16}$$

and let Z_k be an $n \times (n - m)$ matrix, the columns of which form an orthonormal basis for the tangent plane $\mathscr{C}_k$ of (13); i.e.,

$$Z_k' Z_k = I, \qquad Z_k' \nabla h(x_k) = 0. \tag{17}$$

Let also Y_k be an $n \times m$ matrix with columns forming a basis for the subspace spanned by the gradients $\nabla h_1(x_k), \ldots, \nabla h_m(x_k)$. This is the subspace $\mathscr{C}^\perp$ which is orthogonal to $\mathscr{C}_k$

$$\mathscr{C}_k^\perp = \{z \,|\, z'x = 0 \ \forall\, x \in \mathscr{C}_k\} = \{\nabla h(x_k)\xi \,|\, \xi \in R^m\}.$$

Clearly, we have

$$Z_k' Y_k = 0. \tag{18}$$

Actually, we can take $Y_k = \nabla h(x_k)$, but it is possible to obtain other choices via the LQ-factorization of $\nabla h(x_k)$ which yields simultaneously a matrix Z_k satisfying (17) (see Gill and Murray, 1974, *p.* 61). Now, every vector $w \in R^n$ can be written as

$$w = Z_k \xi + Y_k \psi$$

in terms of unique vectors ξ and ψ belonging to R^{n-m} and R^m, respectively. We can write, in this manner,

$$\Delta x = Z_k d_z + Y_k d_y \tag{19}$$

and system (15) can then be written as

$$\nabla^2_{xx} L(x_k, \lambda_k) Z_k d_z + \nabla^2_{xx} L(x_k, \lambda_k) Y_k d_y + \nabla h(x_k)\Delta\lambda = -\nabla_x L(x_k, \lambda_k),$$
$$\nabla h(x_k)' Z_k d_z + \nabla h(x_k)' Y_k d_y = -h(x_k).$$

Since $\nabla h(x_k)' Z_k = 0$, the second equation yields

$$d_y = -[\nabla h(x_k)' Y_k]^{-1} h(x_k). \tag{20}$$

By premultiplying the first equation by Z_k' and by taking into account the fact that $Z_k' \nabla h(x_k) = 0$, we also obtain

(21) $$Z_k' \nabla_{xx}^2 L(x_k, \lambda_k) Z_k d_z = Z_k' [\nabla_{xx}^2 L(x_k, \lambda_k) Y_k [\nabla h(x_k)' Y_k]^{-1} h(x_k) - \nabla f(x_k)].$$

Thus if $Z_k' \nabla_{xx}^2 L(x_k, \lambda_k) Z_k$ is invertible, we can solve for d_z thereby completely determining Δx. A very interesting fact that follows from Eqs. (20) and (21) is that *the vector Δx depends on $\nabla_{xx}^2 L(x_k, \lambda_k)$ only through the product $\nabla_{xx}^2 L(x_k, \lambda_k) Z_k$. Similarly if $\nabla_{xx}^2 L(x_k, \lambda_k)$ is replaced in* (15) *by any matrix H_k such that $Z_k' H_k Z_k$ is invertible, Δx will depend on H_k only through the product $H_k Z_k$.* Regarding the vector $\Delta\lambda$, we see that given Δx it can be determined from the equation

(22) $$\Delta\lambda = -[\nabla h(x_k)' \nabla h(x_k)]^{-1} \nabla h(x_k)' [\nabla_{xx}^2 L(x_k, \lambda_k) \Delta x + \nabla_x L(x_k, \lambda_k)].$$

Now the matrix $Z_k' \nabla_{xx}^2 L(x_k, \lambda_k) Z_k$ may be viewed as the restriction of the Hessian $\nabla_{xx}^2 L(x_k, \lambda_k)$ on the tangent plane $\mathscr{C}_k$. Suppose we add to $\nabla_{xx}^2 L(x_k, \lambda_k)$ a matrix E_k of the form

$$E_k = Z_k \hat{E}_k Z_k',$$

where $\hat{E}_k$ is *diagonal*, thereby forming the matrix

(23) $$H_k = \nabla_{xx}^2 L(x_k, \lambda_k) + Z_k \hat{E}_k Z_k'.$$

Then in view of the fact that $Z_k' Z_k = I$, we have

(24) $$Z_k' H_k Z_k = Z_k' \nabla_{xx}^2 L(x_k, \lambda_k) Z_k + \hat{E}_k.$$

By choosing appropriately $\hat{E}_k$, we can make the matrix H_k positive definite on the tangent plane $\mathscr{C}_k$; i.e.,

$$Z_k' H_k Z_k > 0,$$

and in fact this can be done during the Cholesky factorization process that may be used to solve system (21) similarly as in the unconstrained case considered in Section 1.3.3. At the same time, the Hessian $\nabla_{xx}^2 L(x_k, \lambda_k)$ and its modification H_k operate identically on vectors in the subspace $\mathscr{C}_k^\perp$ spanned by the constraint gradients.

In conclusion, we have shown that by modifying $\nabla_{xx}^2 L(x_k, \lambda_k)$, as in (23), we can obtain a matrix H_k that is positive definite on the tangent plane $\mathscr{C}_k$, and furthermore this can be done conveniently during the factorization process used in solving the system (21). Of course once $\nabla_{xx}^2 L(x_k, \lambda_k)$ is replaced by H_k, we shall obtain a solution $(\Delta x, \Delta\lambda)$ of the system

(25) $$\begin{bmatrix} H_k & \nabla h(x_k) \\ \nabla h(x_k)' & 0 \end{bmatrix} \begin{bmatrix} \Delta x \\ \Delta\lambda \end{bmatrix} = -\begin{bmatrix} \nabla_x L(x_k, \lambda_k) \\ h(x_k) \end{bmatrix}$$

rather than the original system (15). We have yet to demonstrate that some substantive purpose is served by such a modification. As a first step in this direction, we show the following proposition which essentially says that if H_k is positive definite on the tangent plane, the pair (x_k, λ_k) satisfies for all k a relation of the form

$$M(x_k)\nabla_x L(x_k, \lambda_k) = A(x_k, \lambda_k)h(x_k), \tag{26}$$

where A is a continuous $m \times m$ matrix function, and the parameter c is sufficiently large, then the solution $(\Delta x_k, \Delta\lambda_k)$ of system (25) is a direction of descent of the penalty function

$$P_\tau(x, \lambda; c, M) = L(x, \lambda) + \tfrac{1}{2}(c + \tau|\lambda|^2)|h(x)|^2 + \tfrac{1}{2}|M(x)\nabla_x L(x, \lambda)|^2. \tag{27}$$

It appears that in order to construct globally convergent algorithms based on solution of the system (25) and descent of the above penalty function, it is necessary that a condition such as (26) be satisfied by successive iterates (x_k, λ_k). We will subsequently show how one can construct algorithms where condition (26) is automatically satisfied.

Proposition 4.29: Consider the penalty function P_τ of (27) where $\tau \geq 0$ and $M(x)$ is an $m \times n$ twice continuously differentiable matrix function on X^* such that $M(x)\nabla h(x)$ is nonsingular for all $x \in X^*$. Let $\mathscr{H}$ be a compact set of symmetric $n \times n$ matrices, let γ and Γ be some positive scalars, let X be a compact subset of X^*, let Λ be a compact subset of R^m, and let $A(x, \lambda)$ be an $m \times m$ matrix function which is continuous on $X \times \Lambda$. There exist scalars $\bar{c} > 0$ and $\beta > 0$ (depending on $\mathscr{H}$, A, γ, Γ, X, and Λ) such that, for all vectors $x \in X$ and $\lambda \in \Lambda$ and matrices $H \in \mathscr{H}$ satisfying

$$M(x)\nabla_x L(x, \lambda) = A(x, \lambda)h(x), \tag{28}$$

$$\gamma|z|^2 \leq z'Hz \leq \Gamma|z|^2 \qquad \forall\, z \in R^n \quad \text{with} \quad \nabla h(x)'z = 0, \tag{29}$$

the solution $(\Delta x, \Delta\lambda)$, of the system

$$\begin{bmatrix} H & \nabla h(x) \\ \nabla h(x)' & 0 \end{bmatrix} \begin{bmatrix} \Delta x \\ \Delta\lambda \end{bmatrix} = -\begin{bmatrix} \nabla_x L(x, \lambda) \\ h(x) \end{bmatrix}, \tag{30}$$

exists, is unique, and satisfies, for all $c \geq \bar{c}$,

$$\nabla P_\tau(x, \lambda; c, M)' \begin{bmatrix} \Delta x \\ \Delta\lambda \end{bmatrix} \leq -\beta|\nabla P_\tau(x, \lambda; c, M)|^2. \tag{31}$$

Proof: For vectors $x \in X^*$ and $\lambda \in \Lambda$ satisfying (28), let us use the abbreviated notation

$$a = \nabla_x L(x, \lambda), \qquad b = h(x), \qquad N = \nabla h(x), \qquad M = M(x), \tag{32}$$

$$Q = \nabla_x[M(x)\nabla_x L(x, \lambda)], \qquad A = A(x, \lambda). \tag{33}$$

We have, using this notation,

$$\nabla_x P_\tau(x, \lambda; c, M) = a + (c + \tau|\lambda|^2)Nb + QMa, \tag{34}$$

$$\nabla_\lambda P_\tau(x, \lambda; c, M) = b + \tau|b|^2\lambda + N'M'Ma. \tag{35}$$

Consider any $H \in \mathscr{H}$ satisfying (29). Let us denote

$$J = \nabla_x P_\tau(x, \lambda; c, M)'\Delta x + \nabla_\lambda P_\tau(x, \lambda; c, M)'\Delta\lambda, \tag{36}$$

where $(\Delta x, \Delta\lambda)$ is the solution of system (30). (The fact that this solution exists and is unique follows by repetition of the proof of Lemma 1.27.)

Let $p \geq 0$ be a scalar such that the matrix

$$\bar{H} = H + pNN' \tag{37}$$

is positive definite with eigenvalues uniformly bounded above and away from zero over all $x \in X$ and $H \in \mathscr{H}$ satisfying (29). (It is straightforward to show that such a scalar p exists by a minor adaptation of the proof of Lemma 1.25). The second equation of system (30) yields

$$N'\Delta x = -b. \tag{38}$$

By substituting in the first equation and using (37), we obtain

$$\bar{H}\Delta x + N\Delta\lambda = -a - pNb. \tag{39}$$

By multiplying this equation by $N'\bar{H}^{-1}$ and using (38), we obtain

$$\Delta\lambda = (N'\bar{H}^{-1}N)^{-1}[(I - pN'\bar{H}^{-1}N)b - N'\bar{H}^{-1}a]. \tag{40}$$

Substitution in (39) yields finally

$$\begin{aligned} \Delta x = &-[\bar{H}^{-1} - \bar{H}^{-1}N(N'\bar{H}^{-1}N)^{-1}N'\bar{H}^{-1}]a \\ &- \bar{H}^{-1}N(N'\bar{H}^{-1}N)^{-1}b. \end{aligned} \tag{41}$$

We now rewrite the inner product J of (36) using Eqs. (34) and (35). We have

$$J = a'\Delta x + (c + \tau|\lambda|^2)b'N'\Delta x + a'M'Q'\Delta x + b'(I + \tau b\lambda')\Delta\lambda + a'M'MN\Delta\lambda.$$

Using equations (38), (39), and (40), we obtain

$$\begin{aligned} J = {}& a'\Delta x - (c + \tau|\lambda|^2)|b|^2 + a'M'Q'\Delta x + b'(I + \tau b\lambda')(N'\bar{H}^{-1}N)^{-1} \\ & \times [(I - pN'\bar{H}^{-1}N)b - N'\bar{H}^{-1}a] - a'M'M(pNb + a + \bar{H}\Delta x). \end{aligned}$$

By rearranging terms and using (28) and (41), we find that J can be expressed as a sum of two quadratic forms

$$J = R(a, b) - c|b|^2, \tag{42}$$

where

$$\begin{aligned}R(a, b) = & -a'[\bar{H}^{-1} - \bar{H}^{-1}N(N'\bar{H}^{-1}N)^{-1}N'\bar{H}^{-1}]a - a'M'Ma \\ & - b'A'(Q' - M\bar{H})[\bar{H}^{-1} - \bar{H}^{-1}N(N'\bar{H}^{-1}N)^{-1}N'\bar{H}^{-1}]a \\ & - a'M'(Q' - M\bar{H})\bar{H}^{-1}N(N'\bar{H}^{-1}N)^{-1}b \\ & -b'[(I + \tau b\lambda')(N'\bar{H}^{-1}N)^{-1}N'\bar{H}^{-1} + pN'M'M \\ & + (N'\bar{H}^{-1}N)^{-1}N'\bar{H}^{-1}]a \\ & -b'[\tau|\lambda|^2 I - (I + \tau b\lambda')(N'\bar{H}^{-1}N)^{-1}(I - pN'\bar{H}^{-1}N)]b.\end{aligned}$$

(*Note*: A critical step in the above calculation, that uses the assumption (28), is to substitute $b'A'$ in the third term in the right side in place of $a'M'$. This step makes the following argument possible.)

The quadratic form $R(a, b)$ when restricted on the subspace $\{(a, b) \mid b = 0\}$ can be written

$$\begin{aligned}R(a, 0) &= -a'[\bar{H}^{-1} - \bar{H}^{-1}N(N'\bar{H}^{-1}N)^{-1}N'\bar{H}^{-1} + M'M]a \\ &= -\tilde{a}'[I - \tilde{N}(\tilde{N}'\tilde{N})^{-1}\tilde{N}' + \tilde{M}'\tilde{M}]\tilde{a},\end{aligned}$$

where

$$\tilde{a} = \bar{H}^{-1/2}a, \quad \tilde{N} = \bar{H}^{-1/2}N, \qquad \text{and} \qquad \tilde{M} = M\bar{H}^{1/2}.$$

We claim that $R(a, 0)$ is negative definite. Indeed, since both matrices $[I - \tilde{N}(\tilde{N}'\tilde{N})^{-1}\tilde{N}']$ and $\tilde{M}'\tilde{M}$ are positive semidefinite, it follows that $R(a, 0)$ is negative semidefinite. If $R(\bar{a}, 0) = 0$ for some $\bar{a} \neq 0$, then we must have $\bar{a}'[I - \tilde{N}(\tilde{N}'\tilde{N})^{-1}\tilde{N}']\bar{a} = 0$ and $\bar{a}'\tilde{M}'\tilde{M}\bar{a} = 0$. Since $[I - \tilde{N}(\tilde{N}'\tilde{N})^{-1}\tilde{N}']$ is the projection matrix on the subspace $\{z \mid \tilde{N}'z = 0\}$, the first equation shows that $\bar{a}$ belongs to the orthogonal subspace; i.e., $\bar{a} = \tilde{N}\xi$ for some $\xi \in R^m$. Then the equation $\bar{a}'\tilde{M}'\tilde{M}\bar{a} = 0$ yields $\xi'\tilde{N}'\tilde{M}'\tilde{M}\tilde{N}\xi = 0$. Since $\tilde{M}\tilde{N} = MN$ and MN is invertible, we obtain $\xi = 0$ which is a contradiction. Having established negative definiteness of $R(a, 0)$, it follows, using (42) and Lemma 1.25, that for c sufficiently large, we can represent J as a negative definite quadratic form; i.e.,

$$J = [a' \quad b']D(x, \lambda, \bar{H}, c)\begin{bmatrix} a \\ b \end{bmatrix},$$

where $D(x, \lambda, \bar{H}, c)$ is a negative definite matrix. Since the matrix $D(x, \lambda, \bar{H}, c)$ is continuous in all its arguments, it is straightforward to use the previous reasoning and a minor extension of Lemma 1.25 in order to show that there exists $\bar{c} > 0$ such that, for all $c \geq \bar{c}$, $x \in X$, $\lambda \in \Lambda$, and $H \in \mathscr{H}$ satisfying (28) and (29), the eigenvalues of $D(x, \lambda, \bar{H}, c)$ are negative and uniformly bounded above and away from zero. Since, in view of (34) and (35), the square norm of the gradient ∇P_τ can be expressed as a quadratic form in (a, b), it follows that there exists $\beta > 0$ such that (31) holds. Q.E.D.

Let us now consider an algorithm of the form

$$x_{k+1} = x_k + \alpha_k \Delta x_k, \tag{43}$$

$$\lambda_{k+1} = \lambda_k + \alpha_k \Delta \lambda_k, \tag{44}$$

where $(\Delta x_k, \Delta \lambda_k)$ is the solution of the system

$$\begin{bmatrix} H_k & \nabla h(x_k) \\ \nabla h(x_k)' & 0 \end{bmatrix} \begin{bmatrix} \Delta x \\ \Delta \lambda \end{bmatrix} = -\begin{bmatrix} \nabla_x L(x_k, \lambda_k) \\ h(x_k) \end{bmatrix}. \tag{45}$$

α_k is a scalar stepsize based on descent of the exact penalty function $P_\tau(\cdot, \cdot; c, M)$ and is obtained, for example, via the Armijo rule. The matrix H_k is assumed positive definite on the tangent plane $\mathscr{C}_k$ and can be obtained from either the Hessian $\nabla^2_{xx} L(x_k, \lambda_k)$ or a quasi-Newton approximation of it by appropriate modification, if necessary, as described earlier [compare with (23)]. We know from our earlier analysis that if $H_k = \nabla^2_{xx} L(x_k, \lambda_k)$ then for (x_k, λ_k) in a neighborhood of a K–T pair (x^*, λ^*) satisfying Assumption (S) the direction $(\Delta x_k, \Delta \lambda_k)$ is a descent direction for P_τ for sufficiently large c. Proposition 4.29 shows that this is also true even if $H_k \neq \nabla^2_{xx} L(x_k, \lambda_k)$ and (x_k, λ_k) is not near (x^*, λ^*) provided c is sufficiently large and a relation of the form

$$M(x_k) \nabla_x L(x_k, \lambda_k) = A(x_k, \lambda_k) h(x_k) \tag{46}$$

is satisfied for some continuous $m \times m$ matrix A. It is possible to construct a convergent algorithm in which a relation of the form (46) is satisfied by making λ_k depend continuously on x_k via the relation

$$\lambda_k = \hat{\lambda}(x_k), \tag{47}$$

where

$$\begin{aligned} \hat{\lambda}(x) = -\{&\nabla h(x)' M(x)' M(x) \nabla h(x) \\ &+ \tau |h(x)|^2 I\}^{-1} [h(x) + \nabla h(x)' M(x)' M(x) \nabla f(x)]. \end{aligned} \tag{48}$$

Using the above definition we have

$$\begin{aligned} \nabla_\lambda P_\tau[x, \hat{\lambda}(x); c, M] &= h(x) + \tau |h(x)|^2 \hat{\lambda}(x) \\ &\quad + \nabla h(x)' M(x)' M(x) \nabla_x L[x, \hat{\lambda}(x)] \\ &= 0, \end{aligned}$$

so it can be seen that $\hat{\lambda}(x)$ *minimizes* $P_\tau(x, \lambda; c, M)$ *over all* $\lambda \in R^m$. It is also easy to verify via a straightforward calculation that Eq. (46) is satisfied with

$$A(x, \lambda) = -[\nabla h(x)' M(x)']^{-1} [I + \tau \hat{\lambda}(x) h(x)']. \tag{49}$$

Therefore, given any pair (x, λ) with $x \in X^*$, *if we replace* λ *by* $\hat{\lambda}(x)$, *the value of* P_τ *cannot increase*; i.e.,

$$P_\tau[x, \hat{\lambda}(x); c, M] \leq P_\tau(x, \lambda; c, M) \qquad \forall\, x \in X^*, \quad \lambda \in R^m, \tag{50}$$

while at the same time, Eq. (28), *which is sufficient to guarantee the descent property of Proposition* 4.29, *is satisfied.* This leads to the following type of algorithm.

Given $x_k \in X^*$ and $\lambda_k = \hat{\lambda}(x_k)$, compute the solution $(\Delta x_k, \Delta\lambda_k)$ of the system of equations

$$\begin{bmatrix} H_k & \nabla h(x_k) \\ \nabla h(x_k)' & 0 \end{bmatrix} \begin{bmatrix} \Delta x \\ \Delta\lambda \end{bmatrix} = -\begin{bmatrix} \nabla_x L[x_k, \hat{\lambda}(x_k)] \\ h(x_k) \end{bmatrix}, \tag{51}$$

where H_k is positive definite on the tangent plane $\mathscr{C}_k$. Then set

$$x_{k+1} = x_k + \alpha_k \Delta x_k, \tag{52}$$

$$\tilde{\lambda}_{k+1} = \hat{\lambda}(x_k) + \alpha_k \Delta\lambda_k, \tag{53}$$

where α_k is obtained by line search based on descent of the penalty function $P_\tau(\cdot, \cdot; c, M)$ (compare with Proposition 4.29). Then set

$$\lambda_{k+1} = \hat{\lambda}(x_{k+1}), \tag{54}$$

and proceed to the next iteration.

Thus the algorithm above utilizes a two-step procedure at each iteration. Given (x_k, λ_k) with $x_k \in X^*$ and $\lambda_k = \hat{\lambda}(x_k)$, in the first step we obtain by line search a pair $(x_{k+1}, \tilde{\lambda}_{k+1})$ with a lower value of P_τ, and in the second step, we replace $\tilde{\lambda}_{k+1}$ by $\hat{\lambda}(x_{k+1})$ which, in view of (50), also lowers the value of P_τ.

Summarizing the developments so far, we have seen that there are several algorithmic possibilities for minimizing the exact penalty function $P_\tau(x, \lambda; c, M)$ based on solution of the system

$$\begin{bmatrix} H_k & \nabla h(x_k) \\ \nabla h(x_k)' & 0 \end{bmatrix} \begin{bmatrix} \Delta x \\ \Delta\lambda \end{bmatrix} = -\begin{bmatrix} \nabla_x L(x_k, \lambda_k) \\ h(x_k) \end{bmatrix}, \tag{55}$$

where H_k is positive definite on the tangent plane $\mathscr{C}_k$. Given any (x_k, λ_k) sufficiently close to a K–T pair (x^*, λ^*) satisfying Assumption (S), the direction $(\Delta x_k, \Delta\lambda_k)$ obtained from system (55) is a descent direction at (x_k, λ_k) of P_τ if $H_k = \nabla^2_{xx} L(x_k, \lambda_k)$. When far from (x^*, λ^*) or when $H_k \neq \nabla^2_{xx} L(x_k, \lambda_k)$, it may be necessary to replace λ_k by $\hat{\lambda}(x_k)$, given by (48), in order to obtain a descent direction in the same manner. To do this, it is necessary that the penalty parameter c exceeds a certain (generally unknown) threshold. There is complete freedom in choosing the matrix H_k as long as it is positive definite on the tangent plane. Thus H_k may equal the Hessian $\nabla^2_{xx} L(x_k, \lambda_k)$ perhaps modified along the tangent plane, as discussed earlier, or it may equal a quasi-Newton approximation generated, for example, by one of the quasi-Newton formulas discussed in Section 4.4.

A reasonable algorithm seems to be one whereby the direction $(\Delta x_k, \Delta\lambda_k)$ is computed via solution of the system (55) and is tested to determine whether it is a descent direction by computing its inner product with the gradient ∇P_τ. In the case where it is not, the vector $\hat{\lambda}(x_k)$ is computed and λ_k is replaced by $\hat{\lambda}(x_k)$. *It is not necessary to resolve the system* (55) *since the vector* Δx_k *and the vector* $(\lambda_k + \Delta\lambda_k)$ *do not depend on* λ_k [compare with (20)–(22)]. Thus the only additional computation in the case where $(\Delta x_k, \Delta\lambda_k)$ is not a descent direction is the computation of $\hat{\lambda}(x_k)$, and even this need not be difficult since for the typical choice $M = (\nabla h'\nabla h)^{-1}\nabla h$ the computation of $\hat{\lambda}(x_k)$ [compare with (48)] requires the inverse $(\nabla h'\nabla h)^{-1}$ which is normally available from earlier computations during the current iteration [compare with (22)]. In a practical setting, it is of course quite possible that, even after λ_k is replaced by $\hat{\lambda}(x_k)$, a direction of descent is not yet obtained because the parameter c is not sufficiently large. In this case, a reasonable scheme is simply to increase c by multiplication with some scalar until a descent direction is obtained. This can be coupled with an automatic penalty parameter adjustment procedure of the type discussed in Section 4.3.3.

Convergence and Rate of Convergence

It is possible to show various convergence results for specific algorithms of the type described above. These results are based on Proposition 4.29 and the analysis of Chapter 1 and should be routine for the experienced reader.

When the algorithms are combined with the Armijo rule with unity initial stepsize, it is easy to show superlinear convergence to a K–T pair (x^*, λ^*) satisfying Assumption (S) under the condition

$$H_k \to \nabla_{xx}^2 L(x^*, \lambda^*). \tag{56}$$

The proof of this is fairly evident, since we have already shown in the beginning of this section that if $H_k = \nabla_{xx}^2 L(x_k, \lambda_k)$ the algorithm reduces asymptotically to Newton's method for minimizing P_τ.

In some cases where a quasi-Newton scheme is used, it is not reasonable to expect that the condition $H_k \to \nabla_{xx}^2 L(x^*, \lambda^*)$ will be attained in practice. Powell's variable metric scheme, to be discussed in the next section, is a prime example of this situation. For this scheme, one can expect at most that the condition

$$[H_k - \nabla_{xx}^2 L(x^*, \lambda^*)]Z^* \to 0 \tag{57}$$

will be attained, where Z^* is an $n \times (n - m)$ matrix the columns of which form a basis for the tangent plane

$$\mathscr{C}^* = \{z \mid \nabla h(x^*)'z = 0\}.$$

It will be shown later in this section that if this condition holds and if the stepsize α_k in the algorithm

$$x_{k+1} = x_k + \alpha_k \Delta x_k, \qquad \lambda_{k+1} = \lambda_k + \alpha_k \Delta\lambda_k$$

is unity for all k sufficiently large, then the rate of convergence of the sequence $\{x_k\}$ is superlinear. Unfortunately, if only (57) is satisfied in place of (56), it is not possible to guarantee that the condition $\alpha_k = 1$ for all k sufficiently large will be attained as long as we insist on reduction of the exact penalty function P_τ at each iteration. It is possible to bypass this difficulty by modifying the vector $\Delta\lambda_k$ in the following manner:

Consider the iteration

$$x_{k+1} = x_k + \alpha_k \Delta x_k, \qquad \lambda_{k+1} = \lambda_k + \alpha_k \delta_k, \tag{58}$$

where Δx_k together with a vector $\Delta\lambda_k$ solve the system

$$\begin{bmatrix} H_k & \nabla h(x_k) \\ \nabla h(x_k)' & 0 \end{bmatrix} \begin{bmatrix} \Delta x \\ \Delta\lambda \end{bmatrix} = -\begin{bmatrix} \nabla_x L(x_k, \lambda_k) \\ h(x_k) \end{bmatrix}. \tag{59}$$

H_k is positive definite on the tangent plane $\mathscr{C}_k$, and δ_k is defined by

$$\delta_k = -[M(x_k)\nabla h(x_k)]^{-1}[y_k + M(x_k)\nabla_x L(x_k, \lambda_k)], \tag{60}$$

where

$$y_k = \nabla_x[M(x_k)\nabla_x L(x_k, \lambda_k)]'\Delta x_k. \tag{61}$$

Note that the computation of y_k requires the use of second derivatives, but it is possible to estimate y_k accurately by the finite difference scheme

$$y_k \cong t^{-1}[M(x_k + t\Delta x_k)\nabla_x L(x_k + t\Delta x_k, \lambda_k) - M(x_k)\nabla_x L(x_k, \lambda_k)], \tag{62}$$

where t is a small positive scalar.

By essentially repeating the proof of Proposition 4.29, it is possible to show the following result.

Proposition 4.30: Consider the penalty function P_τ of (27) where $\tau \geq 0$ and M is a continuous $m \times n$ matrix such that $M(x)\nabla h(x)$ is nonsingular for each $x \in X^*$. Let $\mathscr{H}$ be a compact set of symmetric $n \times n$ matrices, and γ and Γ be two positive scalars. Also let X be a compact subset of X^*, and Λ be a compact subset of R^m. There exists a scalar $\bar{c} > 0$ and a scalar $\beta > 0$ (depending on $\mathscr{H}$, γ, Γ, X, and Λ) such that for all vectors $x \in X$ and $\lambda \in \Lambda$ and matrices $H \in \mathscr{H}$ satisfying

$$\gamma|z|^2 \leq z'Hz \leq \Gamma|z|^2 \qquad \forall\, z \in R^n \quad \text{with} \quad \nabla h(x_k)'z = 0,$$

the vectors Δx_k and δ_k, defined via the solution of system (59) and Eqs. (60) and (61), satisfy, for all $c \geq \bar{c}$,

$$\nabla P_\tau(x_k, \lambda_k; c, M)' \begin{bmatrix} \Delta x_k \\ \delta_k \end{bmatrix} \leq -\beta|\nabla P_\tau(x_k, \lambda_k; c, M)|^2.$$

Proposition 4.30 asserts that the algorithm (58)–(61) has global descent properties. We now sketch a proof of the fact that if the sequence $\{H_k\}$ is bounded and if the algorithm converges to a K–T pair (x^*, λ^*) satisfying Assumption (S) and the condition $[H_k - \nabla^2_{xx} L(x^*, \lambda^*)]Z^* \to 0$ holds [compare with (57)], then close enough to (x^*, λ^*) the stepsize $\alpha_k = 1$ will be acceptable by the algorithm in the sense that it leads to a reduction of P_τ. To this end, we show that the direction $(\Delta x_k, \delta_k)$ differs from the direction (d_{x_k}, d_{λ_k}) generated by solving the Newton system

$$\begin{bmatrix} \nabla^2_{xx} L(x_k, \lambda_k) & \nabla h(x_k) \\ \nabla h(x_k)' & 0 \end{bmatrix} \begin{bmatrix} d_{x_k} \\ d_{\lambda_k} \end{bmatrix} = - \begin{bmatrix} \nabla_x L(x_k, \lambda_k) \\ h(x_k) \end{bmatrix} \tag{63}$$

by a term that goes to zero faster than $|\nabla L(x_k, \lambda_k)|$; i.e.,

$$\Delta x_k = d_{x_k} + o(|\nabla L(x_k, \lambda_k)|), \tag{64}$$

$$\delta_k = d_{\lambda_k} + o(|\nabla L(x_k, \lambda_k)|). \tag{65}$$

We have that $\Delta x_k (d_{x_k})$ depends on $H_k [\nabla^2_{xx} L(x_k, \lambda_k)]$ only through the matrices $H_k Z_k$ $[\nabla^2_{xx} L(x_k, \lambda_k) Z_k]$, where Z_k is an orthonormal basis matrix for the tangent plane [compare with (20) and (21)]. We can assume, without loss of generality, that $Z_k \to Z^*$ so that (57) yields

$$[H_k - \nabla^2_{xx} L(x^*, \lambda^*)]Z_k \to 0.$$

Since $\Delta x_k = O(|\nabla L(x_k, \lambda_k)|)$ (in view of boundedness of $\{H_k\}$), it is easily seen that (64) holds.

Now the vector d_{λ_k} satisfies

$$\nabla h(x_k) d_{\lambda_k} = -\nabla^2_{xx} L(x_k, \lambda_k) d_{x_k} - \nabla_x L(x_k, \lambda_k),$$

and therefore

$$d_{\lambda_k} = -[M(x_k)\nabla h(x_k)]^{-1}[M(x_k)\nabla^2_{xx} L(x_k, \lambda_k) d_{x_k} + M(x_k)\nabla_x L(x_k, \lambda_k)]. \tag{66}$$

The vector y_k of (61) can be expressed as

$$y_k = [M(x_k)\nabla^2_{xx} L(x_k, \lambda_k) + O(|\nabla L(x_k, \lambda_k)|)]\Delta x_k.$$

Since $\Delta x_k = O(|\nabla L(x_k, \lambda_k)|)$, the previous equation and (60) yield

$$\begin{aligned} \delta_k = -[M(x_k)\nabla h(x_k)]^{-1}[&M(x_k)\nabla^2_{xx} L(x_k, \lambda_k)\Delta x_k \\ &+ o(|\nabla L(x_k, \lambda_k)|) + M(x_k)\nabla_x L(x_k, \lambda_k)]. \end{aligned} \tag{67}$$

By taking into account (64) and comparing (66) and (67), we obtain (65).

We know from our earlier analysis that (d_{x_k}, d_{λ_k}) differs from the Newton direction for minimizing P_τ by a term $o(|\nabla L(x_k, \lambda_k)|)$. In view of (64) and (65), it follows that the same is true for the direction $(\Delta x_k, \delta_k)$. This is sufficient to show that near (x^*, λ^*) the stepsize $\alpha_k = 1$ will be acceptable by the algorithm

(58)–(61) (in the sense that it leads to "sufficient" reduction of the value of P_τ) and that the rate of convergence is superlinear under the conditions stated earlier.

Quasi-Newton Algorithms for Differentiable Exact Penalty Functions Depending Only on x

Let us consider the exact penalty function $\hat{P}$ introduced in Section 4.3.2. We have, for all $x \in X^*$,

$$\hat{P}(x; c) = f(x) + \lambda(x)'h(x) + \tfrac{1}{2}c|h(x)|^2, \tag{68}$$

where

$$\lambda(x) = -[\nabla h(x)'\nabla h(x)]^{-1}\nabla h(x)'\nabla f(x). \tag{69}$$

We saw, in Section 4.3.2, that $\hat{P}$ can be expressed as

$$\hat{P}(x; c) = P[x, \hat{\lambda}(x); c + 1, M] = \min_{\lambda} P(x, \lambda; c + 1, M), \tag{70}$$

where

$$P(x, \lambda; c, M) = L(x, \lambda) + \tfrac{1}{2}c|h(x)|^2 + \tfrac{1}{2}|M(x)\nabla_x L(x, \lambda)|^2, \tag{71}$$

$$M(x) = [\nabla h(x)'\nabla h(x)]^{-1}\nabla h(x)', \tag{72}$$

$$\hat{\lambda}(x) = -h(x) - [\nabla h(x)'\nabla h(x)]^{-1}\nabla h(x)'\nabla f(x). \tag{73}$$

It is thus natural to expect, in view of the analysis given earlier, that algorithms of the type considered so far in this section can also be used for minimizing the penalty function $\hat{P}$. Indeed let us consider an algorithm of the form

$$x_{k+1} = x_k + \alpha_k \Delta x_k, \tag{74}$$

where α_k is a stepsize obtained by descent on $\hat{P}$ (for example via the Armijo rule with unity initial stepsize). Assume that Δx_k together with a vector $\Delta\lambda_k$ solves a system of the form

$$\begin{bmatrix} H_k & \nabla h(x_k) \\ \nabla h(x_k)' & 0 \end{bmatrix} \begin{bmatrix} \Delta x \\ \Delta\lambda \end{bmatrix} = -\begin{bmatrix} \nabla_x L(x_k, \lambda_k) \\ h(x_k) \end{bmatrix}, \tag{75}$$

where λ_k is an arbitrary vector in R^m and H_k is a symmetric $n \times n$ matrix which is positive definite on the tangent plane

$$\mathscr{C}_k = \{z \,|\, \nabla h(x_k)'z = 0\}. \tag{76}$$

We shall show first that Δx_k so obtained is a descent direction of $\hat{P}$ at x_k. To this end, we note the basic fact that Δx_k does not depend on λ_k at all and depends on H_k only through $H_k Z_k$, where Z_k is an $n \times (n - m)$ orthonormal basis matrix for $\mathscr{C}_k$ [compare with (19)–(21) and the related discussion]. We next observe, using (72), that (73) can be written as

$$M(x)\nabla_x L[x, \hat{\lambda}(x)] = -h(x),$$

and hence the condition (28) of Proposition 4.29 is satisfied by all pairs $(x, \hat{\lambda}(x))$, $x \in X^*$, with A equal to $-I$. Proposition 4.29 shows therefore that, for all c sufficiently large, the direction $(\Delta x_k, \Delta\lambda_k)$ is a descent direction of $P(\cdot, \cdot; c + 1, M)$ at $(x_k, \hat{\lambda}(x_k))$. Furthermore we have, for all $x \in X^*$ [compare with (35) and (36) in Section 4.3.2],

$$\nabla_x P[x, \hat{\lambda}(x); c + 1, M] = \nabla\hat{P}(x; c), \tag{77}$$

$$\nabla_\lambda P[x, \hat{\lambda}(x); c + 1, M] = 0. \tag{78}$$

Using these equations in Proposition 4.29, we obtain that, for some $\beta > 0$,

$$\nabla\hat{P}(x_k; c)'\Delta x_k \le -\beta|\nabla\hat{P}(x_k; c)|^2.$$

It follows that Δx_k is a descent direction of $\hat{P}$ at x_k for sufficiently large c. Indeed based on Proposition 4.29, we have proved by the argument above the following result:

Proposition 4.31: Let $\mathscr{H}$ be a compact set of symmetric $n \times n$ matrices, let γ and Γ be two positive scalars, and let X be a compact subset of X^*. There exists a scalar $\bar{c} > 0$ and a scalar $\beta > 0$ (depending on $\mathscr{H}$, γ, Γ, and X) such that for all vectors $x \in X$ and $\lambda \in R^m$ and matrices $H \in \mathscr{H}$ satisfying

$$\gamma|z|^2 \le z'Hz \le \Gamma|z|^2 \qquad \forall\, z \in R^n \quad \text{with} \quad \nabla h(x)'z = 0,$$

the solution $(\Delta x, \Delta\lambda)$ of the system

$$\begin{bmatrix} H & \nabla h(x) \\ \nabla h(x)' & 0 \end{bmatrix} \begin{bmatrix} \Delta x \\ \Delta\lambda \end{bmatrix} = -\begin{bmatrix} \nabla_x L(x, \lambda) \\ h(x) \end{bmatrix}$$

satisfies, for all $c \ge \bar{c}$,

$$\nabla\hat{P}(x; c)'\Delta x \le -\beta|\nabla\hat{P}(x; c)|^2.$$

Proposition 4.31 shows that the algorithm (74), (75) has global convergence properties provided the penalty parameter c is chosen sufficiently large. We shall show that *if $\{H_k\}$ is bounded, the algorithm converges to a K–T pair (x^*, λ^*) satisfying Assumption* (S), *and the condition*

$$\Delta x_k'[H_k - \nabla^2_{xx} L(x^*, \lambda^*)]Z^*/|\Delta x_k| \to 0 \tag{79}$$

holds, where Z^* is an orthonormal basis matrix for the tangent plane $\mathscr{C}^* = \{z \,|\, \nabla h(x^*)'z = 0\}$, *then for sufficiently large k the stepsize $\alpha_k = 1$ is acceptable* (*in the sense that it leads to a "sufficient" reduction of the value of $\hat{P}$*) *and the rate of convergence of the algorithm is superlinear.* To this end, it will be sufficient to show (compare with the proofs of Propositions 1.15 and 1.17) that (79) implies

$$\nabla^2 \hat{P}(x^*; c)\Delta x_k = -\nabla \hat{P}(x_k; c) + o(|x_k - x^*|). \tag{80}$$

In order to show (80), we first observe (using the boundedness of $\{H_k\}$) that we have

$$\Delta x_k = O(|\nabla_x L[x_k, \hat{\lambda}(x_k)]|) + O(|h(x_k)|) = O(|x_k - x^*|). \tag{81}$$

A straightforward calculation shows that

$$\nabla \hat{P}(x_k; c) = \nabla_x L[x_k, \lambda(x_k)] + \nabla\lambda(x_k)h(x_k) + c\nabla h(x_k)h(x_k), \tag{82}$$

$$\nabla^2 \hat{P}(x^*; c) = \nabla_{xx}^2 L(x^*, \lambda^*) + \nabla\lambda(x^*)\nabla h(x^*)' + \nabla h(x^*)\nabla\lambda(x^*)' + c\nabla h(x^*)\nabla h(x^*)'. \tag{83}$$

Hence Eq. (80), which is to be proved, can be written as

$$\begin{aligned}[\nabla_{xx}^2 L(x^*, \lambda^*) &+ \nabla\lambda(x^*)\nabla h(x^*)' + \nabla h(x^*)\nabla\lambda(x^*)' \\ &+ c\nabla h(x^*)\nabla h(x^*)']\Delta x_k \\ &= -[\nabla_x L[x_k, \lambda(x_k)] + \nabla\lambda(x_k)h(x_k) + c\nabla h(x_k)h(x_k)] \\ &+ o(|x_k - x^*|).\end{aligned} \tag{84}$$

Using (81) and the fact that $h(x_k) = -\nabla h(x_k)'\Delta x_k$, we have

$$\nabla\lambda(x_k)h(x_k) = -\nabla\lambda(x_k)\nabla h(x_k)'\Delta x_k = -\nabla\lambda(x^*)\nabla h(x^*)'\Delta x_k + o(|x_k - x^*|), \tag{85}$$

$$\nabla h(x_k)h(x_k) = -\nabla h(x_k)\nabla h(x_k)'\Delta x_k = -\nabla h(x^*)\nabla h(x^*)'\Delta x_k + o(|x_k - x^*|). \tag{86}$$

Using these equations, we see that (84) is equivalent to

$$[\nabla_{xx}^2 L(x^*, \lambda^*) + \nabla h(x^*)\nabla\lambda(x^*)']\Delta x_k + \nabla_x L[x_k, \lambda(x_k)] = o(|x_k - x^*|). \tag{87}$$

From the definition of $\lambda(\cdot)$, we have, for all $x \in X^*$,

$$\nabla h(x)'\nabla_x L[x, \lambda(x)] = \nabla h(x)'\nabla f(x) + \nabla h(x)'\nabla h(x)\lambda(x) = 0. \tag{88}$$

Since $\nabla_x L[x_k, \lambda(x_k)] = O(|x_k - x^*|)$ and $\nabla h(x_k) = \nabla h(x^*) + O(|x_k - x^*|)$, Eq. 88) yields

$$\nabla h(x^*)'\nabla_x L[x_k, \lambda(x_k)] = o(|x_k - x^*|).$$

Also by differentiating (88) at x^*, we obtain

$$\nabla h(x^*)'[\nabla_{xx}^2 L(x^*, \lambda^*) + \nabla h(x^*)\nabla\lambda(x^*)'] = 0.$$

The last two equations yield

$$\nabla h(x^*)'\{[\nabla_{xx}^2 L(x^*, \lambda^*) + \nabla h(x^*)\nabla\lambda(x^*)']\Delta x_k + \nabla_x L[x_k, \lambda(x_k)]\} = o(|x_k - x^*|). \tag{89}$$

We shall now show that

$$(90)\qquad Z^{*\prime}\{[\nabla^2_{xx}L(x^*, \lambda^*) + \nabla h(x^*)\nabla\lambda(x^*)']\Delta x_k + \nabla_x L[x_k, \lambda(x_k)]\} = o(|x_k - x^*|).$$

Since the $n \times n$ matrix

$$[\nabla h(x^*) \quad Z^*]$$

is invertible, Eqs. (89) and (90) will imply the desired relation (87) and hence also (80).

In order to show (90), we note that Eqs. (79) and (81) and the fact that $Z^{*\prime}\nabla h(x^*) = 0$ imply

$$(91)\quad Z^{*\prime}[\nabla^2_{xx}L(x^*, \lambda^*) + \nabla h(x^*)\nabla\lambda(x^*)']\Delta x_k = Z^{*\prime}H_k\Delta x_k + o(|x_k - x^*|).$$

We also have, from the definition of Δx_k,

$$(92)\qquad H_k\Delta x_k + \nabla h(x_k)[\lambda_{k+1} - \lambda(x_k)] = -\nabla_x L[x_k, \lambda(x_k)],$$

where λ_{k+1} is obtained from the solution of system (75). We can write this equation as

$$\lambda_{k+1} - \lambda(x_k) = -[\nabla h(x_k)'\nabla h(x_k)]^{-1}\nabla h(x_k)'\{H_k\Delta x_k + \nabla_x L[x_k, \lambda(x_k)]\},$$

so by also using (81) and the boundedness assumption on $\{H_k\}$, we obtain

$$\lambda_{k+1} - \lambda(x_k) = O(|x_k - x^*|).$$

Using the equation above and the fact that $Z^{*\prime}\nabla h(x_k) = O(|x_k - x^*|)$, we obtain, from (92),

$$(93)\qquad Z^{*\prime}H_k\Delta x_k + Z^{*\prime}\nabla_x L[x_k, \lambda(x_k)] = o(|x_k - x^*|).$$

By combining (91) and (93), we see that (90) holds and therefore our proof of validity of (80) is complete.

To summarize, we have shown that the algorithm (74), (75), coupled with the Armijo rule with unity initial stepsize and descent on $\hat{P}$, has a rate of convergence to a local minimum x^* satisfying (S) which is *superlinear* provided the following three conditions hold:

(a) The penalty parameter c is sufficiently large to ensure that x^* is a strong local minimum of the penalty function $\hat{P}(\cdot\,; c)$.

(b) The sequence $\{H_k\}$ is bounded.

(c) The condition

$$\Delta x_k'[H_k - \nabla^2_{xx}L(x^*, \lambda^*)]Z^*/|\Delta x_k| \to 0$$

holds.

The algorithm (74), (75) has one main disadvantage when compared with the earlier algorithms, which were based on the exact penalty function

$P_\tau(x, \lambda; c, M)$, namely, that each function evaluation requires the computation of $\lambda(x)$ and hence the inverse $[\nabla h(x)'\nabla h(x)]^{-1}$. This however may not be serious, since the solution of system (75) requires $O(n^3)$ operations versus $O(m^3)$ for computation of $[\nabla h(x)'\nabla h(x)]^{-1}$. Also the inverse $[\nabla h(x_k)'\nabla h(x_k)]^{-1}$ may be computed at each iteration as part of the solution of the system (75). Thus additional overhead results only at iterations requiring more than one function call; and we have shown that under circumstances where superlinear convergence is obtained, only one function call per iteration is necessary when near convergence. In any case, it is possible to limit the number of extra evaluations per iteration of $[\nabla h(x)'\nabla h(x)]^{-1}$ to *at most one* by performing instead a line search on the function $P(\cdot, \cdot; c + 1, M)$ of the form

$$x_{k+1} = x_k + \alpha_k \Delta x_k, \qquad \lambda_{k+1} = \hat{\lambda}(x_k) + \alpha_k[\hat{\lambda}(x_k + \Delta x_k) - \hat{\lambda}(x_k)],$$

starting at the pair $(x_k, \hat{\lambda}(x_k))$ for each k. We have already shown that $(\Delta x_k, [\hat{\lambda}(x_k + \Delta x_k) - \hat{\lambda}(x_k)])$ is a descent direction at $(x_k, \hat{\lambda}(x_k))$ of $P(\cdot, \cdot; c + 1, M)$ [Proposition 4.29 and (77), (78)]. Since

$$\hat{P}(x; c) = P[x, \hat{\lambda}(x); c + 1, M] \qquad \forall\, x \in X^*,$$

the preceding analysis shows that if the condition (79) holds, the stepsize $\alpha_k = 1$ will be acceptable by the algorithm near convergence and that the rate of convergence of the algorithm will be again superlinear.

We finally note that the algorithm (74), (75) can similarly be shown to be superlinearly convergent if line search is based on other exact penalty functions of the form

$$\hat{P}_\tau(x; c) = \min_\lambda P_\tau(x, \lambda; c, M). \tag{94}$$

For example, if

$$M(x) = [\nabla h(x)'\nabla h(x)]^{-1}\nabla h(x)',$$

then the minimizing vector in (94) is given by

$$\hat{\lambda}(x) = -(1 + \tau|h(x)|^2)^{-1}\{h(x) + [\nabla h(x)'\nabla h(x)]^{-1}\nabla h(x)'\nabla f(x)\},$$

and a straightforward calculation shows that $\hat{P}_\tau$ takes the form

$$\hat{P}_\tau(x; c) = L[x, \hat{\lambda}(x)] + \tfrac{1}{2}(c + \tau|\hat{\lambda}(x)|^2)|h(x)|^2 + \tfrac{1}{2}|h(x) + \tau|h(x)|^2\hat{\lambda}(x)|^2.$$

For $\tau = 0$, we obtain $\hat{P}_0(x; c) = \hat{P}(x; c - 1)$ where $\hat{P}$ is the penalty function (68). It appears, however, that using a positive scalar τ improves the numerical stability of the resulting algorithm, so the function $\hat{P}_\tau$ of (94) with $\tau > 0$ may offer some advantage over the function $\hat{P}$ of (68). More generally the minimizing vector in (94) is given by

$$\hat{\lambda}(x) = -[\nabla h(x)'M(x)'M(x)\nabla h(x) + \tau|h(x)|^2 I]^{-1} \times [h(x) + \nabla h(x)'M(x)'M(x)\nabla f(x)].$$

When $M(x)$ is defined for all $x \in R^n$ [including those x for which $\nabla h(x)$ does not have full rank, such as for example when $M(x) = \nabla h(x)'$], a choice $\tau > 0$ is particularly interesting since then $\hat{\lambda}(x)$ is defined for all $x \in R^n$ for which either $\nabla h(x)$ has full rank *or* $h(x) \neq 0$. For example, for the two-dimensional problem $\min\{x_1 | x_1^2 + x_2^2 = 1\}$ the function $\hat{P}(x; c)$ of (68) is not defined at the origin. It tends to $-\infty$ and $+\infty$ as x approaches the origin along the directions (1, 0) and (−1, 0) respectively. By contrast this peculiar behavior does not arise when $M(x) = \nabla h(x)$ and $\tau > 0$. In that case $\hat{P}_\tau(x; c)$ is everywhere continuously differentiable.

Some Extensions to Inequality Constraints

We shall develop an extension of the Newton iteration (7), (8) to the inequality constrained problem

(ICP) $\quad$ minimize $f(x)$

$\qquad$ subject to $g_j(x) \leq 0, \quad j = 1, \ldots, r,$

where $f, g_j \in C^3$. The iteration employs an active set strategy and is similar to some of the iterations examined in Section 4.4.3.

We will make use of the exact penalty function [compare with (70) in Section 4.3.3]

$$P_\tau^+(x, \mu; c, \alpha) = f(x) + \tfrac{1}{2}\alpha|\nabla_x L(x, \mu)|^2 + \frac{1}{2(c + \tau|\mu|^2)} \sum_{j=1}^{r} Q_j(x, \mu; c, \alpha, \tau), \tag{95}$$

where

$$Q_j(x, \mu; c, \alpha, \tau) = [\max\{0, \mu_j + 2\alpha\mu_j^2 + (c + \tau|\mu|^2)g_j(x)\}]^2 - (\mu_j + 2\alpha\mu_j^2)^2 - 4\alpha(c + \tau|\mu|^2)\mu_j^2 g_j(x), \qquad j = 1, \ldots, r.$$

Fix $c > 0$, $\tau \geq 0$, and $\alpha > 0$ and define, for each $(x, \mu) \in R^{n+r}$,

$$A^+(x, \mu) = \{j | \mu_j + 2\alpha\mu_j^2 + (c + \tau|\mu|^2)g_j(x) > 0, j = 1, \ldots, r\}, \tag{96}$$

$$A^-(x, \mu) = \{j | j \notin A^+(x, \mu), j = 1, \ldots, r\}. \tag{97}$$

For a given (x, μ), assume (by reordering indices if necessary) that $A^+(x, \mu)$ contains the first p indices where p is an integer with $0 \le p \le r$. Define

$$g_+(x) = \begin{bmatrix} g_1(x) \\ \vdots \\ g_p(x) \end{bmatrix}, \qquad g_-(x) = \begin{bmatrix} g_{p+1}(x) \\ \vdots \\ g_r(x) \end{bmatrix}, \tag{98}$$

$$\mu_+ = \begin{bmatrix} \mu_1 \\ \vdots \\ \mu_p \end{bmatrix}, \qquad \mu_- = \begin{bmatrix} \mu_{p+1} \\ \vdots \\ \mu_r \end{bmatrix}, \tag{99}$$

$$L_+(x, \mu) = f(x) + \mu'_+ g_+(x). \tag{100}$$

We note that p, g_+, g_-, μ_+, μ_-, and L_+ depend on (x, μ), but to simplify notation we do not show explicitly this dependence.

In the extension of Newton's method that we consider, given (x, μ), we denote the next iterate by $(\hat{x}, \hat{\mu})$ where $\hat{\mu} = (\hat{\mu}_1, \ldots, \hat{\mu}_r)$. We also write

$$\hat{\mu}_+ = \begin{bmatrix} \hat{\mu}_1 \\ \vdots \\ \hat{\mu}_p \end{bmatrix}, \qquad \hat{\mu}_- = \begin{bmatrix} \hat{\mu}_{p+1} \\ \vdots \\ \hat{\mu}_r \end{bmatrix}. \tag{101}$$

The iteration, roughly speaking, consists of setting the multipliers of the inactive constraints $[j \in A^-(x, \mu)]$ to zero, and treating the remaining constraints as equalities. More precisely, we set

$$\hat{\mu}_- = 0 \tag{102}$$

and obtain $\hat{x}$, $\hat{\mu}_+$ by solving the system

$$\begin{bmatrix} \nabla^2_{xx} L_+(x, \mu) & \nabla g_+(x) \\ \nabla g_+(x)' & 0 \end{bmatrix} \begin{bmatrix} \hat{x} - x \\ \hat{\mu}_+ - \mu_+ \end{bmatrix} = - \begin{bmatrix} \nabla_x L_+(x, \mu) \\ g_+(x) \end{bmatrix}, \tag{103}$$

assuming, of course, that the matrix on the left above is invertible.

We consider the following combination of the Newton iteration (102), (103) with the Armijo rule and a scaled steepest descent method for minimizing the penalty function $P^+_\tau(\cdot, \cdot; c, \alpha)$ of (95). Let $\sigma \in (0, \frac{1}{2})$, $\beta \in (0, 1)$, $\gamma > 0$, and D be a positive definite matrix. Given (x, μ), the next iterate $(\bar{x}, \bar{\mu})$ is given by

$$\begin{bmatrix} \bar{x} \\ \bar{\mu} \end{bmatrix} = \begin{bmatrix} x \\ \mu \end{bmatrix} + \beta^{\bar{m}} \begin{bmatrix} p_x \\ p_\mu \end{bmatrix}, \tag{104}$$

where $\bar{m}$ is the first nonnegative integer m for which

$$P^+_\tau(x, \mu; c, \alpha) - P^+_\tau(x + \beta^m p_x, \mu + \beta^m p_\mu; c, \alpha) \ge -\sigma \beta^m p' \nabla P^+_\tau(x, \mu; c, \alpha). \tag{105}$$

The direction $p = (p_x, p_\mu)$ is given by the Newton direction, obtained from (102) and (103),

$$p = \begin{bmatrix} p_x \\ p_\mu \end{bmatrix} = \begin{bmatrix} \hat{x} - x \\ \hat{\mu} - \mu \end{bmatrix}, \tag{106}$$

if the matrix on the left in (103) is invertible and

$$-(\hat{x} - x)'\nabla_x P_\tau^+(x, \mu; c, \alpha) - (\hat{\mu} - \mu)'\nabla_\mu P_\tau^+(x, \mu; c, \alpha) \geq \gamma |\nabla P_\tau^+(x, \mu; c, \alpha)|^q, \tag{107}$$

where q is a scalar with $q > 2$. Otherwise

$$p = -D\nabla P_\tau^+(x, \mu; c, \alpha). \tag{108}$$

Based on the results for unconstrained minimization methods developed in Section 1.3, it can be shown that any limit point of a sequence generated by the method described above is a critical point of P_τ^+. There remains to show, similarly as for equality constrained problems, that the direction generated by the Newton iteration (102), (103) approaches asymptotically the Newton direction for minimizing P_τ^+ as (x, μ) approaches a K–T pair (x^*, μ^*) satisfying Assumptions (S^+). A superlinear convergence rate result then follows.

Consider a K–T pair (x^*, μ^*) of (ICP) satisfying Assumption (S^+). In view of the strict complementarity assumption [$\mu_j^* > 0$ if $g_j(x^*) = 0$], for each $c > 0$, $\tau \geq 0$, and $\alpha > 0$, there exists a neighborhood of (x^*, μ^*) within which we have

$$A^+(x, \mu) = A(x^*) = \{j \mid g_j(x^*) = 0, j = 1, \ldots, r\}. \tag{109}$$

Within this neighborhood, the Newton iteration (102), (103) reduces to the Newton iteration for solving the system of necessary conditions

$$\nabla_x L_+(x, \mu) = 0, \qquad g_+(x) = 0,$$

corresponding to the equality constrained problem

$$\begin{aligned} &\text{minimize} \quad f(x) \\ &\text{subject to} \quad g_+(x) = 0. \end{aligned}$$

Based on this fact, it is easy to see that (x^*, μ^*) is a point of attraction of iteration (102), (103), and the rate of convergence is superlinear. Let c, τ, and α be such that $\nabla^2 P_\tau^+(x^*, \mu^*; c, \alpha)$ is positive definite. We shall show that, in a neighborhood of (x^*, μ^*) within which (109) holds, we have

$$\begin{bmatrix} \hat{x} - x \\ \hat{\mu} - \mu \end{bmatrix} = -[H_\tau(x, \mu; c, \alpha)]^{-1}\nabla P_\tau^+(x, \mu; c, \alpha), \tag{110}$$

where $H_\tau(\cdot, \cdot; c, \alpha)$ is a continuous matrix satisfying

$$H_\tau(x^*, \mu^*; c, \alpha) = \nabla^2 P_\tau^+(x^*, \mu^*; c, \alpha). \tag{111}$$

We show this fact for $\tau = 0$.

Consider the $(n + r) \times (n + r)$ matrix

(112) $H =$

$$\left[\begin{array}{c|c|c} \nabla_{xx}^2 L_+ + c\nabla g_+ \nabla g'_+ + \alpha \nabla_{xx}^2 L_+ \nabla_{xx}^2 L_+ & \nabla g_+ + \alpha \nabla_{xx}^2 L_+ \nabla g_+ & \alpha \nabla_{xx}^2 L_+ \nabla g_- + \alpha E \\ \hline \nabla g'_+ + \alpha \nabla g'_+ \nabla_{xx}^2 L_+ & \alpha \nabla g'_+ \nabla g_+ & \alpha \nabla g'_+ \nabla g_- \\ \hline \alpha \nabla g'_- \nabla_{xx}^2 L_+ & \alpha \nabla g'_- \nabla g_+ & \alpha \nabla g'_- \nabla g_- + F \end{array}\right]$$

where all derivatives are evaluated at a point (x, μ) in a neighborhood of (x^*, μ^*) within which (109) holds, the $(r - p) \times (r - p)$ diagonal matrix F is given by

(113) $F =$

$$\begin{bmatrix} -[(1 + 4\alpha\mu_{p+1})(1 + 2\alpha\mu_{p+1})]/c - 4\alpha g_{p+1} & & 0 \\ & \ddots & \\ 0 & & -[(1 + 4\alpha\mu_r)(1 + 2\alpha\mu_r)]/c - 4\alpha g_r \end{bmatrix},$$

and the $n \times (r - p)$ matrix E is given by

$$E = [-\nabla^2 g_{p+1} \nabla_x L - 2\mu_{p+1} \nabla g_{p+1} \,\vdots\, \cdots \,\vdots\, -\nabla^2 g_r \nabla_x L - 2\mu_r \nabla g_r]. \tag{114}$$

The function P^+ can also be written as

$$P^+(x, \mu; c, \alpha) = L_+(x, \mu) + \tfrac{1}{2}c|g_+(x)|^2 + \tfrac{1}{2}\alpha|\nabla_x L(x, \mu)|^2 - \sum_{j=p+1}^{r} \left[\frac{(\mu_j + 2\alpha\mu_j^2)^2}{2c} + 2\alpha\mu_j^2 g_j(x)\right]. \tag{115}$$

By differentiating this expression, we obtain

(116)

$$\nabla P^+(x, \mu; c, \alpha) = \begin{bmatrix} \nabla_x L_+ + c\nabla g_+ g_+ + \alpha \nabla_{xx}^2 L \nabla_x L - 2\alpha \sum_{j=p+1}^{r} \mu_j^2 \nabla g_j \\ g_+ + \alpha \nabla g'_+ \nabla_x L \\ \alpha \nabla g'_- \nabla_x L + F\mu_- \end{bmatrix},$$

where F is given by (113). We now observe that the solution $(\hat{x} - x, \hat{\mu}_+ - \mu_+)$ of the system (103) also satisfies

(117)

$$\left[\begin{array}{c|c} \nabla^2_{xx}L_+ + c\nabla g_+ \nabla g'_+ + \alpha\nabla^2_{xx}L_+\nabla^2_{xx}L_+ & \nabla g_+ + \alpha\nabla^2_{xx}L_+\nabla g_+ \\ \hline \nabla g'_+ + \alpha\nabla g'_+\nabla^2_{xx}L_+ & \alpha\nabla g'_+\nabla g_+ \end{array}\right]\begin{bmatrix} \hat{x} - x \\ \hat{\mu}_+ - \mu_+ \end{bmatrix}$$
$$= -\begin{bmatrix} \nabla_x L_+ + c\nabla g_+ g_+ + \alpha\nabla^2_{xx}L_+\nabla_x L_+ \\ g_+ + \alpha\nabla g'_+\nabla_x L_+ \end{bmatrix}.$$

By using (112)–(114) and (116), (117), it is straightforward to verify that

$$H\begin{bmatrix} \hat{x} - x \\ \hat{\mu}_+ - \mu_+ \\ \hat{\mu}_- - \mu_- \end{bmatrix} = -\nabla P^+(x, \mu; c, \alpha),$$

and hence the vector $(\hat{x}, \hat{\mu})$ generated by (102), (103) satisfies [compare with (110)]

$$\begin{bmatrix} \hat{x} - x \\ \hat{\mu} - \mu \end{bmatrix} = -H^{-1}\nabla P^+(x, \mu; c, \alpha).$$

Denote by H^* the matrix H of (112) evaluated at (x^*, μ^*). Taking into account the fact that $\nabla_x L(x^*, \mu^*) = 0$ and $\mu_j^* = 0, j = p + 1, \ldots, r$, it is easy to verify that

$$H^* = \nabla^2 P^+(x^*, \mu^*; c, \alpha).$$

We have shown therefore that, for $\tau = 0$, (110) and (111) hold with $H(x, \mu; c, \alpha)$ being the matrix (112). The proof for the case where $\tau > 0$ is similar but very tedious as the reader may surmise from the analysis of the case where $\tau = 0$. We shall omit the details.

It is worth noting that if the algorithm (104)–(108) is modified at the expense of a slight loss in reliability, so that the test (107) is replaced by

$$-(\hat{x} - x)'\nabla_x P_\tau^+(x, \mu; c, \alpha) - (\hat{\mu} - \mu)'\nabla_\mu P_\tau^+(x, \mu; c, \alpha) > 0,$$

then, *near a K–T pair (x^*, μ^*) satisfying Assumption* (S^+), *it is not necessary to compute the gradient matrix $\nabla g_-(x)$ corresponding to the inactive constraints.* To see this, note that computation of the Newton direction [compare with (102) and (103)] does not require knowledge of $\nabla g_-(x)$. Next, with the aid of (116), observe that if $\mu_- = 0$ [and hence also $(\hat{\mu}_- - \mu_-) = 0$], then computation of the inner products in (107) and (108) also does not require knowledge of $\nabla g_-(x)$. If the algorithm converges to a K–T pair (x^*, μ^*) satisfying Assumption (S^+), then the Newton iteration will be accepted and the set of inactive constraints will remain the same for all iterations after some index. After this index, we shall have $\mu_- = 0$, and there will be no need for computing $\nabla g_-(x)$ with potentially significant computational savings resulting.

The role of the parameter α in preventing convergence to a local maximum can be observed from the definition of the active constraint set. For inequality constrained problems, local maxima typically have negative Lagrange multipliers associated with active constraints. Now the Newton iteration (103) ignores all constraints j for which [compare with (97)]

$$\mu_j + 2\alpha\mu_j^2 + (c + \tau|\mu|^2)g_j(x) \le 0.$$

This means that if α is sufficiently small, then within a neighborhood of a local maximum–Lagrange multiplier pair (x^*, μ^*) for which strict complementarity holds ($\mu_j^* < 0$ if $g_j(x^*) = 0$) *all* constraints are ignored by the Newton iteration (103) which then becomes an iteration of Newton's method for unconstrained minimization of $f(x)$. Thus even though the method may be initially attracted to a local maximum–Lagrange multiplier pair and may approach it during several iterations while it attempts to reach the feasible region, it has the ability to eventually recognize such local maxima and to take large steps away from them.

We mention also that it may be advantageous to exploit the a priori knowledge that Lagrange multipliers corresponding to inequality constraints are nonnegative. Thus, instead of minimizing P_τ^+ subject to no constraints on (x, μ), it is possible to use special methods that can handle efficiently simple constraints in order to minimize P_τ^+ subject to $\mu \ge 0$ (compare with Section 1.5). This eases the problem of selection of an appropriate value for the parameter α, since by enforcing the constraint $\mu \ge 0$ we preclude the possibility that the method will converge to a K–T pair with a negative Lagrange multiplier such as the usual type of local maximum. When f and g_j are convex functions, then for all x and $\mu \ge 0$, the matrix $\nabla_{xx}^2 L$ is positive semidefinite and the appropriate extension of Proposition 4.15 shows that any positive value of α is suitable. Thus *for convex programming problems, the selection of the parameter α presents no difficulties as long as minimization of P_τ^+ is carried out subject to the constraint $\mu \ge 0$.* This makes the method described above particularly attractive for convex programming problems for which second derivatives of the objective and constraint functions are readily available.

4.5.3 Combinations with Nondifferentiable Exact Penalty Methods—Powell's Variable Metric Approach

As shown in Section 4.4.2 [compare with (26) and Proposition 4.25a], the Newton iteration for solving the system of necessary conditions for (ECP) can be viewed as a special case of the linearization method of Section 4.2 with unity stepsize. The same can be said of the inequality constrained version of Newton's method based on quadratic programming and given in Section 4.4.3. We also saw in Section 4.4.3 a somewhat different type of method for inequality constraints which is based on an active set approach and solution of quadratic programming subproblems of the type appearing in the linearization method [compare with (90) in Section 4.4.3]. It is possible to exploit these relations with the linearization method in an effort to enlarge the region of convergence of Newton-like iterations, and this is the subject of the present section.

Methods that Utilize Second Derivatives

The main idea in such methods is to perform the Newton iteration and test whether some criterion of merit is improved. If so, the results of the iteration are accepted. If not, we fall back to the linearization method. We shall discuss two distinct approaches for the problem

$$\text{(NLP)} \quad \text{minimize} \quad f(x)$$
$$\text{subject to} \quad h_i(x) = 0, \quad g_j(x) \le 0, \quad i = 1, \ldots, m, \quad j = 1, \ldots, r,$$

based on the exact penalty function, of Sections 4.1 and 4.2,

$$f(x) + cP(x) = f(x) + c \max\{0, g_1(x), \ldots, g_r(x), |h_1(x)|, \ldots, |h_m(x)|\},$$

where $c > 0$ is the penalty parameter. *Throughout this section we assume that* $f, h_i, g_j \in C^2$.

First Approach: This method is due to Pschenichny (private communication) and is based on the second active set approach described in Section 4.4.3. Fixed scalars $\delta > 0$ and $\gamma \in (0, 1)$ are selected. Given $x \in R^n$, we consider the quadratic program

$$\text{(QP)}_x \qquad \text{minimize} \quad \nabla f(x)'d + \tfrac{1}{2}|d|^2$$
$$\text{subject to} \quad h_i(x) + \nabla h_i(x)'d = 0, \qquad i = 1, \ldots, m,$$
$$g_j(x) + \nabla g_j(x)'d \le 0, \qquad j \in J_\delta(x),$$

where

$$J_\delta(x) = \{j \mid g_j(x) \ge P(x) - \delta, j = 1, \ldots, r\}. \tag{118}$$

For simplicity, we assume that this problem has at least one feasible solution for every $x \in R^n$ (and hence also a unique optimal solution). Otherwise, modifications of the type described in Section 4.2 must be introduced. Given $x_k \in R^n$ after the kth iteration, let d_k be the optimal solution of (QP)_{x_k} and let $\hat{\lambda}(x_k)$, $\hat{\mu}_j(x_k)$, $j \in J_\delta(x_k)$, be corresponding Lagrange multipliers. Let also $\hat{\mu}_j(x_k) = 0$ for $j \notin J_\delta(x_k)$. Define

$$A(x_k) = \{j \mid \hat{\mu}_j(x_k) > 0\},$$

and assume without loss of generality that $A(x_k)$ contains the first p_k indices, where $p_k \leq r$. Define the $n \times (m + p_k)$ matrix N_k by

$$N_k = [\nabla h_1(x_k), \ldots, \nabla h_m(x_k), \nabla g_1(x_k), \ldots, \nabla g_{p_k}(x_k)].$$

Let

$$E_k = N_k(N_k' N_k)^{-1} N_k' \tag{119}$$

and define

(120)

$$\bar{x}_k = x_k - \{E_k + (I - E_k)\nabla_{xx}^2 L[x_k, \hat{\lambda}(x_k), \hat{\mu}(x_k)]\}^{-1}\nabla_x L[x_k, \hat{\lambda}(x_k), \hat{\mu}(x_k)]$$

if the inverses appearing above exist. Solve $(\mathrm{QP})_{\bar{x}_k}$ and let $\bar{d}_k$ be the corresponding optimal solution. The next point x_{k+1} is obtained as follows:

If the inverses in (119) and (120) exist and

$$|\bar{d}_k| \leq \gamma |d_k|,$$

then

$$x_{k+1} = \bar{x}_k. \tag{121}$$

Otherwise

$$x_{k+1} = x_k + \alpha_k d_k,$$

where the stepsize α_k is obtained as in the linearization method of Section 4.2 based on descent of the exact penalty function $f + cP$, where $c > 0$ is the penalty parameter.

It is easily seen that limit points of the generated sequence $\{x_k\}$ must be either K–T pairs of (NLP) or at least critical points of the exact penalty function $f + cP$. Based on the theory of Section 4.3, it is also easily shown that if the starting point x_0 is sufficiently close to a local minimum x^* of (NLP) satisfying the sufficiency condition (S^+) then x_{k+1} is generated by (120) and (121), for all $k = 0, 1, \ldots,$ and $\{|x_k - x^*|\}$ converges to zero superlinearly.

Second Approach: This approach is basically the linearization method of Section 4.2.2 with the matrices H_k being either equal to $\nabla_{xx}^2 L(x_k, \lambda_k, \mu_k)$, if this is judged appropriate by the algorithm, or equal to some positive definite matrix. (For equality constrained problems, H_k can be taken to be a positive definite modification of $\nabla_{xx}^2 L(x_k, \lambda_k, \mu_k)$ along the tangent plane as discussed in the previous section.) Here λ_k, μ_k are approximations to Lagrange multipliers of the problem obtained for example in the previous iteration. Thus, given x_k after iteration k, the basic method consists of solving the quadratic program

(122)
$$\begin{aligned} &\text{minimize} && \nabla f(x_k)'d + \tfrac{1}{2}d'H_k d \\ &\text{subject to} && h_i(x_k) + \nabla h_i(x_k)'d = 0 \qquad \forall\, i = 1, \ldots, m, \\ & && g_j(x_k) + \nabla g_j(x_k)'d \le 0 \qquad \forall\, j \in J_\delta(x_k), \end{aligned}$$

followed by the iteration

(123)
$$x_{k+1} = x_k + \alpha_k d_k,$$

where $\delta > 0$ is a fixed scalar, $J_\delta(x_k)$ is given by (118), d_k is the unique solution of (122), and α_k is obtained by a line search procedure based on descent of the exact penalty function $f + cP$. For simplicity, we assume that problem (122) has at least one feasible solution for each k and that a suitably large value of the penalty parameter c is known. It is possible, of course, to handle situations where these assumptions are not satisfied as described in Section 4.2.2.

The algorithm should be set up in such a way that near a K–T pair (x^*, μ^*), satisfying Assumption (S$^+$), H_k is chosen to be equal to $\nabla^2_{xx} L(x_k, \lambda_k, \mu_k)$ in which case Eq. (123) is closely related to the Lagrangian method of Section 4.4.3, which is based on the quadratic programming approach. The main additional feature is the introduction of the stepsize α_k which enforces descent of the exact penalty function $f + cP$ and thereby enlarges the region of convergence of the method. If the stepsize α_k turns out to equal unity for all k sufficiently large and the method converges to a local minimum x^* satisfying the sufficiency Assumption (S$^+$), then according to the theory of Section 4.4.3, the rate of convergence is superlinear. Unfortunately, it is impossible to guarantee that a unity stepsize will result in a reduction of the exact penalty function $f + cP$ even when the algorithm is arbitrarily close to a solution. We shall demonstrate this fact later in this section, and we shall discuss possible remedies.

Powell's Variable Metric Approach

Consider again (NLP), the exact penalty function

$$f(x) + cP(x) = f(x) + c \max\{0, g_1(x), \ldots, g_r(x), |h_1(x)|, \ldots, |h_m(x)|\}$$

and the linearization method

(124)
$$x_{k+1} = x_k + \alpha_k d_k,$$

where d_k together with corresponding Lagrange multipliers λ_k, μ_k is the solution of the problem

$$(\text{QP})_0(x_k, H_k, J_k) \qquad \text{minimize} \quad \nabla f(x_k)'d + \tfrac{1}{2}d'H_k d$$

$$\text{subject to} \quad h_i(x_k) + \nabla h_i(x_k)'d = 0 \qquad \forall\, i = 1, \ldots, m,$$

$$g_j(x_k) + \nabla g_j(x_k)'d \leq 0 \qquad \forall\, j \in J_k,$$

and α_k is chosen by one of the line search rules of Section 4.2.1 based on descent of the exact penalty function $f + cP$. We assume for simplicity that problem $(\text{QP})_0(x_k, H_k, J_k)$ has at least one feasible solution—otherwise the algorithm should be modified as in Section 4.2. We also make the assumptions

$$0 < H_k, \qquad J_\delta(x_k) \subset J_k,$$

where $J_\delta(x_k) = \{j \mid g_j(x_k) \geq P(x_k) - \delta\}$ and δ is a fixed positive scalar.

As already discussed in this section, if the starting point x_0 is sufficiently close to a solution x^* satisfying together with Lagrange multipliers λ^*, μ^* Assumption (S^+), H_k is for all k sufficiently close and converges to $\nabla^2_{xx} L(x^*, \lambda^*, \mu^*)$, and the stepsize α_k equals unity for all k, then the resulting method is superlinearly convergent. Note however that since $\nabla^2_{xx} L(x^*, \lambda^*, \mu^*)$ need not be positive definite, if we require that $H_k \to \nabla^2_{xx} L(x^*, \lambda^*, \mu^*)$, then we may violate the positive definiteness requirement on H_k. Powell (1978a) observed that in order to attain superlinear convergence in the linearization algorithm, it is sufficient that

$$\lim_{k \to \infty} [\nabla^2_{xx} L(x^*, \lambda^*, \mu^*) - H_k] Z^* = 0,$$

where Z^* is a matrix of basis vectors for the tangent plane at x^* (compare with the discussion in the previous section). He concluded that it is possible to achieve superlinear convergence by choosing H_k to be for all k a positive definite matrix and showed that this can be done by updating H_k via variable metric formulas utilizing only first derivatives of the objective and constraint functions. He suggested the following updating scheme based on the BFGS formula (see Section 1.3.5)

$$(125) \qquad H_{k+1} = H_k - \frac{H_k p_k p_k' H_k}{p_k' H_k p_k} + \frac{r_k r_k'}{p_k' r_k},$$

where

$$(126) \qquad r_k = \theta_k q_k + (1 - \theta_k) H_k p_k,$$

the vectors p_k and q_k are given by

$$(127) \qquad p_k = x_{k+1} - x_k,$$

$$(128) \qquad q_k = \nabla_x L(x_{k+1}, \lambda_k, \mu_k) - \nabla_x L(x_k, \lambda_k, \mu_k),$$

where λ_k, μ_k are Lagrange multipliers of $(\mathrm{QP})_0(x_k, H_k, J_k)$, and the scalar θ_k is given by

$$\theta_k = \begin{cases} 1 & \text{if} \quad p_k' q_k \geq 0.2 p_k' H_k p_k, \\ \dfrac{0.8 p_k' H_k p_k}{p_k' H_k p_k - p_k' q_k} & \text{if} \quad p_k' q_k < 0.2 p_k' H_k p_k. \end{cases} \tag{129}$$

When $\theta_k = 1$, then from (126), $r_k = q_k$ and the updating formula (125) is the same as the BFGS formula for updating the approximation to $\nabla_{xx}^2 L$. The scalar θ_k is introduced in order to ensure that $p_k' r_k > 0$ which, by Proposition 1.20, ensures that positive definiteness of H_k implies positive definiteness of H_{k+1} via (125). Indeed, with r_k chosen by (126), we have

$$p_k' r_k = \theta_k p_k' q_k + (1 - \theta_k) p_k' H_k p_k,$$

so it is easily seen that if θ_k is chosen by (129) and H_k is positive definite then we have $p_k' r_k > 0$.

There are two main advantages of Powell's algorithm. The first is that there is no need for computation of second derivatives—a typical feature of variable metric methods. The second is that the algorithm maintains positive definiteness of the matrix H_k, and this eliminates the possibility of difficult indefinite quadratic programs arising in the computations as is possible with some of the Newton and quasi-Newton methods of Sections 4.4.3 and 4.4.4.

Powell (1978b) shows that if the stepsize α_k is unity for all k sufficiently large and some additional mild conditions hold, then the rate of convergence of the algorithm to a solution x^* satisfying Assumption (S^+) is superlinear. This is a far from obvious result since the sequence of matrices $\{H_k\}$ typically does not converge to the Hessian of the Lagrangian at the corresponding K–T pair. The result owes its validity primarily to the fact that near convergence many of the steps taken by the algorithm tend to be parallel to the tangent plane at the solution. As a result the variable metric formula (125)–(129) tends to provide a good approximation of the Hessian of the Lagrangian function along this subspace, and this is sufficient to induce superlinear convergence.

On the other hand, we have already mentioned in this section that even when arbitrarily close to x^* it may not be possible to select $\alpha_k = 1$ and still achieve a reduction of the exact penalty function $f + cP$. We proceed to discuss this difficulty.

Rate of Convergence Issues

We first provide an example showing that it may not be possible to select a unity stepsize in algorithms (123) and (124) while achieving a reduction of the exact penalty function $f + cP$, even arbitrarily close to a solution and with an "optimal" scaling matrix H_k. This fact seems to have been first observed by Maratos (1978), (see also Chamberlain *et al.*, 1978).

Example: Consider (NLP) for the case of a single equality constraint $h(x) = 0$, $h: R^n \to R$ (i.e., $m = 1$ and $r = 0$). Let d be the solution of the quadratic program

$$\text{minimize} \quad \nabla f(x)'d + \tfrac{1}{2}d'Hd$$
$$\text{subject to} \quad h(x) + \nabla h(x)'d = 0,$$

where we assume that $\nabla h(x) \neq 0$. Let also λ be the corresponding Lagrange multiplier. We have then

$$\nabla f(x) + \nabla h(x)\lambda + Hd = 0, \tag{130}$$

$$h(x) + \nabla h(x)'d = 0. \tag{131}$$

From the mean value theorem, we have

$$f(x + d) = f(x) + \nabla f(x)'d + \tfrac{1}{2}d'\nabla^2 f(\bar{x})d, \tag{132}$$

$$h(x + d) = h(x) + \nabla h(x)'d + \tfrac{1}{2}d'\nabla^2 h(\tilde{x})d, \tag{133}$$

where $\bar{x}$ and $\tilde{x}$ are points on the line segment connecting x and $(x + d)$. By combining (130), (131), and (132), we obtain

$$f(x + d) = f(x) + \lambda h(x) - d'Hd + \tfrac{1}{2}d'\nabla^2 f(\bar{x})d, \tag{134}$$

while by using (131) in (133), we have

$$h(x + d) = \tfrac{1}{2}d'\nabla^2 h(\tilde{x})d. \tag{135}$$

The last two equations yield

$$\begin{aligned} f(x + d) + c|h(x + d)| - [f(x) + c|h(x)|] \\ = \lambda h(x) - c|h(x)| + \tfrac{1}{2}[d'\nabla^2 f(\bar{x})d + c|d'\nabla^2 h(\tilde{x})d| - 2d'Hd]. \end{aligned} \tag{136}$$

Let (x^*, λ^*) be a local minimum–Langrange multiplier pair satisfying Assumption (S). Assume that x is very close to x^*, but $x \neq x^*$, $h(x) = 0$, and furthermore H is chosen to be the "optimal" scaling matrix $\nabla^2_{xx} L(x^*, \lambda^*)$. Then $d \neq 0$, $\bar{x} \cong x^*$, $\tilde{x} \cong x^*$, and the sign of the expression in the right-hand side of (136) depends on the magnitude of c and the curvature of h. In particular, if $\nabla^2 h(x^*)$ is positive definite (or negative definite), there exists a threshold value $\bar{c}$ such that for all $c \geq \bar{c}$ we have

$$f(x + d) + c|h(x + d)| > f(x) + c|h(x)|,$$

and the Newton step leads to an increase of the exact penalty function $f + cP$. This example reveals also the nature of the difficulty which is that *in moving from x to $(x + d)$ we may attain a decrease of the objective function f*

but also an increase of the penalty $|h|$ *of comparable magnitude*, with a net increase of $f + cP$ for sufficiently large values of the penalty parameter c. As shown by (136), *this situation is more likely to occur when* x *is near the constraint boundary* $[h(x) \cong 0]$ in which case the quadratic term in the right-hand side of (136) dominates. It is interesting to note in this connection that some of the difficulties with establishing the descent property of the Newton direction for differentiable exact penalty functions also occur when x is near the constraint boundary, and this necessitated the introduction of condition (28) in Proposition 4.29.

The phenomenon illustrated in the example above has potentially serious consequences, since it may prevent superlinear convergence of algorithms (123) and (124) even under very favorable circumstances. Two different techniques for overcoming this difficulty are proposed in Chamberlain *et al.* (1979), in Mayne and Polak (1978), and in Gabay (1979). In the first approach, a unity stepsize is accepted even if it does not result in a reduction of the exact penalty function provided additional tests based on descent of the Lagrangian function $L(\cdot, \lambda_k, \mu_k)$ are passed. The overall technique is supplemented by safeguards that ensure satisfactory theoretical convergence properties. The complete details can be found in Chamberlain *et al.* (1979) and in the thesis by Chamberlain (1980).

In the approach of Mayne and Polak (1978) and Gabay (1979), the stepsize search is performed not along the line $\{z \mid z = x_k + \alpha d_k, \alpha \geq 0\}$ but rather along an arc of points which attempts to follow the constraint boundary. We describe this technique for the case of the equality constrained problem

$$\text{(ECP)} \qquad \begin{aligned} &\text{minimize} \quad f(x) \\ &\text{subject to} \quad h(x) = 0. \end{aligned}$$

Similarly as earlier, given x_k, we obtain the solution d_k of the quadratic programming problem

$$\text{(137)} \qquad \begin{aligned} &\text{minimize} \quad \nabla f(x_k)'d + \tfrac{1}{2}d'H_k d \\ &\text{subject to} \quad h(x_k) + \nabla h(x_k)'d = 0. \end{aligned}$$

We then obtain the solution p_k of the quadratic programming problem

$$\text{(138)} \qquad \begin{aligned} &\text{minimize} \quad \tfrac{1}{2}|p|^2 \\ &\text{subject to} \quad h(x_k + d_k) + \nabla h(x_k)'p = 0. \end{aligned}$$

The next point x_{k+1} is given by

$$\text{(139)} \qquad x_{k+1} = x_k + \alpha_k d_k + \alpha_k^2 p_k,$$

where the stepsize α_k is obtained by an Armijo-type line search along the arc $\{x_k + \alpha d_k + \alpha^2 p_k \,|\, \alpha \in [0, 1]\}$. More specifically

$$\alpha_k = \beta^{m_k}, \tag{140}$$

where m_k is the first integer m satisfying

$$f(x_k) + cP(x_k) - [f(x_k + \beta^m d_k + \beta^{2m} p_k) + cP(x_k + \beta^m d_k + \beta^{2m} p_k)] \geq -\sigma \xi_c(x_k; \beta^m d_k) \tag{141}$$

and $\xi_c(x; d)$ is given by [compare with (18) in Section 4.2.1]

$$\xi_c(x; d) = \nabla f(x)'d + c \max\{|h_i(x) + \nabla h_i(x)'d| \,|\, i = 1, \ldots, m\} - cP(x). \tag{142}$$

The scalars β and σ satisfy $\beta \in (0, 1)$ and $\sigma \in (0, \frac{1}{2})$.

It is assumed that the symmetric scaling matrix H_k is uniformly positive definite on the tangent plane; i.e., for some positive scalars γ and Γ and all k, we have

$$\gamma|z|^2 \leq z'H_k z \leq \Gamma|z|^2 \qquad \forall\, z \in R^n \text{ with } \nabla h(x_k)'z = 0. \tag{143}$$

To simplify matters it is also assumed that $\nabla h(x)$ has rank m for all x. This together with (143) implies that both quadratic programs (137) and (138) have a unique optimal solution.

The solution p_k of the quadratic program (138) may be viewed as an approximate Newton step from $(x_k + d_k)$ towards satisfying the constraint $h(x) = 0$. The result that follows can also be proved if (138) is replaced by the quadratic program

$$\begin{aligned} &\text{minimize} \quad \tfrac{1}{2}|p|^2 \\ &\text{subject to} \quad h(x_k + d_k) + \nabla h(x_k + d_k)'p = 0. \end{aligned} \tag{144}$$

The solution of this program can be viewed as a more exact Newton step from $(x_k + d_k)$ towards the constraint surface than the solution p_k of (138). The advantage of using (138) in place of (144) is that the computation of $\nabla h(x_k + d_k)$ is saved. Nonetheless, it is quite possible that, in some situations, using (144) rather than (138) can result in more efficient computation, particularly in the initial stages of the algorithm. Note that the solution p_k of problem (138) can be written explicitly as

$$p_k = -\nabla h(x_k)[\nabla h(x_k)'\nabla h(x_k)]^{-1} h(x_k + d_k) \tag{145}$$

and that the inverse $[\nabla h(x_k)'\nabla h(x_k)]^{-1}$ is normally available as a by-product of the solution of the quadratic program (137).

It is not difficult to modify the proof of Proposition 4.13 and show that, under the preceding assumptions and if c is sufficiently large, all limit points of the sequence $\{x_k\}$ generated by algorithm (137)–(142) are critical points of $f + cP$. We shall leave the verification of this fact to the reader. The following proposition addresses the convergence rate properties of the algorithm.

Proposition 4.32: Let $\{x_k\}$ be a sequence generated by algorithm (137)–(142). Assume that $\{x_k\}$ converges to a local minimum x^* of (ECP) which together with a Lagrange multiplier λ^* satisfies the sufficiency Assumption (S), and furthermore $c > \sum_{i=1}^m |\lambda_i^*|$. Assume also that the sequence $\{H_k\}$ is bounded and satisfies (143) and that

$$\lim_{k\to\infty} [\nabla_{xx}^2 L(x^*, \lambda^*) - H_k] Z^* = 0, \tag{146}$$

where Z^* is an $n \times (n - m)$ matrix of basis vectors for the tangent plane

$$\mathscr{C}^* = \{z \,|\, \nabla h(x^*)'z = 0\}.$$

Then for all k sufficiently large, the stepsize α_k equals unity, and $\{x_k\}$ converges to x^* superlinearly.

The proof of Proposition 4.32 is quite long. For this reason we isolate some of the basic steps in the following lemma.

Lemma 4.33: Let the assumptions of Proposition 4.32 hold, and let λ_k be the Langrange multiplier of the quadratic program (137). Then

(a) $P(x_k + d_k) = O(|d_k|^2)$.
(b) $p_k = O(|d_k|^2)$.
(c) $P(x_k + d_k + p_k) = o(|d_k|^2)$.
(d) There exists a scalar $\gamma > 0$ such that, for all k sufficiently large,

$$\xi_c(x_k; d_k) \le -\gamma |d_k|^2.$$

(e) There holds

$$H_k = \nabla_{xx}^2 L(x_k, \lambda_k) + S_k[H_k - \nabla_{xx}^2 L(x_k, \lambda_k)]S_k + O(1/k),$$

where

$$S_k = \nabla h(x_k)[\nabla h(x_k)'\nabla h(x_k)]^{-1}\nabla h(x_k)'.$$

(f) There holds

$$f(x_k + d_k + p_k) + cP(x_k + d_k + p_k) - f(x_k) - cP(x_k)$$
$$= \xi_c(x_k; d_k) + \tfrac{1}{2} d_k' \nabla_{xx}^2 L(x_k, \lambda_k) d_k + o(|d_k|^2).$$

(g) There holds

$$\xi_c(x_k; d_k) + d_k' \nabla_{xx}^2 L(x_k, \lambda_k) d_k \le o(|d_k|^2).$$

Proof: (a) From Taylor's theorem we have

$$h(x_k + d_k) = h(x_k) + \nabla h(x_k)'d_k + O(|d_k|^2),$$

and since d_k is the solution of (137), we also have $h(x_k) + \nabla h(x_k)'d_k = 0$. The result follows.

(b) Part (b) follows from (145) and part (a).

(c) From Taylor's theorem, we have

$$h(x_k + d_k + p_k) = h(x_k + d_k) + \nabla h(x_k + d_k)'p_k + O(|p_k|^2).$$

Since p_k solves problem (138), we have $h(x_k + d_k) = -\nabla h(x_k)'p_k$ and substitution in the preceding equation yields

$$h(x_k + d_k + p_k) = [\nabla h(x_k + d_k) - \nabla h(x_k)]'p_k + O(|p_k|^2).$$

Since $\nabla h(x_k + d_k) - \nabla h(x_k) = O(|d_k|)$ and from part (b), $p_k = O(|d_k|^2)$, we obtain

$$h(x_k + d_k + p_k) = o(|d_k|^2)$$

from which the result follows.

(d) From (142) and the fact that $h_i(x_k) + \nabla h_i(x_k)'d_k = 0$, we obtain

$$\xi_c(x_k; d_k) = \nabla f(x_k)'d_k - cP(x_k). \tag{147}$$

Let $\bar{c}$ be a scalar such that the matrix $\bar{H}_k$ given by

$$\bar{H}_k = H_k + \bar{c}\nabla h(x_k)\nabla h(x_k)' \tag{148}$$

is positive definite for all k with eigenvalues uniformly bounded below by a positive scalar. [Such a scalar exists in view of the assumption (143) and the boundedness assumption on $\{H_k\}$.] From the necessary conditions for optimality of d_k in problem (137), we have

$$\nabla f(x_k) + \nabla h(x_k)\lambda_k + H_k d_k = 0. \tag{149}$$

Using this equation, (148), and the fact that $\nabla h(x_k)'d_k = -h(x_k)$, we obtain

$$\nabla f(x_k)'d_k - [\lambda_k + \bar{c}h(x_k)]'h(x_k) + d_k'\bar{H}_k d_k = 0.$$

Combining this equation with (147), we obtain

$$\xi_c(x_k; d_k) = [\lambda_k + \bar{c}h(x_k)]'h(x_k) - cP(x_k) - d_k'\bar{H}_k d_k.$$

Using the assumption $c > \sum_{i=1}^m |\lambda_i^*|$ and the fact that $\lambda_k + \bar{c}h(x_k) \to \lambda^*$, we obtain, for sufficiently large k,

$$\xi_c(x_k; d_k) \leq -d_k'\bar{H}_k d_k, \tag{150}$$

and the result follows.

(e) Denote

$$A_k = H_k - \nabla^2_{xx} L(x_k, \lambda_k) - S_k[H_k - \nabla^2_{xx} L(x_k, \lambda_k)]S_k.$$

We have

$$A_k Z^* = [H_k - \nabla^2_{xx} L(x_k, \lambda_k)]Z^* - S_k[H_k - \nabla^2_{xx} L(x_k, \lambda_k)]S_k Z^*.$$

By assumption, $[H_k - \nabla^2_{xx} L(x_k, \lambda_k)]Z^* = O(1/k)$ and $\nabla h(x_k)'Z^* = O(1/k)$. Hence, $S_k Z^* = O(1/k)$, and we obtain

$$A_k Z^* = O(1/k). \tag{151}$$

Also

$$\begin{aligned}\nabla h(x^*)' A_k \nabla h(x^*) = {} & \nabla h(x^*)'[H_k - \nabla^2_{xx} L(x_k, \lambda_k)]\nabla h(x^*) \\ & - \nabla h(x^*)' S_k [H_k - \nabla^2_{xx} L(x_k, \lambda_k)] S_k \nabla h(x^*).\end{aligned}$$

Since $\nabla h(x^*)' S_k = \nabla h(x^*)' + O(1/k)$, we obtain

$$\nabla h(x^*)' A_k \nabla h(x^*) = O(1/k). \tag{152}$$

Every vector $w \in R^n$ can be uniquely decomposed as $w = Z^* y + \nabla h(x^*) z$, so using (151) and (152), we have

$$w' A_k w = [Z^* y + \nabla h(x^*) z]' A_k [Z^* y + \nabla h(x^*) z] = O(1/k).$$

It follows that $A_k = O(1/k)$, and the result is proved.

(f) We have, from Taylor's theorem and part (b),

$$\begin{aligned} f(x_k + d_k + p_k) &= f(x_k) + \nabla f(x_k)'(d_k + p_k) \\ &\quad + \tfrac{1}{2}(d_k + p_k)' \nabla^2 f(x_k)(d_k + p_k) + o(|d_k + p_k|^2) \\ &= f(x_k) + \nabla f(x_k)' d_k + \nabla f(x_k)' p_k \\ &\quad + \tfrac{1}{2} d_k' \nabla^2 f(x_k) d_k + o(|d_k|^2). \end{aligned} \tag{153}$$

Also from (149) and the fact that $h(x_k + d_k) = -\nabla h(x_k)' p_k$, we obtain

$$\begin{aligned} \nabla f(x_k)' p_k &= -p_k' \nabla h(x_k) \lambda_k - p_k' H_k d_k \\ &= h(x_k + d_k)' \lambda_k - p_k' H_k d_k. \end{aligned} \tag{154}$$

Using part (b) and the fact that

$$\begin{aligned} h_i(x_k + d_k) &= h_i(x_k) + \nabla h_i(x_k)' d_k + \tfrac{1}{2} d_k' \nabla^2 h_i(x_k) d_k + o(|d_k|^2) \\ &= \tfrac{1}{2} d_k' \nabla^2 h_i(x_k) d_k + o(|d_k|^2) \end{aligned}$$

in (154), we have

$$\nabla f(x_k)' p_k = \sum_{i=1}^{m} \frac{\lambda_k^i}{2} d_k' \nabla^2 h_i(x_k) d_k + o(|d_k|^2). \tag{155}$$

Combining Eqs. (147), (153), and (155), we have

$$f(x_k + d_k + p_k) = f(x_k) + cP(x_k) + \xi_c(x_k; d_k) + \tfrac{1}{2}d_k' \nabla_{xx}^2 L(x_k, \lambda_k)d_k + o(|d_k|^2).$$

Using part (c), we obtain from this equation the desired result.

(g) From (149), we have

$$\nabla f(x_k)'d_k + d_k' \nabla h(x_k)\lambda_k + d_k' H_k d_k = 0.$$

Using (147) and the fact that $\nabla h(x_k)'d_k = -h(x_k)$ in this equation, we obtain

$$\xi_c(x_k; d_k) + cP(x_k) - h(x_k)'\lambda_k + d_k' H_k d_k = 0.$$

Substituting the expression for H_k in part (e) and using the fact that $\nabla h(x_k)'d_k = -h(x_k)$, we obtain

$$\xi_c(x_k; d_k) + d_k' \nabla_{xx}^2 L(x_k, \lambda_k)d_k = h(x_k)'[\lambda_k + M_k h(x_k)] - cP(x_k) + o(|d_k|^2), \tag{156}$$

where the matrix M_k is given by

$$M_k = [\nabla h(x_k)' \nabla h(x_k)]^{-1} \nabla h(x_k)'[\nabla_{xx}^2 L(x_k, \lambda_k) - H_k]\nabla h(x_k) \times [\nabla h(x_k)' \nabla h(x_k)]^{-1}.$$

Since $\lambda_k + M_k h(x_k) \to \lambda^*$ and $c > \sum_{i=1}^m |\lambda_i^*|$, we have, for sufficiently large k,

$$h(x_k)'[\lambda_k + M_k h(x_k)] - cP(x_k) \le 0. \tag{157}$$

By combining (156) and (157), the result follows. Q.E.D.

We are now ready to complete the proof of Proposition 4.32.

Proof of Proposition 4.32: From parts (d), (f), and (g) of Lemma 4.33, we obtain, for sufficiently large k,

$$\begin{aligned} f(x_k + d_k + p_k) &+ cP(x_k + d_k + p_k) - f(x_k) - cP(x_k) \\ &\le \tfrac{1}{2}\xi_c(x_k; d_k) + o(|d_k|^2) \\ &\le \sigma\xi_c(x_k; d_k) - (\tfrac{1}{2} - \sigma)\gamma|d_k|^2 + o(|d_k|^2) \\ &\le \sigma\xi_c(x_k; d_k). \end{aligned}$$

It follows from the definition (140)–(141) of the stepsize rule that we have $\alpha_k = 1$ for all k sufficiently large.

In order to show the superlinear convergence property of the algorithm, we note that the assumption (146) implies (see the analysis following Proposition 4.31) that

$$|x_k + d_k - x^*| = O(1/k)|x_k - x^*|.$$

It is also easily shown that $d_k = O(|x_k - x^*|)$, and therefore, using part (b) of Lemma 4.33,

$$p_k = O(|x_k - x^*|^2).$$

The two equations above yield

$$\begin{aligned}|x_k + d_k + p_k - x^*| &\le |x_k + d_k - x^*| + |p_k| \\ &= O(1/k)|x_k - x^*| + O(|x_k - x^*|^2) \\ &= O(1/k)|x_k - x^*|.\end{aligned}$$

For k sufficiently large, the stepsize α_k is unity and therefore

$$x_{k+1} = x_k + d_k + p_k.$$

Combining the last two relations, we obtain

$$|x_{k+1} - x^*| \le O(1/k)|x_k - x^*|. \qquad \text{Q.E.D.}$$

We note that if the matrices H_k satisfy the stronger condition

$$\gamma|z|^2 \le z'H_k z \le \Gamma|z|^2 \qquad \forall\, z \in R^n$$

for some positive scalars γ, Γ, in place of (143) then the result of Proposition 4.32 can be proved also for the case where the right side of (141) is replaced by $\beta^m d_k' H_k d_k$ [compare with (150)]. This form of the Armijo rule is consistent with the one of Section 4.2.

While the idea of introducing an additional step towards the constraint surface was motivated by the desire to improve the rate of convergence properties of the algorithm when near a solution, there are indications that this step frequently improves these properties even when far from the solution. The reason is that in order to keep decreasing the value of the exact penalty function $f + cP$ the algorithm must follow closely the constraint surface, particularly for large values of c. The extra step towards the constraint surface helps to achieve this without an excessive number of stepsize reductions and attendant function evaluations at each iteration.

We finally note that it is possible to extend the algorithm just given to inequality constraints by using an active set strategy whereby the active inequality constraints are the ones for which the Lagrange multipliers, obtained from the quadratic program analogous to (137) [compare with (122)], are positive. An alternative approach together with convergence analysis is given in Mayne and Polak (1978).

4.6 Notes and Sources

Notes on Section 4.1: Nondifferentiable exact penalty functions have been analyzed by several authors including Zangwill (1967b), Ermoliev and Shor (1967), Pietrzykowski (1969), Luenberger (1970), Evans *et al.* (1973),

Howe (1973), Bertsekas (1975b), and Dolecki and Rolewicz (1979). Detailed references and a thorough discussion for nonconvex problems is given in Han and Mangasarian (1979). Proposition 4.7 is taken from Mayne and Polak (1978), and the proof of Proposition 4.9 uses an adaptation of an argument in Pschenichny and Danilin (1975, p. 196).

Notes on Section 4.2: The linearization algorithm for minimax and nonlinear programming problems including a global convergence result based on the Armijo rule and descent of the exact penalty function $f + cP$ was first given in Pschenichny (1970). This convergence result is given here as Proposition 4.13. Our proof of this result is new and does not require a Lipschitz assumption on the gradients of the objective and constraints that was necessary in the original proof of Pschenichny. The linearization algorithm was rediscovered in weaker form by Han (1977b), and several related algorithms were given by Mayne and Maratos (1979). Convergence results relating to the linearization algorithm have also been given by Mayne and Polak (1978) and Bazaraa and Goode (1979). The convergence rate of the algorithm is analyzed in detail in Pschenichny and Danilin (1975).

Notes on Section 4.3: The exact penalty functions $P(x, \lambda; c, \alpha)$ and $P(x, \lambda; c, M)$ were introduced by DiPillo and Grippo (1979a). The proofs of all the results of Section 4.3.1 are taken from DiPillo *et al.* (1979) with the exception of Proposition 4.15 which is new. Related penalty functions have been proposed by Boggs and Tolle (1980) and Han and Mangasarian (1981).

The exact penalty function $\hat{P}(x; c)$ of Section 4.3.2 was first introduced by Fletcher (1970) and further discussed in connection with specific algorithms in Fletcher and Lill (1971), Fletcher (1973), Mukai and Polak (1975), Glad and Polak (1979), and McCormick (1978). The line of analysis given here is new and is based on the connection with the penalty functions of DiPillo and Grippo first reported in Bertsekas (1980a).

The algorithms based on second derivatives of Section 4.3.3 are due to DiPillo *et al.* (1979) and Fletcher (1973) with the exception of those algorithms that are based on Newton's method for solving the system $\nabla L(x, \lambda) = 0$, which were first considered in Bertsekas (1980a). The analysis of the penalty parameter choice for the penalty function $\hat{P}(x; c)$ is due to Fletcher (1970), while the corresponding analysis for the penalty functions $P(x, \lambda; c, \alpha)$ and $P(x, \lambda; c, M)$ is new. The main idea of the automatic penalty parameter adjustment schemes is due to Polak (1976) and has been applied by several authors (Mukai and Polak, 1975; Glad and Polak, 1979; Mayne and Maratos, 1979; Mayne and Polak, 1978). The scheme given here is new. An alternative scheme has been given by DiPillo *et al.* (1979).

Notes on Section 4.4 The first extensive work on Lagrangian methods is Arrow *et al.* (1958). Proposition 4.23 is due to Poljak (1970). Analysis relating to first-order methods for inequality constrained convex problems may be found in Zangwill (1969), Maistrovskii (1976), Campos-Filho (1971), Golshtein (1972), and Korpelevich (1976). A class of Lagrangian functions for inequality constrained problems which leads to unconstrained saddle point problems was introduced by Mangasarian (1974, 1975). Some early interesting work on Lagrange multiplier iterations using augmented Lagrangian functions can be found in Miele *et al.* (1971a, b, 1972).

Newton-like and quasi-Newton methods of the Lagrangian type were systematically analyzed only recently. Important works in this area are Garcia-Palomares and Mangasarian (1976), Tapia (1977), Glad (1979), Han (1977a), Biggs (1978), Gabay (1979), and Powell (1978a, b). Early works using quasi-Newton updating formulas for equality constrained problems are Kwakernaak and Strijbos (1972) and Biggs (1972). The proof of local convergence of iteration (45), (46) is due to Glad (1979). Newton's method in the space of primal variables [cf. 48)] was first discussed in Tapia (1977) and Pschenischny and Danilin (1975). Proposition 4.26 is due to Pschenichny and Danilin (1975). The quadratic programming version of Newton's method for inequality constrained problems was first suggested by Wilson (1963). Its convergence rate has been established by Robinson (1974). Corresponding quasi-Newton methods were first proposed by Garcia-Palomares and Mangasarian (1976). Conditions for superlinear convergence of quasi-Newton methods for constrained minimization are given in Boggs, Tolle, and Wang (1982).

Notes on Section 4.5: Some good quasi-Newton algorithms combining Lagrangian methods and multiplier methods are given in Glad (1979).

The results and algorithms of Section 4.5.2 are due to Bertsekas (1980a, b) with the exception of the quasi-Newton algorithm (58)–(62), which was first proposed by Dixon (1980).

Basic references for the material of Section 4.5.3 are Mayne and Polak (1978), Powell (1978a, b), Gabay (1979), Chamberlain (1980), and Chamberlain *et al.* (1979). A related approach is proposed in Coleman and Conn (1980a, b). Proposition 4.32 was proved by Mayne and Polak (1978) and Gabay (1979). Mayne and Polak (1978) have also treated inequality constrained problems.

It is difficult to compare accurately the performance of exact penalty methods, such as those described in Section 4.5, with methods of multipliers discussed in Chapters 2 and 3. The computational evidence available suggests that if relatively good choices of the penalty parameter and the starting point are known, then exact penalty methods are as reliable as multiplier methods and typically require fewer iterations for problems where the sufficiency assumptions (S) or (S^+) are satisfied. On the other hand, the overhead per iteration of exact penalty methods can be significantly higher than the one of multiplier methods—particularly when the dimension of the problem is large. We can however conclude that if good initial information regarding the penalty parameter and the starting point is available, and the dimension of the problem is small, then exact penalty methods hold an advantage over multiplier methods. In the absence of good initial information, multiplier methods tend to be more reliable, easier to "tune," and in the author's experience, often require fewer iterations to converge—particularly when combined with Lagrangian methods as described in Section 4.5.1. We can thus conclude that multiplier methods hold the advantage for problems of large dimension, and for problems where the initial information is of poor quality. The preceding statements should only be viewed as general guidelines, and it should be kept in mind that the relative significance of overhead per iteration depends very much on the computation required for evaluating the function values and gradients needed at each iteration. Furthermore an important factor in comparing the merits of each class of methods is the nature of the application at hand. If repetitive solution of the same problem with minor variations is envisioned, then it may be reasonably assumed that good initial information will eventually become available and this favors the use of an exact penalty method. If only a limited amount of computation needs to be performed after development of the optimization code, one is typically better off using a method of multipliers.

It is also difficult to compare globally convergent Newton and quasi-Newton methods based on differentiable and nondifferentiable penalty functions (Sections 4.5.2 and 4.5.3, respectively). Both types of methods essentially behave identically near a solution when the superlinear convergence property takes effect. Far from a solution, their behavior can be quite different in the sense that in any given iteration the stepsize may have to be reduced by different amounts from its initial value of unity in order to achieve descent for each penalty function. It is significant in this respect that the threshold values for the penalty parameter on any given problem can be greatly different for differentiable and nondifferentiable penalty functions (compare the estimates given in Sections 4.1 and 4.3.3). Methods based on differentiable exact penalty functions require more overhead in view of the fact that they involve first derivatives of the constraints in the penalty function. However this overhead need not include evaluation of second derivatives of objective and constraint functions and is not as much as may appear at first sight (see the discussion of Section 4.5.2). Another aspect of differentiable exact penalty which may constitute a tangible practical disadvantage is the fact that their extensions currently available for handling inequality constraints are not as "clean" as those available for nondifferentiable penalty functions. On the other hand, methods based on differentiable penalty functions have the theoretical advantage that they do not require any modifications such as those of Mayne and Polak (1978), Gabay (1979), and Chamberlain *et al.* (1979) in order to obtain superlinear convergence.

We finally note that in quadratic programming based quasi-Newton schemes, such as Powell's, where an exact penalty function is used as a descent function, one should try to avoid gradient evaluations of the penalty function. For nondifferentiable penalty functions these gradients are of little value even at points where they exist. The computation of a gradient of a differentiable penalty function is undesirable since it involves either exact second derivatives or their finite difference approximations. Therefore one should try to use one-dimensional line search procedures that require function values only. The simplest possibility is to use the Armijo rule with $\sigma = 0$, i.e., a rule that reduces the stepsize by a certain factor until a reduction in the exact penalty function value is observed. While this simplification of the Armijo rule involves a theoretical risk of nonconvergence, this risk, for practical purposes, appears to be negligible.

Chapter 5

Nonquadratic Penalty Functions — Convex Programming

5.1 Classes of Penalty Functions and Corresponding Methods of Multipliers

The quadratic penalty function is the most widely used in practical implementations of methods of multipliers. However, there is occasionally a tangible advantage in using a different penalty function. We describe some situations where this is the case:

(a) It may occur that while the objective function is bounded below along the constraint set, the augmented Lagrangian is unbounded (over the entire space) for every value of the penalty parameter. For example, the augmented Lagrangian for the trivial scalar problem

$$\text{minimize} \quad -x^4$$
$$\text{subject to} \quad x = 0$$

is given by

$$L_c(x, \lambda) = -x^4 + \lambda x + \tfrac{1}{2}c|x|^2.$$

Clearly $L_c(\cdot, \lambda)$ is unbounded below for every c and as a result the unconstrained minimization algorithm used for minimizing $L_c(\cdot, \lambda)$ diverges unless the starting point is close to the unique local minimum of $L_c(\cdot, \lambda)$. This situation can often be corrected by using a penalty function with sufficiently high order of growth. For the preceding example a penalty function of the form

$$\tfrac{1}{2}c(|x|^2 + |x|^5)$$

in place of $\frac{1}{2}c|x|^2$ will resolve the difficulty.

(b) The augmented Lagrangian functions for inequality constraints and some of the approximating functions developed in Chapter 3 do not have continuous second derivatives. On the other hand, the methods most likely to be used for unconstrained minimization of the augmented Lagrangian rely conceptually on continuity of second derivatives. Despite this fact, it appears that for many practical problems the second derivative discontinuities do not have a significant adverse effect on the performance of methods, such as the conjugate gradient method, quasi-Newton methods, and Newton's method. Nonetheless under extreme circumstances, these discontinuities can slow down considerably the rate of convergence of these methods and can be the cause of algorithmic failure. In this case, it is preferable to use a twice continuously differentiable augmented Lagrangian of the type to be introduced shortly.

(c) Multiplier methods corresponding to different types of penalty functions can exhibit drastically different rates of convergence. The speed of convergence may be much faster or much slower depending on the penalty function employed as illustrated in the examples of Section 2.2.4. This perhaps surprising feature, which is not present in ordinary penalty methods, raises the interesting possibility of delineating a penalty function which matches best the problem at hand in terms of computational efficiency.

In what follows in this section, we introduce various classes of penalty functions that are suitable for use in multiplier methods, and develop some of their properties that will be useful for the analysis of subsequent sections.

5.1.1 Penalty Functions for Equality Constraints

Consider the equality constrained problem

$$\text{(ECP)} \qquad \text{minimize } f(x) \qquad \text{subject to } h(x) = 0.$$

We consider the following class of penalty functions:

Class of Penalty Functions P_E: All functions $\phi: R \to R$ having the following properties.

(a) ϕ is continuously differentiable and strictly convex on R.
(b) $\phi(0) = 0$ and $\nabla\phi(0) = 0$.
(c) $\lim_{t\to-\infty} \nabla\phi(t) = -\infty$ and $\lim_{t\to\infty} \nabla\phi(t) = \infty$.

Examples of functions in the class P_E are:

(i) $\phi(t) = \frac{1}{2}t^2$ (quadratic).
(ii) $\phi(t) = \rho^{-1}|t|^\rho$, $\rho > 1$ (ρ-order of growth).
(iii) $\phi(t) = \rho^{-1}|t|^\rho + \frac{1}{2}t^2$, $\rho > 1$.
(iv) $\phi(t) = \cosh(t) - 1$.

We associate with a given penalty function ϕ in the class P_E the augmented Lagrangian function

(1) $$L_c(x, \lambda) = f(x) + \lambda' h(x) + \frac{1}{c} \sum_{i=1}^{m} \phi[ch_i(x)].$$

The first-order method of multipliers corresponding to ϕ consists of sequential unconstrained minimization of the form

(2) $$\text{minimize} \quad L_{c_k}(x, \lambda_k) \qquad \text{subject to} \quad x \in R^n,$$

yielding a vector x_k. Minimization is followed by the multiplier iteration

(3) $$\lambda_{k+1}^i = \lambda_k^i + \nabla\phi[c_k h_i(x_k)], \qquad i = 1, \ldots, m.$$

Note that, for $\phi(t) = \frac{1}{2}t^2$, iteration (3) reduces to $\lambda_{k+1} = \lambda_k + c_k h(x_k)$, and we obtain the quadratic method of multipliers studied in Chapter 2. Similarly as for that method, it is possible to consider inexact minimization of the augmented Lagrangian (1). It is also possible to develop second-order iterations under a second-order differentiability assumption on ϕ. Other variations include the use of a different penalty function and/or penalty parameter for each constraint.

There is a subclass of P_E which admits an analysis which is almost identical to the one for the quadratic penalty function. This is the class of penalty functions ϕ which are twice continuously differentiable with $\nabla^2\phi(0) = 1$. We call such penalty functions *essentially quadratic* since near a solution they behave in essentially the same way as the quadratic penalty function. The entire analysis of Chapter 2 can be shown to hold with minor modifications for essentially quadratic penalty functions as the reader can easily verify. In particular, under Assumption (S), one can prove convergence results for the corresponding multiplier method similar to those for the quadratic method.

The rate of convergence is at least linear if $\{c_k\}$ is bounded above and superlinear if $c_k \to \infty$.

The rate of convergence of the multiplier method corresponding to the penalty function

$$\phi(t) = \rho^{-1}|t|^\rho + \tfrac{1}{2}t^2 \tag{4}$$

is *superlinear* for $\rho \in (1, 2)$ under Assumption (S). We shall demonstrate this fact in Section 5.4 in the context of a convex programming problem. For $\rho \geq 2$, $\phi(t)$ is essentially quadratic and no better than a linear rate of convergence can be expected in general. It may appear that a choice $\rho \in (1, 2)$ would be always preferable but it should be noted that in this case $\phi(t)$ is not twice differentiable at $t = 0$ and in fact $\nabla^2\phi(t)$ tends to ∞ as t tends to zero. This has the effect of making the unconstrained minimization of the augmented Lagrangian ill-conditioned. Thus, the advantage of superlinear convergence of the multiplier iteration may be offset by ill-conditioning difficulties in unconstrained minimization. Nonetheless, we know of problems where the use of the function (4) with $\rho \in (1, 2)$ has yielded better results than the quadratic penalty function. Also, in problems which are solved repetitively with minor variations, it may be possible through the use of good starting points, special powerful unconstrained minimization methods, and "fine-tuning" to reduce significantly the effects of ill-conditioning. Under these circumstances, the method of multipliers that employs the penalty function (4) with $\rho \in (1, 2)$ can substantially outperform the quadratic method.

5.1.2 *Penalty Functions for Inequality Constraints*

Consider the inequality constrained problem

(ICP)
$$\text{minimize} \quad f(x)$$
$$\text{subject to} \quad g(x) \leq 0.$$

We first consider the following class of penalty functions.

Class of Penalty Functions P_{I}: All functions $p: R^2 \to R$ having the following properties.

(a) p is continuous on $R \times [0, +\infty)$, continuously differentiable on $R \times (0, +\infty)$, and possesses for all $t \in R$ the right partial derivative

$$\lim_{\mu \to 0^+} \frac{p(t; \mu) - p(t; 0)}{\mu}.$$

Furthermore, $p(\cdot\,; 0)$ is continuously differentiable with respect to t on R. [The partial derivative with respect to the first argument is denoted by $\nabla_t p(\cdot\,;\cdot)$ and the one with respect to the second argument by $\nabla_\mu p(\cdot\,;\cdot)$.]

(b) $p(t;\cdot)$ is concave on $[0, +\infty)$ for each fixed $t \in R$.

(c) For each $\mu \geq 0$, $p(\cdot\,;\mu)$ is convex on R and satisfies the following strict convexity condition:

If

$$\text{(i) } t_0 > 0 \qquad \text{or} \qquad \text{(ii) } \nabla_t p(t_0;\mu) > 0,$$

then

$$p(t;\mu) - p(t_0;\mu) > (t - t_0)\nabla_t p(t_0;\mu) \qquad \forall\, t \neq t_0.$$

(d) $p(0;\mu) = 0 \quad \forall\, \mu \geq 0$,
(e) $\nabla_t p(0;\mu) = \mu \quad \forall\, \mu \geq 0$,
(f) $\lim_{t\to-\infty} \nabla_t p(t;\mu) = 0 \quad \forall\, \mu \geq 0$,
(g) $\lim_{t\to+\infty} \nabla_t p(t;\mu) = +\infty \quad \forall\, \mu \geq 0$,
(h) $\inf_{t\in R} p(t;\mu) > -\infty \quad \forall\, \mu \geq 0$.

In Fig. 5.1, we show the shape of a typical function in P_{I}. The predominant effect of the multiplier μ is to alter the slope as $p(\cdot\,;\mu)$ passes through the origin [properties (d) and (e)]. For t near zero, $p(t;\mu) \approx \mu t$, but elsewhere the penalty effect dominates. The main consideration is that $p(\cdot\,;\mu)$ passes through the origin with slope μ. As $t \to \infty$, $p(t;\mu)$ grows to infinity with unbounded slope. As $t \to -\infty$, $p(t;\mu)$ approaches or reaches a finite infimum which is less than or equal to zero.

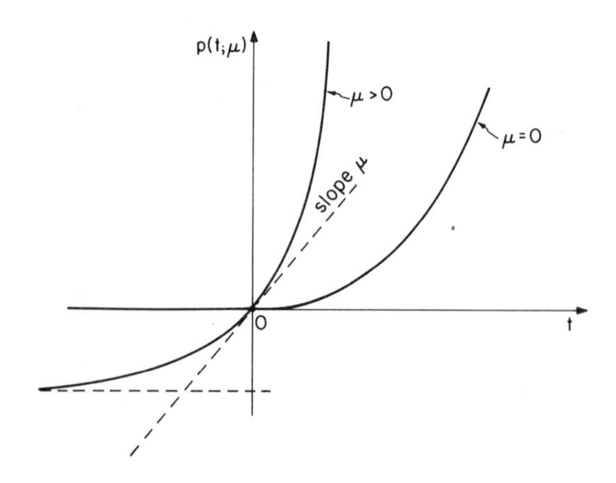

FIG. 5.1 Form of penalty function p in the class P_I

The augmented Lagrangian corresponding to a function $p \in P_{\text{I}}$ is given by

$$L_c(x, \mu) = f(x) + \frac{1}{c} \sum_{j=1}^{r} p[cg_j(x); \mu^j]. \tag{5}$$

The first-order multiplier method corresponding to p consists of sequential unconstrained minimization of the form

$$\begin{aligned} &\text{minimize} \quad L_{c_k}(x, \mu_k) \\ &\text{subject to} \quad x \in X, \end{aligned} \tag{6}$$

yielding a vector x_k. Minimization is followed by the multiplier iteration

$$\mu_{k+1}^j = \nabla_t p[c_k g_j(x_k); \mu_k^j], \qquad j = 1, \ldots, r. \tag{7}$$

Note that iteration (7) is such that the equality

$$\nabla_x L_{c_k}(x_k, \mu_k) = \nabla_x L(x_k, \mu_{k+1})$$

is satisfied for all k, where L is the ordinary Lagrangian function given by $L(x, \mu) = f(x) + \mu' g(x)$. The initial multiplier satisfies $\mu_0 \geq 0$. Note that from Eq. (7) and properties (c), (f), and (g), it follows that $\mu_k \geq 0$ for all k.

Figure 5.2 shows the term $c^{-1}p(ct; \mu)$ in the augmented Lagrangian (5) and the effect of the parameter c in particular. The penalty effect increases with

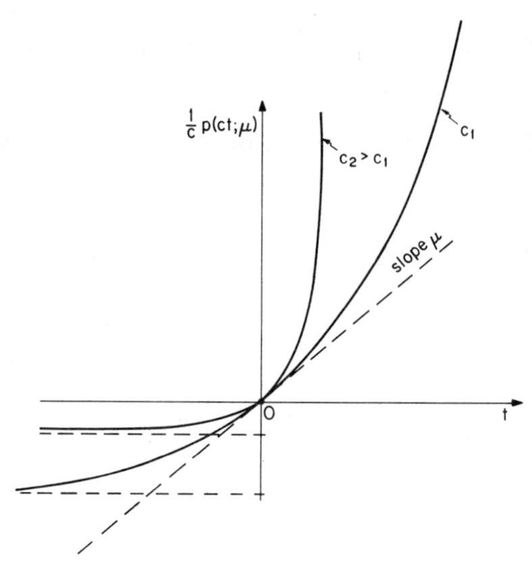

FIG. 5.2 Penalty functions $c_1^{-1}p(c_1 t; \mu)$ and $c_1^{-1}p(c_2 t; \mu)$ for $c_1 < c_2$

increasing c. Indeed for $\mu \geq 0$, the convexity of $p(\cdot\,;\mu)$ and the fact that $p(0;\mu) = 0$ can be used to show that

$$0 < c_1 \leq c_2 \Rightarrow c_1^{-1}p(c_1 t;\mu) \leq c_2^{-1}p(c_2 t;\mu) \qquad \forall\, t \in R, \quad \mu \geq 0.$$

Examples of penalty functions in the class P_{I} are as follows:

Example 1 (Class P_{E}^{+}): This subclass of P_{I} is defined as the class of functions $p\colon R^2 \to R$ of the form

$$p(t;\mu) = \begin{cases} \mu t + \phi(t) & \text{if} \quad \mu + \nabla\phi(t) \geq 0, \\ \min_{\tau \in R} \{\mu\tau + \phi(\tau)\} & \text{otherwise}, \end{cases} \tag{8}$$

where $\phi\colon R \to R$ belongs to the class of penalty functions P_{E} for equality constraints defined in the previous subsection. As an example, if $\phi(t) = \frac{1}{2}t^2$, we obtain the piecewise quadratic function

$$p(t;\mu) = \begin{cases} \mu t + \frac{1}{2}t^2 & \text{if} \quad t \geq -\mu, \\ -\frac{1}{2}\mu^2 & \text{if} \quad t < -\mu. \end{cases} \tag{9}$$

The corresponding augmented Lagrangian (5) can be written as

$$\begin{aligned} L_c(x,\mu) &= f(x) + \frac{1}{c}\sum_{j=1}^{r} p_j[cg_j(x);\mu^j] \\ &= f(x) + \frac{1}{2c}\sum_{j=1}^{r} \{[\max\{0, \mu^j + cg_j(x)\}]^2 - (\mu^j)^2\} \end{aligned}$$

and is identical to the one used for one-sided inequality constraints in Section 3.1. The multiplier iteration (7), corresponding to (9), takes the form

$$\mu_{k+1}^j = \max\{0, \mu_k^j + c_k g_j(x_k)\}, \qquad j = 1, \ldots, r,$$

and again is identical to the one of Section 3.1. More generally, we have, using (8),

$$\nabla_t p(t;\mu) = \begin{cases} \mu + \nabla\phi(t) & \text{if} \quad \mu + \nabla\phi(t) \geq 0, \\ 0 & \text{otherwise}, \end{cases}$$

or equivalently

$$\nabla_t p(t;\mu) = \max\{0, \mu + \nabla\phi(t)\},$$

so the multiplier iteration (7) corresponding to the penalty function $p \in P_{\mathrm{E}}^{+}$ of (8) is given by

$$\mu_{k+1}^j = \max\{0, \mu_k^j + \nabla\phi[c_k g_j(x_k)]\}, \qquad j = 1, \ldots, r. \tag{10}$$

Furthermore the reader can verify that *each function p of the form* (8) *is obtained from the corresponding function $\phi \in P_{\mathrm{E}}$ in the same manner as* (9) *was* *obtained from the quadratic penalty function* $\phi(t) = \frac{1}{2}t^2$ *in Section* 3.1, i.e., *by converting the inequality constraints to equality constraints by using additional variables and by subsequently eliminating these variables from the actual computation.* Thus the class P_E^+ corresponds to multiplier methods for (ICP) after it has been converted to an equality constrained problem. The next example yields penalty functions designed exclusively for inequality constraints. An advantage of such functions is that they lead to twice continuously differentiable augmented Lagrangians. This property cannot be attained with functions in the class P_E^+.

Example 2 (Twice Differentiable): Consider any function $\psi: R \to R$, with $\psi \in C^2$, $\nabla^2\psi(t) > 0$ for all $t \in R$, $\psi(0) = 0$, $\nabla\psi(0) = 1$, $\lim_{t\to-\infty} \psi(t) > -\infty$, $\lim_{t\to-\infty} \nabla\psi(t) = 0$, and $\lim_{t\to\infty} \nabla\psi(t) = \infty$, and any convex function $\xi: R \to R$, with $\xi \in C^2$, $\nabla^2\xi(0) = 0$, $\xi(t) = 0$ for all $t \le 0$, $\xi(t) > 0$ for $t > 0$, and $\lim_{t\to\infty} \nabla\xi(t) = \infty$. Each such pair (ψ, ξ) defines a function $p \in P_I$ by means of

$$p(t; \mu) = \mu\psi(t) + \xi(t).$$

As an example, take

$$\psi(t) = e^t - 1, \qquad \xi(t) = \tfrac{1}{3}[\max\{0, t\}]^3.$$

We have

$$p(t; \mu) = \mu(e^t - 1) + \tfrac{1}{3}[\max\{0, t\}]^3,$$

$$\nabla_t p(t; \mu) = \mu e^t + [\max\{0, t\}]^2, \qquad \partial^2 p(t; \mu)/\partial t^2 = \mu e^t + 2\max\{0, t\}.$$

Another example of a twice differentiable penalty function, which can be evaluated with simple arithmetic operations, is given by

$$p(t; \mu) = \begin{cases} \mu t + \mu t^2 + t^3 & \text{if } t \ge 0, \\ \mu t/(1 - t) & \text{if } t < 0. \end{cases}$$

It is easy to verify that all functions of Examples 1 and 2 satisfy the conditions (a)–(h) and do indeed belong to P_I.

There is another class of penalty functions defined below that is often useful even though the analysis relating to them is not as powerful as the one relating to the class P_I.

Class of Penalty Functions $\hat{P}_I$: All functions $p: R^2 \to R$ of the form

$$p(t; \mu) = \mu\psi(t),$$

where $\psi: R \to R$ is any function such that $\psi \in C^2$, $\nabla^2\psi(t) > 0$ for all $t \in R$, $\psi(0) = 0$, $\nabla\psi(0) = 1$, $\lim_{t\to-\infty} \psi(t) > -\infty$, $\lim_{t\to-\infty} \nabla\psi(t) = 0$, and $\lim_{t\to\infty} \nabla\psi(t) = \infty$.

Note that all functions in the class $\hat{P}_{\mathrm{I}}$ are twice differentiable. A prominent example is the *exponential penalty function*

$$p(t;\mu) = \mu(e^t - 1). \tag{11}$$

Functions in the class $\hat{P}_{\mathrm{I}}$ satisfy all conditions (a)–(h) of the definition of the class P_{I} with the exception of (c) and (g) which are satisfied only for $\mu > 0$.

The augmented Lagrangian corresponding to the class $\hat{P}_{\mathrm{I}}$ is similarly given by

$$L_c(x,\mu) = f(x) + \frac{1}{c}\sum_{j=1}^{r} p[cg_j(x);\mu^j] = f(x) + \frac{1}{c}\sum_{j=1}^{r} \mu^j\psi[cg_j(x)].$$

The first-order multiplier method consists of sequential unconstrained minimization of $L_{c_k}(\cdot,\mu_k)$ over X yielding a vector x_k. Minimization is followed by the multiplier iteration

$$\mu_{k+1}^j = \nabla_t p[c_k g_j(x_k);\mu_k^j] = \mu_k^j \nabla\psi[c_k g_i(x_k)], \qquad j = 1,\ldots,r.$$

The initial multiplier must be positive, i.e., $\mu_0 > 0$. Note that the properties $\lim_{t\to-\infty}\nabla\psi(t) = 0$ and $\nabla^2\psi(t) > 0$ imply that $\nabla\psi(t) > 0$ for all $t \in R$, so it follows that *the sequence $\{\mu_k\}$ generated by the iteration above satisfies $\mu_k > 0$ for all k.*

In the remainder of this chapter, we shall provide a convergence analysis of multiplier methods corresponding to the classes P_{I} and $\hat{P}_{\mathrm{I}}$ as applied to convex programming problems. For this analysis, we shall need a number of properties of the classes P_{I} and $\hat{P}_{\mathrm{I}}$ which we collect in the following proposition.

Proposition 5.1: Let either $p \in P_{\mathrm{I}}$, $\mu \geq 0$, and $t \in R$ or $p \in \hat{P}_{\mathrm{I}}$, $\mu > 0$, and $t \in R$. Then

(a) $\nabla_\mu p(t;\mu) \geq t$,
(b) $t\nabla_t p(t;\mu) \geq p(t;\mu) \geq \mu\nabla_\mu p(t;\mu) \geq \mu t$,
(c) $p(t;\mu) - p(t;\bar{\mu}) \geq t(\mu - \bar{\mu}) \quad \forall\, \bar{\mu} \in [0,\mu]$.
(d) The following five conditions are equivalent:

(d1) $t\nabla_t p(t;\mu) = p(t;\mu)$,
(d2) $p(t;\mu) = \mu t$,
(d3) $p(t;\mu) = 0$,
(d4) $\nabla_t p(t;\mu) = \mu$,
(d5) $t \leq 0$ and $\mu t = 0$.

Proof: We consider only the class P_{I}. The proofs are valid also for the class $\hat{P}_{\mathrm{I}}$ since properties (a)–(h) in the definition of P_{I} are satisfied by $p \in \hat{P}_{\mathrm{I}}$ if $\mu > 0$.

(a, b) Fix $t \in R$ and $\mu \geq 0$. By property (d) and the convexity of $p(\cdot;\mu)$, we have

$$0 = p(0;\mu) \geq p(t;\mu) + (0 - t)\nabla_t p(t;\mu)$$

or

$$t\nabla_t p(t;\mu) \geq p(t;\mu). \tag{12}$$

Similarly, using properties (d) and (e), we obtain

$$p(t;\bar{\mu}) \geq \bar{\mu}t \qquad \forall\, t \in R, \quad \bar{\mu} \geq 0. \tag{13}$$

By concavity of $p(t;\cdot)$, we have

$$p(t;\bar{\mu}) \leq p(t;\mu) + (\bar{\mu} - \mu)\nabla_\mu p(t;\mu) \qquad \forall\, \bar{\mu} \geq 0.$$

Combining the last two inequalities, we obtain

$$\bar{\mu}t \leq p(t;\mu) + (\bar{\mu} - \mu)\nabla_\mu p(t;\mu) \qquad \forall\, \bar{\mu} \geq 0.$$

By setting $\bar{\mu} = 0$, we have

$$\mu\nabla_\mu p(t;\mu) \leq p(t;\mu), \tag{14}$$

while by letting $\bar{\mu} \to \infty$ we obtain

$$t \leq \nabla_\mu p(t;\mu). \tag{15}$$

Combining (12), (14), and (15) we obtain a proof of (a) and (b).

(c) From part (a) we have, for fixed $t \in R$,

$$\int_{\bar{\mu}}^{\mu} \nabla_\mu p(t;\xi)\,d\xi \geq \int_{\bar{\mu}}^{\mu} t\,d\xi$$

or equivalently

$$p(t;\mu) - p(t;\bar{\mu}) \geq t(\mu - \bar{\mu}).$$

(d) To show the equivalence of (d1)–(d5), first assume that (d5) holds. There are two cases to consider:

CASE I. ($t = 0$): In this case, (d1)–(d4) follow immediately from properties (d) and (e).

CASE II. ($t < 0$ and $\mu = 0$): Properties (e) and (f) together with the fact that $\nabla_t p(\cdot;0)$ is nondecreasing (by convexity) yield $\nabla_t p(t;0) = 0$ for $t < 0$. That together with property (d) implies

$$p(t;0) = \nabla_t p(t;0) = \mu = 0.$$

This proves (d1)–(d4).

It will now suffice to show that if (d5) does not hold, then the same is true for (d1)–(d4). If (d5) does not hold then there are two cases.

CASE I. ($t > 0$): The proof of (12) and (13), together with the strict convexity property (c), imply that

$$t\nabla_t p(t;\mu) > p(t;\mu) > \mu t \geq 0. \tag{16}$$

Since $t > 0$, these inequalities imply that (d1)–(d4) do not hold.

CASE II. ($t < 0$ and $\mu > 0$): By property (e), we have $\nabla_t p(0;\mu) = \mu$, and since $\nabla_t p(\cdot\,;\mu)$ is continuous on R, there is an open interval of scalars $\bar{t}$ containing the origin in which $\nabla_t p(\bar{t};\mu) > 0$. By property (c), strict convexity holds for $\bar{t}$ in this interval. Using this fact in the proofs of (12) and (13), we obtain again (16) which shows that (d1)–(d4) do not hold. Q.E.D.

5.1.3 Approximation Procedures Based on Nonquadratic Penalty Functions

The approximation procedure described in Section 3.3 is based on a quadratic penalty function. As a result some of the approximating functions described there, such as the one corresponding to the function (compare with Example 6, Section 3.3)

$$\gamma(t) = \max\{t_1, t_2, \ldots, t_r\}, \tag{17}$$

are not twice differentiable. By using a suitable nonquadratic penalty function, it is possible in some cases to obtain more convenient or twice differentiable approximating functions.

Let $\gamma: R^r \to (-\infty, +\infty]$ be a lower semicontinuous, convex function with $\gamma(t) < \infty$ for at least one $t \in R^r$. Assume further that γ is monotonically nondecreasing in the sense that for any $t_1, t_2 \in R^r$ we have

$$t_1 \leq t_2 \Rightarrow \gamma(t_1) \leq \gamma(t_2). \tag{18}$$

Then for $g: R^n \to R^r$, the problem

$$\begin{aligned} &\text{minimize} \quad \gamma[g(x)] \\ &\text{subject to} \quad x \in X \end{aligned} \tag{19}$$

is equivalent to the problem

$$\begin{aligned} &\text{minimize } \gamma[g(x) - u] \\ &\text{subject to} \quad x \in X, \qquad u \leq 0. \end{aligned}$$

By eliminating the inequality constraints $u \leq 0$ by means of a penalty function $c^{-1}p(cu;\mu)$, $c > 0$ and $\mu \geq 0$, such as the ones considered in the previous section, we obtain the approximating problem

$$\text{(20)} \qquad \begin{aligned} &\text{minimize} \quad Q_c[g(x); \mu] \\ &\text{subject to} \quad x \in X, \end{aligned}$$

where the approximate objective function Q_c is given by

$$\text{(21)} \qquad Q_c(t; \mu) = \min_{u \in R^r} \left\{ \gamma(t - u) + \frac{1}{c} \sum_{i=1}^{r} p(cu_i; \mu^i) \right\}.$$

The general approximation procedure for problems containing several functions of the form $\gamma[g(x)]$ consists of replacing these functions by the approximating functions $Q_c[g(x); \mu]$ wherever they appear, followed by solution of the approximating problem. The process is repeated after suitable updating of the multipliers and the penalty parameter.

As an example, consider the function

$$\text{(22)} \qquad \gamma(t) = \alpha e^{\beta t}, \qquad \alpha > 0$$

[compare with (10) of Section 3.3], and the exponential penalty function [compare with (11)]

$$\text{(23)} \qquad c^{-1} p(cu; \mu) = c^{-1} \mu (e^{cu} - 1).$$

A straightforward calculation yields that the approximating function of (21) is given by

$$Q_c(t; \mu) = \mu^{\beta/(c+\beta)} (\alpha\beta)^{c/(c+\beta)} \frac{c + \beta}{c\beta} e^{c\beta t/(c+\beta)}.$$

The corresponding multiplier iteration is given by

$$\mu_{k+1} = \mu_k^{\beta/(c_k+\beta)} (\alpha\beta)^{c_k/(c_k+\beta)} e^{c_k \beta g(x_k)/(c_k+\beta)}.$$

This last fact can be verified by the reader by adapting the reasoning of Section 3.3 to the present case and by carrying out the straightforward calculations (see also Bertsekas, 1976e).

The exponential penalty function (23) can also be used in connection with the function

$$\text{(24)} \qquad \gamma[g(x)] = \max\{g_1(x), g_2(x), \ldots, g_r(x)\}$$

to yield via (21) the twice differentiable approximating function

$$\text{(25)} \qquad Q_c[g(x); \mu] = \frac{1}{c} \log \left\{ \sum_{i=1}^{r} \mu^i e^{c g_i(x)} \right\}.$$

The corresponding multiplier iteration can be calculated to be

$$\text{(26)} \qquad \mu_{k+1}^i = \mu_k^i e^{c_k g_i(x_k)} \Big/ \sum_{j=1}^r \mu_k^j e^{c_k g_j(x_k)}, \qquad i = 1, \ldots, r.$$

During calculation of $Q_c[g(x); \mu]$, as in (25), it is possible that computer overflow (or underflow) will occur if $cg_i(x)$ is too large (small). This difficulty can be eliminated by computing $Q_c[g(x); \mu]$, using the formula

$$Q_c[g(x); \mu] = \frac{1}{c} \log\left\{ \sum_{i=1}^r \mu_i e_i(x, \mu, c) \right\} + \gamma[g(x)] - \frac{A}{c},$$

with $e_i(x, \mu, c)$ given by

$$e_i(x, \mu, c) = \begin{cases} e^{A - c[\gamma[g(x)] - g_i(x)]} & \text{if } A - c[\gamma[g(x)] - g_i(x)] > -A, \\ 0 & \text{otherwise,} \end{cases}$$

where $\gamma[g(x)]$ is given by (24), and $A > 0$ is a large scalar such that both e^{-A} and e^A lie within the computer's range. Similarly, the updated multipliers μ_{k+1}^i of (26) should be computed by using the formula

$$\mu_{k+1}^i = \mu_k^i e_i(x_k, \mu_k, c_k) \Big/ \sum_{j=1}^r \mu_k^j e_j(x_k, \mu_k, c_k).$$

As an example of using the penalty function (25), consider the following simple method for finding a solution of the system of nonlinear inequalities

$$\text{(27)} \qquad g_i(x) \le 0, \qquad i = 1, \ldots, r.$$

This problem is equivalent to the problem

$$\begin{aligned} &\text{minimize} \quad \gamma[g(x)] \\ &\text{subject to} \quad x \in R^n, \end{aligned}$$

where $\gamma[g(x)]$ is given by (24). Consider a method consisting of sequential unconstrained minimizations of the form [compare with (25)]

$$\text{(28)} \qquad \begin{aligned} &\text{minimize} \quad \frac{1}{c} \log\left\{ \sum_{i=1}^r \mu_k^i e^{c_k g_i(x)} \right\} \\ &\text{subject to} \quad x \in R^n, \end{aligned}$$

where μ_k^i, $i = 1, \ldots, r$ are multipliers satisfying

$$\mu_k^i > 0, \qquad i = 1, \ldots, r, \qquad \sum_{i=1}^r \mu_k^i = 1 \qquad \forall\, k = 0, 1, \ldots,$$

and $c_k > 0$ is a penalty parameter. Let v_k^* be the optimal value of problem (28).

If there exists a feasible solution to the system (27), then it is easily seen that

$$v_k^* \leq 0 \qquad \forall\, k = 0, 1, \ldots,$$

while if there exists a strictly feasible solution $\bar{x}$ with $\gamma[g(\bar{x})] < 0$ then

$$v_k^* < 0 \qquad \forall\, k = 0, 1, \ldots.$$

On the other hand, suppose that there is no feasible solution to the system (27) and assume that

(29) $$c_k \to \infty,$$

and for some $\varepsilon > 0$, we have

(30) $$\mu_k^i \geq \varepsilon \qquad \forall\, i = 1, \ldots, r, \qquad k = 0, 1, \ldots.$$

Then it is easily seen that, for all k sufficiently large, we have $v_k^* > 0$. These observations can be used to show that if the method is operated so that (29) and (30) hold, and for each k, x_k is an optimal solution of problem (28), then

(a) If the system (27) is feasible, every limit point of $\{x_k\}$ is a feasible solution.

(b) If the system (27) is strictly feasible, then there exists an index $\bar{k}$ such that $x_{\bar{k}}$ is a strictly feasible solution.

(c) If the system (27) is infeasible, then there exists an index $\bar{k}$ such that $v_{\bar{k}}^* > 0$ thereby confirming the fact that no feasible solution exists.

We note also that it is possible to show that if the functions g_i, $i = 1, \ldots, r$, are convex then, under a mild assumption, the conclusions (a), (b), and (c) hold even if the conditions (29) and (30) are not enforced (see Proposition 5.12 in Section 5.3).

5.2 Convex Programming and Duality

We consider the following convex programming problem

(CPP) $$\text{minimize } f(x)$$

$$\text{subject to } \quad x \in X, \qquad g_j(x) \leq 0, \quad j = 1, \ldots, r,$$

where we make the following standing assumptions.

Assumption (A1): *The set X is a nonempty convex subset of R^n and the functions $f\colon R^n \to R$, $g_j\colon R^n \to R$, $j = 1, \ldots, r$ are convex over X.*

Assumption (A2): *There exists at least one feasible solution for* (CPP).

Assumption (A3): *The optimal value* f^* *of* (CPP) *is finite, i.e.,*

$$f^* = \inf\{f(x) | x \in X, g_j(x) \leq 0, j = 1, \ldots, r\} > -\infty.$$

It is possible to extend the definition of (CPP) to include linear equality constraints, but, for simplicity, we shall not consider this possibility. The methods we shall discuss together with the corresponding analysis can be suitably extended with essentially trivial modifications.

We shall employ, throughout the remainder of this chapter, the standard terminology of convex analysis. An excellent source for this material is Rockafellar (1970). Thus a function $f: R^n \to [-\infty, \infty]$ is said to be *convex* if the epigraph of f, i.e., the set $\{(x, \rho) | f(x) \leq \rho, x \in R^n, \rho \in R\}$ is convex. We say that f is *proper* if $f(x) > -\infty$ for all $x \in R^n$ and $f(\bar{x}) < \infty$ for at least one $\bar{x} \in R^n$. We say that f is *closed* if it is lower semicontinuous. The *conjugate convex function* of a convex function $f: R^n \to [-\infty, +\infty]$ is defined by

$$f^*(z) = \sup_{x \in R^n} \{z'x - f(x)\}.$$

The function f^* is convex and closed. It is proper if and only if f is proper. If f is closed, then the conjugate of f^* is f. The *subdifferential* $\partial f(x)$ of a convex function f is defined for each $x \in R^n$ by

$$\partial f(x) = \{z | f(\bar{x}) \geq f(x) + z'(\bar{x} - x), \ \forall \bar{x} \in R^n\}.$$

The subdifferential $\partial f(x)$ is a closed (possibly empty) convex set for each x. If f is real valued, then $\partial f(x)$ is nonempty and compact for each x. The preceding discussion is intended to provide only limited orientation, and we shall make frequent references to Rockafellar's text for additional notions and specific results. It is thus necessary that the reader should be somewhat familiar with the contents of this source in order to follow the subsequent development.

We review some known results for (CPP). Consider the ordinary *Lagrangian* function

$$L(x, \mu) = f(x) + \mu' g(x). \tag{1}$$

Definition: A vector $\mu^* \in R^r$ is said to be a *Lagrange multiplier* for (CPP) if $\mu^* \geq 0$ and

$$\inf_{x \in X} L(x, \mu^*) = f^*. \tag{2}$$

We have the following well-known results (see Rockafellar, 1970).

Proposition 5.2: Let μ^* be a Lagrange multiplier for (CPP). Then $x^* \in R^n$ is an optimal solution for (CPP) if and only if the following conditions hold:

$$L(x^*, \mu^*) = \inf_{x \in X} L(x, \mu^*), \tag{3}$$

$$x^* \in X, \qquad g(x^*) \leq 0, \qquad \mu^{*\prime} g(x^*) = 0. \tag{4}$$

Proposition 5.3: The vectors x^* and μ^* form an optimal solution–Lagrange multiplier pair for (CPP) if and only if $x^* \in X$, $\mu^* \geq 0$, and (x^*, μ^*) is a saddle point of the Lagrangian L in the sense

$$L(x^*, \mu) \leq L(x^*, \mu^*) \leq L(x, \mu^*) \qquad \forall\, x \in X, \quad \mu \geq 0. \tag{5}$$

Consider the *dual functional* $d: R^r \to [-\infty, \infty)$ of (CPP) defined by

$$d(\mu) = \begin{cases} \inf\{L(x, \mu) \mid x \in X\} & \text{if } \mu \geq 0, \\ -\infty & \text{otherwise.} \end{cases} \tag{6}$$

The following proposition holds.

Proposition 5.4: (a) If there exists at least one Lagrange multiplier then

$$f^* = \sup_{\mu \geq 0} d(\mu). \tag{7}$$

(b) If (7) holds, then a vector μ^* is a Lagrange multiplier for (CPP) if and only if it is an optimal solution of the *dual problem*

$$\begin{aligned} &\text{maximize} \quad d(\mu) \\ &\text{subject to} \quad \mu \geq 0. \end{aligned} \tag{8}$$

When (7) holds, we say that there is *no duality gap*. It is easily seen that $d(\mu) \leq f^*$ for all $\mu \in R^r$. Therefore the fact that existence of a Lagrange multiplier implies that no duality gap is present follows from the definition of a Lagrange multiplier. Corollary 28.2.1 of Rockafellar's text shows that a sufficient condition for existence of a Lagrange multiplier is the *Slater condition* that there exists an $\bar{x} \in X$ such that $g_j(\bar{x}) < 0$, for all j. *The Slater condition guarantees also that the set of all Lagrange multipliers is compact as well as nonempty* (see Corollary 29.1.5 of Rockafellar's text).

Consider now the *primal functional* $q: R^r \to [-\infty, +\infty]$ defined by

$$q(u) = \inf\{f(x) \mid x \in X, g(x) \leq u\} \qquad \forall\, u \in R^r. \tag{9}$$

The primal and dual functionals are intimately related as we now show. We have, for $\mu \geq 0$,

$$\begin{aligned} d(\mu) &= \inf\{f(x) + \mu' g(x) \mid x \in X\} \\ &= \inf_{u \in R^r} \inf\{f(x) + \mu' g(x) \mid x \in X, g(x) \leq u\} \\ &= \inf_{u \in R^r} \inf\{f(x) + \mu' u \mid x \in X, g(x) \leq u\} \\ &= \inf_{u \in R^r} \{q(u) + \mu' u\} \\ &= -\sup_{u \in R^r} \{(-\mu)' u - q(u)\}, \end{aligned} \tag{10}$$

so that

$$d(\mu) = -q^*(-\mu) \qquad \forall\, \mu \geq 0, \tag{11}$$

where q^* is the conjugate convex function of q.

Now notice that q is monotonically nondecreasing in u in the sense that for all $u \in R^r$ and $\bar{u} \geq 0$ we have $q(u) \geq q(u + \bar{u})$. Hence, if $\mu = (\mu_1, \ldots, \mu_r)'$ is such that $\mu_j < 0$ for some j, we have

$$\inf_{u \in R^r} \{q(u) + \mu' u\} = -\infty.$$

It follows using (6), and (11) that

$$d(\mu) = -q^*(-\mu) \qquad \forall\, \mu \in R^r. \tag{12}$$

The nature of the primal functional provides the key to questions regarding existence of Lagrange multipliers. It is shown in Rockafellar's text (Theorem 29.1) that a vector μ^* is a Lagrange multiplier if and only if $-\mu^* \in \partial q(0)$. If q is closed then we have $f^* = \sup_{\mu \geq 0} d(\mu)$ (see Rockafellar's text, Theorem 30.3). *The primal functional q is in turn closed if X is a closed set and the set of optimal solutions for* (CPP) *is nonempty and compact.* This last fact can be verified by using Theorem 9.2 of Rockafellar's text (see also Theorem 30.4 of the same reference).

The Augmented Lagrangian and the Penalized Dual Functional

Consider now the *augmented Lagrangian* L_c corresponding to a scalar $c > 0$ and a penalty function p, where p belongs to the class P_{I} or the class $\hat{P}_{\text{I}}$ defined in Section 5.1.2. We have

$$L_c(x, \mu) = f(x) + P_c[g(x); \mu], \tag{13}$$

where we use for convenience the notation

$$P_c(z; \mu) = \frac{1}{c} \sum_{j=1}^{r} p(cz_j; \mu_j). \tag{14}$$

Consider the conjugate convex function of $P_c(\cdot\,; \mu)$ defined by

$$P_c^*(s; \mu) = \sup_{z \in R^r} \{z' s - P_c(z; \mu)\} \qquad \forall\, \mu \geq 0. \tag{15}$$

In view of the form (14) it is easily seen that we have

$$P_c^*(s; \mu) = \frac{1}{c} \sum_{j=1}^{r} p^*(s_j; \mu_j) \qquad \forall\, \mu \geq 0, \tag{16}$$

where s_j and μ_j denote the jth coordinates of s and μ respectively, and $p^*(\cdot\,;\mu_j)$ is the conjugate convex function of $p(\cdot\,;\mu_j)$ given by

$$p^*(s_j;\mu_j) = \sup_{t\in R}\{s_j t - p(t;\mu_j)\} \qquad \forall\,\mu_j \geq 0. \tag{17}$$

When p belongs to the class P_E^+ defined in Section 5.1.2, its conjugate can be characterized more precisely. A function $p \in P_E^+$ has the form

$$p(t;\mu_j) = \begin{cases} \mu_j t + \phi(t) & \text{if}\quad \mu_j + \nabla\phi(t) \geq 0,\\ \min\limits_{\tau\in R}\{\mu_j\tau + \phi(\tau)\} & \text{otherwise,}\end{cases} \tag{18}$$

where ϕ belongs to the class P_E of Section 5.1.1. Note that this expression makes sense even if $\mu_j < 0$. From (16) and (17) we obtain via a straightforward calculation for all $\mu_j \in R$

$$p^*(s_j;\mu_j) = \begin{cases} \dfrac{1}{c}\,\phi^*(s_j - \mu_j) & \text{if}\quad s_j \geq 0\\ \infty & \text{otherwise,}\end{cases} \tag{19}$$

and for all $\mu \in R^r$

$$P_c^*(s;\mu) = \begin{cases} \dfrac{1}{c}\displaystyle\sum_{j=1}^{r}\phi^*(s_j - \mu_j) & \text{if}\quad s \geq 0,\\ \infty & \text{otherwise,}\end{cases} \tag{20}$$

where ϕ^* is the conjugate convex function of ϕ defined by

$$\phi^*(y) = \sup_{t\in R}\{yt - \phi(t)\} \qquad \forall\, y \in R. \tag{21}$$

Since, by definition of the class P_E in Section 5.1.1, we have $\phi(t) \geq 0$ for all $t \in R$, $\lim_{t\to-\infty}\nabla\phi(t) = -\infty$, $\lim_{t\to\infty}\nabla\phi(t) = \infty$, $\phi(0) = 0$, $\nabla\phi(0) = 0$, it can be easily seen that

$$0 \leq \phi^*(y) < \infty \qquad \forall\, y \in R,$$
$$\min_{y\in R}\phi^*(y) = \phi^*(0) = 0.$$

Since ϕ is strictly convex and differentiable, it follows from Theorem 26.3 of Rockafellar (1970) that ϕ^* is also strictly convex, and differentiable. Finally the facts that ϕ is the convex conjugate of ϕ^* and ϕ is real-valued imply that $\lim_{t\to\infty}\nabla\phi^*(t) = -\infty$ and $\lim_{t\to\infty}\nabla\phi^*(t) = \infty$. The conclusion is that $\phi \in P_E$ *if and only if* $\phi^* \in P_E$. We shall make frequent use of the properties of ϕ^* just shown.

The *penalized dual functional*, denoted d_c, is defined on the set $\{\mu \mid \mu \geq 0\}$ by

$$d_c(\mu) = \inf\{L_c(x,\mu) \mid x \in X\} \qquad \forall\,\mu \geq 0. \tag{22}$$

If $p \in P_E^+$ this definition also makes sense for every $\mu \in R^r$ (not just for $\mu \geq 0$), and for $p \in P_E^+$ we shall view d_c in what follows as a function defined by (22) for *all* $\mu \in R^r$. A calculation similar to the one in (10) yields

$$d_c(\mu) = \inf_{u \in R^r} \{q(u) + P_c(u; \mu)\}, \tag{23}$$

where $q(\cdot)$ is the primal functional of (CPP).

The following proposition provides some basic facts.

Proposition 5.5. Let $c > 0$.

(a) Assume that $\mu \geq 0$ and $p \in P_I$, or $\mu > 0$ and $p \in \hat{P}_I$, or $\mu \in R^r$ and $p \in P_E^+$. The conjugate convex function $P_c^*(\cdot\, ; \mu)$ satisfies

$$0 \leq P_c^*(s; \mu) < \infty \qquad \forall\, s \geq 0, \tag{24}$$

$$P_c^*(s; \mu) = \infty \qquad \text{if} \quad s_j < 0 \quad \text{for some} \quad j = 1, \ldots, r. \tag{25}$$

Furthermore $P_c^*(\cdot\, ; \mu)$ is strictly convex on the set $\{s \,|\, s \geq 0\}$.

(b) Assume that $\mu \geq 0$ and $p \in P_I$, or $\mu > 0$ and $p \in \hat{P}_I$, or $\mu \in R^r$ and $p \in P_E^+$. Assume also that $f^* = \sup_{\mu \geq 0} d(\mu)$. Then

$$d_c(\mu) = \max_{s \in R^r} \{d(s) - P_c^*(s; \mu)\}, \tag{26}$$

and the maximum above is attained at a unique point $s(\mu, c) \geq 0$. Furthermore if the infimum in the definition

$$d_c(\mu) = \inf_{x \in X} L_c(x, \mu) \tag{27}$$

is attained at a point $x(\mu, c)$ (not necessarily unique) we have

$$s(\mu, c) = \nabla_z P_c[g[x(\mu, c)]; \mu], \tag{28}$$

where $s(\mu, c)$ is the unique point attaining the maximum in (26).

(c) Assume that $p \in P_E^+$, and that $f^* = \sup_{\mu \geq 0} d(\mu)$. Then d_c is continuously differentiable on R^r and

$$\frac{\partial d_c(\mu)}{\partial \mu_j} = \nabla \phi^*[s_j(\mu, c) - \mu_j] \qquad \forall\, \mu \in R^r, \quad j = 1, \ldots, r, \tag{29}$$

where $s_j(\mu, c)$ is the jth coordinate of the vector $s(\mu, c)$ defined in (b) above.

Proof: (a) From the properties of P_I, $\hat{P}_I$, and P_E^+ it can be seen that in all cases $p(0; \mu_j) \leq 0$. Therefore from (17) we have $p^*(s_j; \mu_j) \geq 0$ for all $s_j \geq 0$. Also in all cases $\lim_{t \to -\infty} \nabla_t p(t, \mu) = 0$ and $\lim_{t \to \infty} \nabla_t p(t; \mu) = \infty$. It follows that the supremum in (17) is attained if $s_j > 0$, while for $s_j = 0$ we have $p^*(0; \mu_j) = -\inf_{t \in R} p(t; \mu_j) < \infty$. Hence $p^*(s_j; \mu_j) < \infty$ for all $s_j \geq 0$ and (24) follows.

Since in all cases $p(\cdot\,; \mu_j)$ is nondecreasing and bounded below, (17) yields $p^*(s_j; \mu_j) = \infty$ if $s_j < 0$, from which (25) follows.

In all cases $p(\cdot\,; \mu_j)$ is a real-valued, differentiable convex function. Therefore by Theorem 26.3 of Rockafellar (1970), we obtain that $p^*(\cdot\,; \mu_j)$ is strictly convex on the set of s_j for which the subdifferential of $p^*(\cdot\,; \mu_j)$ with respect to s_j is nonempty. Since $p^*(\cdot\,; \mu_j)$ is defined on the real line it follows that $p_c^*(\cdot\,; \mu_j)$ is strictly convex on $\{s_j | p^*(s_j; \mu_j) < \infty\}$. Therefore, by (24), $P_c^*(\cdot\,; \mu)$ is strictly convex on the set $\{s | s \geq 0\}$.

(b) Assumptions (A1)–(A3) and the fact that $f^* = \sup_{\mu \geq 0} d(\mu)$ imply that the functions q, q^*, and d are proper convex functions. Furthermore $P_c(\cdot\,; \mu)$ is real-valued. These facts guarantee that assumption (a) of the Fenchel duality theorem (Rockafellar, 1970, Theorem 31.1) is satisfied. It follows from (23) and the conclusion of this theorem that

$$d_c(\mu) = \max_{s \in R^r}\{-q^*(-s) - P_c^*(s; \mu)\}, \tag{30}$$

and that the maximum is attained at some point $s(\mu, c)$. This point must be unique in view of the fact that $P_c^*(\cdot\,; s)$ is strictly convex and real-valued on $\{s | s \geq 0\}$ as shown in part (a). Equation (26) follows from (12) and (30).

If $x(\mu, c)$ attains the infimum in (27), then the vector $u(\mu, c) = g[x(\mu, c)]$ attains the infimum in (23). We have, using (23) and (30),

$$\begin{aligned}
&q[u(\mu, c)] + P_c[u(\mu, c); \mu] \\
&\quad = -q^*[-s(\mu, c)] - P_c^*[s(\mu, c); \mu] \\
&\quad = -\sup_{u \in R^r}\{-s(\mu, c)'u - q(u)\} - \sup_{u \in R^r}\{s(\mu, c)'u - P_c(u; \mu)\} \\
&\quad \leq s(\mu, c)'u(\mu, c) + q[u(\mu, c)] - s(\mu, c)'u + P_c(u; \mu) \qquad \forall\, u \in R^r.
\end{aligned}$$

Hence

$$P_c(u; \mu) \geq P_c[u(\mu, c); \mu] + s(\mu, c)'[u - u(\mu, c)] \qquad \forall\, u \in R^r,$$

so $s(\mu, c)$ is a subgradient of $P_c(\cdot\,; \mu)$ at $u(\mu, c)$. Since $P_c(\cdot\,; \mu)$ is differentiable and $u(\mu, c) = g[x(\mu, c)]$ we obtain (28).

(c) Since $\sup_{\mu \geq 0} d(\mu) = f^*$ and $f^* > -\infty$ [by Assumption (A3)] it follows from (24) and (26) that

$$-\infty < d_c(\mu) \leq f^* \qquad \forall\, \mu \in R^r. \tag{31}$$

Hence d_c is real-valued and as a result the subdifferential $\partial d_c(\mu)$ is nonempty and compact for all $\mu \in R^r$. Fix $\mu \in R^r$ and let $w \in \partial d_c(\mu)$. Then for all $\bar{\mu} \in R^r$, we have, using (26),

$$\begin{aligned}
d[s(\mu, c)] - P_c^*[s(\mu, c); \bar{\mu}] &\leq d_c(\bar{\mu}) \leq d_c(\mu) + w'(\bar{\mu} - \mu) \\
&= d[s(\mu, c)] - P_c^*[s(\mu, c); \mu] + w'(\bar{\mu} - \mu) \qquad \forall\, \bar{\mu} \in R^r.
\end{aligned} \tag{32}$$

Using (20) we obtain

$$P_c^*[s(\mu, c); \bar{\mu}] = \frac{1}{c} \sum_{j=1}^{r} \phi^*[s_j(\mu, c) - \bar{\mu}_j],$$

$$P_c^*[s(\mu, c); \mu] = \frac{1}{c} \sum_{j=1}^{r} \phi^*[s_j(\mu, c) - \mu_j].$$

Combining the above two equations and (32) we obtain for all j and $\bar{\mu}_j \in R$

$$\phi^*[s_j(\mu, c) - \bar{\mu}_j] \geq \phi^*[s_j(\mu, c) - \mu_j] + (-w_j)(\bar{\mu}_j - \mu_j).$$

This implies that $(-w_j)$ is a subgradient of the function $h_j(\mu_j) = \phi^*[s_j(\mu, c) - \mu_j]$ at μ_j. Hence

$$w_j = \nabla\phi^*[s_j(\mu, c) - \mu_j] \qquad \forall\, j = 1, \ldots, r,$$

and (29) follows. Q.E.D.

The differentiability property of d_c, shown in Proposition 5.5c, does not hold for $p \notin P_E^+$, since d_c is not even defined outside the set $\{\mu | \mu \geq 0\}$. It is possible, however, to show that if the assumptions of Proposition 5.5c are satisfied and $p \in P_I$ or $p \in \hat{P}_I$ then the penalized dual functional d_c is continuously differentiable on the set $\{\mu | \mu > 0\}$. The proof of this is very similar to the proof of Proposition 5.5c and is left for the reader.

We can now prove the following proposition relating optimal solutions and Lagrange multipliers of (CPP) with minimizing points of the augmented Lagrangian and optimal solutions of the following penalized dual problem

$$\begin{aligned} &\text{maximize} \quad d_c(\mu) \\ &\text{subject to} \quad \mu \geq 0. \end{aligned}$$

Proposition 5.6. Let $c > 0$, and assume that $p \in P_I$ or $p \in \hat{P}_I$.

(a) Assume $f^* = \sup_{\mu \geq 0} d(\mu)$. Then the set of maximizing points of both d and d_c over $\{\mu | \mu \geq 0\}$ coincides with the set of Lagrange multipliers of (CPP).

(b) Assume that $p \in P_I$ and let μ^* be a Lagrange multiplier for (CPP). Then a vector x^* is an optimal solution of (CPP) if and only if it minimizes $L_c(\cdot, \mu^*)$ over $x \in X$.

(c) Assume that $p \in P_I$. Then (x^*, μ^*) is an optimal solution–Lagrange multiplier pair of (CPP) if and only if $x^* \in X$, $\mu^* \geq 0$, and (x^*, μ^*) is a saddle point of L_c in the sense that

$$L_c(x^*, \mu) \leq L_c(x^*, \mu^*) \leq L_c(x, \mu^*) \qquad \forall\, x \in X, \quad \mu \geq 0. \tag{33}$$

If in addition $p \in P_E^+$, then (x^*, μ^*) is an optimal solution–Lagrange multiplier pair if and only if $x^* \in X$, $\mu^* \in R^r$, and

$$L_c(x^*, \mu) \le L_c(x^*, \mu^*) \le L_c(x, \mu^*) \qquad \forall\, x \in X, \quad \mu \in R^r. \tag{34}$$

Proof: (a) We first note that by a similar argument as the one used in the proof of Proposition 5.5 we can show that the relations

$$0 \le P_c^*(s; \mu) \qquad \forall\, s \ge 0 \tag{35}$$

$$d_c(\mu) = \max_{s \in R^r}\{d(s) - P_c^*(s; \mu)\} \tag{36}$$

hold for all $\mu \ge 0$ (i.e., even if $p \in \hat{P}_I$ and $\mu_j = 0$ for some j). Furthermore the maximum in (36) is attained for some $s(\mu, c) \ge 0$ (not necessarily unique if $p \in \hat{P}_I$ and $\mu_j = 0$ for some j). Therefore using also Proposition 5.1b

$$d(\mu) \le d_c(\mu) \le f^* \qquad \forall\, \mu \ge 0. \tag{37}$$

From Proposition 5.4 and (37) it follows that if μ^* is a Lagrange multiplier it must maximize both d and d_c over $\{\mu \mid \mu \ge 0\}$, while if μ^* maximizes d it must be a Lagrange multiplier.

Let $\mu^* \ge 0$ maximize d_c over $\{\mu \mid \mu \ge 0\}$. We will show that μ^* is a Lagrange multiplier. Indeed in view of (37) and the fact that $f^* = \sup_{\mu \ge 0} d(\mu)$ we have

$$f^* = d_c(\mu^*) = d[s(\mu^*, c)] - P_c^*[s(\mu^*, c); \mu^*]. \tag{38}$$

From (35), (37), (38) and the fact that $f^* > -\infty$ we obtain

$$d[s(\mu^*, c)] = f^* \tag{39}$$

and

$$P_c^*[s(\mu^*, c); \mu^*] = 0. \tag{40}$$

From (40) we obtain

$$\sup_{t \in R}\{t s_j(\mu^*, c) - \frac{1}{c}\, p(ct; \mu_j^*)\} = 0 \qquad \forall\, j = 1, \ldots, r$$

and

$$c t s_j(\mu^*, c) \le p(ct; \mu_j^*) \qquad \forall\, j = 1, \ldots, r, \quad t \in R. \tag{41}$$

Since $p(0; \mu_j^*) = 0$, it follows from (41) that $s_j(\mu^*, c)$ is a subgradient of $p(\cdot\,; \mu_j^*)$ at $t = 0$. But $\nabla_t p(0; \mu_j^*) = \mu_j^*$ for all $\mu_j^* \ge 0$, so it follows that

$$s_j(\mu^*, c) = \mu_j^* \qquad \forall\, j = 1, \ldots, r.$$

From (39) and Proposition 5.4 it follows that μ^* is a Lagrange multiplier.

(b) We have by part (a)

$$f^* = d_c(\mu^*) = \inf_{x \in X} L_c(x, \mu^*) \leq L_c(x^*, \mu^*) \tag{42}$$
$$= f(x^*) + P_c(x^*; \mu^*) \qquad \forall\, x^* \in X.$$

If x^* is an optimal solution of (CPP) we have $f^* = f(x^*)$ and $P_c(x^*; \mu^*) \leq 0$. It follows from (42) that x^* minimizes $L_c(\cdot, \mu^*)$ over X.

Conversely, assume that x^* minimizes $L_c(\cdot, \mu^*)$ over X. Using Proposition 5.1b and the fact that μ^* is a Lagrange multiplier we have

$$L_c(x^*, \mu^*) \geq L(x^*, \mu^*) \geq d(\mu^*) = d_c(\mu^*) = \inf_{x \in X} L_c(x, \mu^*). \tag{43}$$

Since x^* minimizes $L_c(\cdot, \mu^*)$ over X, equality must hold throughout in (43). Therefore

$$P_c[g(x^*); \mu^*] = \mu^{*\prime} g(x^*)$$

and, using Proposition 5.1d, it follows that

$$g(x^*) \leq 0, \qquad \mu^{*\prime} g(x^*) = 0. \tag{44}$$

Since equality holds in (43) we have

$$L(x^*, \mu^*) = d(\mu^*) = \inf_{x \in X} L(x, \mu^*). \tag{45}$$

Using (44), (45), and Proposition 5.2 it follows that x^* is an optimal solution of (CPP).

(c) If (x^*, μ^*) is an optimal solution–Lagrange multiplier pair we have by part (b)

$$L_c(x^*, \mu^*) \leq L_c(x, \mu^*) \qquad \forall\, x \in X. \tag{46}$$

Also, since $g(x^*) \leq 0$, we have $P_c[g(x^*); \mu] \leq 0$ for all $\mu \geq 0$. Therefore

$$L_c(x^*, \mu) \leq f(x^*) = L_c(x^*, \mu^*) \qquad \forall\, \mu \geq 0. \tag{47}$$

From (46) and (47) we obtain (33). If $p \in P_{\mathrm{E}}^+$, then we have $P_c[g(x^*); \mu] \leq 0$ for all $\mu \in R^r$ and similarly (34) follows.

Conversely, assume that $x^* \in X$, $\mu^* \geq 0$ and (33) is satisfied. Then

$$L_c(x^*, \mu^*) = \sup_{\mu \geq 0} L_c(x^*, \mu) \geq \sup_{\mu \geq 0} L(x^*, \mu) \tag{48}$$
$$= \begin{cases} f(x^*) & \text{if } \ g(x^*) \leq 0 \\ +\infty & \text{otherwise.} \end{cases}$$

Therefore we must have $g(x^*) \leq 0$ and it follows that

$$P_c[g(x^*); \mu^*] \leq P_c[g(x^*); 0] = 0.$$

Hence

$$(49) \qquad L_c(x^*, \mu^*) = f(x^*) + P_c[g(x^*); \mu^*] \le f(x^*) = L_c(x^*, 0).$$

Using (33) it follows that equality holds throughout in (49), and therefore

$$(50) \qquad P_c[g(x^*); \mu^*] = 0.$$

From Proposition 5.1d we obtain

$$(51) \qquad g(x^*) \le 0, \qquad \mu^{*\prime} g(x^*) = 0.$$

We have from (50) and (33)

$$(52) \quad f(x^*) = L_c(x^*, \mu^*) = \inf_{x \in X} L_c(x, \mu^*)$$

$$\le \inf_{\substack{x \in X \\ g(x) \le 0}} L_c(x, \mu^*) \le f(x) \qquad \forall\, x \in X, \quad g(x) \le 0.$$

Combining (51) and (52) we obtain that x^* is an optimal solution of (CPP).

Denote $u^* = g(x^*)$. Then from (52) we have that u^* attains the infimum in the equation

$$(53) \qquad f^* = \inf_{u \in R^r} \{q(u) + P_c(u; \mu^*)\}.$$

It follows that q is proper and, since $P_c(\,\cdot\,; \mu^*)$ is real-valued, application of the Fenchel duality theorem (Rockafellar, 1970, Theorem 31.1) yields for some vector s^*

$$(54) \quad q(u^*) + P_c(u^*; \mu^*) = -q^*(-s^*) - P_c^*(s^*; \mu^*)$$

$$= -\sup_{u \in R^r}\{-u's^* - q(u)\} - \sup_{u \in R^r}\{u's^* - P_c(u; \mu^*)\}.$$

From (54) we obtain

$$q(u^*) + P_c(u^*; \mu^*) \le u^{*\prime}s^* + q(u^*) - u's^* + P_c(u; \mu^*) \qquad \forall\, u \in R^r,$$

or

$$P_c(u^*; \mu^*) + (u - u^*)'s^* \le P_c(u; \mu^*) \qquad \forall\, u \in R^r.$$

Hence s^* is a subgradient of $P_c(\,\cdot\,; \mu^*)$ at u^*. In view of (50) and Proposition 5.1d we obtain

$$(55) \qquad s^* = \mu^*.$$

From (54) we also obtain

$$(56) \quad q(u^*) + P_c(u^*; \mu^*) \le u's^* + q(u) - u^{*\prime}s^* + P_c(u^*; \mu^*) \qquad \forall\, u \in R^r.$$

Combining (55) and (56) we obtain

$$q(u^*) + u^{*\prime}\mu^* \le q(u) + u'\mu^* \qquad \forall\, u \in R^r,$$

or equivalently

$$q(u^*) + u^{*\prime}\mu^* = \inf_{x \in X} L(x, \mu^*).$$

Since $q(u^*) = f^*$ and $u^{*\prime}\mu^* = 0$ by (51) it follows that μ^* is a Lagrange multiplier for (CPP).

It is easily seen that, if $p \in P_{\mathrm{E}}^+$, then, for all $\mu \in R^r$, we have $L_c(x^*, \mu^+) \ge L_c(x^*, \mu)$ where μ^+ is the vector with coordinates $\max\{0, \mu_j\}$, $j = 1, \ldots, r$ with strict inequality holding if $\mu_j < 0$ for some j. Therefore (34) implies that $\mu^* \ge 0$, and the preceding proof shows that (x^*, μ^*) is an optimal solution–Lagrange multiplier pair. Q.E.D.

Note that, by Proposition 5.6b, it is sufficient for optimality that x^* minimizes $L_c(\cdot, \mu^*)$ over X. By contrast, it is not enough for a vector x^* to minimize $L(\cdot, \mu^*)$ over X. The additional conditions $g(x^*) \le 0$ and $\mu^{*\prime}g(x^*) = 0$ are also needed (compare with Proposition 5.2). Proposition 5.6c shows that when $X = R^n$ and $p \in P_{\mathrm{E}}^+$, the search for local minima–Lagrange multiplier pairs can be reduced to a search for *unconstrained* saddle points of L_c. We refer to the paper by Mangasarian (1975) for related algorithms and analysis. Note that the results of parts (b) and (c) of Proposition 5.6 do not hold if $p \in \hat{P}_{\mathrm{I}}$. Nonetheless this fact does not seem to impair the utility of the class $\hat{P}_{\mathrm{I}}$ for algorithmic purposes as will be discussed in the next section.

5.3 Convergence Analysis of Multiplier Methods

The algorithms that we consider are based on exact or approximate minimization of the augmented Lagrangian

$$L_c(x, \mu) = f(x) + P_c[g(x); \mu] = f(x) + \frac{1}{c}\sum_{j=1}^{r} p[cg_j(x); \mu_j],$$

where $p \in P_{\mathrm{I}}$ or $p \in \hat{P}_{\mathrm{I}}$.

Throughout this and the next section, we shall adopt the following assumption in addition to the assumptions (A1)–(A3) made in the beginning of Section 5.2.

Assumption (A4): *The set X is closed and* (CPP) *has a nonempty and compact optimal solution set denoted X^*. Furthermore the set of all Lagrange multipliers for* (CPP) *denoted M^* is nonempty and compact.*

Actually some of the subsequent results can be obtained under assumptions weaker than (A4). The reader can easily identify these results, and thus we prefer to assume (A4) at the outset so as to avoid overburdening the presentation.

We consider two algorithms, denoted A and B, employing exact and inexact minimization of $L_c(\cdot, \mu)$, respectively.

Algorithm A (Exact Minimization): Select an initial penalty parameter $c_0 > 0$ and an initial multiplier μ_0 satisfying $\mu_0 \geq 0$ if $p \in P_{\mathrm{I}}$ and $\mu_0 > 0$ if $p \in \hat{P}_{\mathrm{I}}$.

STEP 1: Given μ_k and c_k, find an x_k solving the problem

$$\text{minimize} \quad L_{c_k}(x, \mu_k) \qquad \text{subject to} \quad x \in X.$$

STEP 2: Set

$$\mu_{k+1}^j = \nabla_t p[c_k g_j(x_k); \mu_k^j], \qquad j = 1, \ldots, r. \tag{1}$$

Select $c_{k+1} \geq c_k$, and return to Step 1.

Notice that if $p \in P_{\mathrm{I}}$, we have, from (1), $\mu_k \geq 0$ for all k, while if $p \in \hat{P}_{\mathrm{I}}$ we have $\mu_k > 0$ for all k. Also note that (1) can be written as

$$\mu_{k+1} = \nabla_z P_{c_k}[g(x_k); \mu_k]. \tag{2}$$

This equation together with Proposition 5.5 [compare with (26) and (28)] imply that μ_{k+1} *is the unique point attaining the maximum in the equation*

$$d_{c_k}(\mu_k) = \max_{s \in R^r} \{d(s) - P_{c_k}^*(s; \mu_k)\}. \tag{3}$$

A geometric interpretation of this fact for the case where $p \in P_{\mathrm{E}}^+$ is given in Fig. 5.3.

In practice, the minimization in Step 1 of Algorithm A should be carried out only approximately. Not only this is necessary in order for the algorithm to be implementable, but in addition it usually results in substantial computational savings. We provide below an implementable version of the algorithm which employs inexact minimization. For $c > 0$ and $\mu \geq 0$, consider the convex function $\bar{L}_c(\cdot, \mu)$ given by

$$\bar{L}_c(x, \mu) = \begin{cases} L_c(x, \mu) & \text{if} \quad x \in X, \\ \infty & \text{if} \quad x \notin X. \end{cases} \tag{4}$$

Denote by $\Delta_x \bar{L}_c(x, \mu)$ the element of minimum Euclidean norm of the subdifferential (with respect to x) $\partial_x \bar{L}_c(x, \mu)$ for every $x \in R^n$ for which $\partial_x \bar{L}_c(x, \mu)$ is nonempty. We have

$$|\Delta_x \bar{L}_c(x, \mu)| = \min_{z \in \partial_x \bar{L}_c(x, \mu)} |z|. \tag{5}$$

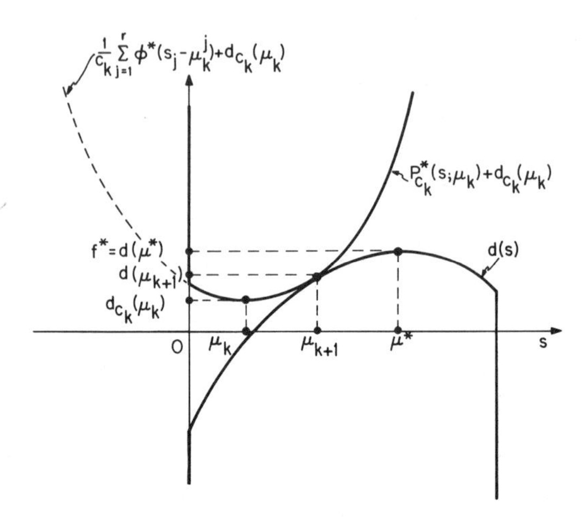

FIG. 5.3 Geometric interpretation of the multiplier iteration for $p \in P_E^+$

Note that $\Delta_x \bar{L}_c(x, \mu)$ is just the ordinary gradient $\nabla_x L_c(x, \mu)$, if $L_c(\cdot, \mu)$ is differentiable and x lies in the interior of X. In the following algorithm, minimization of $L_{c_k}(\cdot, \mu_k)$ over X is terminated at a point x_k where $|\Delta_x \bar{L}_{c_k}(x_k, \mu_k)|$ is sufficiently small. This type of stopping criterion makes sense only if an algorithm is available that produces sequences $\{z_i\}$ with the property $\lim_{i\to\infty} |\Delta_x \bar{L}_{c_k}(z_i, \mu_k)| = 0$. Such algorithms are available if $X = R^n$ and $f, g_j \in C^1$. They are also available in other situations, for example if X is specified by upper and lower bounds on the coordinates of x (see Section 1.5).

Algorithm B (Inexact Minimization): Select a sequence $\{\eta_k\}$ with $\eta_k \geq 0$ for all k, $\eta_k \to 0$, an initial penalty parameter $c_0 > 0$, and an initial multiplier μ_0 satisfying $\mu_0 \geq 0$ if $p \in P_I$ and $\mu_0 > 0$ if $p \in \hat{P}_I$.

Step 1: Given μ_k and c_k, find an x_k satisfying

$$|\Delta_x \bar{L}_{c_k}(x_k, \mu_k)|^2 \leq \eta_k \sum_{j=1}^{r} \left\{ \nabla_t p[c_k g_j(x_k); \mu_k^j] g_j(x_k) - \frac{1}{c_k} p[c_k g_j(x_k); \mu_k^j] \right\} \tag{6}$$

Step 2: Set

$$\mu_{k+1}^j = \nabla_t p[c_k g_j(x_k); \mu_k^j], \qquad j = 1, \ldots, r. \tag{7}$$

Select $c_{k+1} \geq c_k$, and return to Step 1.

Again from (7), we obtain for all k, $\mu_k \geq 0$ if $p \in P_I$ and $\mu_k > 0$ if $p \in \hat{P}_I$. Also from Proposition 5.1b, it follows that the right-hand side of the stopping criterion (6) is nonnegative. When $\eta_k = 0$ in (6), then x_k minimizes $L_{c_k}(\cdot, \mu_k)$ over X. Since $\eta_k \to 0$, the possibly inexact minimization indicated by (6) is asymptotically exact. We shall demonstrate shortly that when $\eta_k > 0$, then under a fairly mild assumption a vector x_k satisfying (6) can be obtained by means of a finite process. This assumption is stated below and will be in effect only for the results relating to Algorithm B.

Assumption (A5) (For Algorithm B): *There exists a positive scalar α such that, for all $x, \bar{x} \in R^n$, and $z \in \partial f(\bar{x})$,*

$$f(x) \geq f(\bar{x}) + z'(x - \bar{x}) + \tfrac{1}{2}\alpha|x - \bar{x}|^2. \tag{8}$$

In the remainder of this chapter every result for Algorithm A *assumes* (A1)–(A4), *while every result for Algorithm* B *assumes* (A1)–(A4) *and* (A5). The following proposition shows that Step 1 in both Algorithms A and B can be carried out. For the proof of the proposition, we shall need the notion of a direction of recession. Let $h: R^n \to (-\infty, +\infty]$ be a closed, proper convex function. A vector $z \in R^n$, $z \neq 0$, determines a direction in R^n, namely the direction of the ray emanating from the origin and passing through z. For $x \in R^n$ such that $h(x) < \infty$, the one-dimensional function $\eta(t) = h(x + tz)$, $t \in R$, is a cross section of h along the direction z passing through x. The direction z is a called a *direction of recession* of h if $\eta(t)$ is nonincreasing over the entire real line. It can be shown that *the set of minimizing points of h (i.e., the set $\{\bar{x} \mid h(\bar{x}) = \inf_{x \in R^n} h(x)\}$) is nonempty and compact if and only if h has no directions of recession* (Rockafellar, 1970, Corollary 8.7.1, Theorem 27.1d). Another relevant fact is that if for some $\bar{\alpha} \in R$ the level set $\{x \mid h(x) \leq \bar{\alpha}\}$ is nonempty and compact then all level sets $\{x \mid h(x) \leq \alpha\}$, $\alpha \in R$, are compact. The *recession function of h* denoted $h0^+$ is defined by

$$h0^+(z) = \lim_{t \to \infty} \frac{h(x + tz) - h(x)}{t} = \lim_{t \downarrow 0} th\left(\frac{z}{t}\right) \qquad \forall\, z \in R^n, \tag{9}$$

where x is *any* vector such that $h(x) < \infty$ (see Rockafeller, 1970, Theorem 8.5, Corollary 8.5.2). Thus $h0^+(z)$ does not depend on x as long as $h(x) < \infty$. It may be shown that the direction $z \neq 0$ is a direction of recession of h at *every* x for which $h(x) < \infty$ if and only if $h0^+(z) \leq 0$. [In fact this statement constitutes an equivalent definition of a direction of recession (see Rockafellar, 1970).] Part of Assumption (A4) is that (CPP) *has a nonempty and compact solution set, which is equivalent to assuming that the functions $g_1, \ldots, g_r$ and the function $\bar{f}$ given by*

$$\bar{f}(x) = \begin{cases} f(x) & \text{if } x \in X, \\ \infty & \text{if } x \notin X \end{cases} \tag{10}$$

have no common direction of recession. This can be seen from the fact that for any collection $h_1, \ldots, h_m$ of closed proper convex functions for which the sum $(h_1 + \cdots + h_m)$ is not identically $+\infty$, we have

$$(h_1 + \cdots + h_m)0^+ = h_1 0^+ + \cdots + h_m 0^+, \tag{11}$$

(see Rockafellar, 1970, Theorem 9.3). Our earlier assertion follows if we apply this formula to the function $\bar{f}$ and the functions $\delta_1, \ldots, \delta_r$ given by

$$\delta_j(x) = \begin{cases} 0 & \text{if } g_j(x) \le 0 \\ \infty & \text{if } g_j(x) > 0, \end{cases} \qquad j = 1, \ldots, r. \tag{12}$$

Proposition 5.7: (a) Let $p \in P_{\mathrm{I}}$ and $\mu \ge 0$ or $p \in \hat{P}_{\mathrm{I}}$ and $\mu > 0$. Under (A1)–(A4), the set of vectors minimizing $L_c(\cdot, \mu)$ over X is nonempty and compact for every $c > 0$. If in addition (A5) holds, then this set consists of a single vector.

(b) For each k in Algorithm B, if $\eta_k > 0$, μ_k is not a Lagrange multiplier, $\{z_i\}$ is a sequence converging to the unique minimizing point of $\bar{L}_{c_k}(\cdot, \mu_k)$, and $\Delta_x \bar{L}_{c_k}(z_i, \mu_k) \to 0$, then there exists a vector $x_k \in \{z_1, z_2, \ldots\}$ satisfying the stopping criterion (6).

Proof: (a) We shall show that for every $c > 0$ the function $\bar{L}_c(\cdot, \mu)$ given by (4) has no direction of recession. In view of the earlier discussion, this implies that the set of minimizing points of $L_c(\cdot, \mu)$ over X is nonempty and compact. We need to compute the recession function of $\bar{L}_c(\cdot, \mu)$. We have, using (10) and (11),

$$\bar{L}_c 0^+(z, \mu) = \bar{f}0^+(z) + \frac{1}{c} \sum_{j=1}^{r} h_j 0^+(z), \tag{13}$$

where h_j is given by

$$h_j(x) = p[cg_j(x); \mu_j], \qquad j = 1, \ldots, r.$$

Using (9), we have

$$h_j 0^+(z) = \lim_{t \to \infty} \frac{p[cg_j(x + tz); \mu_j] - p[cg_j(x); \mu_j]}{t}. \tag{14}$$

Suppose z is a direction of recession of g_j. Then

$$g_j(x + tz) \le g_j(x) \qquad \forall\, t \ge 0.$$

Using the properties of p, we have, for all $t \ge 0$

$$-\infty < \inf_u p(cu; \mu_j) \le p[cg_j(x + tz); \mu_j] \le p[cg_j(x); \mu_j].$$

It follows that the limit in (14) is zero. Now suppose z is not a direction of recession of g_j. By (9),

$$h_j 0^+(z) = \lim_{t \downarrow 0} tp\left[cg_j\left(\frac{z}{t}\right); \mu_j\right] = \lim_{t \downarrow 0} tp\left[ctg_j\left(\frac{z}{t}\right)\Big/ t; \mu_j\right]. \tag{15}$$

Since z is not a direction of recession of g_j we have $g_j0^+(z) = \lim_{t\downarrow 0} tg_j(z/t) > 0$, so there exist $\bar{t} > 0$ and $\bar{\alpha} > 0$ such that $tg_j(z/t) \geq \bar{\alpha}$ for all $t \in (0, \bar{t}]$. Then, from (15),

$$h_j0^+(z) \geq \lim_{t\downarrow 0} tp(c\bar{\alpha}/t;\, \mu_j) = p0^+(c\bar{\alpha};\, \mu_j).$$

Since $\lim_{u\to\infty} p(cu;\, \mu_j) = \infty$, it follows that $p0^+(c\bar{\alpha};\, \mu_j) = \infty$, so $h_j0^+(z) = \infty$. We have thus shown than

$$h_j0^+(z) = \begin{cases} 0 & \text{if} \quad z \text{ is a direction of recession of } g_j, \\ \infty & \text{if} \quad z \text{ is not a direction of recession of } g_j. \end{cases}$$

Using (13), we have

$$\bar{L}_c0^+(z, \mu) = \begin{cases} \bar{f}0^+(z) & \text{if} \quad z \text{ is a direction of recession of each } g_j, \\ & \qquad j = 1, \ldots, r, \\ \infty & \text{otherwise,} \end{cases}$$

Thus, $\bar{L}_c0^+(z, \mu) > 0$ for all $z \neq 0$ [equivalently, the set of minimizing points of $\bar{L}_c(\cdot, \mu)$ is nonempty and compact] if and only if $\bar{f}$ and $g_1, \ldots, g_r$ have no common direction of recession. As stated earlier, this is equivalent to (CPP) having a nonempty and compact solution set. This proves part (a) except for the last assertion.

If (A5) holds, then it is easily seen that we have, for all $x, \bar{x} \in R^n$, $c > 0$, and $z \in \partial_x \bar{L}_c(\bar{x}, \mu)$,

$$\bar{L}_c(x, \mu) \geq \bar{L}_c(\bar{x}, \mu) + z'(x - \bar{x}) + \tfrac{1}{2}\alpha|x - \bar{x}|^2.$$

If $\bar{x}$ minimizes $\bar{L}_c(\cdot, \mu)$ then $0 \in \partial_x \bar{L}_c(\bar{x}, \mu)$, so by taking $z = 0$ in the preceding relation, we obtain

$$\bar{L}_c(x, \mu) \geq \bar{L}_c(\bar{x}, \mu) + \tfrac{1}{2}\alpha|x - \bar{x}|^2 \qquad \forall\, x \in R^n,$$

from which the uniqueness of the minimizing point follows.

(b) If μ_k is not a Lagrange multiplier and $\bar{z}$ is the minimizing point of $\bar{L}_{c_k}(\cdot, \mu_k)$ (and also the limit of $\{z_i\}$), then we cannot have both $g_j(\bar{z}) \leq 0$ for all j and $\sum_{j=1}^r g_j(\bar{z})\mu_k^j = 0$. By using Proposition 5.1(b) and (d), we obtain

$$\lim_{i\to\infty} \sum_{j=1}^r \left\{\nabla_t p[c_k g_j(z_i);\, \mu_k^j] g_j(z_i) - \frac{1}{c_k} p[c_k g_j(z_i);\, \mu_k^j]\right\}$$

$$= \sum_{j=1}^r \left\{\nabla_t p[c_k g_j(\bar{z});\, \mu_k^j] g_i(\bar{z}) - \frac{1}{c_k} p[c_k g_j(\bar{z});\, \mu^j]\right\} > 0.$$

Since $\Delta_x \bar{L}_{c_k}(z_i, \mu_k) \to 0$ and $\eta_k > 0$, we obtain that the stopping criterion (6) will be satisfied for sufficiently large i. Q.E.D.

We turn now to proving that the vectors μ_k generated by Algorithms A and B eventually ascend the ordinary dual functional d. This fact leads to the interpretation of Algorithms A and B as primal–dual methods.

Proposition 5.8. If $\{(x_k, \mu_k)\}$ is a sequence generated by Algorithm A or B then:

(a) For Algorithm A, we have

$$d(\mu_k) \le d_{c_k}(\mu_k) \le d(\mu_{k+1}) \qquad \forall\, k = 0, 1, \ldots \tag{16}$$

with strict inequality throughout if μ_k is not a Lagrange multiplier.

(b) For Algorithm B and all k such that $\eta_k < 2\alpha$, we have

$$d(\mu_k) \le L_{c_k}(x_k, \mu_k) \le d(\mu_{k+1}) \tag{17}$$

with strict inequality throughout if μ_k is not a Lagrange multiplier.

Proof: We will prove part (b). A similar (in fact simpler) argument proves also part (a). In view of Eq. (7) defining μ_{k+1}, it is easy to show that

$$\Delta_x \bar{L}_{c_k}(x_k, \mu_k) = \Delta_x \bar{L}(x_k, \mu_{k+1}) \tag{18}$$

where

$$\bar{L}(x, \mu) = \begin{cases} L(x, \mu) & \text{if} \quad x \in X, \\ \infty & \text{if} \quad x \notin X. \end{cases}$$

From (A5) we have, for all $x \in R^n$ and $z \in \partial_x \bar{L}(x_k, \mu_{k+1})$,

$$\bar{L}(x, \mu_{k+1}) \ge \bar{L}(x_k, \mu_{k+1}) + z'(x - x_k) + \tfrac{1}{2}\alpha|x - x_k|^2,$$

from which by taking infima with respect to x, we find that

$$d(\mu_{k+1}) \ge \bar{L}(x_k, \mu_{k+1}) - (1/2\alpha)|\Delta_x \bar{L}(x_k, \mu_{k+1})|^2. \tag{19}$$

The stopping rule (6) can also be written as

$$|\Delta_x \bar{L}_{c_k}(x_k, \mu_k)|^2 \le \eta_k\{L(x_k, \mu_{k+1}) - L_{c_k}(x_k, \mu_k)\}. \tag{20}$$

Combining (18)–(20), we obtain

$$\begin{aligned} \bar{L}(x_k, \mu_{k+1}) - d(\mu_{k+1}) &\le \frac{1}{2\alpha}|\Delta_x \bar{L}(x_k, \mu_{k+1})|^2 \\ &= \frac{1}{2\alpha}|\Delta_x \bar{L}_{c_k}(x_k, \mu_k)|^2 \\ &\le \frac{\eta_k}{2\alpha}\{L(x_k, \mu_{k+1}) - L_{c_k}(x_k, \mu_k)\} \\ &\le L(x_k, \mu_{k+1}) - L_{c_k}(x_k, \mu_k). \end{aligned} \tag{21}$$

Since $\bar{L}(x_k, \mu_{k+1}) = L(x_k, \mu_{k+1})$, we obtain

$$L_{c_k}(x_k, \mu_k) \le d(\mu_{k+1}).$$

We also have, by the definition of d and the properties of p,

$$d(\mu_k) \le L(x_k, \mu_k) \le L_{c_k}(x_k, \mu_k), \tag{22}$$

so the proof of (17) is complete. If μ_k is not a Lagrange multiplier, then Proposition 5.1 (b) and (d) and the definition of μ_{k+1} imply that $L(x_k, \mu_{k+1}) > L_{c_k}(x_k, \mu_k)$, so in view of $\eta_k < 2\alpha$, the last inequality in (21) is strict. Similarly, the right inequality in (22) is strict, so strict inequality is obtained throughout in (17). Q.E.D.

Corollary 5.9: A sequence $\{\mu_k\}$ generated by Algorithm A or B is bounded.

Proof: Since $\eta_k \to 0$, there exists an index $\bar{k}$ such that $\eta_k < 2\alpha$ for all $k \ge \bar{k}$. By Proposition 5.8, for all $k \ge \bar{k} + 1$, μ_k belongs to the level set $\{\mu \mid d(\mu) \ge d(\mu_{\bar{k}+1})\}$ and $d(\mu_{\bar{k}+1}) \ge L_{c_{\bar{k}}}(x_{\bar{k}}, \mu_{\bar{k}}) > -\infty$. Since the set of maximizing points of d [i.e., the set of Lagrange multipliers of (CPP)] is compact by (A4), the same is true for all level sets of the form $\{\mu \mid d(\mu) \ge \beta\}$ where $\beta > -\infty$ (Rockafellar, 1970, Corollary 8.7.1). Thus, the level set $\{\mu \mid d(\mu) \ge d(\mu_{\bar{k}+1})\}$ is compact, and it follows that $\{\mu_k\}$ is bounded. Q.E.D.

We continue the convergence analysis by first considering the case where $p \in P_1$.

Proposition 5.10: Let $p \in P_1$. A sequence $\{x_k\}$ generated by Algorithm A or B is bounded.

Proof: Let $\{\mu_k\}$ be the corresponding multiplier sequence generated by the algorithm. By Corollary 5.9, $\{\mu_k\}$ is bounded, so there exists $M > 0$ such that $0 \le \mu_k^j \le M$ for all k and j. Using the properties of p, we have

$$\frac{1}{c_k} p(c_k t; \mu_k^j) \ge \frac{1}{c_k} p(c_k t; 0) \ge \frac{1}{c_0} p(c_0 t; 0) \qquad \forall\, t \ge 0,$$

$$\frac{1}{c_k} p(c_k t; \mu_k^j) \ge \frac{1}{c_k} p(c_k t; M) \ge \frac{1}{c_0} p(c_0 t; M) \ge \frac{1}{c_0} \inf_\tau p(\tau; M) \qquad \forall\, t < 0.$$

Using these inequalities and the fact that $p(t; 0) = 0$ for all $t < 0$ and $\inf_\tau p(\tau; M) \le 0$, we have, for all j and k,

$$\frac{1}{c_k} p(c_k t; \mu_k^j) \ge \frac{1}{c_0} \{p(c_0 t; 0) + \inf_\tau p(\tau; M)\} \qquad \forall\, t \in R.$$

Hence

$$\bar{L}_{c_k}(x, \mu_k) \geq \bar{f}(x) + \frac{1}{c_0} \sum_{j=1}^{r} \left\{ p[c_0 g_j(x); 0] + \inf_\tau p(\tau; M) \right\} \tag{23}$$

$$= \bar{L}_{c_0}(x, 0) + \frac{r}{c_0} \inf_\tau p(\tau; M) \qquad \forall\, x \in R^n.$$

Now the function $\bar{L}_{c_0}(x, 0)$ has no direction of recession, as shown in the proof of Proposition 5.7, and hence it has bounded level sets. By Proposition 5.8, we have that, for k sufficiently large, $\bar{L}_{c_k}(x_k, \mu_k) = L_{c_k}(x_k, \mu_k) \leq d(\mu_{k+1}) \leq f^*$. Using (23), it follows that x_k belongs to the level set

$$\left\{ x \mid \bar{L}_{c_0}(x, 0) \leq f^* - \frac{r}{c_0} \inf_\tau p(\tau; M) \right\}.$$

Hence $\{x_k\}$ is bounded. Q.E.D.

The following is the main convergence result for Algorithms A and B and for p chosen within the class P_{I}.

Proposition 5.11: Let $p \in P_{\mathrm{I}}$. Every limit point of a sequence $\{(x_k, \mu_k)\}$ generated by Algorithm A or B is an optimal solution–Lagrange multiplier pair of (CPP). Furthermore at least one such limit point exists and we have $\lim_{k \to \infty} f(x_k) = \lim_{k \to \infty} d(\mu_k) = f^*$.

Proof: Let $(\bar{x}, \bar{\mu})$ be a limit point of a subsequence $\{(x_k, \mu_k)\}_K$. We first show that $\bar{x}$ is feasible. Indeed since X is closed and $x_k \in X$ for all k, we have $\bar{x} \in X$, so, if $\bar{x}$ is infeasible, there must exist $j \in \{1, \ldots, r\}$, $\delta > 0$, and an index $\bar{k}$ such that $g_j(x_k) \geq \delta$ for all $k \in K$ with $k \geq \bar{k}$. For such k, we have, by Proposition 5.1c,

$$\begin{aligned} L_{c_k}(x_k, \mu_k) - L(x_k, \mu_k) &\geq c_k^{-1} p[c_k g_j(x_k); \mu_k^j] - \mu_k^j g_j(x_k) \\ &\geq c_k^{-1} p[c_k g_j(x_k); 0] \geq c_0^{-1} p[c_0 \delta; 0] > 0. \end{aligned} \tag{24}$$

We may assume without loss of generality that $\eta_k < 2\alpha$ for all $k \geq \bar{k}$. Therefore by Proposition 5.8b we have $d(\mu_k) \leq d(\mu_{k+1}) \leq f^*$ for all $k \geq \bar{k}$. It follows that we must have $\{d(\mu_{k+1}) - d(\mu_k)\} \to 0$, so (17) and (22) imply

$$\{L_{c_k}(x_k, \mu_k) - L(x_k, \mu_k)\} \to 0. \tag{25}$$

This contradicts (24), and therefore $\bar{x}$ is feasible.

Using Proposition 5.1b, we have, for all j,

$$c_k^{-1} p[c_k g_j(x_k); \mu_k^j] \geq c_0^{-1} p[c_0 g_j(x_k); \mu_k^j] \geq \mu_k^j g_j(x_k),$$

so (25) implies

$$c_0^{-1} p[c_0 g_j(\bar{x}); \bar{\mu}^j] = \bar{\mu}^j g_j(\bar{x}) \qquad \forall\, j = 1, \ldots, r.$$

Using Proposition 5.1d, we obtain $\bar{\mu}^j g_j(\bar{x}) = 0$ for all j. Hence we have, by using (17) and (22),

$$\max_{\mu} d(\mu) \geq \lim_{k\to\infty} d(\mu_k) = \lim_{k\to\infty} L(x_k, \mu_k) = f(\bar{x}) + \sum_{j=1}^{r} \bar{\mu}^j g_i(\bar{x}) = f(\bar{x}).$$

Since $\bar{x}$ is feasible, we also have $f(\bar{x}) \geq f^* = \max_\mu d(\mu)$, so it follows that $f(\bar{x}) = f^*$, $\bar{x}$ is optimal, and $\lim_{k\to\infty} d(\mu_k) = d(\bar{\mu}) = f^*$. The existence of least one limit point follows from the boundedness of $\{(x_k, \mu_k)\}$ (compare with Corollary 5.9 and Proposition 5.10). Q.E.D.

The preceding proposition establishes that first-order multiplier iterations based on penalty functions from the class P_{I} have satisfactory convergence properties. Unfortunately when the penalty function p is chosen from the class $\hat{P}_{\mathrm{I}}$, the convergence results available are not as powerful. The main reason is that $p(t; 0)$ is zero for $t > 0$, and this affects materially the proofs of Propositions 5.10 and 5.11. The author's extensive computational experience with penalty functions in $\hat{P}_{\mathrm{I}}$ (particularly the exponential penalty function) suggests, however, that their convergence properties are as good in practice as those of penalty functions in the class P_{I}. The following analysis supports the validity of this observation.

Let S be the set of all subsets of $\{1, \ldots, r\}$. For any index set $J \in S$ consider the function d_J defined by

$$d_J(\mu) = \inf_{x\in X} \left\{ f(x) + \sum_{j\in J} \mu^j g_j(x) \right\} \qquad \forall\, \mu \geq 0, \tag{26}$$

and $d_J(\mu) = -\infty$ if $\mu^j < 0$ for some $j \in J$. Clearly d_J is the dual functional corresponding to the problem

$$(\mathrm{CPP})_J \qquad \text{minimize} \quad f(x)$$
$$\text{subject to} \quad x \in X, \qquad g_j(x) \leq 0, \qquad j \in J.$$

This problem is the same as (CPP) except that the inequality constraints $g_j(x) \leq 0, j \notin J$, have been eliminated. The corresponding dual optimal value is

$$d_J^* = \sup_{\mu \geq 0} d_J(\mu). \tag{27}$$

It is easily seen from (26) that we have, for all $J_1, J_2 \in S$,

$$d_{J_1}^* \leq d_{J_2}^* \qquad \text{if} \quad J_1 \subset J_2.$$

In particular,

$$d_J^* \leq \sup_{\mu\geq 0} d(\mu) = f^*. \tag{28}$$

Define

$$\tilde{S} = \{J \in S \mid -\infty < d_J^* < f^*\}. \tag{29}$$

We shall prove the convergence of Algorithms A and B for the case where $p \in \hat{P}_{\mathrm{I}}$ under the following assumption.

Assumption (A6) (For Penalty Functions $p \in \hat{P}_{\mathrm{I}}$): *For each $J \in \tilde{S}$, and $j \in J$ there do not exist any two vectors $\bar{\mu} \geq 0$ and $\tilde{\mu} \geq 0$ maximizing d_J and such that $\bar{\mu}^j = 0$ and $\tilde{\mu}^j > 0$.*

Note that (A6) is not a very restrictive assumption. It is satisfied in particular if each of the problems $(\mathrm{CPP})_J$, $J \in \tilde{S}$, has a unique Lagrange multiplier vector. We do not know whether it is possible to relax this assumption and still be able to prove the result of the following proposition.

Proposition 5.12: Let $p \in \hat{P}_{\mathrm{I}}$ and assume that (A6) holds. Then every limit point of a sequence $\{\mu_k\}$ generated by Algorithm A or B is a Lagrange multiplier of (CPP). Furthermore at least one such limit point exists and we have $\lim_{k\to\infty} f(x_k) = \lim_{k\to\infty} d(\mu_k) = f^*$.

Proof: We have, for all k,

$$\begin{aligned} L_{c_k}(x_k, \mu_k) - L(x_k, \mu_k) &= \sum_{j=1}^{r} \left\{\frac{1}{c_k} p[c_k g_j(x_k); \mu_k^j] - \mu_k^j g_j(x_k)\right\} \\ &\geq \sum_{j=1}^{r} \left\{\frac{1}{c_0} p[c_0 g_j(x_k); \mu_k^j] - \mu_k^j g_j(x_k)\right\} \\ &= \frac{1}{c_0} \sum_{j=1}^{r} \mu_k^j \{\psi[c_0 g_j(x_k)] - c_0 g_j(x_k)\}. \end{aligned} \tag{30}$$

Similarly, as in the proof of Proposition 5.11 [compare with (25)], we have, for some $\bar{k}$,

$$d(\mu_k) \leq L(x_k, \mu_k) \leq L_{c_k}(x_k, \mu_k) \leq d(\mu_{k+1}) \leq f^* \qquad \forall\, k \geq \bar{k}, \tag{31}$$

$$\{L_{c_k}(x_k, \mu_k) - L(x_k, \mu_k)\} \to 0. \tag{32}$$

Since for all k and j, we have $\mu_k^j > 0$ and $\psi[c_0 g_j(x_k)] - c_0 g_j(x_k) \geq 0$ [compare with Proposition 5.1b], we obtain, from (30) and (32),

$$\mu_k^j \{\psi[c_0 g_j(x_k)] - c_0 g_j(x_k)\} \to 0 \qquad \forall\, j = 1, \ldots, r. \tag{33}$$

Using the properties of ψ, it is easy to see that (33) implies

$$\mu_k^j g_j(x_k) \to 0 \qquad \forall\, j = 1, \ldots, r. \tag{34}$$

Combining (31) and (34), we obtain

$$\lim_{k\to\infty} d(\mu_k) = \lim_{k\to\infty} L(x_k, \mu_k) = \lim_{k\to\infty} f(x_k) \leq f^*. \tag{35}$$

If $\lim_{k\to\infty} f(x_k) = f^*$, we are done, so in the remainder of the proof we assume that $\lim_{k\to\infty} f(x_k) < f^*$.

Since, by Corollary 5.9, $\{\mu_k\}$ is bounded, it has at least one limit point. We shall first show that in order to prove the proposition it is sufficient to prove the following statement (S).

(S) If for some limit point $\bar{\mu}$ of $\{\mu_k\}$ and index j we have $\bar{\mu}^j = 0$, then $\lim_{k\to\infty} \mu_k^j = 0$.

Indeed if (S) holds, then we can extract a subsequence $\{\mu_k\}_K$ converging to some vector $\bar{\mu}$ such that, for all j,

$$\bar{\mu}^j = 0 \Rightarrow \limsup_{\substack{k\to\infty \\ k\in K}} g_j(x_k) \le 0, \tag{36}$$

$$\bar{\mu}^j > 0 \Rightarrow \lim_{\substack{k\to\infty \\ k\in K}} g_j(x_k) = 0. \tag{37}$$

To see this, note that if $\tilde{\mu}^j = 0$ for some limit point $\tilde{\mu}$ of $\{\mu_k\}$, then by (S), $\lim_{k\to\infty} \mu_k^j = 0$. Thus, we have $\mu_k^j > \mu_{k+1}^j = \mu_k^j \nabla\psi[c_k g_j(x_k)]$ for an infinite number of indices k. This implies that $\nabla\psi[c_k g_j(x_k)] < 1$ or equivalently, $g_j(x_k) < 0$ for an infinite number of indices, so a subsequence $\{\mu_k\}_K$ convergent to some vector $\bar{\mu}$ can be chosen satisfying (36). Relation (37) must also be satisfied in view of (34). Now consider, for all $k \in K$, the vector u_k with coordinates

$$u_k^j = \max\{0, g_j(x_k)\} \qquad \forall\, j = 1, \ldots, r. \tag{38}$$

We then have

$$q(u_k) = \inf\{f(x) \mid x \in X, g_j(x) \le u_k^j, j = 1, \ldots, r\} \le f(x_k),$$

where q is the primal functional. From (36)–(38) we have $\{u_k\}_K \to 0$, and since q is lower semicontinuous [in view of the assumption that the set of optimal solution of (CPP) is nonempty and compact—see the discussion in Section 5.2], we have

$$f^* = q(0) \le \liminf_{\substack{k\to\infty \\ k\in K}} q(u_k) \le \lim_{\substack{k\to\infty \\ k\in K}} f(x_k).$$

Combining this relation with (35), we obtain $\lim_{k\to\infty} d(\mu_k) = \lim_{k\to\infty} f(x_k) = f^*$. Taking into account (36) and (37), we conclude that the proposition will be proved if we can show statement (S).

To prove (S), we argue by contradiction. Suppose there exists an index $\bar{j}$ and two subsequences $\{\mu_k\}_{\bar{K}}$ and $\{\mu_k\}_{\tilde{K}}$ converging to $\bar{\mu}$ and $\tilde{\mu}$, respectively, such that

$$\bar{\mu}^{\bar{j}} = 0, \qquad \tilde{\mu}^{\bar{j}} > 0.$$

Consider the index sets

$$J = \left\{ j \,\middle|\, \limsup_{\substack{k\to\infty \\ k\in K}} g_j(x_k) \le 0,\ \limsup_{\substack{k\to\infty \\ k\in \tilde{K}}} g_j(x_k) \le 0 \right\} \qquad J^+ = J \cup \{\bar{j}\}$$

In view of (34), we have

$$\lim_{\substack{k\to\infty \\ k\in \tilde{K}}} g_{\bar{j}}(x_k) = 0,$$

and hence,

$$\limsup_{\substack{k\to\infty \\ k\in \tilde{K}}} g_j(x_k) \le 0 \qquad \forall\, j \in J^+.$$

By using an argument similar to the one given earlier starting with (38), we have

$$d_J(\tilde{\mu}) = d_J(\bar{\mu}) = \max_{\mu\ge 0} d_J(\mu) = d_J^* = \lim_{k\to\infty} f(x_k),$$

$$d_{J^+}(\tilde{\mu}) = \max_{\mu\ge 0} d_{J^+}(\mu) = d_{J^+}^* = \lim_{k\to\infty} f(x_k).$$

Thus $d_J^* = d_{J^+}^*$ and since $\bar{\mu}^{\bar{j}} = 0$, we also have

$$d_{J^+}^* = d_J^* = d_J(\bar{\mu}) = \inf_{x\in X} \left\{ f(x) + \sum_{j\in J} \bar{\mu}^j g_j(x) \right\}$$

$$= \inf_{x\in X} \left\{ f(x) + \sum_{j\in J^+} \bar{\mu}^j g_j(x) \right\} = d_{J^+}(\bar{\mu}).$$

So both $\bar{\mu}$ and $\tilde{\mu}$ maximize d_{J^+} over $\mu \ge 0$, while for the index $\bar{j} \in J^+$ we have $\bar{\mu}^{\bar{j}} = 0$ and $\tilde{\mu}^{\bar{j}} > 0$. By our earlier assumption, $\lim_{k\to\infty} f(x_k) < f^*$. So $J^+ \in \tilde{S}$, and we obtain a contradiction of (A6). Thus we have proved (S), and by the earlier discussion, the proof of the proposition is complete. Q.E.D.

Special Results for the Quadratic Penalty Function

The quadratic penalty function is given by

$$p(t_j; \mu_j) = \begin{cases} \mu_j t_j + \frac{1}{2} t_j^2 & \text{if } \mu_j + t_j \ge 0, \\ -\frac{1}{2}\mu_j^2 & \text{if } \mu_j + t_j < 0. \end{cases} \tag{39}$$

It belongs to P_E^+ and a fortiori to P_I, and it has already been considered in Chapter 3. The conjugate convex function of $\phi(t) = \frac{1}{2}t^2$ is $\phi^*(y) = \frac{1}{2}y^2$ as the reader can easily verify. Thus we have, by using Proposition 5.5,

$$P_c^*(s; \mu) = \frac{1}{2c} \sum_{j=1}^{r} |s_j - \mu_j|^2 = \frac{1}{2c} |s - \mu|^2 \qquad \forall\, s \ge 0, \tag{40}$$

and

(41) $$d_c(\mu) = \max_{s \in R^r}\left\{d(s) - \frac{1}{2c}|s - \mu|^2\right\}.$$

If $s(\mu, c)$ is the unique maximizing point in the equation above we have from Proposition 5.5

$$\nabla d_c(\mu) = c^{-1}[s(\mu, c) - \mu], \qquad s(\mu, c) = \nabla_z P_c[g[x(\mu, c)]; \mu].$$

where $x(\mu, c)$ is any vector minimizing $L_c(\cdot, \mu)$ over X. Using (2) and the relations above, we obtain, for any sequence $\{\mu_k\}$ generated by Algorithm A and all k,

(42) $$\mu_{k+1} = s(\mu_k, c_k),$$

(43) $$\nabla d_{c_k}(\mu_k) = c_k^{-1}(\mu_{k+1} - \mu_k).$$

Thus

$$\mu_{k+1} = \mu_k + c_k \nabla d_{c_k}(\mu_k),$$

and *each iteration of Algorithm* A *may be viewed as a fixed stepsize steepest ascent iteration aimed at maximizing* d_{c_k}.

For any $\bar{\mu} \in R^r$ consider the quadratic function

$$h(\mu) = d[s(\bar{\mu}, c)] - (1/2c)|s(\bar{\mu}, c) - \mu|^2.$$

It satisfies

(44) $$d_c(\bar{\mu}) = h(\bar{\mu})$$

and

(45) $$d_c(\mu) \geq h(\mu) \qquad \forall\, \mu \in R^r.$$

These two properties imply that

(46) $$\nabla d_c(\bar{\mu}) = \nabla h(\bar{\mu}).$$

Since $h(\mu)$ is quadratic with Hessian $-c^{-1}I$, we have

(47) $$h(\mu) = h(\bar{\mu}) + \nabla h(\bar{\mu})'(\mu - \bar{\mu}) - (1/2c)|\mu - \bar{\mu}|^2.$$

By combining (44)–(47), we obtain

(48) $$d_c(\mu) \geq d_c(\bar{\mu}) + \nabla d_c(\bar{\mu})'(\mu - \bar{\mu}) - (1/2c)|\mu - \bar{\mu}|^2 \qquad \forall\, \mu, \bar{\mu} \in R^r.$$

This relation yields a short proof of convergence of the generalized multiplier iteration

(49) $$\mu_{k+1} = \mu_k + \alpha_k \nabla d_c(\mu_k),$$

where α_k is a stepsize satisfying

(50) $$\delta c \leq \alpha_k \leq (2 - \delta)c$$

and δ is any scalar with $0 < \delta \leq 1$ (compare with the analysis of Section 2.3.1).

Proposition 5.13: Let p be the quadratic penalty function (39). Then the sequence $\{\mu_k\}$ generated by iteration (49) is bounded and each of its limit points is a Lagrange multiplier for (CPP).

Proof: From (48)–(50) we have, for all k,

$$
\begin{aligned}
(51) \qquad d_c(\mu_{k+1}) &\geq d_c(\mu_k) + \alpha_k |\nabla d_c(\mu_k)|^2 - (\alpha_k^2/2c)|\nabla d_c(\mu_k)|^2 \\
&= d_c(\mu_k) + [\alpha_k(2c - \alpha_k)/2c]|\nabla d_c(\mu_k)|^2 \\
&\geq d_c(\mu_k) + \tfrac{1}{2}c\delta^2 |\nabla d_c(\mu_k)|^2.
\end{aligned}
$$

Hence $d_c(\mu_{k+1}) \geq d_c(\mu_k)$ for all k, and $\{\mu_k\}$ belongs to the compact set $\{\mu | d_c(\mu) \geq d_c(\mu_0)\}$. Thus $\{\mu_k\}$ is bounded. Since $f^* \geq d_c(\mu_{k+1})$, from (51) we also obtain $|\nabla d_c(\mu_k)| \to 0$ so all limit points of $\{\mu_k\}$ maximize d_c and are thus Lagrange multipliers for (CPP). Q.E.D.

There is also an analog of Proposition 5.13 for an algorithm involving inexact minimization of the augmented Lagrangian which can be found in Bertsekas (1975a).

Another interesting fact regarding the quadratic penalty function is that for sequences $\{\mu_k\}$ generated by the corresponding Algorithms A and B we can assert that they converge to a unique limit point [even though (CPP) may have more than one Lagrange multiplier]. Actually such a statement can be made under other conditions—for example, each time we can assert that for some $q > 0$ and $\beta \in (0, 1)$, we have

$$(52) \qquad |\mu_{k+1} - \mu_k| \leq q\beta^k.$$

If (52) is satisfied, then $\{\mu_k\}$ can be easily shown to be a Cauchy sequence and therefore must converge to a unique limit. Conditions under which (52) is satisfied will be derived in the next section (compare with Propositions 5.22 and 5.24 and Lemma 5.17). It is quite interesting however that these conditions are not necessary when the penalty function is quadratic.

Propostion 5.14: Let p be the quadratic penalty function (39). Then a sequence $\{\mu_k\}$ generated by Algorithm A converges to a Lagrange multiplier of (CPP).

Proof: It was shown earlier [compare with (42)] that μ_{k+1} equals the unique maximizing point $s(\mu_k, c_k)$ in (41). Thus we have

$$d(\mu_{k+1}) - (1/2c)|\mu_{k+1} - \mu_k|^2 \geq d(s) - (1/2c)|s - \mu_k|^2 \qquad \forall\, s \in R^r.$$

For any s such that $d(s) \geq d(\mu_{k+1})$, the relation above yields

$$|\mu_{k+1} - \mu_k|^2 \leq |s - \mu_k|^2,$$

so μ_{k+1} is the projection of μ_k on the level set $\{\mu | d(\mu) \geq d(\mu_{k+1})\}$, and we have

$$(\mu_{k+1} - \mu_k)'(s - \mu_{k+1}) \geq 0 \qquad \forall\, s \in \{\mu | d(\mu) \geq d(\mu_{k+1})\}.$$

This yields

$$\begin{aligned} |s - \mu_k|^2 &= |s - \mu_{k+1}|^2 + 2(s - \mu_{k+1})'(\mu_{k+1} - \mu_k) + |\mu_{k+1} - \mu_k|^2 \\ &\geq |s - \mu_{k+1}|^2 \qquad \forall\, s \in \{\mu | d(\mu) \geq d(\mu_{k+1})\}. \end{aligned}$$

In particular, for every Lagrange multiplier $\bar{\mu}$, we have

$$|\mu_{k+1} - \bar{\mu}| \leq |\mu_k - \bar{\mu}| \qquad \forall\, k = 0, 1, \ldots. \tag{53}$$

By Proposition 5.11, every limit point of $\{\mu_k\}$ is a Lagrange multiplier and at least one limit point exists. It follows that $\{\mu_k\}$ can have at most one limit point, since if $\bar{\mu}$ and $\tilde{\mu}$ were two distinct limit points then we could find indices $\bar{k}$ and $\tilde{k} > \bar{k}$ such that

$$|\mu_{\bar{k}} - \bar{\mu}| < \tfrac{1}{2}|\bar{\mu} - \tilde{\mu}|, \qquad |\mu_{\tilde{k}} - \tilde{\mu}| < \tfrac{1}{2}|\bar{\mu} - \tilde{\mu}|,$$

thereby obtaining

$$|\mu_{\tilde{k}} - \bar{\mu}| \geq |\bar{\mu} - \tilde{\mu}| - |\mu_{\tilde{k}} - \tilde{\mu}| > \tfrac{1}{2}|\bar{\mu} - \tilde{\mu}| > |\mu_{\bar{k}} - \bar{\mu}|$$

and violating (53). Q.E.D.

5.4 Rate of Convergence Analysis

In the rate of convergence analysis of this section *we restrict attention to methods utilizing penalty functions in the class* P_{E}^{+}, so that

$$P_c(z; \mu) = \frac{1}{c} \sum_{j=1}^{r} p(cz_j; \mu_j), \tag{1}$$

where

$$p(t_j; \mu_j) = \begin{cases} \mu_j t_j + \phi(t_j) & \text{if} \quad \mu_j + \nabla\phi(t_j) \geq 0, \\ \min_{\tau \in R} \{\mu_j \tau + \phi(\tau)\} & \text{otherwise} \end{cases} \tag{2}$$

for some penalty function $\phi \in P_{\mathrm{E}}$. There are some convergence rate results available for methods utilizing other penalty functions but these are fragmentary and they will not be presented.

We shall be interested in the rate at which a sequence $\{\mu_k\}$ generated by Algorithm A or B converges to the set M^* of Lagrange multiplier vectors of (CPP). Denote

$$\|\mu - M^*\| = \min_{\mu^* \in M^*} |\mu - \mu^*|. \tag{3}$$

We examine the convergence of $\{\mu_k\}$ to the set M^* in terms of the distance $\|\mu_k - M^*\|$. Note that, from the convergence result of Proposition 5.11, we have

$$\|\mu_k - M^*\| \to 0.$$

We shall make use of the following additional assumptions:

Assumption (A7): *There exist scalars $M_2 \geq M_1 > 0$ and $\rho > 1$ such that for some open interval N_0 containing zero*

$$M_1 |t|^{\rho-1} \leq |\nabla\phi(t)| \leq M_2 |t|^{\rho-1} \qquad \forall\, t \in N_0, \tag{4}$$

where ϕ corresponds to p as in (2).

Assumption (A8): *There exists a δ-neighborhood* $(\delta > 0)$

$$S(M^*; \delta) = \{\mu \,|\, \text{there exists } \mu^* \in M^* \text{ with } |\mu - \mu^*| < \delta\}, \tag{5}$$

a scalar $\gamma > 0$, and a scalar $q \geq 1$ such that the dual functional d satisfies

$$d(\mu) \leq \max_{w \in R^r} d(w) - \gamma\|\mu - M^*\|^q \qquad \forall\, \mu \in S(M^*; \delta). \tag{6}$$

Assumption (A7) may be explained as a growth assumption on ϕ. Roughly speaking, it states that in a neighborhood of zero, $\phi(t)$ behaves like $|t|^\rho$. Similarly (A8) is a growth assumption on the dual functional d. It says that in a neighborhood of the maximum set M^*, $d(\mu)$ grows (downward) at least as fast as $\gamma\|\mu - M^*\|^q$. This assumption is much weaker than regularity assumptions which require d to be twice differentiable with negative definite Hessian at a unique maximum (compare with Section 2.3). In fact (A8) does not require once differentiability of d or even finiteness of d in the entire neighborhood $S(M^*; \delta)$.

We assume throughout this section that (A1)–(A4) *and* (A7) *and* (A8) *hold. In all the results where explicit reference is made to Algorithm* B *we also assume that* (A5) *holds.*

Preliminary Analysis

We first introduce some notation and conventions and subsequently prove a few lemmas which set the stage for the proof of the main propositions.

For each $\mu \in R^r$ we denote by $\hat{\mu}$ the unique projection of μ on M^*; i.e.,

$$\hat{\mu} = \arg\min_{\mu^* \in M^*} |\mu - \mu^*| \qquad \forall\, \mu \in R^r. \tag{7}$$

When considering results relating to Algorithm B, we use the notation

$$v_k = \eta_k/2\alpha, \qquad k = 0, 1, \ldots. \tag{8}$$

To simplify statements of results, *we assume without essential loss of generality (since* $\eta_k \to 0$*) that for some* $\bar{v} > 0$ *we have*

$$0 \le v_k \le \bar{v} < 1 \qquad \forall\, k = 0, 1, \ldots. \tag{9}$$

The results of all subsequent lemmas and propositions, where v_k *and* $\bar{v}$ *appear, hold also with* $v_k \equiv 0$ *for the case of Algorithm* A.

Consider the conjugate convex function $P_c^*(\cdot\,; \mu)$ of the penalty function $P_c(\cdot\,; \mu)$. As shown in Section 5.2 we have, for all $\mu \in R^n$,

$$P_c^*(s; \mu) = \begin{cases} (1/c) \sum_{j=1}^r \phi^*(s_j - \mu_j) & \text{if} \quad s \ge 0 \\ \infty & \text{otherwise.} \end{cases} \tag{10}$$

For a sequence $\{\mu_k\}$ generated by Algorithm A or B, denote by u_k the vector with coordinates

$$u_k^j = (1/c_k)\nabla\phi^*(\mu_{k+1}^j - \mu_k^j), \qquad j = 1, \ldots, r. \tag{11}$$

Note that

$$u_k = \nabla_s P_{c_k}^*(\mu_{k+1}; \mu_k) \qquad \text{if} \quad \mu_{k+1} > 0, \tag{12}$$

and more generally u_k is a subgradient of $P_{c_k}^*(\mu_{k+1}; \mu_k)$ with respect to the first argument. In terms of Fig. 5.3, u_k *can be identified with a support hyperplane to the graphs of* $P_{c_k}^*(\cdot\,; \mu_k) + d_{c_k}(\mu_k)$ *and* $d(\cdot)$ *at the "point of contact" corresponding to* μ_{k+1}. We shall derive an alternative characterization of u_k. We have, for all j,

$$\phi(c_k u_k^j) = \sup_{t \in R}\{c_k u_k^j t - \phi^*(t)\}.$$

In view of (11), it follows that $(\mu_{k+1}^j - \mu_k^j)$ attains the supremum above so

$$\begin{aligned} \phi(c_k u_k^j) &= c_k u_k^j(\mu_{k+1}^j - \mu_k^j) - \phi^*(\mu_{k+1}^j - \mu_k^j) \\ &= c_k u_k^j(\mu_{k+1}^j - \mu_k^j) - \sup_t\{t(\mu_{k+1}^j - \mu_k^j) - \phi(t)\}. \end{aligned}$$

It follows that

$$\phi(c_k u_k^j) \le c_k u_k^j(\mu_{k+1}^j - \mu_k^j) - t(\mu_{k+1}^j - \mu_k^j) + \phi(t) \qquad \forall\, t \in R,$$

or equivalently

$$\phi(t) \ge \phi(c_k u_k^j) + (t - c_k u_k^j)(\mu_{k+1}^j - \mu_k^j) \qquad \forall\, t \in R.$$

Hence $(\mu_{k+1}^j - \mu_k^j)$ is a subgradient of ϕ at $c_k u_k^k$, and since ϕ is differentiable we have

$$\mu_{k+1}^j = \mu_k^j + \nabla\phi(c_k u_k^j), \qquad j = 1, \ldots, r. \tag{13}$$

Now for both Algorithms A and B, we have, as shown earlier [Section 5.1.2, Eq. (10)],

$$\mu_{k+1}^j = \max\{0, \mu_k^j + \nabla\phi[c_k g_j(x_k)]\}, \qquad j = 1, \ldots, r. \tag{14}$$

By comparing (13) and (14) and taking into account that ϕ is strictly convex, it is easily seen that

$$u_k^j = \max\{g_j(x_k), \tau_k^j/c_k\}, \qquad j = 1, \ldots, r, \tag{15}$$

where

$$\tau_k^j = \arg\min_\tau\{\mu_k^j\tau + \phi(\tau)\}, \qquad j = 1, \ldots, r. \tag{16}$$

We now develop some preliminary results through a series of lemmas.

Lemma 5.15: For all $\mu \in R^r$, $c > 0$, and $x \in R^n$, we have

$$L_c(x, \mu) = \max_s\{L(x, s) - P_c^*(s; \mu)\}. \tag{17}$$

Furthermore if $\{(x_k, \mu_k)\}$ is a sequence generated by Algorithm A or B, we have, for all k.

$$L_{c_k}(x_k, \mu_k) = L(x_k, \mu_{k+1}) - P_c^*(\mu_{k+1}; \mu_k), \tag{18}$$

where P_c^* is given by (10).

Proof: In view of (10) and the fact that $P_c^*(\cdot; \mu)$ is the conjugate convex function of $P_c(\cdot; \mu)$, we have

$$L_c(x, \mu) = f(x) + P_c[g(x); \mu] = f(x) + \max_s\{s'g(x) - P_c^*(s; \mu)\} \tag{19}$$

$$= \max_s\{L(x, s) - P_c^*(s; \mu)\},$$

where the maximum is attained by strict convexity of $P_c^*(\cdot; \mu)$.

We have, by definition for all k,

$$\mu_{k+1} = \nabla_z P_{c_k}[g(x_k); \mu_k],$$

so $g(x_k)$ attains the maximum in the equation

$$P_{c_k}^*(\mu_{k+1}; \mu_k) = \sup_z\{\mu_{k+1}'z - P_{c_k}(z; \mu_k)\}.$$

It follows that

$$\mu_{k+1}'g(x_k) - P_{c_k}^*(\mu_{k+1}; \mu_k) = P_{c_k}[g(x_k); \mu_k]$$

$$= \sup_s\{s'g(x_k) - P_{c_k}^*(s; \mu)\}.$$

Therefore μ_{k+1} attains the maximum in (19) when $x = x_k$, $c = c_k$, and $\mu = \mu_k$, and (18) follows. Q.E.D.

Lemma 5.16: Let $\{\mu_k\}$ be a sequence generated by Algorithm A or B. For all k sufficiently large, we have

$$0 \le \gamma \|\mu_{k+1} - M^*\|^q \le P^*_{c_k}(\hat{\mu}_k; \mu_k) - (1 - v_k) P^*_{c_k}(\mu_{k+1}; \mu_k), \tag{20}$$

where $\hat{\mu}_k$ is the projection of μ_k on M^* [compare with (7)], and $v_k = \eta_k/2\alpha$ [compare with (8)]. In addition, $|\mu_{k+1} - \mu_k| \to 0$.

Proof: Using Lemma 5.15 and the fact that $\max_\mu d(\mu) = d(\hat{\mu}_k) \le L(x_k, \hat{\mu}_k)$, we have

$$\begin{aligned} L_{c_k}(x_k, \mu_k) &= L(x_k, \mu_{k+1}) - P^*_{c_k}(\mu_{k+1}; \mu_k) \\ &\ge L(x_k, \hat{\mu}_k) - P^*_{c_k}(\hat{\mu}_k; \mu_k) \\ &\ge \max_\mu d(\mu) - P^*_{c_k}(\hat{\mu}_k; \mu_k). \end{aligned} \tag{21}$$

The stopping rule for Algorithm B can also be written

$$|\Delta_x \bar{L}_{c_k}(x_k, \mu_k)|^2 \le \eta_k \{L(x_k, \mu_{k+1}) - L(x_k, \mu_k)\},$$

while we have

$$\Delta_x \bar{L}_{c_k}(x_k, \mu_k) = \Delta_x \bar{L}(x_k, \mu_{k+1}).$$

Combining the two relations above with (18), we obtain

$$|\Delta_x \bar{L}(x_k, \mu_{k+1})|^2 \le \eta_k P^*_{c_k}(\mu_{k+1}; \mu_k). \tag{22}$$

We have already shown [compare with (19) in Section 5.3] that

$$L(x_k, \mu_{k+1}) \le d(\mu_{k+1}) + (1/2\alpha)|\Delta_x \bar{L}(x_k, \mu_{k+1})|^2. \tag{23}$$

The last two relations yield

$$L(x_k, \mu_{k+1}) \le d(\mu_{k+1}) + v_k P^*_{c_k}(\mu_{k+1}; \mu_k).$$

Combining this relation with (21) and (6) (which is in effect for k sufficiently large), we obtain

$$\begin{aligned} \max_\mu d(\mu) - P^*_{c_k}(\hat{\mu}_k; \mu_k) + P^*_{c_k}(\mu_{k+1}; \mu_k) &\le L(x_k, \mu_{k+1}) \\ &\le \max_\mu d(\mu) - \gamma \|\mu_{k+1} - M^*\|^q \\ &\quad + v_k P^*_{c_k}(\mu_{k+1}; \mu_k) \end{aligned}$$

from which (20) follows. Also from (20) and the expression (10) for P_c^*, we have

$$(1 - v_k) \sum_{j=1}^{r} \phi^*(\mu_{k+1}^j - \mu_k^j) \leq \sum_{j=1}^{r} \phi^*(\hat{\mu}_k^j - \mu_k^j).$$

Since $(\hat{\mu}_k^j - \mu_k^j) \to 0$ and v_k is bounded away from unity, we obtain $\phi^*(\mu_{k+1}^j - \mu_k^j) \to 0$ for all j which implies $|\mu_{k+1} - \mu_k| \to 0$. Q.E.D.

Lemma 5.17: Let $\{\mu_k\}$ be a sequence generated by Algorithm A or B. There exists a scalar M_0 such that for all k

$$|\mu_{k+1} - \mu_k| \leq M_0 \|\mu_k - M^*\|.$$

Proof: From (4),

$$M_1 |t|^{\rho-1} \leq |\nabla\phi(t)| \leq M_2 |t|^{\rho-1} \qquad \forall\, t \in N_0,$$

and by integration

$$(M_1/\rho)|t|^\rho \leq \phi(t) \leq (M_2/\rho)|t|^\rho \qquad \forall\, t \in N_0.$$

Hence for any scalar s,

$$\sup_{t \in N_0}\left\{st - \frac{M_1}{\rho}|t|^\rho\right\} \geq \sup_{t \in N_0}\{st - \phi(t)\} \geq \sup_{t \in N_0}\left\{st - \frac{M_2}{\rho}|t|^\rho\right\}.$$

Let $[-\alpha, \alpha] \subset N_0$, $\alpha > 0$. Then if $|s| \leq M_1\alpha^{\rho-1}$, the suprema above are attained and by the definition of the conjugate convex function, we obtain

$$(24) \qquad (1/\sigma M_2^{\sigma-1})|s|^\sigma \leq \phi^*(s) \leq (1/\sigma M_1^{\sigma-1})|s|^\sigma, \qquad |s| \leq M_1\alpha^{\rho-1},$$

where σ is the conjugate exponent of ρ defined by $\sigma^{-1} + \rho^{-1} = 1$ or equivalently

$$\sigma = \rho/(\rho - 1).$$

Since $|\mu_{k+1} - \mu_k| \to 0$ (by Lemma 5.16) and $|\mu_k - \hat{\mu}_k| \to 0$ (by Proposition 5.11), we have that for all j and all k sufficiently large both $|\mu_{k+1}^j - \mu_k^j|$ and $|\mu_k^j - \hat{\mu}_k^j|$ are less than $M_1\alpha^{\rho-1}$. Applying (10), (20), and (24), we obtain

$$\begin{aligned}
\frac{1}{c_k \sigma M_2^{\sigma-1}} \sum_{j=1}^{r} |\mu_{k+1}^j - \mu_k^j|^\sigma &\leq \frac{1}{c_k} \sum_{j=1}^{r} \phi^*(\mu_{k+1}^j - \mu_k^j) \\
&= P_{c_k}^*(\mu_{k+1}; \mu_k) \leq (1 - v_k)^{-1} P_{c_k}^*(\hat{\mu}_k; \mu_k) \\
&\leq \frac{1}{(1 - v_k) c_k \sigma M_1^{\sigma-1}} \sum_{j=1}^{r} |\hat{\mu}_k^j - \mu_k^j|^\sigma.
\end{aligned}$$

Hence

$$|\mu_{k+1} - \mu_k|_\sigma \leq (M_2/M_1)^{1/\rho}(1 - v_k)^{(1-\rho)/\rho}|\hat{\mu}_k - \mu_k|_\sigma,$$

where $|\cdot|_\sigma$ denotes the l_σ-norm. Passing to the standard norm $|\cdot|$, via the topological equivalence theorem for all norms on R^r, and using the fact that $v_k \le \bar{v} < 1$, we obtain, for k sufficiently large and some $M_0 > 0$,

$$|\mu_{k+1} - \mu_k| \le M_0 |\hat{\mu}_k - \mu_k| = M_0 \|\mu_k - M^*\|.$$

By increasing M_0 to a sufficiently high value, we obtain that this relation holds for all k. Q.E.D.

Lemma 5.18: Let $\{\mu_k\}$ be a sequence generated by Algorithm A or B. For all k sufficiently large, we have

$$\text{(25)} \quad M_1 |c_k u_k|^{\rho-1} \le |\mu_{k+1} - \mu_k| \le r^{(2-\rho)/2} M_2 |c_k u_k|^{\rho-1} \qquad \text{if} \quad \rho \le 2$$

and

$$\text{(26)} \quad r^{(2-\rho)/2} M_1 |c_k u_k|^{\rho-1} \le |\mu_{k+1} - \mu_k| \le M_2 |c_k u_k|^{\rho-1} \qquad \text{if} \quad \rho \ge 2,$$

where u_k is given by (11), r is the number of constraints, and M_1 and M_2 are as in Assumption (A7).

Proof: Using (13) and the fact that $|\mu_{k+1} - \mu_k| \to 0$ (Lemma 5.16), we obtain $\nabla\phi(c_k u_k^j) \to 0$. It follows by the continuity and strict monotonicity of $\nabla\phi$ that $c_k u_k^j \to 0$. Hence, $c_k u_k^j \in N_0$ for k sufficiently large. Applying (A7) and (13), we have, for k sufficiently large,

$$M_1 |c_k u_k^j|^{\rho-1} \le |\nabla\phi(c_k u_k^j)| = |\mu_{k+1}^j - \mu_k^j| \le M_2 |c_k u_k^j|^{\rho-1}.$$

By squaring and summing over j, we obtain

$$M_1^2 \sum_{j=1}^r |c_k u_k^j|^{2(\rho-1)} \le |\mu_{k+1} - \mu_k|^2 \le M_2^2 \sum_{j=1}^r |c_k u_k^j|^{2(\rho-1)}.$$

Now it is easy to prove that if $0 < \rho - 1 \le 1$ then

$$\left(\sum_{j=1}^r |c_k u_k^j|^2\right)^{\rho-1} \le \sum_{j=1}^r |c_k u_k^j|^{2(\rho-1)} \le r^{2-\rho}\left(\sum_{j=1}^r |c_k u_k^j|^2\right)^{\rho-1}.$$

Combining the last two relations we obtain

$$M_1^2 |c_k u_k|^{2(\rho-1)} \le |\mu_{k+1} - \mu_k|^2 \le M_2^2 r^{2-\rho} |c_k u_k|^{2(\rho-1)}$$

from which, by taking square roots, (25) follows. If $1 \le \rho - 1$, we have

$$r^{2-\rho}\left(\sum_{j=1}^r |c_k u_k^j|^2\right)^{\rho-1} \le \sum_{j=1}^r |c_k u_k^j|^{2(\rho-1)} \le \left(\sum_{j=1}^r |c_k u_k^j|^2\right)^{\rho-1},$$

and similarly we obtain

$$r^{2-\rho} M_1^2 |c_k u_k|^{2(\rho-1)} \le |\mu_{k+1} - \mu_k|^2 \le M_2^2 |c_k u_k|^{2(\rho-1)}$$

Again by taking square roots, we obtain (26). Q.E.D.

The following technical lemma provides some useful estimates.

Lemma 5.19: Let $\{\mu_k\}$ be a sequence generated by Algorithm A or B. For all k sufficiently large we have

$$(27) \qquad \gamma\|\mu_{k+1} - M^*\|^q - v_k P^*_{c_k}(\mu_{k+1}; \mu_k) \le |u_k|\,\|\mu_{k+1} - M^*\|,$$

$$(28) \quad \|\mu_{k+1} - M^*\|^2 + \gamma c_k\|\mu_{k+1} - M^*\|^q - v_k c_k P^*_{c_k}(\mu_{k+1}; \mu_k) \le \|\mu_{k+1} - c_k u_k - M^*\|\,\|\mu_{k+1} - M^*\|.$$

Proof: Take k sufficiently large so that $\mu_{k+1} \in S(M^*; \delta)$. Then applying (A8), (22), (23), and the fact that $\max_\mu d(\mu) = d(\hat{\mu}_{k+1}) \le L(x_k, \hat{\mu}_{k+1})$, we obtain

$$\begin{aligned}(29) \quad \gamma\|\mu_{k+1} - M^*\|^q &\le \max_\mu d(\mu) - d(\mu_{k+1}) \\ &\le L(x_k, \hat{\mu}_{k+1}) - L(x_k, \mu_{k+1}) \\ &\quad + (1/2\alpha)|\Delta_x \bar{L}(x_k, \mu_{k+1})|^2 \\ &= g(x_k)'(\hat{\mu}_{k+1} - \mu_{k+1}) + (1/2\alpha)|\Delta_x \bar{L}(x_k, \mu_{k+1})|^2. \\ &\le g(x_k)'(\hat{\mu}_{k+1} - \mu_{k+1}) + v_k P^*_{c_k}(\mu_{k+1}; \mu_k).\end{aligned}$$

We have $\hat{\mu}_{k+1} \ge 0$, while from (14)–(16) we obtain that if $g_j(x_k) < u_k^j$ then $\mu_{k+1}^j = 0$ while if $\mu_{k+1}^j > 0$ then $g_j(x_k) = u_k^j$. It follows that

$$g(x_k)'(\hat{\mu}_{k+1} - \mu_{k+1}) \le u_k'(\hat{\mu}_{k+1} - \mu_{k+1}),$$

so (29) yields

$$\begin{aligned}(30) \qquad \gamma\|\mu_{k+1} - M^*\|^q &\le u_k'(\hat{\mu}_{k+1} - \mu_{k+1}) + v_k P^*_{c_k}(\mu_{k+1}; \mu_k) \\ &\le |u_k|\,\|\mu_{k+1} - M^*\| + v_k P^*_{c_k}(\mu_{k+1}; \mu_k).\end{aligned}$$

This proves (27).

By multiplying with c_k both sides of the first inequality in (30) and adding $(\mu_{k+1} - \mu^*)'(\mu_{k+1} - \hat{\mu}_{k+1})$, we obtain, for each $\mu^* \in M^*$,

$$\begin{aligned}(\mu_{k+1} - \mu^*)'(\mu_{k+1} - \hat{\mu}_{k+1}) &+ \gamma c_k\|\mu_{k+1} - M^*\|^q \\ &\le (\mu_{k+1} - \mu^* - c_k u_k)'(\mu_{k+1} - \hat{\mu}_{k+1}) + v_k c_k P^*_{c_k}(\mu_{k+1}; \mu_k) \\ &\le |\mu_{k+1} - \mu^* - c_k u_k|\,\|\mu_{k+1} - M^*\| + v_k c_k P^*_{c_k}(\mu_{k+1}; \mu_k).\end{aligned}$$

Since $\hat{\mu}_{k+1}$ is the projection of μ_{k+1} on M^*, we have

$$\begin{aligned}(\mu_{k+1} - \mu^*)'&(\mu_{k+1} - \hat{\mu}_{k+1}) \\ &\ge |\mu_{k+1} - \hat{\mu}_{k+1}|^2 = \|\mu_{k+1} - M^*\|^2 \qquad \forall\, \mu^* \in M^*,\end{aligned}$$

so the last two inequalities yield

$$\|\mu_{k+1} - M^*\|^2 + \gamma c_k\|\mu_{k+1} - M^*\|^q - v_k c_k P^*_{c_k}(\mu_{k+1}; \mu_k) \le |\mu_{k+1} - \mu^* - c_k u_k|\,\|\mu_{k+1} - M^*\|.$$

By taking the minimum over $\mu^* \in M^*$ we obtain (28). Q.E.D.

Convergence Rate of Algorithm A (Exact Minimization)

Proposition 5.20 (Superlinear Convergence): Assume $q > 1$ and denote

$$w = \frac{1}{(\rho - 1)(q - 1)}.$$

If $\{\mu_k\}$ is a sequence generated by Algorithm A, $\mu_k \notin M^*$ for all k, and $w > 1$, then

$$\limsup_{k \to \infty} \frac{\|\mu_{k+1} - M^*\|}{\|\mu_k - M^*\|^w} < \infty;$$

i.e., the rate of convergence is superlinear of order at least w.

Proof: Apply Lemmas 5.17 and 5.18 together with (27) (with $v_k = 0$). For sufficiently large k,

$$\begin{aligned} M_0\|\mu_k - M^*\| &\geq |\mu_{k+1} - \mu_k| \geq \overline{M}_1|c_k u_k|^{\rho-1} \\ &\geq \overline{M}_1(\gamma c_k\|\mu_{k+1} - M^*\|^{q-1})^{\rho-1}, \end{aligned}$$

where $\overline{M}_1 = M_1 \min\{1, r^{(2-\rho)/2}\}$. Hence

$$\|\mu_{k+1} - M^*\| \leq (1/\gamma c_k)^{1/(q-1)}(M_0/\overline{M}_1)^w\|\mu_k - M^*\|^w$$

from which the result follows. Q.E.D.

Proposition 5.21 (Finite Convergence): Assume $q = 1$. If $\{\mu_k\}$ is a sequence generated by Algorithm A, then there exists an index $\bar{k}$ such that $\mu_k \in M^*$ for all $k \geq \bar{k}$.

Proof: From (27) (with $v_k = 0$), we obtain for all k sufficiently large

$$0 \leq (|u_k| - \gamma)\|\mu_{k+1} - M^*\|. \tag{31}$$

From Lemma 5.16, we have $|\mu_{k+1} - \mu_k| \to 0$, so (11) yields $|u_k| \to 0$. Since $\gamma > 0$, we have $|u_k| - \gamma < 0$ for all k sufficiently large so the only way (31) can hold is if $\|\mu_{k+1} - M^*\| = 0$ for all k sufficiently large. Q.E.D.

Proposition 5.22 (Linear Convergence): Assume that $\rho = 2$ in (A7) and $q = 2$ in (A8). Assume further that ϕ is twice continuously differentiable in a neighborhood of the origin and $\nabla^2\phi(0) = 1$. If $\{\mu_k\}$ is a sequence generated by Algorithm A and $\mu_k \notin M^*$ for all k, then

$$\limsup_{k \to \infty} \frac{\|\mu_{k+1} - M^*\|}{\|\mu_k - M^*\|} \leq \frac{1}{1 + \gamma\bar{c}} \quad \text{if} \quad \lim_{k \to \infty} c_k = \bar{c} < \infty,$$

and

$$\lim_{k \to \infty} \frac{\|\mu_{k+1} - M^*\|}{\|\mu_k - M^*\|} = 0 \quad \text{if} \quad \lim_{k \to \infty} c_k = \infty.$$

Proof: By Taylor's theorem and using the fact that $\nabla^2\phi(0) = 1$, we have

$$\nabla\phi(c_k u_k^j) = c_k u_k^j + o(c_k u_k^j),$$

where $o(\cdot)$ is such that $\lim_{\alpha\to 0} o(\alpha)/\alpha = 0$. Hence from (13),

$$\mu_{k+1}^j = \mu_k^j + \nabla\phi(c_k u_k^j) = \mu_k^j + c_k u_k^j + o(c_k u_k^j)$$

or equivalently

$$\mu_{k+1} - \hat{\mu}_k - c_k u_k = \mu_k - \hat{\mu}_k + o(c_k u_k).$$

Combining this equation with (28) (with $v_k = 0$ and $q = 2$), we obtain

$$(32)\quad (1 + \gamma c_k)\|\mu_{k+1} - M^*\| \le |\mu_{k+1} - \hat{\mu}_k - c_k u_k| \le |\mu_k - \hat{\mu}_k| + o(c_k u_k) = \|\mu_k - M^*\| + o(c_k u_k).$$

By Lemma 5.17 and (25), we have

$$|c_k u_k| \le M_1^{-1}|\mu_{k+1} - \mu_k| \le (M_0/M_1)\|\mu_k - M^*\|,$$

so (32) yields

$$\|\mu_{k+1} - M^*\| \le (1 + \gamma c_k)^{-1}[\|\mu_k - M^*\| + o(\|\mu_k - M^*\|)].$$

From this the result follows. Q.E.D.

Interpretation of Results

The last three propositions show that the rate of convergence of Algorithm A is primarily determined by the two scalars q and ρ, introduced in Assumptions (A7) and (A8). The scalar q depends on the rate at which the dual

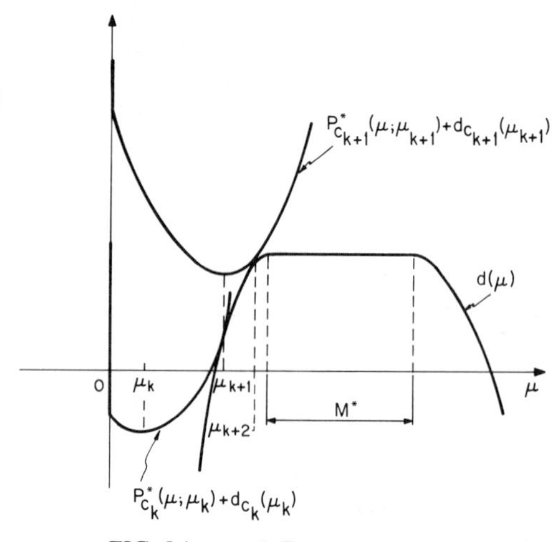

FIG. 5.4 $q \simeq 1$, Fast convergence

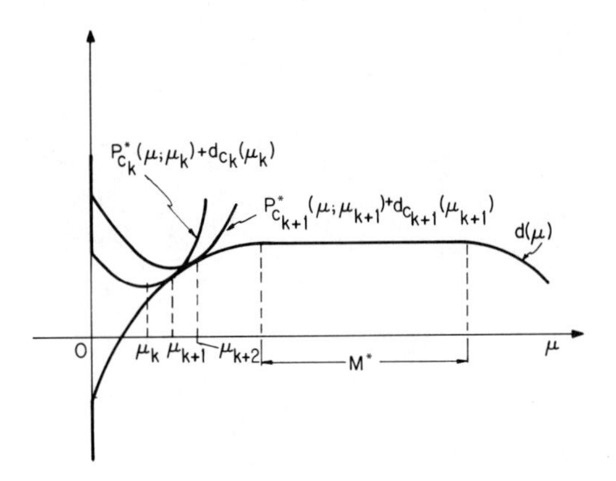

FIG. 5.5 $q \gg 1$, Slow convergence

functional d grows (downward). If the graph of d is sharply pointed ($q \simeq 1$), the rate of convergence is fast. If d is relatively "flat" near the optimal set M^* (q: large), the rate of convergence is slow. These observations are illustrated in Figs. 5.4 and 5.5, where we have chosen ϕ (and hence also ϕ^*) to be quadratic. At the same time, the rate of growth of the penalty function ϕ is equally important in determining the rate of convergence. When ρ is large, then ϕ grows slowly and ϕ^* grows rapidly near the origin. As a result, the rate of convergence is poor as shown in Fig. 5.6. Conversely, when ρ is small then ϕ grows rapidly, ϕ^* grows slowly, and the rate of convergence is fast as shown in Fig. 5.7.

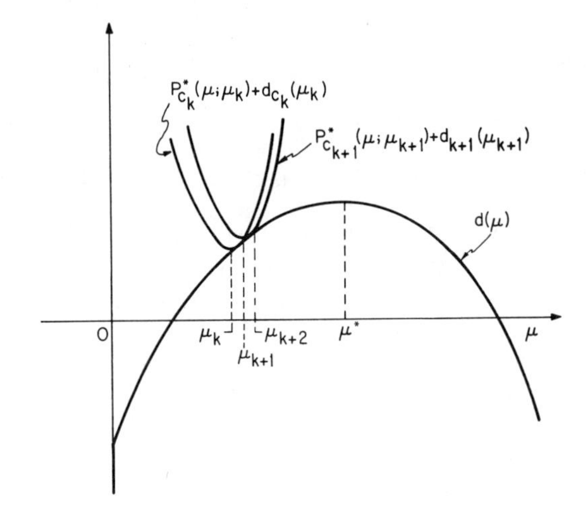

FIG. 5.6 $\rho \gg 1$, Slow convergence

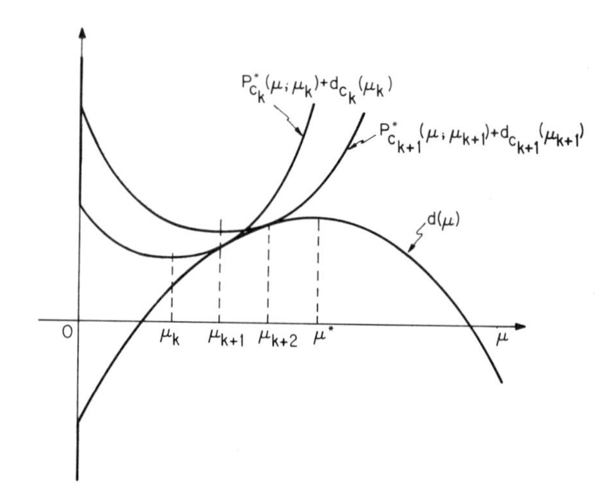

FIG. 5.7 $\rho \simeq 1$, Fast convergence

An extreme case occurs when $q = 1$. Then by Proposition 5.21, convergence occurs in a finite number of iterations. In this case, the dual functional d has a "corner" all around its boundary. More precisely, if μ is any vector in $S(M^*; \delta)$ and $\hat{\mu}$ is its projection on M^*, then (A8) (with $q = 1$) implies

$$d(\mu) \le d(\hat{\mu}) - \gamma|\mu - \hat{\mu}|.$$

For every $w \in \partial d(\mu)$, we have $d(\hat{\mu}) \le d(\mu) + w'(\hat{\mu} - \mu)$ so $w'(\hat{\mu} - \mu) \ge d(\hat{\mu}) - d(\mu) \ge \gamma|\mu - \hat{\mu}|$. Hence,

$$|w||\hat{\mu} - \mu| \ge w'(\hat{\mu} - \mu) \ge \gamma|\mu - \hat{\mu}| \qquad \forall\, w \in \partial d(\mu), \quad \mu \in S(M^*; \delta).$$

It follows that, for $q = 1$, (A8) implies

$$|w| \ge \gamma \qquad \forall\, w \in \partial d(\mu), \quad \mu \in S(M^*; \delta), \quad \mu \notin M^*,$$

so all subgradients at points near but outside M^* must have a norm exceeding γ. We show in Fig. 5.8 a situation where $q = 1$, and illustrate the process of finite convergence. A typical case is when d is polyhedral as, for example, where (CPP) is a linear (more generally polyhedral) program. Then it is straightforward to show that (A8) is satisfied with $q = 1$ so convergence of Algorithm A is obtained in a finite number of steps for every $p \in P_{\text{E}}^+$.

Some other conclusions from Propositions 5.20 and 5.22 are that when $q < 2$ the quadratic penalty function leads to a superlinear rate of convergence while if $q = 2$ the convergence rate is at least linear (superlinear if

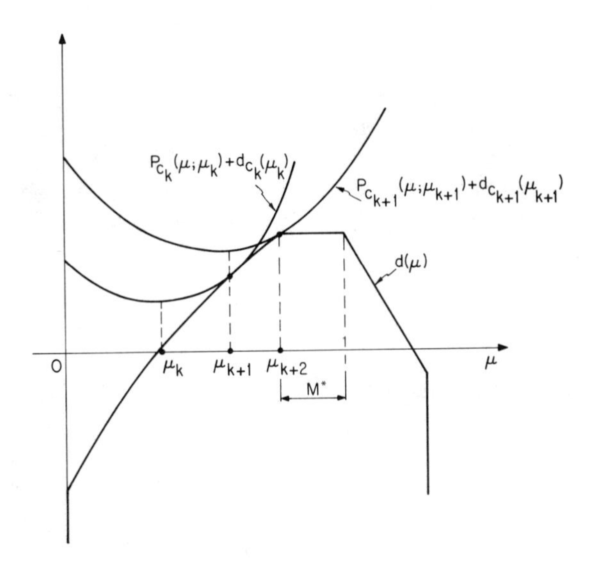

FIG. 5.8 $q = 1$, Finite convergence

$c_k \to \infty$). When $q > 2$ then it is possible to show by example that the quadratic penalty function leads to a convergence rate which is sublinear in general. Thus, when $q > 2$ the only way one can achieve a linear or superlinear convergence rate is by choosing a penalty function with $\rho \in (1, 2)$. More generally Proposition 5.20 shows that *any* order of convergence can be achieved by a suitable choice of the penalty function. There is a price for this however. When $\rho < 2$, ϕ is not twice differentiable at the origin and the minimization of the augmented Lagrangian $L_c(\cdot, \mu)$ may pose difficulties due to ill-conditioning. As a result rapid convergence of $\{\mu_k\}$ is achieved at the expense of ill-conditioning the unconstrained minimization. However, in situations where one repeatedly solves the same basic problem with minor variations, one may be able to "fine tune" the algorithm by choosing $\{c_k\}$ and $\{\eta_k\}$ in a near optimal fashion. Since good estimates of the solution are already known, ill-conditioning may not be a problem, and then one can exploit the superior convergence rate of the order $\rho < 2$ penalty without incurring undue cost in computing the unconstrained minima. It is worth pointing out that our results imply that the order $\rho < 2$ penalty functions lead to fast convergence only after the method is near convergence. When far from the solution, Fig. 5.7 indicates that convergence may be slow unless the penalty function ϕ contains, implicitly or explicitly, terms of the form $|t|^{\rho_1}$ where $\rho_1 \geq 2$. For this reason penalty functions of the form $\phi(t) = |t|^\rho + |t|^{\rho_1}$, with $1 < \rho < 2$ and $2 \leq \rho_1$, seem to be preferable to functions of the form $\phi(t) = |t|^\rho$ with $1 < \rho < 2$.

Comparison with Penalty Methods

If the multipliers μ_k are not updated in Algorithm A but rather are held fixed at a constant value, then an ordinary exterior penalty method is obtained. We derive the rate of convergence of this method and show that it is typically inferior to the one of the method of multipliers. For convenience we assume that $\mu_k \equiv 0$ so the penalty method consists of a sequence of minimizations of the form

$$\text{(33)} \qquad \begin{aligned} &\text{minimize} \quad L_{c_k}(x, 0) \\ &\text{subject to} \quad x \in X \end{aligned}$$

for a sequence $\{c_k\}$ such that

$$\text{(34)} \qquad 0 < c_k < c_{k+1} \qquad \forall\, k = 0, 1, \ldots, \qquad c_k \to \infty.$$

The multiplier update formula is not used but it is still relevant, as it provides a sequence $\{\tilde{\mu}_k\}$ of Lagrange multiplier estimates

$$\tilde{\mu}_k^j = \nabla_t p[c_k g_j(x_k); 0], \qquad j = 1, \ldots, r,$$

where x_k is a solution of problem (33). We shall derive an estimate of $\|\tilde{\mu}_k - M^*\|$.

By Proposition 5.5, $\tilde{\mu}_k$ is the unique vector attaining the maximum in the expression

$$\max_{s \in R^r} \{d(s) - P_{c_k}^*(s; 0)\},$$

so for any $\mu^* \in M^*$, we have

$$\text{(35)} \qquad d(\mu^*) - P_{c_k}^*(\mu^*; 0) \le d(\tilde{\mu}_k) - P_{c_k}^*(\tilde{\mu}_k; 0).$$

The sequence $\{\tilde{\mu}_k\}$ is bounded since d has bounded level sets and $d(\tilde{\mu}_k) \ge d_{c_k}(0) \ge d_{c_0}(0)$ (by Proposition 5.8). By taking limits as $c_k \to \infty$ in (35), we obtain using (10)

$$d(\tilde{\mu}_k) \to d(\mu^*), \qquad \|\tilde{\mu}_k - M^*\| \to 0.$$

It follows that for c_k sufficiently large we have $\tilde{\mu}_k \in S(M^*; \delta)$, so by using (A8), we have

$$\text{(36)} \qquad d(\tilde{\mu}_k) \le \max_{\mu} d(\mu) - \gamma\|\tilde{\mu}_k - M^*\|^q = d(\mu^*) - \gamma\|\tilde{\mu}_k - M^*\|^q.$$

Combining (35) and (36), we obtain

$$\text{(37)} \qquad \gamma\|\tilde{\mu}_k - M^*\|^q \le P_{c_k}^*(\mu^*; 0) - P_{c_k}^*(\tilde{\mu}_k; 0).$$

Since for $s \ge 0$, the function

$$c_k P_{c_k}^*(s; 0) = \sum_{j=1}^{r} \phi^*(s_j)$$

is real valued and convex, it is Lipschitz continuous on bounded sets, so if N is the Lipschitz constant, for the set $\{\mu | d(\mu) \geq d_{c_0}(0)\}$, we have

$$|P_{c_k}^*(\mu^*; 0) - P_{c_k}^*(\tilde{\mu}_k; 0)| \leq (N/c_k)|\mu^* - \tilde{\mu}_k|. \tag{38}$$

Combining (37) and (38), we obtain for all c_k sufficiently large

$$\gamma \|\tilde{\mu}_k - M^*\|^q \leq (N/c_k)|\tilde{\mu}_k - \mu^*| \qquad \forall\, \mu^* \in M^*. \tag{39}$$

Taking the infimum over $\mu^* \in M^*$ we obtain, for $q > 1$,

$$\|\tilde{\mu}_k - M^*\| \leq (N/c_k\gamma)^{1/(q-1)}. \tag{40}$$

By comparing this estimate with the estimates of Propositions 5.20 and 5.22, we see that the convergence rate properties of the penalty method are not as attractive as those of multiplier methods. Indeed the rate of convergence of the penalty method depends on the rate at which the parameter c_k is increased. Note that (40) indicates that for q near unity the convergence is faster. The order of growth ρ of the penalty function does not enter the estimate (40), and indeed it appears that the choice of the penalty function is immaterial unless the multiplier update formula is utilized.

An interesting situation occurs when $q = 1$. In this case, we obtain, from (39),

$$0 \leq (N - c_k\gamma)\|\tilde{\mu}_k - M^*\|,$$

so it follows that

$$\tilde{\mu}_k \in M^* \qquad \text{if} \quad c_k > N/\gamma.$$

Thus, *when* $q = 1$, *the penalty method* (33), (34) *yields a Lagrange multiplier of* (CPP) *for* c_k *sufficiently large. In particular, this occurs when the dual functional is polyhedral.* This situation is illustrated in Fig. 5.9.

Convergence Rate of Algorithm B (Inexact Minimization)

Proposition 5.23 (Superlinear Convergence): Assume $q > 1$ and denote

$$w = \frac{\rho}{(\rho - 1)q}$$

If $\{\mu_k\}$ is generated by Algorithm B, $\mu_k \notin M^*$ for all k, and $w > 1$, then

$$\limsup_{k \to \infty} \frac{\|\mu_{k+1} - M^*\|}{\|\mu_k - M^*\|^w} < \infty.$$

Proof: By Lemma 5.18, we have

$$|\mu_{k+1} - \mu_k| \geq \overline{M}_1 |c_k u_k|^{\rho - 1},$$

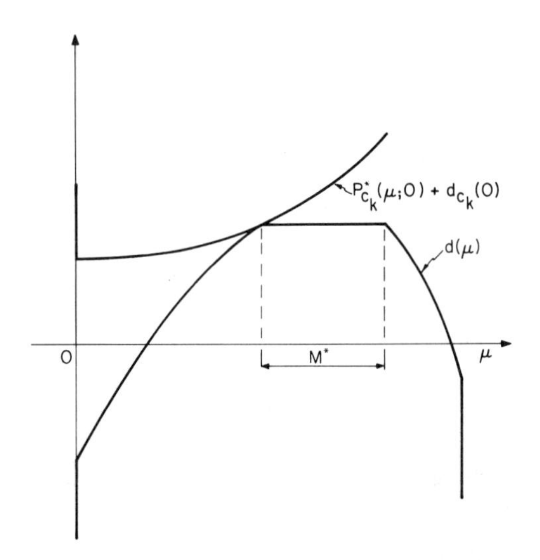

FIG. 5.9 Finite convergence of the penalty method for c_k sufficiently large

where $\overline{M}_1 = M_1 \min\{1, r^{(2-\rho)/2}\}$. Using this inequality together with (27), we obtain

$$\left(\frac{1}{\overline{M}_1}|\mu_{k+1} - \mu_k|\right)^{\sigma-1} \geq |c_k u_k| \geq \gamma c_k \|\mu_{k+1} - M^*\|^{q-1} - \frac{c_k v_k P^*_{c_k}(\mu_{k+1}; \mu_k)}{\|\mu_{k+1} - M^*\|}, \tag{41}$$

where $\sigma = \rho/(\rho - 1)$. From (24), we have for large k

$$\begin{aligned} c_k P^*_{c_k}(\mu_{k+1}; \mu_k) &= \sum_{j=1}^{r} \phi^*(\mu^j_{k+1} - \mu^j_k) \\ &\leq \frac{1}{\sigma M_1^{\sigma-1}} \sum_{j=1}^{r} |\mu^j_{k+1} - \mu^j_k|^\sigma \leq \frac{D}{\sigma M_1^{\sigma-1}} |\mu_{k+1} - \mu_k|^\sigma, \end{aligned}$$

where $D = \max\{1, r^{(2-\sigma)/2}\}$. Combining this with (41), we obtain

$$\left(\frac{1}{\overline{M}_1}|\mu_{k+1} - \mu_k|\right)^{\sigma-1} \geq \gamma c_k \|\mu_{k+1} - M^*\|^{q-1} - \frac{D v_k}{\sigma M_1^{\sigma-1}} \frac{|\mu_{k+1} - \mu_k|^\sigma}{\|\mu_{k+1} - M^*\|}.$$

Equivalently,

$$\begin{aligned} \gamma M_1^{\sigma-1} \|\mu_{k+1} - M^*\|^q &\leq \frac{1}{c_k} \|\mu_{k+1} - M^*\| \left(\frac{M_1}{\overline{M}_1} |\mu_{k+1} - \mu_k|\right)^{\sigma-1} \\ &\quad + \frac{D v_k}{\sigma c_k} |\mu_{k+1} - \mu_k|^\sigma \end{aligned} \tag{42}$$

By Lemma 5.17, we have

$$|\mu_{k+1} - \mu_k| \le M_0 \|\mu_k - M^*\|, \tag{43}$$

while using the triangle inequality

$$\begin{aligned}\|\mu_{k+1} - M^*\| \le |\mu_{k+1} - \hat{\mu}_k| &\le |\mu_{k+1} - \mu_k| + |\mu_k - \hat{\mu}_k| \\ &= |\mu_{k+1} - \mu_k| + \|\mu_k - M^*\|.\end{aligned}$$

The last two relations yield

$$\|\mu_{k+1} - M^*\| \le (1 + M_0)\|\mu_k - M^*\|. \tag{44}$$

Combining (42)–(44),

$$\gamma M_1^{\sigma-1}\|\mu_{k+1} - M^*\|^q \le \left[\frac{1 + M_0}{c_k}\left(\frac{M_0 M_1}{\overline{M}_1}\right)^{\sigma-1} + \frac{Dv_k}{\sigma c_k} M_0^\sigma\right]\|\mu_k - M^*\|^\sigma.$$

Given that $\sigma = \rho/(\rho - 1)$ we have, for k sufficiently large,

$$\|\mu_{k+1} - M^*\|/\|\mu_k - M^*\|^{\rho/(\rho-1)q} \le M,$$

where M is some scalar, and the proposition is proved. Q.E.D.

Note from Proposition 5.23 that the order of convergence of Algorithm B $[\rho/(\rho - 1)q]$ is smaller than the one of Algorithm A $[1/(\rho - 1)(q - 1)]$. It is possible however to increase the order of convergence of Algorithm B up to $1/(\rho - 1)(q - 1)$ provided a mechanism is introduced that forces the scalar η_k in the stopping rule to decrease sufficiently fast. It can be shown that if the stopping rule of Algorithm B is of the form

$$|\Delta_x \bar{L}_{c_k}(x_k, \mu_k)|^2 \le \eta_k \sum_{j=1}^{r} \left\{\nabla_t p[c_k g_j(x_k); \mu_k^j] g_j(x_k) - \frac{1}{c_k} p[c_k g_j(x_k); \mu_k^j]\right\},$$

where

$$\eta_k = \min\left\{\bar{\eta}_k, B \sum_{j=1}^{r} |\nabla_t p[c_k g_j(x_k); \mu_k^j] - \mu_k^j|^\beta\right\}$$

and $\{\bar{\eta}_k\}$, B, and β satisfy

$$0 \le \bar{\eta}_{k+1} \le \bar{\eta}_k \quad \forall\, k = 0, 1, \ldots, \quad \bar{\eta}_k \to 0, \quad B > 0, \quad \beta \ge \frac{1 - (\rho - 1)(q - 1)}{(\rho - 1)(q - 1)},$$

then the order of convergence of Algorithm B is restored to $1/(\rho - 1)(q - 1)$—the same as for Algorithm A. A proof of this fact is given in Kort and Bertsekas (1976, pp. 287–288).

Proposition 5.24 (Linear Convergence): Assume that $\rho = 2$ in (A7) and $q = 2$ in (A8). Assume further that ϕ is twice continuously differentiable in a neighborhood of the origin and $\nabla^2\phi(0) = 1$. If $\{\mu_k\}$ is a sequence generated by Algorithm B and $\mu_k \notin M^*$ for all k, then

$$\limsup_{k\to\infty} \frac{\|\mu_{k+1} - M^*\|}{\|\mu_k - M^*\|} \le \frac{1}{1+\gamma\bar{c}} \qquad \text{if} \quad \lim_{k\to\infty} c_k = \bar{c} < \infty,$$

and

$$\lim_{k\to\infty} \frac{\|\mu_{k+1} - M^*\|}{\|\mu_k - M^*\|} = 0 \qquad \text{if} \quad \lim_{k\to\infty} c_k = \infty.$$

Proof: As in the proof of Proposition 5.22, we have

$$\mu_{k+1} - \hat{\mu}_k - c_k u_k = \mu_k - \hat{\mu}_k + o(c_k u_k).$$

Using this equation together with (28) (for $q = 2$), we obtain

$$\begin{aligned} (1+\gamma c_k)\|\mu_{k+1} - M^*\|^2 &- c_k v_k P^*_{c_k}(\mu_{k+1};\mu_k) \\ &\le |\mu_{k+1} - \hat{\mu}_k - c_k u_k|\,\|\mu_{k+1} - M^*\| \\ &\le \|\mu_k - M^*\|\,\|\mu_{k+1} - M^*\| + o(c_k u_k)\|\mu_{k+1} - M^*\|. \end{aligned} \tag{45}$$

By Lemma 5.17 and (25),

$$|c_k u_k| \le (M_0/M_1)\|\mu_k - M^*\|. \tag{46}$$

By using (20) to upper-bound $P^*_{c_k}(\mu_{k+1};\mu_k)$, substituting in (45), and using also (46), we obtain

$$\begin{aligned} \left(1 + \frac{\gamma c_k}{1-v_k}\right)\|\mu_{k+1} - M^*\|^2 &- \frac{c_k v_k}{1-v_k} P^*_{c_k}(\hat{\mu}_k;\mu_k) \\ &\le \|\mu_k - M^*\|\,\|\mu_{k+1} - M^*\| + \|\mu_{k+1} - M^*\| o(\|\mu_k - M^*\|). \end{aligned}$$

Using (24) (for $\sigma = 2$), we also have

$$c_k P^*_{c_k}(\hat{\mu}_k;\mu_k) = \sum_{j=1}^{r} \phi^*(\hat{\mu}_k^j - \mu_k^j) \le \frac{1}{2M_1}\sum_{j=1}^{r} |\hat{\mu}_k^j - \mu_k^j|^2 = \frac{1}{2M_1}\|\mu_k - M^*\|^2.$$

The last two inequalities yield

$$\begin{aligned} \left(1 + \frac{\gamma c_k}{1-v_k}\right)\|\mu_{k+1} - M^*\|^2 &- \frac{1}{2M_1}\frac{v_k}{1-v_k}\|\mu_k - M^*\|^2 \\ &\le \|\mu_k - M^*\|\,\|\mu_{k+1} - M^*\| + \|\mu_{k+1} - M^*\| o(\|\mu_k - M^*\|). \end{aligned}$$

Dividing through by $\|\mu_k - M^*\|^2$, we obtain

$$R_k(\alpha_k R_k - \gamma_k) \le \beta_k, \tag{47}$$

where

$$R_k = \|\mu_{k+1} - M^*\|/\|\mu_k - M^*\|,$$

$$\alpha_k = 1 + \frac{\gamma c_k}{1 - v_k}, \qquad \beta_k = \frac{1}{2M_1}\frac{v_k}{1 - v_k}, \qquad \gamma_k = 1 + \frac{o(\|\mu_k - M^*\|)}{\|\mu_k - M^*\|}.$$

Since $\beta_k \to 0$, from (47) we must have either $R_k \to 0$ or else

$$\limsup_{k\to\infty} (a_k R_k - \gamma_k) \le 0.$$

If $\lim_{k\to\infty} c_k = \bar{c} < \infty$, it follows in either case that

$$\limsup_{k\to\infty} R_k \le \frac{\lim_{k\to\infty} \gamma_k}{\lim_{k\to\infty} \alpha_k} = \frac{1}{1 + \gamma\bar{c}}.$$

If $c_k \to \infty$ then $\alpha_k \to \infty$, $\gamma_k \to 1$, and it follows that $R_k \to 0$. Q.E.D.

Note that under the assumptions of Proposition 5.24, the bound on the convergence ratio

$$\|\mu_{k+1} - M^*\|/\|\mu_k - M^*\|$$

is identical for both Algorithms A and B, so the ultimate speed of convergence is unaffected by the fact that minimization of $L_{c_k}(\cdot, \mu_k)$ is not exact. This of course, is true for the particular stopping rule utilized in Algorithm B. When other stopping rules are used, then there is no guarantee that this property will be maintained. In fact it is possible to construct examples (see Bertsekas, 1975c) where the (natural) stopping rule

$$|\Delta_x \bar{L}_{c_k}(x_k, \mu_k)|^2 \le \varepsilon_k, \qquad 0 < \varepsilon_{k+1} < \varepsilon_k, \quad \varepsilon_k \to 0,$$

is used, and the assumptions of Proposition 5.24 are satisfied, but the convergence rate of the corresponding algorithm is sublinear.

5.5 Conditions for Penalty Methods to be Exact

It was shown in the previous section (Proposition 5.21) that the method of multipliers with exact minimization under certain (rather restrictive) assumptions yields a Lagrange multiplier of (CPP) in a finite number of iterations. One extra minimization will be required in order to obtain an optimal solution of (CPP) [compare with Proposition 5.6b]. On the other hand, it is possible to obtain under much less restrictive assumptions an optimal solution of (CPP) provided we use a nondifferentiable penalty function. We developed the relevant theory for nonconvex problems in

Section 4.1. In this section, we develop a generalized version of this theory for convex problems.

We consider (CPP) and assume throughout that *Assumptions* (A1)–(A3) *of Section* 5.2 *are in effect. We also assume that* $f^* = \sup_{\mu \geq 0} d(\mu)$. We consider penalty functions $p\colon R^r \to R$ which are real valued convex and satisfy

(1) $$p(t) = 0 \qquad \forall\, t \leq 0,$$

(2) $$p(t) > 0 \qquad \text{if} \quad t_j > 0 \quad \text{for some } j = 1, \ldots, r.$$

It is easily seen that (1) implies that the conjugate

$$p^*(s) = \sup_t \{s't - p(t)\}$$

satisfies

(3) $$p^*(s) \geq 0 \qquad \forall\, s \in R^r.$$

Typical conjugate convex pairs p and p^* that are of interest within the context of this section are shown in Fig. 5.10.

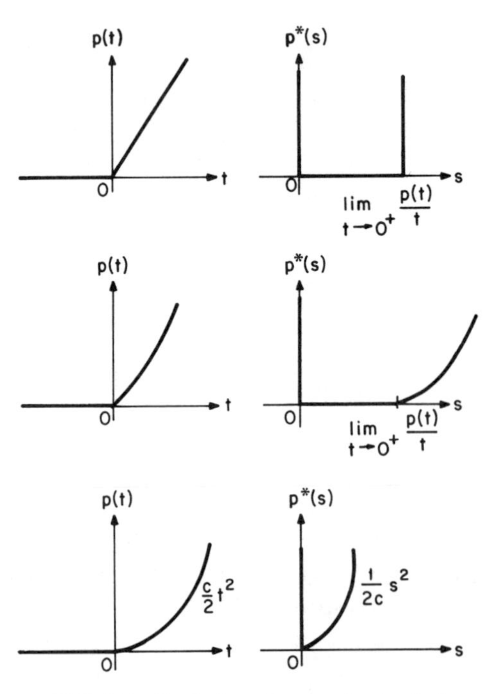

FIG. 5.10 Typical conjugate pairs p and p^*

Consider the problem

$$\text{(4)} \qquad \begin{aligned} &\text{minimize} \quad f(x) + p[g(x)] \\ &\text{subject to} \quad x \in X. \end{aligned}$$

It is easy to verify, by essentially repeating the proof of Proposition 5.7a, that if *Assumption* (A.4) *holds then problem* (4) *has a nonempty and compact solution set.* We are interested in deriving conditions under which optimal solutions of problem (4) are also optimal solutions of (CPP). The situation can be visualized by using Fenchel's duality theory (Rockafellar, 1970, Theorem 31.1) to write

$$\text{(5)} \qquad \begin{aligned} \inf_{x \in X} \{f(x) + p[g(x)]\} &= \inf_{u} \{q(u) + p(u)\} \\ &= \max_{s \in R^r} \{d(s) - p^*(s)\}, \end{aligned}$$

where q is the primal functional of (CPP) [compare with (9) in Section 5.2] and d is the dual functional. Notice the similarity of Eq. (5) with Eqs. (23) and (26) of Section 5.2 (compare with Proposition 5.5). It can be easily shown (see also the proof of Proposition 5.5b) that our assumptions guarantee that condition (a) of (Rockafellar, 1970, Theorem 31.1) is satisfied, and this in turn implies that *the maximum in* (5) *is attained* (even though this maximum may equal $-\infty$). The maximization in (5) is illustrated in Fig. 5.11 where the scalar $\tilde{f}$ is defined by

$$\tilde{f} = \inf_{x \in X} \{f(x) + p[g(x)]\}.$$

It can be seen that in order for problem (4) to have the same optimal value as (CPP), it is necessary for the conjugate p^* to be "flat" along a sufficiently large "area." This is formalized in the following proposition.

Proposition 5.25: Assume that (A1)–(A3) hold and that

$$f^* = \sup_{\mu \geq 0} d(\mu).$$

(a) In order for some optimal solution of problem (4) to be an optimal solution of (CPP) it is necessary that there exists a Lagrange multiplier $\bar{\mu}$ of (CPP) for which

$$\text{(6)} \qquad t'\bar{\mu} \leq p(t) \qquad \forall\, t \in R^r.$$

(b) In order for problem (4) and (CPP) to have exactly the same optimal solutions, it is sufficient that

$$\text{(7)} \qquad t'\bar{\mu} < p(t) \qquad \forall\, t \in R^r \text{ with } t_j > 0 \text{ for some } j$$

for some Lagrange multiplier $\bar{\mu}$ of (CPP).

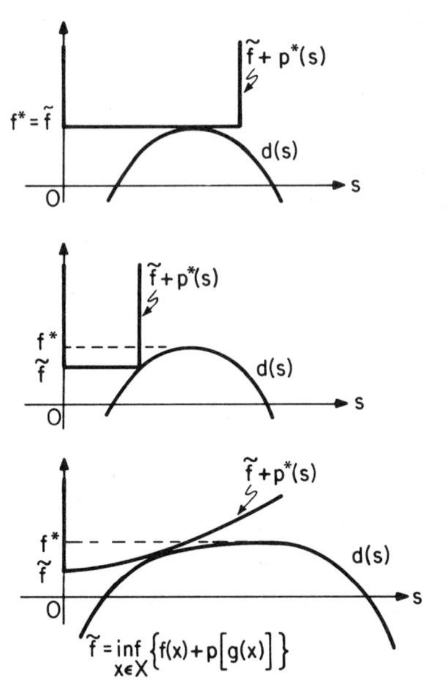

FIG. 5.11 Geometric interpretation of the maximization in (5)

Proof: (a) Let $\bar{x}$ be a common optimal solution of problem (4) and (CPP). Since $\bar{x}$ is feasible for (CPP), we have $f^* = f(\bar{x})$ and $p[g(\bar{x})] = 0$. Using (5) we obtain

$$f^* = f(\bar{x}) + p[g(\bar{x})] = \max_{s \in R^r} \{d(s) - p^*(s)\}.$$

Let $\bar{\mu}$ be any vector attaining the maximum above. Then

$$f^* = d(\bar{\mu}) - p^*(\bar{\mu}) \leq \sup_{\mu} d(\mu) - p^*(\bar{\mu}) = f^* - p^*(\bar{\mu}). \tag{8}$$

Hence $p^*(\bar{\mu}) \leq 0$ and since from (3) we have $p^*(s) \geq 0$ for all s, it follows that

$$p^*(\bar{\mu}) = 0. \tag{9}$$

It follows from (8) that $d(\bar{\mu}) = f^*$, so $\bar{\mu}$ is a Lagrange multiplier. We can rewrite (9) as

$$\sup_t \{\bar{\mu}'t - p(t)\} = 0$$

so

$$t'\bar{\mu} - p(t) \leq 0 \qquad \forall\, t \in R^r$$

implying (6).

(b) If $\bar{x}$ is an optimal solution of (CPP), then by (1), (7), and the definition of a Lagrange multiplier, we have, for all $x \in X$,

$$f(\bar{x}) + p[g(\bar{x})] = f(\bar{x}) = f(\bar{x}) + \bar{\mu}'g(\bar{x})$$
$$\leq f(x) + \bar{\mu}'g(x) \leq f(x) + p[g(x)].$$

Hence, $\bar{x}$ is also a solution of problem (4).

Conversely, if $\bar{x}$ is a solution of problem (4), then $\bar{x}$ is either a feasible point in which case it is also a solution of (CPP) (in view of $p[g(x)] = 0$ for all feasible points x), or it is infeasible in which case $g_j(\bar{x}) > 0$ for some j. Then by using (7) we have that there exists an $\varepsilon > 0$ such that

$$\bar{\mu}'g(\bar{x}) + \varepsilon < p[g(\bar{x})]. \tag{10}$$

Let $\tilde{x}$ be a feasible solution of (CPP) such that $f(\tilde{x}) \leq f^* + \varepsilon$. Since $p[g(\tilde{x})] = 0$ and $f^* = \inf_{x \in X} \{f(x) + \bar{\mu}'g(x)\}$ we obtain

$$f(\tilde{x}) + p[g(\tilde{x})] = f(\tilde{x}) \leq f^* + \varepsilon \leq f(\bar{x}) + \bar{\mu}'g(\bar{x}) + \varepsilon. \tag{11}$$

By combining Eqs. (10) and (11) we obtain

$$f(\tilde{x}) + p[g(\tilde{x})] < f(\bar{x}) + p[g(\bar{x})],$$

which contradicts the fact that $\bar{x}$ is an optimal solution of problem (4). Hence problem (4) and (CPP) have exactly the same solutions. Q.E.D.

As an application of Proposition 5.25, consider the penalty function

$$p(t) = c \sum_{j=1}^{r} \max\{0, t_j\},$$

where $c > 0$. Clearly it satisfies (1) and (2). Condition (6) can be written as

$$\sum_{j=1}^{r} \bar{\mu}_j t_j \leq c \sum_{j=1}^{r} \max\{0, t_j\} \qquad \forall\, t \in R^r$$

and is clearly equivalent to

$$\bar{\mu}_j \leq c \qquad \forall\, j = 1, \ldots, r.$$

Similarly, condition (7) is equivalent to

$$\bar{\mu}_j < c \qquad \forall\, j = 1, \ldots, r.$$

More generally, consider the case where

$$p(t) = \sum_{j=1}^{r} p_j(t_j),$$

where $p_j\colon R \to R$ are convex real-valued penalty functions satisfying

$$p_j(t) = 0 \quad \forall\, t \leq 0, \qquad p_j(t) > 0 \quad \forall\, t > 0. \tag{12}$$

Then condition (6) can be written as

$$\sum_{j=1}^{r} \bar{\mu}_j t_j \leq \sum_{j=1}^{r} p_j(t_j) \qquad \forall\, t \in R^r$$

and is clearly equivalent to

$$\bar{\mu}_j t_j \leq p_j(t_j) \qquad \forall\, t_j \in R, \quad j = 1, \ldots, r.$$

In view of (12) and the convexity of p_j, this condition can be easily seen to be equivalent to

$$\bar{\mu}_j \leq \lim_{t_j \to 0+} \frac{p_j(t_j)}{t_j} \qquad \forall\, j = 1, \ldots, r.$$

Similarly condition (7) is equivalent to

$$\bar{\mu}_j < \lim_{t_j \to 0+} \frac{p_j(t_j)}{t_j} \qquad \forall\, j = 1, \ldots, r.$$

Consider also for $c > 0$ the penalty function

$$p(t) = c \max\{0, t_1, \ldots, t_r\}$$

discussed in Section 4.1. Condition (6) can be written as

$$\sum_{j=1}^{r} \bar{\mu}_j t_j \leq c \max\{0, t_1, \ldots, t_r\} \qquad \forall\, t \in R^r,$$

and a little thought shows that it is equivalent to

$$\sum_{j=1}^{r} \bar{\mu}_j \leq c.$$

Similarly, condition (7) is equivalent to

$$\sum_{j=1}^{r} \bar{\mu}_j < c.$$

The results of Proposition 5.25 for this penalty function should be compared with the results of Section 4.1.

5.6 Large Scale Separable Integer Programming Problems and the Exponential Method of Multipliers

We noted in Section 5.4 that multiplier methods are fully applicable to linear or polyhedral programs and furthermore give convergence in a finite number of iterations if the penalty function used belongs to the class P_E^+ and the minimization of the augmented Lagrangian is carried out exactly (Proposition 5.21). Despite their finite convergence property, it seems that multiplier methods are typically *not* competitive with the simplex method for solution of linear programs of *small* dimension. There are however large dimensional linear programs with special structure for which the simplex method is hopelessly time consuming and for which other methods designed for nondifferentiable optimization are much more effective. In this section we discuss the application of approximation methods for nondifferentiable optimization to an interesting class of polyhedral programs.

An important feature of nondifferentiable optimization methods is that, by contrast with the simplex method, they are not oriented towards moving from one extreme point of the feasible set to a neighboring extreme point. As a result, they are typically not guaranteed to solve the problem in a finite number of iterations. However, because they are not constrained to follow a (usually conservative) path consisting of adjacent extreme points they are often capable of locating rather quickly a good approximation to an optimal solution. In most applications, this is sufficient for practical purposes. The potential of nondifferentiable optimization methods for solving important classes of polyhedral problems arising for example via a duality transformation in integer programming was appreciated early in the Soviet Union following the development of the subgradient method (Shor, 1964; Poljak, 1969b) and space dilation methods (Shor, 1970; Shor and Jourbenko, 1971). Considerable interest in nondifferentiable optimization was also generated several years later in the West (Held and Karp, 1970; Held *et al.*, 1974) and new methods such as the ε-subgradient method (Bertsekas and Mitter, 1971, 1973; Lemarechal, 1974), conjugate subgradient methods (Wolfe, 1975; Lemarechal, 1975), and other descent methods (Goldstein, 1977, Mifflin, 1977) were developed (see Auslender, 1976, and Shapiro, 1979, for an extensive account).

The approximation methods for nondifferentiable optimization discussed in Sections 3.3 and 5.1.3 provide an interesting alternative for solving polyhedral optimization problems arising in integer programming via a duality transformation. One of their advantages versus subgradient-type methods is that, in addition to solving the nondifferentiable optimization problem at hand, they provide additional information in the form of the multipliers entering the approximation formulas. These multipliers often turn out to be extremely valuable in generating a good suboptimal solution of the original integer programming problem. In this section, we describe this methodology as applied to an important class of integer programming problems. It has been recently successful in solving problems involving several thousands of integer variables that have resisted solution for many years using other methods (see Bertsekas *et al.*, 1981).

The approximation approach to be described is based on the exponential penalty function as in Section 5.1.3. By contrast with penalty functions in the class P_E^+, the use of the exponential function does not lead to a finitely convergent algorithm for polyhedral problems. Its main advantage is that it leads to twice differentiable approximating problems, and it appears that this is a very important factor when the problem to be approximated is polyhedral.

A Class of Integer Programming Problems

We consider the following (primal) integer programming problem

$$\text{(PIP)} \quad \text{minimize} \quad \sum_{i=1}^{I} f_i(x_i, n_i)$$

$$\text{subject to} \quad \sum_{i=1}^{I} h_i(x_i, n_i) \leq b \qquad n_i \in N_i, \quad x_i \in X_i(n_i), \quad i = 1, \ldots, I,$$

where for each i, n_i is an integer variable constrained to take values in a bounded integer set N_i, and for each $n_i \in N_i$, x_i is a vector in R^{p_i} constrained to take values in a bounded polyhedron $X_i(n_i)$ which depends on n_i. *The real-valued function $f_i(\cdot, n_i)$ is assumed to be concave (for example linear) on $X_i(n_i)$ for all i and $n_i \in N_i$*. Also, for each i and $n_i \in N_i$, the function $h_i(\cdot, n_i)$ maps R^{p_i} into R^m. The vector $b \in R^m$ is given. *We assume that each component of the function $h_i(\cdot, n_i)$ is concave on $X_i(n_i)$.*

One may interpret (PIP) as a problem of finding a minimum cost production schedule by I production units while meeting the "demand" constraints implied by $\sum_{i=1}^{I} h_i(x_i, n_i) \leq b$. Each set $X_i(n_i)$ may be viewed as a "production region" within which the production cost is $f_i(x_i, n_i)$. In this way, a broad variety of production cost functions is allowed including discontinuous, concave, and piecewise linear convex functions. An example is shown in Fig. 5.12, where x_i is a scalar and there are four production regions.

A dynamic version of (PIP) is obtained by considering a time horizon of T periods ($T > 1$) and "setup" costs for passing from one production region to another at the beginning of each time period. The problem is

$$(1) \quad \text{minimize} \quad \sum_{t=1}^{T} \sum_{i=1}^{I} \{f_{it}(x_{it}, n_{it}) + s_{it}(n_{i,t-1}, n_{it})\}$$

$$\text{subject to} \quad \sum_{i=1}^{I} h_{it}(x_{it}, n_{it}) \leq b_t, \qquad t = 1, \ldots, T,$$

$$n_{it} \in N_{it}(n_{i,t-1}), \qquad x_{it} \in X_{it}(n_{it}), \quad i = 1, \ldots, I, \quad t = 1, \ldots, T.$$

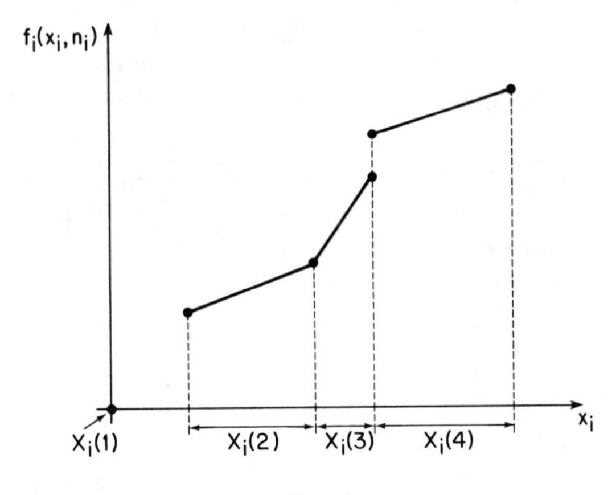

FIG. 5.12

Here, for each i and t, f_{it}, h_{it}, and X_{it} satisfy similar assumptions as f_i, h_i, and X_i in (PIP), the integer constraint set $N_{it}(n_{i,t-1})$ for n_{it} depends on $n_{i,t-1}$, and $s_{it}(n_{i,t-1}, n_{it})$ represents the setup cost for passing from the production region $n_{i,t-1}$ to the region n_{it} at time period t.

We now introduce a dual problem for (PIP) and its dynamic version.

The Dual Problem

For $\mu \geq 0$ define

$$d(\mu) = \min_{\substack{n_i \in N_i \\ x_i \in X_i(n_i)}} \left\{ \sum_{i=1}^{I} [f_i(x_i, n_i) + \mu' h_i(x_i, n_i)] - \mu' b \right\}. \tag{2}$$

The dual problem corresponding to (PIP) is

(DIP)
$$\text{maximize} \quad d(\mu)$$
$$\text{subject to} \quad \mu \geq 0.$$

Similarly for $\mu = (\mu_1, \ldots, \mu_T)$, where for all t, $\mu_t \in R^m$, $\mu_t \geq 0$, define

$$\tilde{d}(\mu) = \min_{\substack{n_{it} \in N_{it}(n_{i,t-1}) \\ x_{it} \in X_{it}(n_{it})}} \left\{ \sum_{t=1}^{T} \sum_{i=1}^{I} [f_{it}(x_{it}, n_{it}) + s_{it}(n_{i,t-1}, n_{it}) + \mu_t' h_{it}(x_{it}, n_{it})] - \mu_t' b_t \right\}. \tag{3}$$

The dual problem corresponding to the dynamic problem (1) is

$$\text{maximize} \quad \tilde{d}(\mu) \tag{4}$$
$$\text{subject to} \quad \mu \geq 0.$$

The value of the dual functional $d(\mu)$ can be calculated relatively easily. In view of the separable nature of the objective and constraint functions, the minimization in (2) can be carried out separately for each i. Because $f_i(\cdot, n_i)$ and the components $h_i(\cdot, n_i)$ have been assumed concave over $X_i(n_i)$, it is sufficient to carry out each of the separate minimizations in (2) over the extreme points of $X_i(n_i)$. In what follows we shall implicitly assume that these extreme points are readily available. If $x_i(j, n_i)$, $j = 1, \ldots, j_{n_i}$, are the extreme points of $X_i(n_i)$, we can write $d(\mu)$ as

$$(5)\qquad d(\mu) = \sum_{i=1}^{I} \min_{\substack{n_i \in N_i \\ j=1,\ldots,j_{ni}}} \{f_i[x_i(j, n_i), n_i] + \mu' h_i[x_i(j, n_i), n_i]\} - \mu' b.$$

Similarly, given the extreme points of $X_{it}(n_{it})$ for each i and t, it is possible to carry out the minimization in (3) separately for each i by means of a simple dynamic programming recursion. In what follows *we shall restrict ourselves for simplicity to* (PIP) *but the analysis to be given fully extends to its dynamic version* (1) (see Bertsekas *et al.*, 1981).

One of the most common approaches for solving integer programs such as (PIP) is the so-called *Lagrangian relaxation* method (Geoffrion, 1974; Shapiro, 1979) which is based on solution of the dual problem. The dual optimal value provides a lower bound to the optimal value of the original integer program. This lower bound together with the solution of the dual problem is used in turn (possibly in conjunction with the branch-and-bound technique) to provide a good approximate solution of the original problem. Verification of the quality of this solution is based on its cost (which is an upper bound to the optimal cost) and the lower bound obtained from the dual problem.

An approach of this type is typically successful in solving problems of large size only if the following two conditions are met:

(a) The difference between the optimal values of the primal and dual problems (the duality gap) is relatively small.

(b) The method used for solving the dual problem provides sufficient information for generating a nearly optimal feasible solution of the primal problem.

It turns out that the *duality gap for* (PIP) *is typically quite small* (*in relative terms*) *if the number of separable terms I is large, and in fact becomes smaller as I increases.* We shall demonstrate this fact under some conditions [see Assumption (A) that follows] in the next subsection.

Regarding the possibility of generating a good suboptimal solution of (PIP), the solution of the following problem is of particular interest.

The Relaxed Problem

Consider a problem which is the same as (PIP) except that instead of choosing, for each i, an integer $n_i \in N_i$ and a vector $x_i \in X_i(n_i)$ we choose a *probability distribution* over all the extreme points of the sets $X_i(n_i)$, $n_i \in N_i$. In other words, for each i, we enlarge the feasible set $\{(x_i, n_i) \mid n_i \in N_i, x_i \in X_i(n_i)\}$ to include all *randomized decisions*. If $x_i(j, n_i)$, $j = 1, \ldots, j_{n_i}$, denotes the extreme points of $X_i(n_i)$ and $p_i(j, n_i)$ are the corresponding probabilities, the relaxed version of (PIP) is stated as

$$\text{(RIP)} \quad \text{minimize} \quad \sum_{i=1}^{I} \sum_{n_i \in N_i} \sum_{j=1}^{j_{n_i}} p_i(j, n_i) f_i[x_i(j, n_i), n_i]$$

$$\text{subject to} \quad \sum_{i=1}^{I} \sum_{n_i \in N_i} \sum_{j=1}^{j_{n_i}} p_i(j, n_i) h_i[x_i(j, n_i), n_i] \le b,$$

$$\sum_{n_i \in N_i} \sum_{j=1}^{j_{n_i}} p_i(j, n_i) = 1, \qquad i = 1, \ldots, I,$$

$$p_i(j, n_i) \ge 0, \qquad i = 1, \ldots, I, \quad n_i \in N_i, \quad j = 1, \ldots, j_{n_i}.$$

There is also a relaxed problem for the dynamic problem (1) (see Bertsekas *et al.*, 1981), which admits an interpretation consistent with the theory of relaxed optimal control.

The duality gap estimate of the next section and the subsequent analysis is based on the following assumption which can be expected to hold for many problems of practical interest:

Assumption (A): Given any feasible solution

$$\{p_i(j, n_i) \mid i = 1, \ldots, I, n_i \in N_i, j = 1, \ldots, j_{n_i}\}$$

of (RIP) there exists a feasible solution $\{(\bar{n}_i, \bar{x}_i) \mid i = 1, \ldots, I\}$ of (PIP) such that

$$h_i(\bar{x}_i, \bar{n}_i) \le \sum_{n_i \in N_i} \sum_{j=1}^{j_{n_i}} p_i(j, n_i) h_i[x_i(j, n_i), n_i] \qquad \forall\, i = 1, \ldots, I.$$

We note that (RIP) is a linear program in the variables $p_i(j, n_i)$, $i = 1, \ldots, I$, $n_i \in N_i$, and $j = 1, \ldots, j_{n_i}$. If we write, using (5), the dual problem (DIP) in the equivalent form

$$\text{maximize} \quad \sum_{i=1}^{I} z_i - \mu' b$$

$$\text{subject to} \quad \mu \ge 0, \qquad f_i[x_i(j, n_i), n_i] + \mu' h_i[x_i(j, n_i), n_i] \ge z_i,$$

$$i = 1, \ldots, I, \quad n_i \in N_i, \quad j = 1, \ldots, j_{n_i},$$

then we find that (RIP) *and* (DIP) *are dual linear programs and therefore have the same optimal value.* Furthermore, every feasible solution of (RIP) in which all probabilities $p_i(j, n_i)$ are either zero or unity correspond to a feasible solution of (PIP). Therefore *the duality gap can alternatively be viewed as the decrease in optimal cost obtained by allowing randomized decisions.*

In view of the fact that (RIP) has I equality constraints and m inequality constraints (except for the nonnegativity constraints), we have that in any basic solution of (RIP) there can be at most $(m + I)$ nonzero probabilities $p_i(j, n_i)$. Since for each i at least one probability $p_i(j, n_i)$ must be nonzero, it follows (assuming $I > m$) that, for at least $(I - m)$ indices i, all the probabilities $p_i(j, n_i)$ are either zero or unity, and for only a maximum of m indices it is possible to have two or more probabilities $p_i(j, n_i)$ being nonzero. This suggests that, if I is much larger than m, then it should be possible to devise heuristic rules for modifying an optimal solution of (RIP) to obtain a feasible solution of (PIP) with value that is relatively close to the optimal value of (RIP). The value of this feasible solution of (PIP) can then be compared with the optimal value of (RIP) or equivalently, the optimal value of (DIP). If these values are sufficiently close, the feasible solution will be accepted as final. Otherwise one has to proceed with the branch-and-bound technique. Actually the procedure we have described amounts to examining the first node of the branch-and-bound tree in the context of the Lagrangian relaxation process.

We claim that *if I is much larger than m and use is made of the solution of the relaxed problem, then an excellent suboptimal solution of* (SIP) *can typically be obtained at the very first node of the branch-and-bound tree.* We shall demonstrate this fact via the computational example given in Section 5.1.3. The papers by Lauer *et al.* (1981) and Bertsekas *et al.* (1981) provide computational results substantiating this claim for the corresponding methodology as applied to dynamic problems of the form (1) arising in electric power system scheduling.

The main advantage that the approximation method of Section 5.1.3 offers over subgradient-type methods is that, when used to solve the dual problem (DIP), it simultaneously provides an optimal solution of the relaxed problem (RIP). This relaxed solution can then be used to generate a good suboptimal solution of (PIP). We note that the capability of solving simultaneously both the dual and the relaxed problem is also shared by the simplex method. It is unclear whether in a specific instance of (PIP) it is preferable to use the approximation method over the simplex method. The advantages of the approximation method manifest themselves primarily in the context in the dynamic problem (1) for which the simplex method quickly becomes unwieldy as the number of time periods T increases.

5.6.1 *An Estimate of the Duality Gap*

The estimate of the duality gap to be derived applies to a broad class of problems that includes (PIP) as a special case. It is therefore worthwhile as well as convenient to develop this estimate in a general setting.

Consider the following problem:

$$\text{(P)} \qquad \text{minimize} \quad \sum_{i=1}^{I} f_i(x_i)$$

$$\text{subject to} \quad x_i \in X_i, \qquad \sum_{i=1}^{I} h_i(x_i) \leq b,$$

where I is a positive integer, b is a given vector in R^m (m is a positive integer), X_i is a subset of R^{p_i} (p_i is a positive integer for each i), and f_i: conv$(X_i) \to R$ and h_i: conv$(X_i) \to R^m$ are given functions defined on the convex hull of X_i denoted conv(X_i). We assume the following:

Assumption (A1): There exists at least one feasible solution of problem (P).

Assumption (A2): For each i, the subset of R^{p_i+m+1}

$$\{(x_i, h_i(x_i), f_i(x_i)) \mid x_i \in X_i\}$$

is compact.

Assumption (A2) implies that X_i is compact. It is satisfied whenever X_i is compact and both f_i and h_i are continuous on X_i. It is also satisfied for the special case of the integer program (PIP) described earlier. Note that no convexity assumptions are made on f_i, h_i, or X_i.

For each i, define the function $\tilde{f}_i$: conv$(X_i) \to R$ by

$$\text{(6)} \quad \tilde{f}_i(\tilde{x}) = \inf\left\{ \sum_{j=1}^{p_i+1} \alpha^j f_i(x^j) \;\middle|\; \tilde{x} = \sum_{j=1}^{p_i+1} \alpha^j x^j,\ x^j \in X_i,\ \sum_{j=1}^{p_i+1} \alpha^j = 1,\ \alpha^j \geq 0 \right\} \quad \forall\, \tilde{x} \in \text{conv}(X_i).$$

The function $\tilde{f}_i$ may be viewed as a "convexified" version of f_i on conv(X_i). Figure 5.13 shows an example of f_i and the corresponding $\tilde{f}_i$, where X_i consists of the union of an interval and a single point. Similarly, define the function $\tilde{h}_i$: conv$(X_i) \to R^m$ by

$$\text{(7)} \quad \tilde{h}_i(\tilde{x}) = \inf\left\{ \sum_{j=1}^{p_i+1} \alpha^j h_i(x^j) \;\middle|\; \tilde{x} = \sum_{j=1}^{p_i+1} \alpha^j x^j,\ x^j \in X_i,\ \sum_{j=1}^{p_i+1} \alpha^j = 1,\ \alpha^j \geq 0 \right\} \quad \forall\, \tilde{x} \in \text{conv}(X_i),$$

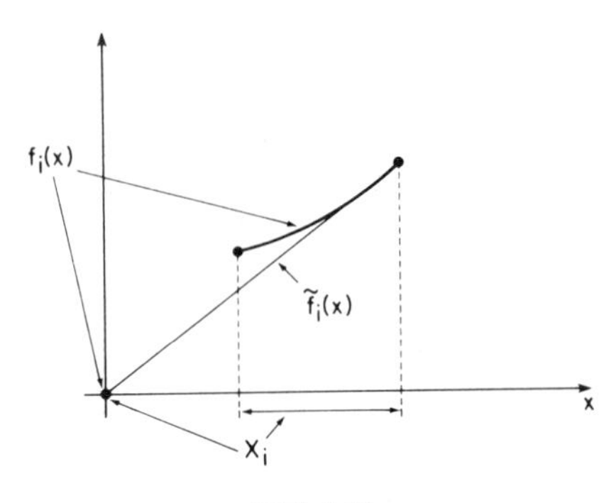

FIG. 5.13

where the infimum is taken separately for each of the m coordinates of the function h_i. Note that if f_i is a convex function on(X_i) and (X_i) is convex, then $f_i = \tilde{f}_i$. A similar statement can be made concerning h_i and $\tilde{h}_i$.

Our third assumption corresponds to Assumption (A) regarding (PIP).

Assumption (A3): For each i, given any vector $\tilde{x}$ in $\text{conv}(X_i)$, there exists $x \in X_i$ such that $h_i(x) \leq \tilde{h}_i(\tilde{x})$.

Note that (A3) is satisfied if X_i is convex and each component of h_i is a convex function. It can be expected to hold in many problems of practical interest.

Define for each i the function $\hat{f}_i: \text{conv}(X_i) \to R$ by

$$\hat{f}_i(\tilde{x}) = \inf\{f_i(x) \mid h_i(x) \leq \tilde{h}_i(\tilde{x}),\ x \in X_i\} \qquad \forall\, \tilde{x} \in \text{conv}(X_i). \tag{8}$$

Note that, by (A3), the constraint set for the minimization indicated in (8) is nonempty. Our estimate of the duality gap is given in terms of the scalars

$$\rho_i = \sup\{\hat{f}_i(x) - \tilde{f}_i(x) \mid x \in \text{conv}(X_i)\}. \tag{9}$$

Since we have, for all $x \in \text{conv}(X_i)$,

$$\hat{f}_i(x) \leq \sup\{f_i(x_i) \mid x_i \in X_i\}, \qquad \tilde{f}_i(x) \geq \inf\{f_i(x_i) \mid x_i \in X_i\},$$

it follows that an easily obtainable overestimate of ρ_i is

$$\rho_i \leq \sup\{f_i(x_i) \mid x_i \in X_i\} - \inf\{f_i(x_i) \mid x_i \in X_i\}.$$

Figures 5.14–5.17 show the scalar ρ_i for X_i consisting of the union of an interval and a single point, and for specific cases of f_i and h_i.

Consider now the dual problem

$$\text{(D)} \qquad \text{maximize} \quad d(\mu) = \inf_{\substack{x_i \in X_i \\ i=1,\ldots,I}} \left\{ \sum_{i=1}^{I} [f_i(x_i) + \mu' h_i(x_i)] - \mu' b \right\}$$

$$\text{subject to} \quad \mu \geq 0.$$

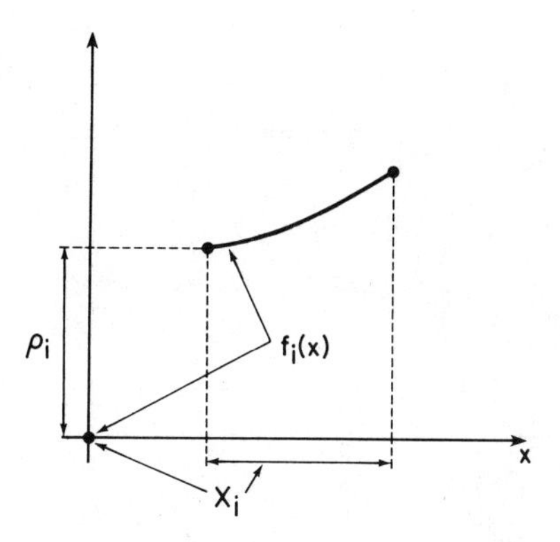

FIG. 5.14 $h_i(x) = -x, \rho_i > 0$

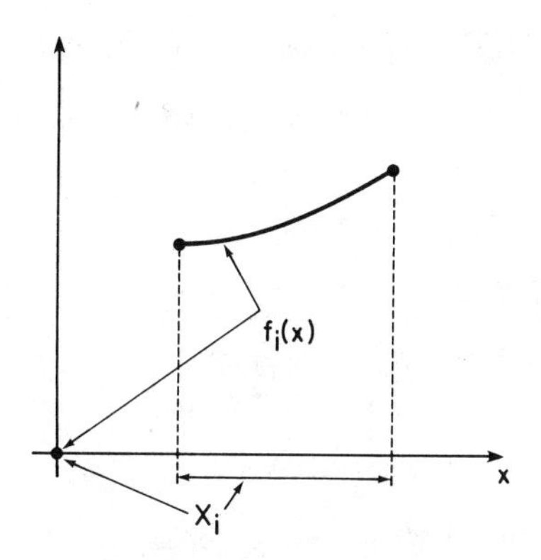

FIG. 5.15 $h_i(x) = x, \rho_i = 0$

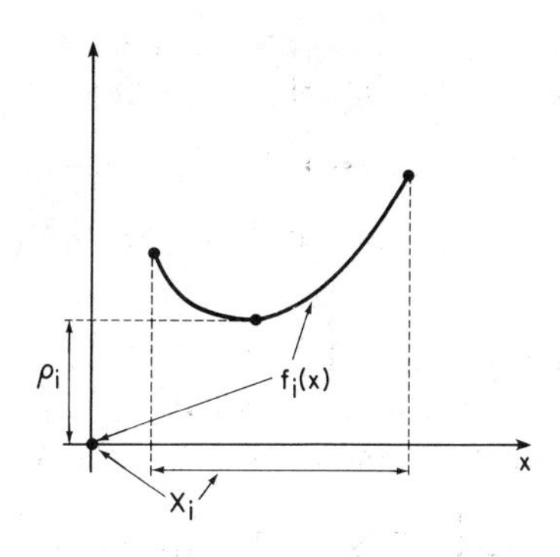

FIG. 5.16 $h_i(x) = -x, \rho_i > 0$

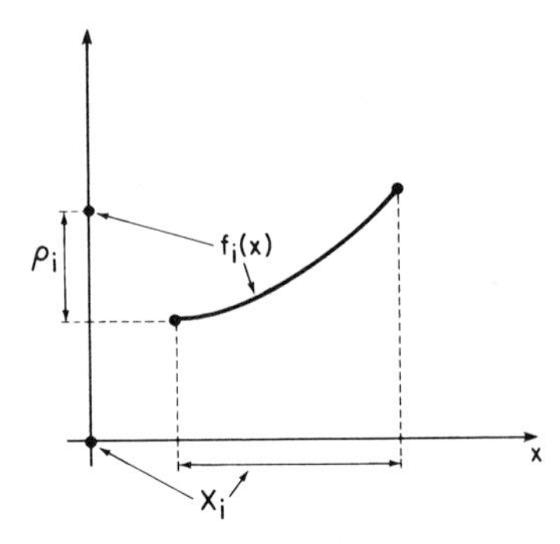

FIG. 5.17 $h_i(x) = x, \rho_i > 0$

Let inf(P) and sup(D) denote the optimal values of the primal and dual problems respectively. We have the following result.

Proposition 5.26: Under Assumptions (A1)–(A3) there holds

(10) $$\inf(P) - \sup(D) \leq (m + 1)E,$$

where

(11) $$E = \max\{\rho_i \mid i = 1, \ldots, I\}.$$

Proof: Define the sets

(12) $$Y_i = \{y_i \mid y_i = [h_i(x_i), f_i(x_i)], x_i \in X_i\}, \qquad i = 1, \ldots, I,$$

and their vector sum

(13) $$Y = Y_1 + Y_2 + \cdots + Y_I.$$

In view of (A2), Y, conv(Y), and Y_i, conv(Y_i), $i = 1, \ldots, I$, are all compact sets. By definition of Y, we have

(14) $$\inf(P) = \min\{w \mid \text{there exists } (z, w) \in Y \text{ with } z \leq b\}.$$

By using (A1), (A2), and a standard duality argument (see Magnanti *et al.*, 1976; Shapiro, 1979, p. 150) we can also show that

(15) $$\sup(D) = \min\{w \mid \text{there exists } (z, w) \in \text{conv}(Y) \text{ with } z \leq b\}.$$

We now use the following theorem (see Ekeland and Temam, 1976, Appendix I).

Shapley–Folkman Theorem: Let Y_i, $i = 1, \ldots, I$, be a collection of subsets of R^{m+1}. Then for every $y \in \text{conv}(\sum_{i=1}^{I} Y_i)$, there exists a subset $I(y) \subset \{1, \ldots, I\}$ containing at most $(m + 1)$ indices such that

$$y \in \left[\sum_{i \notin I(y)} Y_i + \sum_{i \in I(y)} \text{conv}(Y_i)\right].$$

Now let $(\bar{z}, \bar{w}) \in \text{conv}(Y)$ be such that [compare with (15)]

$$\bar{w} = \sup(\text{D}), \qquad \bar{z} \le b. \tag{16}$$

By applying the Shapley–Folkman theorem to the set $Y = \sum_{i=1}^{I} Y_i$ given by (12) and (13), we have that there exists a subset $\bar{I} \subset \{1, \ldots, I\}$, with at most $(m + 1)$ indices, and vectors

$$(\bar{b}_i, \bar{w}_i) \in \text{conv}(Y_i), \quad i \in \bar{I}, \qquad \bar{x}_i \in X_i, \quad i \notin \bar{I},$$

such that [compare with (16)]

$$\sum_{i \notin \bar{I}} h_i(\bar{x}_i) + \sum_{i \in \bar{I}} \bar{b}_i = \bar{z} \le b, \tag{17}$$

$$\sum_{i \notin \bar{I}} f_i(\bar{x}_i) + \sum_{i \in \bar{I}} \bar{w}_i = \sup(\text{D}). \tag{18}$$

Using the Carathéodory theorem for representing elements of the convex hull of a set, we have that, for each $i \in \bar{I}$, there must exist vectors $x_i^1, \ldots, x_i^{m+2} \in X_i$ and scalars $\alpha_i^1, \ldots, \alpha_i^{m+2}$ such that

$$\sum_{j=1}^{m+2} \alpha_i^j = 1, \qquad \alpha_i^j \ge 0, \quad j = 1, \ldots, m + 2,$$

$$\bar{b}_i = \sum_{j=1}^{m+2} \alpha_i^j h_i(x_i^j), \qquad \bar{w}_i = \sum_{j=1}^{m+2} \alpha_i^j f_i(x_i^j).$$

Using the definition of $\tilde{f}_i$, $\tilde{h}_i$, and ρ_i [compare with (6)–(9)], we have

$$\bar{b}_i \ge \tilde{h}_i\left(\sum_{j=1}^{m+2} \alpha_i^j x_i^j\right), \tag{19}$$

$$\bar{w}_i \ge \tilde{f}_i\left(\sum_{j=1}^{m+2} \alpha_i^j x_i^j\right) \ge \hat{f}_i\left(\sum_{j=1}^{m+2} \alpha_i^j x_i^j\right) - \rho_i. \tag{20}$$

By combining (16)–(20), we obtain

$$\sum_{i \notin \bar{I}} h_i(\bar{x}_i) + \sum_{i \in \bar{I}} \tilde{h}_i\left(\sum_{j=1}^{m+2} \alpha_i^j x_i^j\right) \le b \tag{21}$$

$$\sum_{i \notin \bar{I}} f_i(\bar{x}_i) + \sum_{i \in \bar{I}} \hat{f}_i\left(\sum_{j=1}^{m+2} \alpha_i^j x_i^j\right) \le \sup(\text{D}) + \sum_{i \in \bar{I}} \rho_i. \tag{22}$$

Given any $\varepsilon > 0$ and $i \in \bar{I}$, we can find [using (A3)] a vector $\bar{x}_i \in X_i$ such that [compare with (8)]

$$f_i(\bar{x}_i) \le \hat{f}\left(\sum_{j=1}^{m+2} \alpha_i^j x_i^j\right) + \varepsilon, \quad h_i(\bar{x}_i) \le \tilde{h}_i\left(\sum_{j=1}^{m+2} \alpha_i^j x_i^j\right).$$

These relations together with (21) and (22) yield

$$\sum_{i=1}^{I} h_i(\bar{x}_i) \le b, \tag{23}$$

$$\sum_{i=1}^{I} f_i(\bar{x}_i) \le \sup(\mathrm{D}) + \sum_{i \in \bar{I}} (\rho_i + \varepsilon), \tag{24}$$

Since by (23), $(\bar{x}_1, \ldots, \bar{x}_I)$ is a feasible vector for (P), we have $\inf(\mathrm{P}) \le \sum_{i=1}^{I} f_i(\bar{x}_i)$, and (24) yields

$$\inf(\mathrm{P}) \le \sup(\mathrm{D}) + \sum_{i \in \bar{I}} (\rho_i + \varepsilon). \tag{25}$$

Since ε is arbitrary, $\bar{I}$ contains at most $(m + 1)$ elements, and $E = \max\{\rho_i \mid i = 1, \ldots, I\}$, (25) proves the desired estimate (10). Q.E.D.

The significance of Proposition 5.26 lies in the fact that the estimate $(m + 1)E$ depends only on m and E but not on I. Thus if we consider instead of problem (P), the problem

$$\text{minimize} \quad \frac{1}{I} \sum_{i=1}^{I} f_i(x_i)$$

$$\text{subject to} \quad x_i \in X_i, \qquad \sum_{i=1}^{I} h_i(x_i) \le b,$$

the objective function of which represents "average cost per term," the duality gap estimate becomes

$$\inf(\mathrm{P}) - \sup(\mathrm{D}) \le \frac{m + 1}{I} E.$$

Thus the duality gap goes to zero as $I \to \infty$. Otherwise stated, *if the optimal value of problem* (P) *is proportional to* I, *the ratio of the duality gap over the optimal value goes to zero as* $I \to \infty$.

5.6.2 Solution of the Dual and Relaxed Problems

We consider solution of the dual problem (DIP) via the approximation method based on the exponential penalty function [compare with Section 5.1.3, Eqs. (25) and (26)]. Taking into account the expression (5) for the dual functional, we form an *approximate dual functional* defined by

$$d_c(\mu; p) = -\frac{1}{c} \sum_{i=1}^{I} \log\left\{ \sum_{n_i \in N_i} \sum_{j=1}^{j_{n_i}} p_i(j, n_i) e^{-c a_i(\mu, j, n_i)} \right\} - \mu' b, \tag{26}$$

where

$$a_i(\mu, j, n_i) = f_i[x_i(j, n_i), n_i] + \mu' h_i[x_i(j, n_i), n_i], \tag{27}$$

c is a positive scalar parameter, and the multipliers $p_i(j, n_i)$ are positive scalars satisfying

$$\sum_{n_i \in N_i} \sum_{j=1}^{j_{n_i}} p_i(j, n_i) = 1, \qquad i = 1, \ldots, I. \tag{28}$$

The approximation method consists of sequential solution of approximate dual problems of the form

$$\begin{aligned} &\text{maximize} \quad d_{c_k}(\mu; p^k) \\ &\text{subject to} \quad \mu \geq 0 \end{aligned} \tag{29}$$

followed by multiplier iterations of the form

$$p_i^{k+1}(j, n_i) = p_i^k(j, n_i) e^{-c_k a_i(\mu_k, j, n_i)} \Bigg/ \sum_{\bar{n}_i \in N_i} \sum_{\bar{j}=1}^{j_{\bar{n}_i}} p_i^k(\bar{j}, \bar{n}_i) e^{-c_k a_i(\mu_k, \bar{j}, \bar{n}_i)}, \tag{30}$$

where μ_k solves problem (29). The initial multipliers $p_i^0(j, n_i)$ are strictly positive and satisfy (28), and the penalty parameter sequence $\{c_k\}$ satisfies $0 < c_k \leq c_{k+1}$ for all k.

We note that the approximate dual problem (29) is twice continuously differentiable and thus can be solved by means of a constrained version of Newton's method—for example, the one discussed in Section 1.5. The generated sequence $\{\mu_k\}$ can be expected to converge to an optimal solution of the dual problem (compare with Proposition 5.12), and the multiplier sequences $\{p_i^k(j, n_i)\}$ can also be expected to converge to limits $\bar{p}_i(j, n_i)$ satisfying [compare with (28)]

$$\bar{p}_i(j, n_i) \geq 0, \qquad i = 1, \ldots, I, \quad n_i \in N_i, \quad j = 1, \ldots, j_{n_i},$$

$$\sum_{n_i \in N_i} \sum_{j=1}^{j_{n_i}} \bar{p}_i(j, n_i) = 1, \qquad i = 1, \ldots, I.$$

Furthermore, by applying Proposition 5.12 and using the fact that the relaxed problem (RIP) and the dual problem (DIP) are dual linear programs, we find that *the set of multipliers* $\{\bar{p}_i(j, n_i) \mid i = 1, \ldots, I,\ n_i \in N_i,\ j = 1, \ldots, j_{ni}\}$ *is an optimal solution of the relaxed problem.* Thus *the approximation method can be expected to solve simultaneously both the dual and the relaxed problems.*

We now demonstrate via example how the solution of the relaxed problem can be used to generate a good suboptimal solution of (PIP).

Example: Consider the problem

$$\text{minimize} \quad \sum_{i=1}^{I} f_i(x_i) \tag{31}$$

$$\text{subject to} \quad \sum_{i=1}^{I} x_i \geq b,$$

where b is a given scalar, for each i, x_i is a scalar taking values in a set of the form

$$X_i = \{0\} \cup [\alpha_i, \beta_i],$$

and $f_i: X_i \to R$ has the form shown in Fig. 5.18. Clearly this problem is a special case of (PIP). There are three production regions ($n_i \in \{1, 2, 3\}$) and the corresponding production sets are

$$X_i(1) = \{0\}, \qquad X_i(2) = [\alpha_i, \tfrac{1}{2}(\alpha_i + \beta_i)], \qquad X_i(3) = [\tfrac{1}{2}(\alpha_i + \beta_i), \beta_i].$$

Thus, there is a total of five extreme points $x_i(j, n_i)$, for each i,

$$x_i(1, 1) = 0, \quad x_i(1, 2) = \alpha_i,$$
$$x_i(2, 2) = \tfrac{1}{2}(\alpha_i + \beta_i), \quad x_i(1, 3) = \tfrac{1}{2}(\alpha_i + \beta_i), \quad x_i(2, 3) = \beta_i.$$

Suppose that after applying the approximation method, we obtain a solution $\{\bar{p}_i(j, n_i)\}$ of the relaxed problem. We shall describe a reasonable procedure for generating a feasible solution of problem (31). The main idea is to assign, for each i, a production region $\bar{n}_i$ on the basis of the probabilities $\bar{p}_i(j, n_i)$ and then choose optimally x_i within the corresponding region. The procedure is guaranteed to generate a feasible solution assuming that problem (31) has at least one such solution.

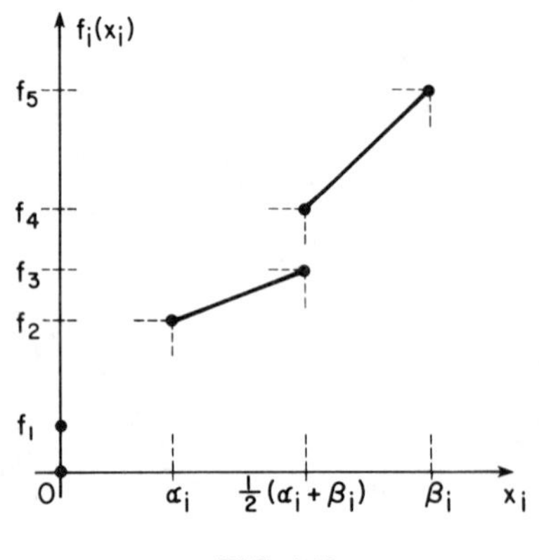

FIG. 5.18

Choose $\gamma = 0.5$. For all i, set

$$\bar{n}_i = 1 \quad \text{if} \quad \bar{p}_i(1, 1) > \gamma,$$
$$\bar{n}_i \neq 1 \quad \text{if} \quad \bar{p}_i(1, 1) \leq \gamma.$$

Then test whether the condition

$$\sum_{\substack{i=1 \\ \bar{n}_i \neq 1}}^{I} \beta_i \geq b \tag{32}$$

is satisfied. If not, increase γ by 0.1 and repeat the procedure until (32) is satisfied. Let $\bar{I} = \{i \mid \bar{n}_i \neq 1\}$, and choose $\delta = 0$. For all $i \in \bar{I}$, set

$$\bar{n}_i = 2 \quad \text{if} \quad \bar{p}_i(1, 2) + \bar{p}_i(2, 2) > \bar{p}_i(1, 3) + \bar{p}_i(2, 3) + \delta,$$
$$\bar{n}_i = 3 \quad \text{if} \quad \bar{p}_i(1, 2) + \bar{p}_i(2, 2) \leq \bar{p}_i(1, 3) + \bar{p}_i(2, 3) + \delta$$

Define $I_2 = \{i \mid \bar{n}_i = 2\}$ and $I_3 = \{i \mid \bar{n}_i = 3\}$, and test whether the condition

$$\sum_{i \in I_2} \frac{\alpha_i + \beta_i}{2} + \sum_{i \in I} \beta_i \geq b \tag{33}$$

is satisfied. If not, increase δ by 0.1, and repeat the procedure until (33) is satisfied. Then solve the (trivial) linear program

$$\begin{aligned} &\text{minimize} \quad \sum_{i=1}^{I} f_i(x_i) \\ &\text{subject to} \quad \sum_{i=1}^{I} x_i \geq b, \qquad x_i \in X_i(\bar{n}_i), \quad i = 1, \ldots, I, \end{aligned} \tag{34}$$

to obtain a feasible solution $\{\bar{x}_i \mid i = 1, \ldots, I\}$ of problem (31). It can be easily seen that the procedure for selecting $\bar{n}_i$ described above guarantees that problem (34) has at least one feasible solution.

In Table 5.1, we provide some computational results using this procedure and randomly generated problems. The table shows the number of production units I, and the ratio $(UB - LB)/LB$, where LB is the best lower bound obtained from solution of the dual problem, and UB is the value of the feasible solution of the primal problem (31) generated via the procedure described above. Each entry represents an average over five randomly generated problems with $b = 2I$ and α_i, β_i, $f_1, \ldots, f_5$ (refer to Fig. 5.18) chosen as

$$f_1 = r_1, \quad f_2 = 1.5r_2, \quad f_3 = f_2 + 1.5r_3, \quad f_4 = f_3 + 0.5r_4,$$
$$f_5 = f_4 + 2r_5, \quad \alpha_i = 0.5r_6, \quad \beta_i = \alpha_i + 3r_7,$$

where $r_1, \ldots, r_7$ are scalars chosen according to a uniform probability distribution from the interval $[0, 1]$. It can be seen from Table 5.1 that the relative difference between the upper and lower bound decreases as I increases, which is consistent with the estimate of the duality gap of Proposition 5.26.

TABLE 5.1

Estimates of Relative Duality Gap for Various Problem Sizes

I	$\frac{UB-LB}{LB}$
10	9.56×10^{-2}
50	2.21×10^{-2}
100	1.17×10^{-2}
150	0.574×10^{-2}
200	0.359×10^{-2}
250	0.309×10^{-2}
300	0.187×10^{-2}

5.7 Notes and Sources

Notes on Section 5.1: The class of penalty functions P_E and $\hat{P}_I$ were first proposed in Kort and Bertsekas (1972), while the class P_I was proposed in Kort and Bertsekas (1973). Rockafellar (1971) proposed earlier the use of the augmented Lagrangian based on the quadratic penalty function $\phi(t) = (1/2)t^2$. The use of the exponential penalty function in approximation procedures for minimax problems was suggested in Bertsekas (1977). For an application of this function in optimization of multicommodity network flows, see Vastola (1979), and for an application in electric diode network analysis, see Bertsekas (1976e). Convergence analysis for nonconvex problems relating to this function and the associated method of multipliers is provided in Nguyen and Strodiot (1979). The application of the exponential penalty function to the problem of finding a feasible point of a system of inequalities was suggested by Schnabel (1980).

Notes on Section 5.2: Most of the results of this section were shown by Rockafellar (1973a) for the case where ϕ is quadratic. They were extended in Kort and Bertsekas (1973, 1976) and Kort (1975a).

Notes on Section 5.3: The algorithms of this section together with Propositions 5.7–5.11 are due to Kort and Bertsekas (1973, 1976). Some additional material is given in the Ph.D. thesis by Kort (1975a). A convergence analysis of the algorithm corresponding to the quadratic penalty function was also given independently by Rockafellar (1973b). The inexact minimization algorithm suggested by Rockafellar (1973b) uses a different stopping rule which can be translated into a gradient-based stopping rule such as ours, as shown later in Rockafellar (1976c). Relations of the methods of multipliers with the proximal point algorithm were shown in Rockafellar (1976a,b). The same references describe a modified version of the method which is suitable for problems where Assumption (A5) is not satisfied. Proposition 5.12 is an improved version of an unpublished result due to Kort and Bertsekas. Proposition 5.13 was given in Bertsekas (1975a), but the inequality (48) on which its proof rests is due to Rockafellar (1973a). Proposition 5.14 is due to Rockafellar (1973b).

Notes on Section 5.4: The convergence rate analysis of this section is due to Kort and Bertsekas (1976) with the exception of the result of Proposition 5.21. This result together with the finite convergence property of the penalty method for c sufficiently large are new in the general form given here but stem from independent work of Poljak and Tretjakov (1974) and Bertsekas (1975b, 1976c) regarding polyhedral convex programming problems and the quadratic penalty function. A convergence rate analysis of the proximal point algorithm (which includes as a special case the method of multipliers with a quadratic penalty function) is given in Luque (1981). This work also considers situations where the rate is sublinear.

Notes on Section 5.5: Proposition 5.25 is a generalized version of a result given in Bertsekas (1975b) but descends from earlier works on specific exact nondifferentiable penalty functions referenced in Section 4.6.

Notes on Section 5.6: The material in this section is new and was developed as the monograph was being written. It is based on a method for power system scheduling described in Bertsekas *et al.* (1981) and Lauer *et al.* (1981). The analysis of the duality gap bears similarity with the one of Aubin and Ekeland (1976). Our estimate is based on different assumptions and is considerably sharper than the one in that reference.

References

The following abbreviations have been used in the reference list.

M.P.	Math. Programming	S.J.C.	SIAM J. on Control
M.S.	Management Science	S.J.C.O.	SIAM J. on Control and Optimization
J.O.T.A.	J. Opt. Th. & Appl.	O.R.	Operations Research

Armijo, L. (1966). Minimization of functions having continuous partial derivatives. *Pacific J. Math.* **16**, 1–3.

Arrow, K. J., and Solow, R. M. (1958). Gradient methods for constrained maxima with weakened assumptions. In "Studies in Linear and Nonlinear Programming" (K. J. Arrow, L. Hurwitz, and H. Uzawa, eds.), pp. 166–176. Stanford Univ. Press, Stanford, California.

Arrow, K. J., Hurwicz, L., and Uzawa, H., eds. (1958). "Studies in Linear and Nonlinear Programming." Stanford Univ. Press, Stanford, California.

Arrow, K. J., Gould, F. J., and Howe, S. M. (1973). A general saddle point result for constrained minimization. *M.P.* **5**, 225–234.

Aubin, J. P., and Ekeland, I. (1976). Estimates of the duality gap in nonconvex optimization. *Math. Oper. Res.* **1**, 225–245.

Auslender, A. (1976). "Optimization: Methodes Numeriques." Mason, Paris.

Avriel, M. (1976). "Nonlinear Programming: Analysis and Methods." Prentice-Hall, Englewood Cliffs, New Jersey.

Bazaraa, M. S., and Goode, J. J. (1979). "An Extension of Armijo's Rule to Minimax and Quasi-Newton Methods for Constrained Optimization," Indust. Systems Engr. Rep. Georgia Inst. of Tech., Atlanta, Georgia. *European J. of Operational Research*, to appear.

Beale, E. M. L. (1972). A derivation of conjugate gradients. *In* "Numerical Methods for Nonlinear Optimization" (F. A. Lootsma, ed.), pp. 39–43. Academic Press, New York.

Bertsekas, D. P. (1973). Convergence rate of penalty and multiplier methods. *Proc. 1973 IEEE Confer. Decision Control, San Diego, Calif.*, pp. 260–264.

Bertsekas, D. P. (1974a). Partial conjugate gradient methods for a class of optimal control problems. *IEEE Trans Automat. Control* **19**, 209–217.

Bertsekas, D. P. (1974b). Nondifferentiable optimization via approximation. *Proc. Annual Allerton Confer. Circuit System Theory, 12th, Allerton Park, Ill.* pp. 41–52. Also *in* "Mathematical Programming Study 3" (M. Balinski and P. Wolfe, eds.). pp. 1–25. North-Holland Publ., Amsterdam, 1975.

Bertsekas, D. P. (1974c). On the Goldstein–Levitin–Poljak gradient projection method. *Proc. 1974 IEEE Decision Control Conf., Phoenix, Ariz.*, pp. 47–52. Also *IEEE Trans. Automat. Control* **21**, 174–184 (1976).

Bertsekas, D. P. (1975a). Multiplier methods for convex programming. *IEEE Trans. Automat. Control* **20**, 385–388.

Bertsekas, D. P. (1975b). Necessary and sufficient conditions for a penalty method to be exact. *M.P.* **9**, 87–99.

Bertsekas, D. P. (1975c). Combined primal-dual and penalty methods for constrained minimization. *S.J.C.* **13**, 521–544.

Bertsekas, D. P. (1976a). On penalty and multiplier methods for constrained optimization. *S.J.C.O.* **14**, 216–235.

Bertsekas, D. P. (1976b). Multiplier methods: A survey. *Automatica—J. IFAC* **12**, 133–145.

Bertsekas, D. P. (1976c). Newton's method for linear optimal control problems. *Proc. IFAC Symp. Large-Scale Systems, Udine, Italy* pp. 353–359.

Bertsekas, D. P. (1976d). Minimax methods based on approximation. *Proc. 1976 Johns Hopkins Confer. Inform. Sci. Systems, Baltimore, Md.* pp. 363–365.

Bertsekas, D. P. (1976e). A new algorithm for solution of resistive networks involving diodes. *IEEE Trans. Circuits and Systems* **23**, 599–608.

Bertsekas, D. P. (1977). Approximation procedures based on the method of multipliers. *J.O.T.A.* **23**, 487–510.

Bertsekas, D. P. (1978). On the convergence properties of second order methods of multipliers. *J.O.T.A* **25**, 443–449.

Bertsekas, D. P. (1979a). A convergence analysis of the method of multipliers for nonconvex constrained optimization. Presented at *Proc. Workshop Augmented Lagrangians, IIASA, Vienna.*

Bertsekas, D. P. (1979b) Convexification procedures and decomposition algorithms for large-scale nonconvex optimization problems. *J.O.T.A.* **29**, 169–197.

Bertsekas, D. P. (1980a). "Enlarging the region of convergence of Newton's method for constrained optimization," LIDS Rep. R-985. MIT, Cambridge, Massachusetts. Also *J.O.T.A.* **36**, (1982) 221–252.

Bertsekas, D. P. (1980b). Variable metric methods for constrained optimization based on differentiable exact penalty functions. *Proc. Allerton Confer. Comm. Control Comput., Allerton Park, Ill.* pp. 584–593.

Bertsekas, D. P. (1980c). "Projected Newton methods for optimization problems with simple constraints," LIDS Rep. R-1025. MIT, Cambridge, Massachusetts. Also *S.J.C.O.* **20**, (1982), pp. 221–246.

Bertsekas, D. P., and Gafni, E. (1983). "Projected Newton Methods and Optimization of Multicommodity Flows," *IEEE Trans. Automat. Control,* AC-28, pp. 1090–1096.

Bertsekas, D. P., and Mitter, S. K. (1971). Steepest descent for optimization problems with nondifferentiable cost functionals. *Proc. Annual Princeton Confer. Inform. Sci. Systems, 5th, Princeton, N.J.* pp. 347–351.

Bertsekas, D. P., and Mitter, S. K. (1973). A descent numerical method for optimization problems with nondifferentiable cost functionals. *S.J.C.* **11**, 637–652.

Bertsekas, D. P., Lauer, G. S., Sandell, N. R., Jr., and Posbergh, T. A. (1981). Optimal short term scheduling of large-scale power systems. *Proc. 1981 IEEE Confer. Decision Control, San Diego, Calif.* pp. 432-443, *IEEE Trans. Automat. Control,* Vol. AC-28, 1983, pp. 1–11.

Biggs, M. C. (1972). Constrained minimization using recursive equality quadratic programming. *In* "Numerical Methods for Nonlinear Optimization" (F. A. Lootsma, ed.), pp. 411–428. Academic Press, New York.

Biggs, M. C. (1978). On the convergence of some constrained minimization algorithms based on recursive quadratic programming. *J. Inst. Math. Appl.* **21**, 67–81.

Boggs, P. T., and Tolle, J. W. (1980). Augmented Lagrangians which are quadratic in the multiplier. *J.O.T.A.* **31**, 17–26.

Boggs, P. T., Tolle, J. W., and Wang, P. (1982). On the local convergence of quasi-Newton methods for constrained optimization. *S.J.C.O.* **20**, 161–171.

Brayton, R. K., and Cullum, J. (1979). An algorithm for minimizing a differentiable function subject to box constraints and errors. *J.O.T.A.* **29**, 521–558.

Brent, R. P. (1972). "Algorithms for Minimization without Derivatives." Prentice-Hall, Engelwood Cliffs, New Jersey.

Broyden, C. G. (1970). The convergence of a class of double rank minimization algorithms. *J. Inst. Math. Appl.* **6**, 76–90.

Broyden, C. G. (1972). Quasi-Newton methods. *In* "Numerical Methods for Unconstrained Optimization" (W. Murray, ed.), pp. 87–106. Academic Press, New York.

Broyden, C. G., Dennis, J., and Moré, J. J. (1973). On the local and superlinear convergence of quasi-Newton methods. *J. Inst. Math. Appl.* **12**, 223–245.

Brusch, R. B. (1973). A rapidly convergent method for equality constrained function minimization. *Proc. 1973 IEEE Confer. Decision Control, San Diego, Calif.* pp. 80–81.

Buys, J. D. (1972). Dual algorithms for constrained optimization. Ph.D. Thesis, Rijksuniversiteit de Leiden.

Campos-Filho, A. S. (1971). Numerical computation of optimal control sequences. *IEEE Trans. Automat. Control* **16**, 47–49.

Cannon, M. D., Cullum, C. D., and Polak, E. (1970). "Theory of Optimal Control and Mathematical Programming," McGraw-Hill, New York.

Chamberlain, R. M. (1979). Some examples of cycling in variable metric methods for constrained minimization. *M.P.* **16**, 378–383.

Chamberlain, R. M. (1980). The theory and application of variable metric methods to constrained optimization problems. Ph.D. Thesis, Univ. of Cambridge, Cambridge, England.

Chamberlain, R. M., Lemarechal, C., Pedersen, H. C., and Powell, M. J. D. (1979). The watchdog technique for forcing convergence in algorithms for constrained optimization. *Internat. Symp. Math. Programming, 10th, Montreal, Math. Programming Stud.* **16**, (1982), to appear.

Coleman, T. F., and Conn, A. R. (1980a). "Nonlinear Programming via an Exact Penalty Function: Asymptotic Analysis," Comput. Sci. Dept., Rep. CS-80-30. Univ. of Waterloo, *M.P.*, to appear.

Coleman, T. F., and Conn, A. R. (1980b). "Nonlinear Programming via an Exact Penalty Function: Global Analysis," Comput. Sci. Dept., Rep. CS-80-31. Univ. of Waterloo, *M.P.*, to appear.

Daniel, J. W. (1971). "The Approximate Minimization of Functionals." Prentice-Hall, Englewood Cliffs, New Jersey.

Davidon, W. C. (1959). "Variable Metric Method for Minimization," R and D Rep. ANL-599 (Ref.). U.S. At. Energy Commission, Argonne Nat. Lab., Argonne, Illinois.

Dembo, R. S., Eisenstadt, S. C., and Steihaug, T. (1980). "Inexact Newton Methods," Working Paper, Ser. B, No. 47. Yale Sch. Organ. Management, New Haven, Connecticut.

Dennis, J. E., and Moré, J. J. (1974). A characterization of superlinear convergence and its application to quasi-Newton methods. *MC* **28**, 549–560.

Dennis, J. E., and Moré, J. J. (1977). Quasi-Newton methods: motivation and theory. *SIAM Rev.* **19**, 46–89.

DiPillo, G., and Grippo, L. (1979a). A new class of augmented Lagrangians in nonlinear programming. *S.J.C.O.* **17**, 618–628.

DiPillo, G., and Grippo, L. (1979b). "An Augmented Lagrangian for Inequality Constraints in Nonlinear Programming Problems," Rep. 79–22. Ist. Automat., Univ. di Roma.

DiPillo, G., Grippo, L., and Lampariello, F. (1979). A method for solving equality constrained optimization problems by unconstrained minimization. *Proc. IFIP Confer. Optim. Tech., 9th, Warsaw* pp. 96–105.

Dixon, L. C. W. (1972a). Quasi-Newton algorithms generate identical points. *M.P.* **2**, 383–397.

Dixon, L. C. W. (1972b). Quasi-Newton algorithms generate identical points. *M.P.* **3**, 346–358.

Dixon, L. C. W. (1980). "On the Convergence Properties of Variable Metric Recursive Quadratic Programming Methods," Numer. Optim. Centre, Rep. No. 110. Hatfield Polytechnic, Hatfield, England.

Dolecki, S., and Rolewicz, S. (1979). Exact penalties for local minima. *S.J.C.O.* **17**, 596–606.

Dunn, J. C. (1980). Newton's method and the Goldstein Step Length Rule for constrained minimization problems. *S.J.C.O.* **18**, 659–674.

Dunn, J. C. (1981). Global and asymptotic convergence rate estimates for a class of projected gradient processes. *S.J.C.O.* **19**, 368–400.

Ekeland, I., and Teman, R. (1976). "Convex Analysis and Variational Problems." North-Holland Publ., Amsterdam.

Ermoliev, Y. M., and Shor, N. Z. (1967). On the minimization of nondifferentiable functions." *Kibernetika* (*Kiev*) **3**, 101–102.

Evans, J. P., Gould, F. J., and Tolle, J. W. (1973). Exact penalty functions in nonlinear programming. *Math. Programming* **4**, 72–97.

Everett, H. (1963). Generalized Lagrange multiplier method for solving problems of optimal allocation of resources. *O.R.* **11**, 399–417.

Fadeev, D. K., and Fadeeva, V. N. (1963). "Computational Methods of Linear Algebra." Freeman, San Francisco, California.

Fiacco, A. V., and McCormick, G. P. (1968). "Nonlinear Programming: Sequential Unconstrained Minimization Techniques." Wiley, New York.

Fletcher, R. (1970). A class of methods for nonlinear programming with termination and convergence properties. *In* "Integer and Nonlinear Programming" (J. Abadie, ed.), pp. 157–173. North-Holland Publ., Amsterdam.

Fletcher, R. (1973). A class of methods for nonlinear programming: III. Rates of convergence. *In* "Numerical Methods for Nonlinear Optimization" (F. A. Lootsma, ed.), pp. 371–381. Academic Press, New York.

Fletcher, R. (1975). An ideal penalty function for constrained optimization. *In* "Nonlinear Programming 2" (O. Mangasarian, R. Meyer, and S. Robinson, eds.) pp. 121–163. Academic Press, New York.

Fletcher, R., and Freeman, T. L. (1977). A modified Newton method for minimization, *J.O.T.A.* **23**, 357–372.

Fletcher, R., and Lill, S. (1971). A class of methods for nonlinear programming: II. computational experience. In "Nonlinear Programming" (J. B. Rosen, O. L. Mangasarian, and K. Ritter, eds.), pp. 67–92. Academic Press, New York.

Fletcher, R., and Powell, M. J. D. (1963). A rapidly convergent descent algorithm for minimization. *Comput. J.* **6**, 163–168.

Fletcher, R., and Reeves, C. M. (1964). Function minimization by conjugate gradients. *Comput. J.* **7**, 149–154.

Gabay, D. (1979). Reduced quasi-Newton methods with feasibility improvement for nonlinearly constrained optimization. *Math. Programming Stud.* **16**, (1982), to appear.
Garcia-Palomares, U. M. (1975). Superlinearly convergent algorithms for linearly constrained optimization. *In* "Nonlinear Programming 2" (O. L. Mangasarian, R. R. Meyer, and S. M. Robinson, eds.), pp. 101–119. Academic Press, New York.
Garcia-Palomares, U. M., and Mangasarian, O. L. (1976). Superlinearly convergent quasi-Newton algorithms for nonlinearly constrained optimization problems. *M.P.* **11**, 1–13.
Geoffrion, A. (1974). Lagrangian relaxation for integer programming. *Math. Programming Stud.* **2**, 82–114.
Gill, P. E., and Murray, W. (1972). Quasi-Newton methods for unconstrained optimization. *J. Inst. Math. Appl.* **9**, 91–108.
Gill, P. E., and Murray, W. (1974). "Numerical Methods for Constrained Optimization." Academic Press, New York.
Gill, P. E., Murray, W., and Wright, M. H. (1981). "Practical Optimization." Academic Press, New York.
Glad, T. (1979). Properties of updating methods for the multipliers in augmented Lagrangians. *J.O.T.A.* **28**, 135–156.
Glad, T., and Polak, E. (1979). A multiplier method with automatic limitation of penalty growth. *M.P.* **17**, 140–155.
Goldfarb, D. (1970). A family of variable-metric methods derived by variational means. *Math. Comp.* **24**, 23–26.
Goldfarb, D. (1980). Curvilinear path steplength algorithms for minimization which use directions of negative curvature. *M.P.* **18**, 31–40.
Goldstein, A. A. (1962). Cauchy's method of minimization. *Numer. Math.* **4**, 146–150.
Goldstein, A. A. (1964). Convex programming in Hibert Space. *Bull. Amer. Math. Soc.* **70**, 709–710.
Goldstein, A. A. (1966). Minimizing functionals on normed linear spaces. *S.J.C.* **4**, 81–89.
Goldstein, A. A. (1974). On gradient projection. *Proc. Annual Allerton Confer., 12th, Allerton Park, Ill.* pp. 38–40.
Goldstein, A. A. (1977). Optimization of Lipschitz continuous functions. *M.P.* **13**, 14–22.
Goldstein, A. A., and Price, J. B. (1967). An effective algorithm for minimization. *Numer. Math.* **10**, 184–189.
Golshtein, E. G. (1972). A generalized gradient method for finding saddle points. *Matecon* **8**, 36–52.
Greenstadt, J. (1970). Variations on variable metric methods. *Math. Comput.* **24**, 1–18.
Haarhoff, P. C., and Buys, J. D. (1970). A new method for the optimization of a nonlinear function subject to nonlinear constraints. *Comput. J.* **13**, 178–184.
Han, S. P. (1977a). Dual variable metric algorithms for constrained optimization. *S.J.C.O.* **15**, 135–194.
Han, S. P. (1977b). A globally convergent method for nonlinear programming. *J.O.T.A.* **22**, 297–309.
Han S. P., and Mangasarian, O. L. (1979). Exact penalty functions in nonlinear programming. *M.P.* **17**, 251–269.
Han, S. P., and Mangasarian, O. L. (1981). A dual differentiable exact penalty function. Computer Sciences Tech. Rep. #434, Univ. of Wisconsin, Madison.
Held, M., and Karp, R. M. (1970). The Traveling Salesman Problem and minimum spanning trees. *O.R.* **18**, 1138–1162.
Held, M., Wolfe, P., and Crowder, H. (1974). Validation of subgradient optimization. *M.P.* **6**, 62–88.
Hestenes, M. R. (1966). "Calculus of Variations and Optimal Control Theory." Wiley, New York.

Hestenes, M. R. (1969). Multiplier and gradient methods. *J.O.T.A.* **4**, 303–320.

Hestenes, M. R. (1975). "Optimization Theory: The Finite Dimensional Case." Wiley, New York.

Hestenes, M. R. (1980). "Conjugate Direction Methods in Optimization." Springer-Verlag, Berlin and New York.

Hestenes, M. R., and Stiefel, E. L. (1952). Methods of conjugate gradients for solving linear systems. *J. Res. Nat. Bur. Standards Sect. B* **49**, 409–436.

Howe, S. (1973). New conditions for exactness of a simple penalty function. *S.J.C.* **11**, 378–381.

Jijtontrum, K. (1980). Accelerated convergence for the Powell/Hestenes multiplier methods. *M.P.* **18**, 197–214.

Kantorovich, L. V. (1945). On an effective method of solution of extremal problems for a quadratic functional. *Dokl. Akad. Nauk SSSR* **48**, 483–487.

Klessig, R., and Polak, E. (1972). Efficient implementation of the Polak–Ribiere conjugate gradient algorithms. *S.J.C.* **10**, 524–549.

Korpelevich, G. M. (1976). The extragradient method for finding saddle points and other problems. *Matecon* **12**, 35–49.

Kort, B. W. (1975a). Combined primal-dual and penalty function algorithms for nonlinear programming. Ph.D. Thesis, Stanford Univ., Stanford. California.

Kort, B. W. (1975b). Rate of convergence of the method of multipliers with inexact minimization. *In* "Nonlinear Programming 2" (O. Mangasarian, R. Meyer, and S. Robinson, eds.), pp. 193–214. Academic Press, New York.

Kort, B. W., and Bertsekas, D. P. (1972). A new penalty function method for constrained minimization. *Proc 1972 IEEE Confer. Decision Control, New Orleans. La.* pp. 162–166.

Kort, B. W., and Bertsekas, D. P. (1973). Multiplier methods for convex programming. *Proc. 1073 IEEE Conf. Decision Control, San Diego, Calif.* pp. 428–432.

Kort, B. W., and Bertsekas, D. P. (1976). Combined primal-dual and penalty methods for convex programming. *S.J.C.O.* **14**, 268–294.

Kwakernaak, H., and Strijbos, R. C. W. (1972). Extremization of functions with equality constraints. *M.P.* **2**, 279–295.

Lasdon, L. S. (1970). "Optimization Theory for Large Systems." Macmillan, New York.

Lauer, G., Bertsekas, D. P., Sandell, N., and Posbergh, T. (1981). Optimal solution of large-scale unit commitment problems. *IEEE Trans. Power Systems Apparatus.* **101**, 79–86.

Lemarechal, C. (1974). An algorithm for minimizing convex functions." *In* "Information Processing '74" (J. L. Rosenfeld, ed.), pp. 552–556. North-Holland Publ., Amsterdam.

Lemarechal, C. (1975). An extension of Davidon methods to nondifferentiable problems. *Math. Programming Stud.* **3**, 95–109.

Lenard, M. L. (1973). Practical convergence conditions for unrestrained optimization. *M.P.* **4**, 309–325.

Lenard, M. L. (1976). Convergence conditions for restarted conjugate gradient methods with inaccurate line searches. *M.P.* **10**, 32–51.

Lenard, M. L. (1979). A computational study of active set strategies in nonlinear programming with linear constraints. *M.P.* **16**, 81–97.

Levitin, E. S., and Poljak, B. T. (1965). Constrained minimization methods. *Z. Vyčisl. Mat. i Mat. Fiz.* **6**. 787–823. Engl. transl. in *C.M.M.P.* **6**, 1–50 (1966).

Luenberger, D. G. (1979). "Optimization by Vector Space Methods." Wiley, New York.

Luenberger, D. G. (1970). Control problems with kinks. *IEEE Trans. Automat. Control* **15**, 570–575.

Luenberger, D. G. (1973). "Introduction to Linear and Nonlinear Programming." Addison-Wesley, Reading, Massachusetts.

Luque, J. R. (1981). "Asymptotic convergence analysis of the proximal point algorithm." LIDS Rep. P-1142, M.I.T., Cambridge, Mass. To appear in *S.J.C.O.*

McCormick, G. P. (1969). Anti-zig-zagging by bending. *M.S.* **15**, 315–319.
McCormick, G. P. (1978). An idealized exact penalty function. *In* "Nonlinear Programming 3" (O. Mangasarian, R. Meyer, and S. Robinson, eds.) pp. 165–195. Academic Press, New York.
McCormick, G. P., and Ritter, K. (1972). Methods of conjugate directions versus quasi-Newton methods. *M.P.* **3**, 101–116.
Magnanti, T. L. Shapiro, J. F., and Wagner, M. H. (1976). Generalized linear programming solves the dual. *M.S.* **22**, 1195–1203.
Maistrovskii, G. D. (1976). Gradient methods for finding saddle points. *Matecon* **12**. 3–22.
Mangasarian, O. L. (1969). "Nonlinear Programming." Prentice-Hall, Englewood Cliffs. New Jersey.
Mangasarian, O. L. (1974). Unconstrained methods in optimization. *Proc. Allerton Conf. Circuit System Theory, 12th, Univ. Ill., Urbana* pp. 153–160.
Mangasarian, O. L. (1975). Unconstrained Lagrangians in nonlinear programming. *S.J.C.* **13**, 772–791.
Maratos, N. (1978). Exact penalty function algorithms for finite dimensional and control optimization problems. Ph.D. Thesis, Imperial College Sci. Tech., Univ. of London.
Mayne, D. Q., and Maratos, N. (1979). A first order exact penalty function algorithm for equality constrained optimization problems. *M.P.* **16**, 303–324.
Mayne, D. Q., and Polak, E. (1978). "A Superlinearly Convergent Algorithm for Constrained Optimization Problems," Res. Rep. 78–52. Dept. Comput. Control, Imperial College, London. *Math. Programming Stud.* **16**, (1982), to appear.
Miele, A., Cragg, E. E., Iyer, R. R., and Levy, A. V. (1971a). Use of the augmented penalty function in mathematical programming problems, Part I. *J.O.T.A.* **8**, 115–130.
Miele, A., Cragg, E. E., and Levy, A. V. (1971b). Use of the augmented penalty function in mathematical programming problems, Part II. *J.O.T.A.* **8**, 131–153.
Miele, A., Moseley, P. E., and Cragg, E. E. (1972). On the method of multipliers for mathematical programming problems. *J.O.T.A.* **10**, 1–33.
Mifflin, R. (1977). An algorithm for constrained optimization with semismooth functions. *Math. Oper. Res.* **2**, 191–207.
Moré, J. J., and Sorensen, D. C. (1979). On the use of directions of negative curvature in a modified Newton method. *M.P.* **16**, 1–20.
Mukai, H., and Polak, E. (1975). A quadratically convergent primal-dual algorithm with global convergence properties for solving optimization problems with equality constraints. *M.P.* **9**, 336–349.
Murray. W. (1972). Second derivative methods. *In* "Numerical Methods for Unconstrained Optimization" (W. Murray, ed.), pp. 57–71. Academic Press, New York.
Nguyen, V. H., and Strodiot, J. J. (1979). On the convergence rate of a penalty function method of exponential type. *J.O.T.A.* **27**, 495–508.
Oren, S. S. (1973). Self-scaling variable metric algorithms without line search for unconstrained minimization. *Math. Comput.* **27**, 873–885.
Oren, S. S. (1974). Self-scaling variable metric algorithm, Part II. *M.S.* **20**, 863–874.
Oren, S. S. (1978). A combined variable metric conjugate gradient algorithm for a class of large-scale unconstrained minimization problems. *In* "Optimization Techniques, Part II" (J. Stoer, ed.), pp. 107–115. Springer-Verlag, Berlin and New York.
Oren S. S., and Luenberger, D. G. (1974). Self-scaling variable metric algorithm, Part I. *M.S.* **20**, 845–862.
Oren, S. S., and Spedicato, E. (1976). Optimal conditioning of self-scaling variable metric algorithms. *M.P.* **10**, 70–90.
Ortega, J. M., and Rheinboldt, W. C. (1970). "Iterative Solution of Nonlinear Equations in Several Variables." Academic Press, New York.
Papavassilopoulos, G. (1977). Algorithms for a class of nondifferentiable problems. M.S. Thesis, Dept. Electr. Engr., Univ. of Illinois, Urbana. Also *J.O.T.A.* **34**, (1981), 41–82.

Pietrzykowski, T. (1969). An exact potential method for constrained maxima. *SIAM J. Numer. Anal.* **6**, 269–304.

Polak, E. (1971). "Computational Methods in Optimization: A Unified Approach." Academic Press, New York.

Polak, E. (1976). On the global stabilization of locally convergent algorithms. *Automatica—J. IFAC* **12**, 337–342.

Polak, E., and Ribiere, G. (1969). Note sur la convergence de méthodes de directions conjugées. *Rev. Fr. Inform. Rech. Oper.* **16-R1**, 35–43.

Polak, E., and Tits, A. L. (1979). "A Globally Convergent, Implementable Multiplier Method with Automatic Penalty Limitation," Memo No. UCB/ERL M79/52. Electron. Res. Lab., Univ. of California, Berkeley. *Applied Math. and Optim.* **6**, (1980), 335–360.

Poljak, B. T. (1963). Gradient methods for the minimization of functionals. *Ž. Vyčisl. Mat. i Mat. Fiz.* **3**, 643–653.

Poljak, B. T. (1969a). The conjugate gradient method in extremal problems. *Ž. Vyčisl. Mat. i Mat. Fiz.* **9**, 94–112.

Poljak, B. T. (1969b). Minimization of unsmooth functionals. *Ž. Vyčisl. Mat. i Mat. Fiz.* **9**, 14–29.

Poljak, B. T. (1970). Iterative methods using Lagrange multipliers for solving extremal problems with constraints of the equation type. *Ž. Vyčisl. Mat. i Mat. Fiz.* **10**, 1098–1106.

Poljak, B. T. (1971). The convergence rate of the penalty function method. *Ž. Vyčhisl. Mat. i Mat. Fiz.* **11**, 3–11.

Poljak, B. T. (1979). On Bertsekas' method for minimization of composite functions. *Internat. Symp. Systems Opt. Analysis* (A. Bensoussan and J. L. Lions, eds.). pp. 179–186. Springer-Verlag, Berlin and New York.

Poljak, B. T., and Tretjakov, N. V. (1973). The method of penalty estimates for conditional extremum problems. *Ž. Vyčhisl. Mat. i Mat. Fiz.* **13**, 34–46.

Poljak, B. T., and Tretjakov, N. V. (1974). An iterative method for linear programming and its economic interpretation. *Matecon* **10**, 81–100.

Powell, M. J. D. (1964). An efficient method for finding the minimum of a function of several variables without calculating derivatives. *Comput. J.* **7**, 155–162.

Powell, M. J. D. (1969). A method for nonlinear constraints in minimization problems. *In* "Optimization" (R. Fletcher, ed.), pp. 283–298. Academic Press, New York.

Powell, M. J. D. (1970). A new algorithm for unconstrained optimization. *In* "Nonlinear Programming" (J. B. Rosen, O. L. Mangasarian, and K. Ritter, eds.), pp. 31–65. Academic Press, New York.

Powell, M. J. D. (1971). On the convergence of the variable metric algorithm. *J. Inst. Math. Appl.* **7**, 21–36.

Powell, M. J. D. (1973). On search directions for minimization algorithms. *M.P.* **4**, 193–201.

Powell, M. J. D. (1977). Restart procedures for the conjugate gradient method. *M.P.* **12**, 241–254.

Powell, M. J. D. (1978a). Algorithms for nonlinear constraints that use Lagrangian functions. *M.P.* **15**, 224–248.

Powell, M. J. D. (1978b). The convergence of variable metric methods for nonlinearly constrained optimization calculations. In "Nonlinear Programming 3" (O. L. Mangasarian, R. Meyer, and S. Robinson, eds.), pp. 27–63. Academic Press, New York.

Pschenichny, B. N. (1970). Algorithms for the general problem of mathematical programming. *Kibernetika* (*Kiev*), **6**, 120–125.

Pschenichny, B. N., and Danilin, Y. M. (1975). "Numerical Methods in Extremal Problems." MIR, Moscow. (Engl. transl., 1978.)

Ritter, K. (1973). A superlinearly convergent method for minimization with linear inequality constraints. *M.P.* **4**, 44–71.

Robinson, S. M. (1974). Perturbed Kuhn–Tucker points and rates of convergence for a class of nonlinear programming algorithms. *M.P.* **7**, 1–16.

Rockafellar, R. T. (1970). "Convex Analysis." Princeton Univ. Press, Princeton, New Jersey.

Rockafellar, R. T. (1971). New applications of duality in convex programming. *Proc. Confer. Probab., 4th, Brasov, Romania*, pp. 73–81.

Rockafellar, R. T. (1973a). A dual approach to solving nonlinear programming problems by unconstrained optimization. *M.P.* **5**, 354–373.

Rockafellar, R. T. (1973b). The multiplier method of Hestenes and Powell applied to convex programming. *J.O.T.A.* **12**, 555–562.

Rockafellar, R. T. (1974). Augmented Lagrange multiplier functions and duality in nonconvex programming. *S.J.C.* **12**, 268–285.

Rockafellar, R. T. (1976a). Monotone operators and the proximal point algorithm. *S.J.C.O.* **14**, 877–898.

Rockafellar, R. T. (1976b). Augmented Langrangians and applications of the proximal point algorithm in convex programming. *Math. Oper. Res.* **1**, 97–116.

Rockafellar, R. T. (1976c). Solving a nonlinear programming problem by way of a dual problem. *Symp. Mat.* **27**, 135–160.

Rupp, R. D. (1972). Approximation of the Classical Isoperimetric Problem. *J.O.T.A.* **9**, 251–264.

Sargent, R. W. H., and Sebastian, D. J. (1973). On the convergence of sequential minimization algorithms. *J.O.T.A.* **12**, 567–575.

Schnabel, R. B. (1980). Determining feasibility of a set of nonlinear inequality constraints. *Math. Programming Stud.* **16**, (1982), to appear.

Shanno, D. F. (1970). Conditioning of quasi-Newton methods for function minimization. *Math. Comput.*, **24**, 647–656.

Shanno, D. F. (1978a). Conjugate gradient methods with inexact searches. *Math. Oper. Res.* **3**, 244–256.

Shanno, D. F. (1978b). On the convergence of a new conjugate gradient algorithm. *SIAM J. Numer. Anal.* **15**, 1247–1257.

Shapiro J. E. (1979). "Mathematical Programming Structures and Algorithms." Wiley, New York.

Shor, N. Z. (1964). On the structure of algorithms for the numerical solution of planning and design problems. Thesis. Kiev.

Shor, N. Z. (1970). Utilization of the operation of space dilation in the minimization of convex functions. *Kiberbetika* (*Kiev*) **6**, 6–12.

Shor, N. Z., and Jourbenko, N. G. (1971). A method of minimization using space dilation in the direction of two successive gradients. *Kibernetika* (*Kiev*) **7**, 51–59.

Sorenson, H. W. (1969). Comparison of some conjugate direction procedures for functions minimization. *J. Franklin Inst.* **288**, 421–441.

Stephanopoulos, G., and Westerberg, A. W. (1975). The use of Hestenes' method of multipliers to resolve dual gaps in engineering system optimization. *J.O.T.A.* **15**, 285–309.

Stoilow, E. (1977). The augmented Lagrangian method in two-level static optimization. *Arch. Automat. Telemech.* **22**, 219–237.

Tapia, R. A. (1977). Diagonalized multiplier methods and quasi-Newton methods for constrained minimization. *J.O.T.A.* **22**, 135–194.

Tapia, R. A. (1978). Quasi-Newton methods for equality constrained optimization: Equivalence of existing methods and a new implementation. *In* "Nonlinear Programming 3" (O. L. Mangasarian, R. R. Meyer, and S. M. Robinson, eds.), pp. 125–164. Academic Press, New York.

Vastola, K. S. (1979). A numerical study of two measures of delay for network routing. M.S. Thesis. Dept. Electr. Engr., Univ. of Illinois, Champaign-Urbana.

Watanabe, N., Nishimura, Y., and Matsubara, M. (1978). Decomposition in large system optimization using the method of multipliers. *J.O.T.A.* **25**, 181–193.

Wierzbicki, A. P. (1971). A penalty function shifting method in constrained static optimization and its convergence properties. *Arch. Automat. Telemech.*, **16**, 395–416.

Wilson, R. B. (1963). A simplicial algorithm for concave programming. Ph.D. Thesis, Grad. Sch. Business Admin., Harvard Univ., Cambridge, Massachusetts.

Wolfe, P. (1969). Convergence conditions for ascent methods. *SIAM Rev.*, **11**, 226–235.

Wolfe, P. (1975). A method of conjugate subgradients for minimizing nondifferentiable functions. *Math. Programming Stud.* **3**, 145–173.

Zangwill, W. I. (1967a). Minimizing a function without calculating derivatives. *Comput. J.* **10**, 293–296.

Zangwill, W. I. (1967b). Nonlinear programming via penalty functions. *M.S.* **13**, 344–358.

Zangwill, W. I. (1969). "Nonlinear Programming." Prentice-Hall, Englewood Cliffs. New Jersey.

Index

O

P

Q

R

S

T

U

Computer Science and Applied Mathematics

A SERIES OF MONOGRAPHS AND TEXTBOOKS

Editor

Werner Rheinboldt

University of Pittsburgh

HANS P. KÜNZI, H. G. TZSCHACH, AND C. A. ZEHNDER. Numerical Methods of Mathematical Optimization: With ALGOL and FORTRAN Programs, Corrected and Augmented Edition

AZRIEL ROSENFELD. Picture Processing by Computer

JAMES ORTEGA AND WERNER RHEINBOLDT. Iterative Solution of Nonlinear Equations in Several Variables

AZARIA PAZ. Introduction to Probabilistic Automata

DAVID YOUNG. Iterative Solution of Large Linear Systems

ANN YASUHARA. Recursive Function Theory and Logic

JAMES M. ORTEGA. Numerical Analysis: A Second Course

G. W. STEWART. Introduction to Matrix Computations

CHIN-LIANG CHANG AND RICHARD CHAR-TUNG LEE. Symbolic Logic and Mechanical Theorem Proving

C. C. GOTLIEB AND A. BORODIN. Social Issues in Computing

ERWIN ENGELER. Introduction to the Theory of Computation

F. W. J. OLVER. Asymptotics and Special Functions

DIONYSIOS C. TSICHRITZIS AND PHILIP A. BERNSTEIN. Operating Systems

ROBERT R. KORFHAGE. Discrete Computational Structures

PHILIP J. DAVIS AND PHILIP RABINOWITZ. Methods of Numerical Integration

A. T. BERZTISS. Data Structures: Theory and Practice, Second Edition

N. CHRISTOPHIDES. Graph Theory: An Algorithmic Approach

ALBERT NIJENHUIS AND HERBERT S. WILF. Combinatorial Algorithms

AZRIEL ROSENFELD AND AVINASH C. KAK. Digital Picture Processing

SAKTI P. GHOSH. Data Base Organization for Data Management

DIONYSIOS C. TSICHRITZIS AND FREDERICK H. LOCHOVSKY. Data Base Management Systems

JAMES L. PETERSON. Computer Organization and Assembly Language Programming

WILLIAM F. AMES. Numerical Methods for Partial Differential Equations, Second Edition

ARNOLD O. ALLEN. Probability, Statistics, and Queueing Theory: With Computer Science Applications

ELLIOTT I. ORGANICK, ALEXANDRA I. FORSYTHE, AND ROBERT P. PLUMMER. Programming Language Structures

ALBERT NIJENHUIS AND HERBERT S. WILF. Combinatorial Algorithms. Second edition.

JAMES S. VANDERGRAFT. Introduction to Numerical Computations

AZRIEL ROSENFELD. Picture Languages, Formal Models for Picture Recognition

ISAAC FRIED. Numerical Solution of Differential Equations

ABRAHAM BERMAN AND ROBERT J. PLEMMONS. Nonnegative Matrices in the Mathematical Sciences

BERNARD KOLMAN AND ROBERT E. BECK. Elementary Linear Programming with Applications

CLIVE L. DYM AND ELIZABETH S. IVEY. Principles of Mathematical Modeling

ERNEST L. HALL. Computer Image Processing and Recognition

ALLEN B. TUCKER, JR. Text Processing: Algorithms, Languages, and Applications

MARTIN CHARLES GOLUMBIC. Algorithmic Graph Theory and Perfect Graphs

GABOR T. HERMAN. Image Reconstruction from Projections: The Fundamentals of Computerized Tomography

WEBB MILLER AND CELIA WRATHALL. Software for Roundoff Analysis of Matrix Algorithms

ULRICH W. KULISCH AND WILLARD L. MIRANKER. Computer Arithmetic in Theory and Practice

LOUIS A. HAGEMAN AND DAVID M. YOUNG. Applied Interative Methods

I. GOHBERG, P. LANCASTER AND L. RODMAN. Matrix Polynomials.

AZRIEL ROSENFELD AND AVINASH C. KAK. Digital Picture Processing, Second Edition, Vol. 1, Vol. 2

DIMITRI P. BERTSEKAS. Constrained Optimization and Lagrange Multiplier Methods

In preparation

LEONARD UHR. Algorithm-Structured Computer Arrays and Networks: Architectures and Processes for Images, Percepts, Models, Information

GÖTZ ALEFELD AND JURGEN HERZBERGER. Introduction to Interval Computations. Translated by Jon Rockne

FRANÇOISE CHATELIN. Spectral Approximation of Linear Operators